Forest Seed and Seedling Cultivation
林木种苗培育学

Editor-in-Chief: Liu Yong

刘 勇 主编

Forest Seed and Seedling Cultivation is a textbook about the theory and technology of tree seed production and seedling cultivation in forest establishment. It is also a required textbook for the core curriculum of forestry and urban forestry specialties.

图书在版编目（CIP）数据

林木种苗培育学 = Forest Seed and Seedling Cultivation：英文 / 刘勇主编. -- 北京：中国林业出版社, 2024. 10. -- （高等院校外国留学生教育系列教材）. -- ISBN 978-7-5219-2974-4

Ⅰ．S723.1

中国国家版本馆 CIP 数据核字第 2024HJ8920 号

责任编辑：肖基浒
封面设计：睿思视思视觉设计

出版发行　中国林业出版社
　　　　　（100009, 北京市西城区刘海胡同 7 号, 电话 83143562）
电子邮箱　jiaocaipublic@163.com
网　　址　https://www.cfph.net
印　　刷　北京印刷集团有限责任公司
版　　次　2024 年 10 月第 1 版
印　　次　2024 年 10 月第 1 次印刷
开　　本　787mm×1092mm　1/16
印　　张　30.5
字　　数　990 千字
定　　价　96.00 元

Editorial Team of *Forest Seed and Seedling Cultivation*

Editor-in-Chief:
    Liu Yong, Professor, Beijing Forestry University

Associate Editors:
    Li Guolei, Professor, Beijing Forestry University
    Hou Zhixia, Professor, Beijing Forestry University

Editorial Team Members:
    Bai Shulan, Professor, Inner Mongolia Agricultural University
    Cao Banghua, Professor, Shandong Agricultural University
    Fu Xiangxiang, Professor, Nanjing Forestry University
    Guo Shujuan, Professor, Beijing Forestry University
    Lin Na, Lecturer, South China Agricultural University
    Lu Xiujun, Professor, Shenyang Agricultural University
    Peng Zuodeng, Professor, Beijing Forestry University
    Shen Hailong, Professor, Northeast Forestry University
    Wang Aifang, Associate Professor, Hebei Agricultural University
    Wang Jun, Professor, Beijing Forestry University
    Wang Weiwei, Engineer, Institute of Botany, Chinese Academy of Sciences
    Wei Xiaoli, Professor, Guizhou University
    Ying Yeqing, Professor, Zhejiang Agricultural and Forestry University
    Zhang Gang, Professor, Hebei Agricultural University
    Zhao Hewen, Professor, Beijing Agricultural College
    Zheng Yushan, Professor, Fujian Agriculture and Forestry University
    Zhu Yan, Assistant Researcher, Chinese Academy of Forestry

Preface

Forest seed and seedlings are the basic materials of tree planting. Adequate quantities and good quality seeds and seedlings guarantee the success of artificial afforestation and land greening. The term artificial afforestation means direct seeding and planting has been implemented rather than a natural regeneration that occurs when seed is dispersed without human involvement. The forest coverage rate in China has increased from 8.6% at the beginning of the founding of the People's Republic of China in the 1940s to 24% in 2023. The artificial forest accounts for a large proportion, and the production of forest seed and seedlings has made a great contribution to this improvement. Tree seed and seedling science has important knowledge about the theory and technology of seed and seedling cultivation, which is the core course of forestry education. This textbook has been written for this course.

Tree seed and seedling science belongs to the discipline of silviculture. China has an impressively long history of practices related to seed and seedling cultivation. The oracle bone inscriptions in the Yin Dynasty and Shang Dynasty included characters such as "Nursery". The *Book of Fansheng* in the Western Han Dynasty and the *Qi Min Yao Shu* in the Northern Wei Dynasty contain detailed descriptions of seed collection, seed germination acceleration, sowing, cutting, nursery tending, and management. Tree seed and seedling science became a part of modern science and technology in the first half of the 20th century; however, it did not have significant development until the newly found People's Republic of China recognized and prioritized the importance of regenerating forests and green areas. *Science of Afforestation*, compiled by the East and Central China Cooperation Group in 1959 and by the afforestation teaching and research group of Beijing Forestry College in 1961, has become a milestone of silviculture in New China. *Science of Afforestation*, edited by Sun Shixuan in 1981, reflects the scientific and technological achievements before 1978; at that time China started to reform and open up to the world. In 2001, *Silviculture*, edited by Shen Guofang, reflected the experience and knowledge accumulated in forest cultivation in China in the 20th century and was responsible for changing the textbook name *Science of Afforestation* to *Silviculture*. This current textbook builds on the foundation of *Science of Afforestation* and *Silviculture*.

The content of this textbook includes the establishment of a nursery, collection of germplasm resources, seed production of improved varieties, seedling cultivation, nursery management, and more, reflecting the basic theory, knowledge, and technology of forest seed and seedling production in China. In recent years, China has promoted seed and seedlings as an industry, which has greatly promoted the development of nursery technology. Readers who need a deeper understanding of some or all of the content can consult the source list of references and additional reading materials at the end of each chapter.

Teachers engaged in forest seed and seedling science in major agricultural and forestry universities across the country participated in the compilation of this textbook, *Forest Seed and Seedling Cultivation*. Their many years of teaching experience has contributed to the clear expression of the knowledge of forest seed and seedlings, making it easier for readers to understand.

The publication of this textbook is strongly supported by Chinese Forestry Publishing House, the College of Forestry in Beijing Forestry University. R. Kasten Dumroese, Research Plant Physiologist in the Forest Service's Rocky Mountain Research Station, USA, provided a lot of help; Candace J. Akins, Freelance Science Editor, USA, polished the text of the English manuscript. Graduate students Guo Huanhuan, Pei Ziqi, and Wang Yanchao participated in searching data, sorting out charts, proofreading text, and so forth: The authors express heartfelt thanks to them!

Liu Yong
2024. 4. 18

Contents

Preface

Chapter 1 Development of Forest Seed and Seedling Science and Technique System ······ 1
    1.1 Development and Evolution of Forest Seed and Seedling Science ················ 1
    1.2 Characteristics of Forest Seed and Seedling Cultivation ······················ 3
    1.3 Theoretical Basis and Technical System of Directional Seedling Cultivation ······ 4
    1.4 Overview of Seedling Industry Development in China ························ 11

Chapter 2 Establishment of Forest Nurseries ················ 17
    2.1 Site Selection of Forest Nursery ················ 17
    2.2 Planning and Design of Forest Nursery ················ 19

Chapter 3 Tree Germplasm Resources and Breeding of New Varieties ················ 46
    3.1 Significance of Tree Germplasm Resources ················ 46
    3.2 Characteristics of Tree Germplasm Resources ················ 48
    3.3 Collection and Conservation of Tree Germplasm Resources ················ 49
    3.4 Breeding and Protection of New Tree Varieties ················ 61

Chapter 4 Production of Improved Propagation Material of Forest Trees ················ 73
    4.1 Provenance Selection and Seed Regionalization ················ 73
    4.2 Production Base of Improved Tree Species ················ 77
    4.3 Tree Fruiting Law, Seed Collection, and Processing ················ 85
    4.4 Collection and Processing of Asexual Propagation Materials ················ 99

Chapter 5 Propagative Materials Storage and Seed Quality Evaluation for Tree Species ················ 103
    5.1 Seed Storage Physiology and Storage Method ················ 103
    5.2 Storage Physiology and Storage Method of Vegetative Propagules ················ 111
    5.3 Quality Evaluation of Tree Seeds ················ 115

Chapter 6 Seedling Growth and Physiology ················ 143
    6.1 Seedling Types, Characteristics, and Seedling Age ················ 143
    6.2 Seedling Growth Rhythm ················ 146
    6.3 Factors Affecting Seedling Growth ················ 153

Chapter 7 Nursery Soil Management and Seedling Protection ················ 169
    7.1 Soil Fertility Management ················ 169
    7.2 Water Management ················ 189
    7.3 Symbiotic Fungi Inoculation and Management ················ 194
    7.4 Weed Management ················ 210
    7.5 Pest and Disease Control ················ 224

Chapter 8 Bare-rooted Seedling Cultivation ... 231
- 8.1 Seed and Seedbed Preparation ... 231
- 8.2 Seedling-raising Pattern and Seedbed Preparation ... 250
- 8.3 Sowing Season and Sowing Quantity ... 252
- 8.4 Sowing Method ... 255
- 8.5 Nursery Soil and Seedling Management ... 258

Chapter 9 Container Seedling Cultivation ... 274
- 9.1 Characteristics of Container Seedling Production ... 274
- 9.2 Seedling Containers ... 275
- 9.3 Seedling Matrix ... 285
- 9.4 Sowing and Seedling Management ... 289
- 9.5 Greenhouse and Environment Control ... 301

Chapter 10 Vegetative Propagation ... 304
- 10.1 Cutting ... 304
- 10.2 Grafting ... 320
- 10.3 Tissue Culture ... 336

Chapter 11 Cultivation of Transplants and Large-size Seedlings ... 355
- 11.1 Bare-Root Transplantation ... 355
- 11.2 Transplantation and Cultivation with Soil Boll ... 364
- 11.3 Large-size Container Seedling Cultivation ... 373

Chapter 12 Seedling Pruning ... 386
- 12.1 Biological Basis of Pruning Seedlings ... 386
- 12.2 Principles of Pruning ... 388
- 12.3 Method of Pruning Seedlings ... 392

Chapter 13 Seedling Quality Evaluation and Seedling Lifting ... 411
- 13.1 Seedling Quality Assessment ... 411
- 13.2 Seedling Investigation Method ... 430
- 13.3 Seedling Lifting, Packing, and Transporting ... 436

Chapter 14 Nursery Management ... 453
- 14.1 Nursery Positioning and Target Management ... 453
- 14.2 Human Resource Management of Nursery ... 455
- 14.3 Production Management of Nursery ... 457
- 14.4 Marketing Management ... 464
- 14.5 Nursery Files Management ... 469

Chapter 1 Development of Forest Seed and Seedling Science and Technique System

Liu Yong

Chapter Summary: This chapter introduces the development of forest seedling science, its characteristics, the technical system, the problems and prospects of the development of China's seedling industry, and the teaching system of forest seedling science. The purpose is to give the reader an overall view of the teaching content, teaching methods, and the status of the seedling industry, which plays an important guiding role for learning this course material.

1.1 Development and Evolution of Forest Seed and Seedling Science

China has an impressively long history of practices related to seedling cultivation. For example, Emperor Xuanyuan sowed all kinds of seeds of grasses and trees at the right time; that is, tree propagation by sowing the seeds at the right time. The oracle bone inscriptions in the Shang Dynasty included characters such as "Yuan" and "Pu". Yuan refers to the planting place of fruit trees, economic trees, and vegetables; Pu means a piece of land for planting fruit trees, vegetables, flowers, and other plants surrounded by a fence on the open space around the house. Chuang Tzu, a thinker in the Spring and Autumn Period, once served as an official of a lacquer tree garden, meaning he was a forest ranger of a lacquer forest, indicating that tree planting held considerable status at that time. Subsequently, from the Qin, Han, Three Kingdoms to Wei, Jin, and Southern and Northern Dynasties, a wealth of information on seedling cultivation was gradually accumulated. The *Book of Fansheng* in Western Han Dynasty, *Qi Min Yao Shu* in Northern Wei Dynasty, and *Qun Fang Pu* in Ming Dynasty contain detailed descriptions of seed collection, seed germination acceleration, sowing, cutting, nursery tending, and management. However, the accumulation of this knowledge and experience is still in a scattered state, unable to form a complete system.

Seedling cultivation has become a part of modern science and technology and is related to the history of the industrial revolution of Europe. The industrial revolution began in the 18th century; many scholars believe it contributed to the destruction of forests. With the deterioration of the ecological environment, the restoration of forests is of critical importance to mankind. The first textbook of silviculture (a branch of forestry dealing with the development and care of forests) was compiled by German Hager R. in the 18th century, and it marked the birth of silviculture. Germany became the originator of modern forestry. In the first half of the 20th century, many monographs on forest cultivation appeared in the United States, Britain, Japan, and the Soviet Union and were introduced into China one after another. In 1933, *Afforestation Essentials* and *Afforestation of Tree Species*, published by Chen Rong, became the foundation work of silviculture in China. *Afforestation*, compiled by the East and Central China Cooperation Group in 1959 and by the afforestation teaching and research group of Beijing Forestry College in 1961, has become a milestone of silviculture in New China. *Afforestation*, edited by Sun Shixuan in 1981, reflects the scientific and technological achievements before 1978; at that time China started to reform and open up to the world. In 2001, *Silviculture*, edited by Shen Guofang, reflected the experience and knowledge accumulated in forest cultivation in China in the 20th century and was responsible for changing the name "afforestation" to "silviculture".

Seed and seedling cultivation is an important aspect of silviculture; however, the seeds and seedlings themselves are the final product and form the seedling industry in the market. Therefore, with the development of silviculture, seed and seedling cultivation has gradually formed a complete knowledge and technology system with the goal of cultivating seedlings for afforestation, and the textbooks such as *Forest Seed and Seedling Manual*, *Forest Seed and Seedling Science*, and others have been published successively since 1982.

At the same time, seed and seedling cultivation is also an important part of garden and horticulture. In the 16th and 17th century, a large number of nurseries appeared in France and eventually extended to the whole of Europe, gradually forming a seed and seedling industry. The *Gardener's Guide* appeared in Belgium in 1366, and the first glass greenhouse was built in 1598. The Royal Society of Horticulture was established in Britain in 1804. In the 19th century, the seedling industry expanded to the United States, and in its early stages it focused primarily on breeding and grafting fruit trees but then extended to ornamental plants and afforestation tree species. In recent decades, the seed and seedling industry has been evolving in most developed countries, and the United States has become the largest seed and seedling producer. Although China's seed and seedling industry started relatively late, it has developed to a large scale and with rapid momentum. The knowledge and technology for cultivating garden and horticultural plants gradually formed a system, and the teaching materials of *Garden Nursery* and *Garden Nursery Science* were published in 1982 and 1988, respectively.

So far, the seed and seedling sciences have developed in two major areas: Forest seed and seedlings, and landscape seed and seedlings. With the rapid development of global urbanization, the urban ecological environment is deteriorating, and this urbanization process in China is occurring even more quickly given government efforts. The ecological environmental problems have become very prominent, which has spawned an increased interest in urban forestry, with the main goal being the improvement of the urban ecological environment. Urban tree cultivation technology has received unprecedented attention, and accordingly, tree seed and seedling science, which considers both forest seed and seedlings and landscape seed and seedlings, has developed quickly.

1.2 Characteristics of Forest Seed and Seedling Cultivation

Seed and seedling cultivation, strictly speaking, should be called forest or tree seed and seedling cultivation, because the seed and seedlings here refer to the seed and seedlings of trees (trees and shrubs), instead of crops or herbs. It integrates the theory and technology of forestry and landscape seed and seedling cultivation, with the following characteristics.

1.2.1 More Diversified Cultivation Objectives

The afforestation site is characterized by a complex environment, including natural environment, artificial environment, and various biological environments dominated by human beings. The various needs of people, in terms of how the land is used, will inevitably lead to a range of requirements for the different types of forest, trees, and green land, and their naturally distinct seedling demands. Therefore, various objectives must be considered for seed and seedling cultivation, such as small-size seedlings with strong resistance suitable for ecological forest construction in mountainous areas, seedlings with colorful leaves suitable for forest construction in urban scenic areas, large-size seedlings suitable for office and residential areas, and so forth.

1.2.2 Highlight Non-standardized Production Mode

Forest seed and seedling science emphasizes the standardization of seedlings. Through the formulation of seedling standards and technical regulations, unified technical measures, unified production process, and unified seedling management are adopted for the specific type of seed-lings, and finally, standardized and uniform seedlings are produced. Forestry production requires large areas of artificial forest construction, with a large demand for the same seedling type and for consistent quality of the standardized seedling to ensure good survival rates. Inconsistent quality of seedlings will seriously affect the survival effect of afforestation.

By contrast, urban seedling cultivation has a greater diversity of cultivation objectives, which means the demand for each kind of seedlings is not necessarily great. Hence, it is difficult to achieve standardized production, and sometimes cultivation measures must be formulated according to the situation of individual seedlings. For example, before the 2008 Beijing Olympic Games, the tall seedlings of evergreen conifers were in short supply, with the prices soaring. So, some nursery directors in Beijing have encouraged technicians to work out pruning plans for each seedling that needs to be pruned in the field. With the diversification of public demand, the non-standard production mode will need greater attention.

1.2.3 Diversity of Nursery Functions

Nurseries in peri-urban areas are not only the permanent planning green space of the city but also the production base of urban tree seedlings. Its cultivation goals require greater abundance and a greater variety of seedlings. Functions of the nursery are also expanding: Some of them become the urban park with production functions so that people can buy seedlings and flowers, while also enjoying various kinds of trees, flowers, and plants (Figure 1-1). Other peri-urban nurseries serve the function of science education, so that citizens and students can understand the knowledge related to green and seedling production chains. And still others cooperate with the research institutes to carry out scientific studies related to seedling cultivation, becoming scientific research and teaching bases. With the further improvement of public living standards, the demand for a range of nursery functions will become more diverse.

Figure 1-1　A colorful garden-style city nursery (Photo by Liu Yong).

1.3　Theoretical Basis and Technical System of Directional Seedling Cultivation

　　Directional seedling cultivation means that based on the seedling order in the nursery, the site conditions where trees will be planted, and the purpose of users, the corresponding technical measures need to be adopted so that the seeds and seedlings meet the requirements of users in color, shape, physiology, resistance, and other aspects. The end goal is to match seedlings with the site to ensure successful afforestation performance. Given the diversity of cultivation goals and non-standard production modes, adopting the directional seedling cultivation is a good approach.

　　Matching seedlings with the site means to look at the circumstances from the perspective of the seedling users, which requires full consideration of the characteristics of the specific seedlings, the afforestation site conditions, and the nursery facilities. You will need to stipulate explicitly the type, size, and physiological characteristics of seedlings in the afforestation design. Ultimately, the directional seedling cultivation will need to align with the design requirements to ensure healthy target seedlings for afforestation.

Directional seedling cultivation is similar to the target seedling concept in the American forest nursery community. Both concepts have the purpose of combining nursery seedling cultivation with afforestation to prevent what can be considered blind nursery seedling cultivation. For example, in the past, cultivation may not have taken into account the intended future use of seedlings, which often led to structural surplus or structural shortage in quantity; a mismatch of seedling type, size, quality, and physiological characteristics with the planting site conditions; and the risk of improper seedling selection with insufficient regard for the planting goals and site conditions—any of which seriously affected the afforestation performance.

1.3.1 Basis of Directional Seedling Cultivation

To realize the directional seedling cultivation, three basic conditions are needed. First, various seedling cultivation techniques may be available. After years of development, a variety of seedling cultivation techniques have been formed. For example, seed-accelerating germination technology includes water soaking, low temperature stratification, high temperature stratification, variable temperature stratification, sulfuric acid soaking, plant hormone treatment, electrostatic field, osmotic regulation, potassium permanganate treatment, rare earth treatment, and many more. Regulating methods to avoid seedling stress include technologies of nutrition control, water control, chemical control, biological control, mechanical control, and so forth. Technical diversity has laid a foundation for directional seedling cultivation.

Second, the plasticity of seedlings is a factor. Plasticity means the capacity for being altered by varying environmental conditions. Seedlings are plastic in morphology, physiology, and resistance. Cultivation by some combination of seedling techniques, such as water stress treatment, fertilization, growth inhibitors, mycorrhizal fungi, photoperiod, and so forth, will result in seedlings that differ greatly in morphology, physiology, vigor, resistance, and survival rate of planting. For example, periodic water stress to the seedlings, to a certain extent, at the end of summer is conducive to the formation of apical buds to go to dormancy, improving hardiness and enhancing resistance; the short-day treatment in summer can induce fast-growing seedlings to form the terminal buds, promote hardiness, enhance stress resistance, and facilitate the afforestation and planting in the late summer or autumn; and fertilization technology can be used to promote seedling growth and increase biomass. In addition, under the premise of not causing the seedlings to overgrow, fertilizing in autumn can increase the content of nutrient elements within the seedlings to prepare them for growth after afforestation. Therefore, the plasticity of seedlings makes the directional seedling cultivation possible.

Third, nurseries and the end users of the plants are partners. "Direction" comes from the perspective of users, which determines the seedling type, size, physiological status, resistance characteristics, and so on, according to the afforestation purpose and site conditions. The nursery organizes the production in terms of variety selection, seedling cultivation technology, and seedling vigor protection, so they can provide seedlings that meet the user requirements according to the time and the site. After planting the seedlings, users observe their survival and growth situations and provide feedback to the nursery. According to the feedback information, the nursery adjusts the corresponding seedling cultivation techniques to cultivate more suitable seedlings for users. In this process, the nursery intervened in the afforestation purpose of users and site evaluation, while the user participated in the determination of the seedling type and seedling cultivation technology, both of which benefited the other, allowing them to become partners. Only in this way can the seedlings be cultivated directionally and suitably for the site conditions, ensuring a successful afforestation effect.

1.3.2 Technical System of Directional Seedling Cultivation

The technical mechanics of directional seedling cultivation is an information circulation system. First, users determine the purpose of tree planting, evaluate the site conditions, and negotiate with the nursery to determine the required seedling characteristics, such as tree species, variety, seedling type, age, size, physiology, resistance, and so on. Then, the nursery carries out variety selection or introduction; improves seed propagation, seed production, and seedling cultivation; regulates the growth environment according to the characteristics of required seedlings in the seedling cultivation process; and finally produces the seedlings for users. Users observe their survival and growth conditions after planting and provide feedback to the nursery, which will improve and perfect the cultivation technology in the next round of production (Figure 1-2).

1.3.2.1 Purpose and Site Conditions of Tree Planting

Characteristics of directed seedlings depend on the purpose of tree planting and the site conditions, both of which can vary widely. The characteristics of seedlings include tree species, quality, type, age, and potential resistance to adapting to the site conditions. If the site is in a city, the city is constructed for human living, and where people are found, culture is found; therefore, tree planting should conform to the city's history and culture to realize the planting purpose. For the construction of large-scale ecological protection forests in mountainous areas, seedlings of smaller size can be used; if plantings are intended for mountainous areas with worse site conditions, the tree species with strong resistance should be selected, and container seedlings can be used; for greenery around an urban government office building, seedlings with greater size should be selected. From the perspective of visual aesthetics, color matching should also be considered: Select not only evergreen trees but also colored-leaf trees to make spring, summer, and autumn colorful. For urban road greening, the safe passage of vehicles and pedestrians must be planned for, so large-size seedlings are needed, and a sufficient, clear bole height is required. For a double-decker bus to pass, the clear bole height should be more than 5 m (4.8 m at least above the ground is required for the standard double-decker). For riverside sites with high groundwater level, water-resistant poplar, willow, and *metasequoia* can be selected; pines and cypresses can be used in dry areas with low groundwater level. In terms of cultural heritage, tree species selection is directly related to human life, emotions, and tastes. For example, poplar grows fast with a tall tree shape and dense crown, which is a good tree species for shade and cool as well as for the wood. Thus, it is well-loved by the

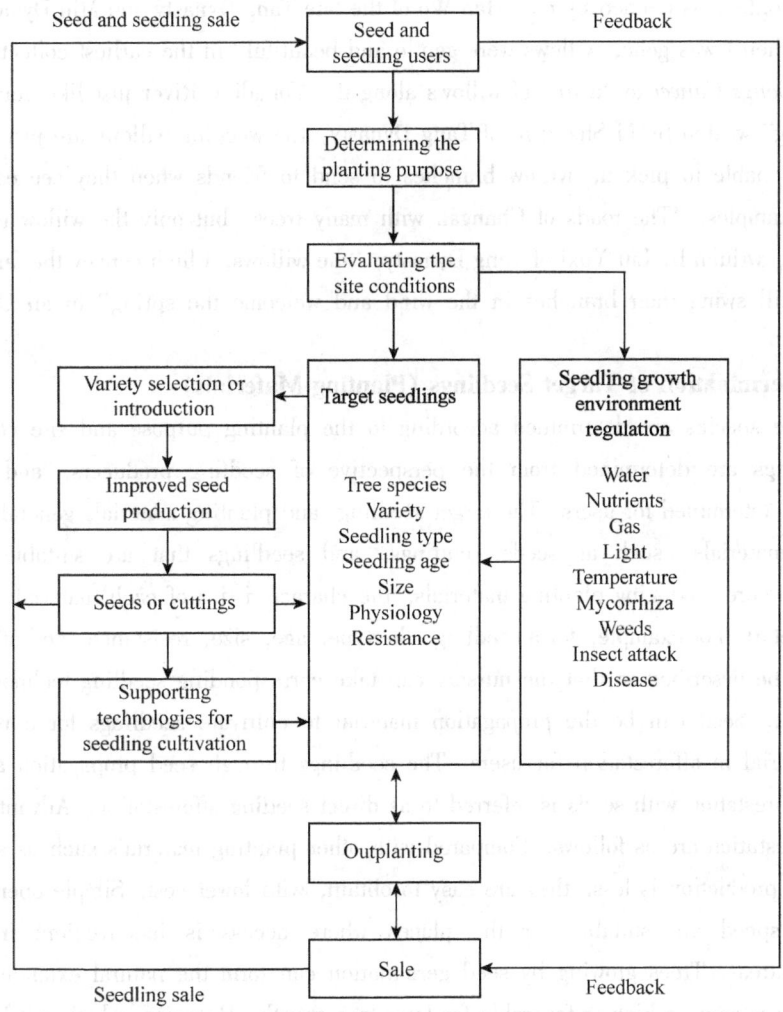

Figure 1-2 Schematic diagram of technical system of directional seedling cultivation.

people and planted around the house and beside the village road. "Farewell in the Pavilion in Jinling", written by the famous poet Li Bai of the Tang Dynasty, mentions "three poplars are facing to the gate of the pavilion"; in the poem of "Ci Yidu", Dai Liang of the Ming Dynasty mentioned the same scenery of "outside the city, the poplar is in the side of the pavilion". However, the aspen in the poplar family is listed as a cemetery tree species in the "Nineteen Ancient Poems". Liang Renxiao in Tang Dynasty organized and built Penglai Palace, and he selected the poplar as a green tree species. When he introduced the tree species to General Qibiheli, he told him that the poplar grew fast and it would make Penglai Palace full of trees. Qibiheli did not agree, responding that "the aspen usually represented the sadness and depression", which means that the aspen belongs to the cemetery, not the palace. Liang Renxiao understood perfectly. He ordered that all poplars be uprooted and the Chinese parasol tree (*Firmiana platanifolia*) be planted instead. Willows thrive in a wet environment and do well when planted beside rivers and roads. The branches of weeping willows drape gracefully downward and are accessible, so it is called "the whole body is weak so as to hang to people", as penned by poet Han Wo of the late Tang Dynasty and Min Dynasty. From "in retrospect, when I was gone, willows were gentle and beautiful" in the earliest collection of poems, *the Book of Songs-Caiwei* to "a line of willows along the Yongding River just like your long hair in passed Spring" written by Li Shangyin of Tang Dynasty, the weeping willows are just like farewell, so it is fashionable to pick up willow branches to send to friends when they see each other off. Additional examples: "The roads of Changan with many trees, but only the willow responsible for the farewell", written by Liu Yuxi of Tang Dynasty; "the willows, which convey the farewell feelings of human, still swing their branches in the wind and welcome the spring" in an Opera of Yuan Dynasty.

1.3.2.2 Determination of Target Seedlings (Planting Materials)

After the species are determined according to the planting purpose and site conditions, the target seedlings are determined from the perspective of seedling producers, and the planting materials are determined for users. The target seedlings and planting materials generally refer to the propagation materials, such as seeds, cuttings, and seedlings that are suitable for the site conditions. Before producing planting materials, the characteristics of each material should be described in detail. For example, for a seedling, the type, age, size, resistance, and other characteristics should be described so that the nursery can take corresponding seedling technology measures for production. Seed can be the propagation material to cultivate seedlings for a nursery or the planting material in afforestation for users. The seedlings through seed propagation are called the seedling; afforestation with seeds is referred to as direct seeding afforestation. Advantages of direct seeding afforestation are as follows. Compared with other planting materials such as seedlings, the work of seed production is less, they are easy to obtain, with lower cost. Simple operation and fast afforestation speed are suitable for the places where access is inconvenient in the remote mountainous areas. Trees growing by seed germination can form the natural extended root system with strong resistance, which is favorable for later tree growth. However, whether it is necessary to use seeds for afforestation depends on tree species, site conditions, and planting purpose. The species suitable for direct seeding afforestation include oaks (*Quercus* spp.), Chinese pine (*Pinus tabuliformis*), masson pine (*Pinus massoniana*), Yunnan pine (*Pinus yunnanensis*), and some shrub species. For example, Beijing used airplanes to sow *Pinus tabuliformis* seeds in mountainous areas, and Chongqing used them to sow *Pinus massoniana*, cypress (*Cupressus* spp.), loblolly pine (*Pinus taeda*), black locus (*Robinia pseudoacacia*), slash pine (*Pinus elliottii*), Chinese privet (*Ligustrun lucidum*), Chinese guger tree (*Schima superba*), Formosa sweet gum (*Liquidambar formosana*), and other tree species, all obtaining fine results. Bare-rooted seedlings are cultivated in the field and raised as bare-root, meaning they are not with containers. This method is a traditional seedling type with relatively simple technology and low cost, but a part of the root system is cut off during the lifting of seedlings. When the root system is exposed for a certain period from lifting of seedlings to planting, it reduces seedling vitality and affects the survival rate of planting. Therefore, the cultivation method is suitable only for the trees that are hardy and in better site conditions. Containerized seedlings are cultivated in a container with culture medium. In the process of cultivation, the culture medium and root system form the root plug; therefore, the root system is protected by the container and the root plug, and the seedling vitality allows for a high survival rate of planting. Compared to bare-rooted seedlings, container seedlings have been raised by a new technique that produces seedlings more suitable for poor site conditions. Asexual propagation seedlings are cultivated by using the vegetative organs, such as roots, stems, leaves, and buds, or described another way, by using plant tissues, cells, and protoplasts. It is characterized by the adoption of vegetative organs as reproductive materials, without pollination and formation of seeds. So, asexual propagation seedlings are able to inherit maternal characters by using this important way to cultivate excellent varieties and have great significance in the selection of new tree varieties. For example, in the process of smoketree (*Cotinus coggygria*) seedling cultivation in the Ming Tombs nursery in Beijing, a mutant seedling with purple leaf color was found. If the seeds were used for propagation, the leaf color would turn back to the usual green; however, the purple leaves can be maintained by using the branches of the variation seedlings for cutting propagation, so as to cultivate a new variety of *Cotinus coggygria* var. *purpurens*.

Large-size seedlings have been transplanted many times, and their diameter at breast height (DBH) is more than 8-10 cm. Because of their larger size, they can form the landscape after planting, which is a common type of greening seedlings for urban roads, parks, offices, and residential areas. With the development of forestry production and science and technology, more and more types of seedlings are available. According to propagation materials of seedlings, they can be divided into seeded seedlings and vegetative propagation seedlings. According to seedling cultivation methods, they can be divided into bare-rooted seedlings and containerized seedlings. According to seedling cultivation times, they can be divided into 1-year-old seedlings and perennial seedlings. According to whether the seedlings are transplanted in the cultivation period, they can be divided into transplanting seedlings (bed-changing seedling) and retaining seedlings. According to seedling environments, they can be divided into test tube seedlings, greenhouse seedlings, and field seedlings. On the basis of seedling size, they can be divided into standard seedlings (seedlings with national quality standards) and large-size seedlings. Because no unified classification standard is available, when determining the seedling type of planting materials in detail, there may be a cross of multiple classification methods, which has already appeared in production. For example, the transplanting container seedling is the container seedling formed by transplanting bare-rooted seedlings into the container; the container transplanting seedling (plug + 1) is the bare-rooted seedling that grows for one year after being outplanted from the container to the field; the tissue culture field seedling is the bare-rooted seedling outplanted from the tissue culture to the field.

1.3.2.3 Gene and Sex Determination of Target Seedlings

The target seedlings, or planting materials, will grow under specific site conditions and adapt to a specific environment in the future. Therefore, their reproductive materials should meet the requirements in genetics and gender. The provenance represents the genetic makeup of the tree species, which refers to the geographical source of seeds or other reproductive materials. The same tree species in a specific natural environment for a long time will inevitably form genetic characteristics and geographical variation to adapt to local conditions. If the conditions between planting site and provenance differ significantly, trees would grow poorly, even all die, which has occurred in China. For example, after the successful introduction of Xinjiang walnut (*Juglans regia*) by Beijing in 1958, more than 10 other provinces (autonomous regions) in China competed to introduce Xinjiang walnut seeds, with a high seed quantity of 10×10^4 kg, but many locations had unsuitable climatic conditions, resulting in huge losses. Generally speaking, the local provenance is most suitable for climate and soil conditions, so the local seeds should be used as much as possible to cultivate seedlings and plants in that same area. If no local tree species are available, however, and trees need to be transported in from other places, some tree species can refer to the *China Forest Seed Division*. For the tree species that are not included in the division, seeds should be selected as much as possible to match the local climate and soil conditions. The gender of the target seedling or tree species is also important. When using asexual propagation methods to cultivate seedlings, because the propagation materials are the vegetative organs (roots, stems, leaves, buds, and so forth), the gender of the seedling is the same as that of the seed tree that provides the propagation materials. The gender of a tree species being planted into a place is important to know. Male trees release pollen from their flowers, and female trees receive the pollen to reproduce sexually. Since no pollination occurs in male plants, you want to avoid a stand of trees made up entirely of male plants. But without male plants, female plants will not bear fruit. In contrast, bearing too much fruit can be also a problem. Poplars and willows are important urban greening trees in China. Both of these trees are dioecious; that is, female and male parts grow on completely different trees. When the male flowers mature, pollination occurs, without catkins and flying batting, whereas when the female flowers mature or the fruit matures, the poplar and willow catkins bear the seeds to fly away and distribute elsewhere.

In previous years, a large number of female plants were planted in many cities without considering the gender, which leads to urban environmental pollution by catkins of poplar and willow. Therefore, the genes and gender of the target seedlings need to be prudently determined at the beginning of seedling cultivation.

1.3.2.4 Supporting Technologies for the Directional Seedling Cultivation

The seedling technology adopted for the target seedling should relate to the factors that influence the survival and growth of the seedling in the planting area. Therefore, in site evaluation, determine the limiting factors of the site conditions that affect the survival and growth of seedlings. The nursery can then arrange corresponding technical measures in seedling production to strengthen seedling resistance in some aspects and ensure the success of afforestation.

For example, on the sunny slopes in the northern mountains of China, water is the main factor limiting the survival and growth of seedlings given the intense light and rapid water loss. So, water control measures should be adopted in seedling cultivation, and drought-resistance training to seedlings should be carried out to enhance their ability to resist drought. While in high altitudes or high-latitude shady slopes, low temperature becomes the main limiting factor; therefore, you need to take measures such as nutrition loading to enhance the cold resistance ability of seedlings. For planting Chinese pine and Yunnan pine in an area with poor soil, mycorrhizal inoculation can greatly improve the survival and growth of seedlings.

1.3.2.5 Target Seedlings Outplanting or Planting Time

Outplanting is the last step in directional seedling cultivation and an important stage of protecting the seedling quality. Improper measures at this time will reduce the vitality of seedlings or even cause seedling death. The appropriate lifting time of seedlings is determined according to the planting time. The lifting methods of seedlings are selected according to the characteristics of the tree species and the seedling size. The appropriate packaging and transportation methods are adopted according to the transportation distance. These decisions must be selected carefully so as to ensure the vitality of the seedlings and the survival rate of the planting.

1.3.2.6 Information Feedback

Nursery sales should collect feedback from the users on the pros and cons of the seedlings, so that the nursery can correspondingly improve the seedling cultivation technology to better meet the requirements of users.

1.4 Overview of Seedling Industry Development in China

The seedling industry in China is still in the early stages of development and is in the process of transformation from planned economy to market economy. In the past, seedlings were produced by an improved seed base and nurseries belonging to public institutions, which organized production according to the national plan. But now the industry must transform into an enterprise with independent legal persons and with production goals and methods that align to the market demand. And indeed, many modern seedling enterprises of independent legal persons are emerging, as well as many farmers who are using their contracted land to manage seedlings, thus forming a huge seedling production group. Correspondingly, the huge demand of ecological civilization construction, the rapid development of cities, and the improvement of human living standards has contributed to the formation of the seedling industry.

1.4.1 Opportunity Period of Seedling Industry

Because the country attaches great importance to the growth of ecological civilization, many ecological construction projects are in development for the future. Demand for seedlings is still heavy, which is of great significance to promote the development of seedling industry. From the perspective of land greening, the existing forest coverage rate in China is 21.63%. General guidelines suggest that to achieve a better environmental protection, the forest coverage rate should be at least 31%, which means the country still has an area of 960 000 km^2 (96 000 000 hm^2) that needs to be greened.

According to the rough calculation of 2000 plants per hectare, the number of seedlings needed is 192 billion. The current greening rate is increasing by approximately 1% per year, and the number of seedlings needed each year is 19.2 billion, which will take at least 10 years to complete.

From the perspective of urbanization construction, promoting urbanization has become an important task of our country. The urbanization rate in 2015 was 56%; the urbanization rate in developed countries generally exceeds 70%. If our target is 70%, we need approximately 14% more development, which is expected to take at least 10-20 years to achieve. Based on the experience of Japan, as long as the urbanization is advancing steadily, the investment in municipal landscape engineering industry can maintain a rapid growth. Based on this, domestic urban greening also has a flourishing period of growth that will take 10-20 years at least. From the perspective of beautiful China, the state has upgraded the "ecological civilization construction" to the height of the general layout of national construction and development. China has transformed the past "quaternity" which includes economic construction, political construction, cultural construction, and social construction to a "five-sphere integrated plan" striving to construct a "beautiful China". Under the background of "beautiful China" environmental construction, such as air control, soil remediation, and water area control is constantly strengthening. It is reasonable to think that in the future, the seedling industry in China will be improved by more specific policy support, more powerful government support, and more abundant financial guarantees. In brief, the next 10-20 years will still be a prosperous period for the rapid development of the seedling industry. Nonetheless, the development trend is not a straight, increasing line. The line may go up and down, with a change period of approximately every 5 years.

1.4.2 Development Trend of Seedling Industry

1.4.2.1 Improved Breeding

An improved variety is approved by the national or provincial seed and seedling management department or is tested by practice. An improved variety must have better ornamental appeal, resistance, and stronger market vitality. In seedling production, selecting the improved varieties is the first and the most important thing, which is directly related to the profit of the nursery. According to data provided by the National Forestry and Grassland Administration, 2776 improved tree species have been approved and extended. The utilization rate of improved varieties of major tree species for afforestation in China has increased from 20% in 1999 to 51% in 2015, and for some commercial forests and economic forests, the utilization rate of improved varieties has reached 70%, with the expectation that it will increase to 75% in 2020.

1.4.2.2 Diversification of Varieties

In the past, afforestation needs only to be green, because of the low expectations in the early days of this work. But with the improvement of living standards, human demands are more and more diverse. Plants not only need to be green but also colorful and beautiful, which makes afforestation pay more attention to diversity and functional tree species, and the development trend of variety diversification can be roughly divided into the following aspects.

(1) Novel Varieties

One of the highlights of the seedling market is the rapid development of colored-leaf trees. In the process of urban and rural greening, developed countries attach great importance to the collection and use of colored-leaf trees, and at present, the colored-leaf trees with obvious seasonal hues account for approximately 30% of the seedling market. But the proportion of colored-leaf trees in China is less than 10%, and even in the big cities, the main landscape belt is less than 20%. Colored-leaf trees can be roughly divided into red leaves, yellow leaves, blue leaves, purple leaves, and floral leaves.

Because the urban greening engineering institutions are eager to see the immediate effect, the large-size colored-leaf seedlings were adopted generally, so the colored-leaf trees with big seedlings sell well, but the tree quantity is far from enough. A good way to increase the number of large-size colored-leaf trees is to use grafting. For example, select a large-size seedling that is not a colored-leaf tree; cut the top of the seedling; then graft a colored-leaf tree of the same genus onto it. Colored-leaf seedlings whose DBH is more than 8 cm sell for a good price and are in demand, such as *Sophora japonica* cv. Jinye, *Sophora japonica* cv. Golden Stem, *Ulmus pumila* cv. Jinye, gold-edged box elder (*Acer negundo* cv. Aurea), and Chinese ash (*Fraxinus chinensis*).

Containerized seedlings of colored-leaf trees will become a new favorite in the market. Container seedlings have many advantages, such as developed root system, high survival rate of outplanting, complete plant crown, transplanting without seasonal restrictions, superior color performance, convenient transportation, and more, and container seedlings have been widely used in landscaping. The market share of container seedlings with colored leaves is larger in foreign countries, while in the domestic market, container seedlings of colored-leaf trees have just started, and the demand for them is growing.

(2) Stress-resistant Varieties

Many places that need afforestation are located in adverse environments for plants, so the tree species with strong resistance to a variety of stresses, protective effect, and the adaptability to poor site conditions are particularly needed.

(3) Low Manual Maintenance Varieties

Some varieties are well-suited for special green spaces, such as water-saving, slowly growing, and low manual maintenance tree species on a roof.

(4) Functional Varieties

More and more attention has been paid to functional tree species. With the increase of national environmental control efforts, the market of functional tree species will be growing. Some institutions in Beijing have carried out the selection of tree species that absorb fine, inhalable particles ($PM_{2.5}$) to improve air quality standards. In addition, the market is improving for economic forest species that meet the needs of new rural construction and ecological forest construction, with ecological benefits.

(5) Modeling Tree Species

At present, the application of modeling trees in landscape greening in China is not well-known, and few landscape modeling trees are on the market. In the important scenes of developed countries, the modeling tree embodies an intense national tradition and modern cultural atmosphere. It may be one of the basic trends of seedling industry development in the future; communities will skillfully apply the traditional culture and modern art to the modeling and allocation of landscape trees. In the trees suitable for modeling, deciduous trees include elm, hackberry, Chinese flowering crabapple, Hubei ash (*Fraxinus hupehensis*), crape myrtle, and others; evergreen trees include Chinese littleleaf box, citrus, *Podocarpus*, gurgeon stopper, pittosporum, Chinese holly, and others.

(6) Precious Tree Species

Precious tree species refer to rare and endangered plants, which are also national key protected plants. In the past these trees were not regularly used in landscape engineering. With the development of science and technology, and the improvement of economic level, the progress of artificial propagation of precious trees is accelerating. For example, Chinese yew (*Taxus chinensis*) is a first-class national protected plant. In the past, most of the *Taxus chinensis* were found in the wild, and people were not allowed to dig, transport, and trade them. But now, the market of *Taxus chinensis* bred artificially has begun to form, large-size seedlings have emerged, and the number of small-size seedlings is quite considerable.

(7) Native Tree Species

Native plants are those that are native or originating from a local area, that have experienced a long process of evolution in the local area, and so they are suitable for the local habitat conditions. Their physiological, genetic, and morphological characteristics are suitable for the local natural conditions, with stronger adaptability, such as *Sophora japonica*, *Ulmus pumila*, Siberian apricot (*Prunus sibirica*), Chinese chastetree (*Vitex negundo* var. *heterophylla*) in North China. With the further attention to afforestation of native plants, it has become an important trend for nurseries to adopt native plants to cultivate seedlings. A recent investigation of American nurseries indicated that the tree species in many nurseries have changed significantly, moving away from conifers and including more broad-leaved trees, shrubs, and even herbs. Around 2012, the East Tennessee Nursery in Delano, Tennessee, USA, carried 5 pine species, 17 oak species, and other broad-leaved species. In the Wildlife Habitat Nursery in Princeton, Idaho, USA, local broad-leaved trees, shrubs, and grasses are mainly cultivated to be used for vegetation restoration and afforestation in wetland riverbanks and other such places. More than 300 trees species, including grasses and shrubs, are cultivated in the Plants of the Wild Nursery in Tekoa, Washington, USA. A diversity of tree species and local plantings have become the direction of nursery development.

1.4.2.3 Facilities and Industrialization of Seedling Cultivation

With the rapid development of science and technology, industrialized seedling cultivation is on the rise. This approach adopts the enterprisation and scale to organize seedling production and management, based on advanced seedling facilities and equipment. Industrialized seedling cultivation is the large-scale seedling cultivation mode, which adopts modern biological technology, soilless cultivation technology, environment control technology, and information management technology to achieve standardized production including specialization, mechanization, and automation to realize high-quality seedling production with high efficiency and stability in controlled environmental conditions (Figure 1-3).

Compared with traditional seedling cultivation methods, industrialized seedling cultivation technology has the following advantages: Less seed consumption and smaller floor area; shortening the seedling age out of the nursery and saving seedling cultivation time; reducing the occurrence of diseases and insect pests; improving seedling production efficiency and reducing costs; and being conducive to the enterprise management and rapid promotion of new technologies to achieve continuous production. In addition, the facilities of seedling cultivation can reduce costs and improve efficiency, and without facilities, there will be no real standardization.

Figure 1-3 A modern greenhouse is one of the important facilities for container seedlings(Photo by Liu Yong).

1.4.2.4 Specialization

The traditional nursery is only the location for seedling cultivation, therefore, as long as a piece of land is available, a person can enter into the seedling industry. Because of outdated, "backward" management methods and management concepts, however, improved nursery benefits cannot be guaranteed. Currently, a modern nursery has to be a modern enterprise, which organizes production under the guidance of science and refinement. Such improvements require a large investment on a national scale. An up-to-date nursery requires the operators of the nursery determine the target orientation of the enterprise according to market capacity. They also must select or introduce varieties according to market demand, organize production methods, control costs and the input-output cycle according to commodity standards, carry out strict product quality inspection, and finally open up sale channels. Only by specializing every link can we obtain the best interests under the premise of ensuring the quality of seedlings.

1.4.2.5 Networking

Nursery sales are often a headache for nursery managers, and finding suitable seedlings is often a problem for afforestation institutions or individuals. The fundamental cause is that the seedling production became divorced from marketing, with the divergence between the seedlings used in construction designing and the seedling production. The emergence of the internet provides an ideal opportunity to solve this problem. The nursery releases its products on the internet, and users pick out and buy the products on the internet as well, which has brought convenience to many people. With the further development of the network, additional conveniences in production and marketing will continue to emerge. At this time, nursery producers should not miss the opportunity and should actively participate in the online network. Deeply integrating the seedling production, sales, and information feedback with the network will ensure success.

Questions for Review

1. In addition to the three aspects pointed out in the chapter, what are the important characteristics of tree seedling cultivation to you?

2. In the technical system of directional seedling cultivation, which aspects should be done well to cultivate target seedlings?

3. According to the cultivation goals of a variety of seedlings, please make all the seedling techniques sum up into a block diagram.

4. Please think about how to develop the large nursery and the small nursery, respectively, from the existing problems and development trends of the seedling industry.

5. This course adopts the question-oriented, participatory teaching approach. How do you cultivate your ability to ask questions? And find answers?

References and Additional Readings

CHENG F Y, 2012. Nursery of landscape plants[M]. Beijing: China Forestry Publishing House.
DUMROESE R K, LUNA T, LANDIS T D, 2009. Nursery manual for native plants: A guide for tribal nurseries [M]. Washington, DC: US Department of Agriculture, Forest Service.
LANDIS T D, DUMROESE R K, HAASE D L, 2010. The container tree nursery manual. Volume 7, seedling processing, storage, and outplanting. Agriculture Handbook 674[M]. Washington, DC: US Department of Agriculture, Forest Service.
LIU Y, LI G L, ZHU Y, 2013. Current situation and enlightenment of American tree seedling cultivation technology [J]. World Forestry Research, 26(4): 75-80.
SHEN H L, 2009. Seedling cultivation[M]. Beijing: China Forestry Publishing House.
WULIYASI, LIU Y, 2004. Theories of directed seedling cultivation for reforestation varieties[J]. Journal of Beijing Forestry University, 26(4): 85-90.
ZHAI M P, SHEN G F, 2016. Silviculture[M]. 3rd edition. Beijing: China Forestry Publishing House.

Chapter 2 Establishment of Forest Nurseries

Peng Zuodeng

Chapter Summary: This chapter provides an overview of forest nursery establishment, explaining site selection, position of nursery construction, scale of nursery construction, and criteria for nursery construction. Preparatory work that should be carried out for nursery planning and construction includes land survey, natural condition survey, and socioeconomic condition survey. Content of nursery planning and design is explained, including general planning, nursery engineering design, nursery cultivation process and technology design, organization and operation of management, investment estimate and nursery cost estimation, and more.

A forest nursery is a certain area of land used to meet the purpose of breeding seedlings; hence, the nursery is an important place for seedling production. To facilitate the development of seedling production activities and efficiently develop and utilize the land resources, it is necessary to conduct scientific and reasonable overall planning and design for the land. Often reconstruction and expansion of an old nursery can meet the needs of production development. The construction of the nursery and the overall scientific organization of seedling production are important links of forestry production. Nursery planning and design is related to nursery production and management and long-term development, not only will it directly affect the future nursery products—seedling yield and quality, outplanting performance—but also the economic, ecological, and social benefits of forest cultivation.

2.1 Site Selection of Forest Nursery

The selection of a nursery site is the basis of nursery establishment. The quality of the nursery site is directly related to the quality, yield, and cost of nursery seedlings. High quality and robust nursery trees can be cultivated only by selecting appropriate nursery sites and by strengthening management. If the nursery site is not selected properly, it will bring irreparable losses to the seedling production in the future. It is very important to choose a good nursery site whether to establish a permanent nursery or a temporary nursery.

Many factors determine whether the nursery site is suitable or not. You must carry out detailed investigation and studies on various conditions. After thinking fully about the strategic planning of local forestry development, determine the positioning of the planned nursery; the type and construction scale and standards; the local natural, social, and economic status; and the regional location and natural and social conditions of the nursery. Generally, requirements are lower for the land of a temporary nursery, while higher requirements are needed for a permanent nursery. In the specific selection of nursery sites, the following aspects are the main considerations.

Location. To select a nursery site, the main seedling application areas, the center or the nearby afforestation region should be considered. To meet the conditions of seedling production, it is necessary to be close to the main traffic line. In this way, seedling damage caused by long-distance transport can be reduced and the survival rate of afforestation can be improved. In addition, the materials to raise seedlings can be transported to the nursery at appropriate times, and the seedlings can be transported to the planting land in the shortest time. To facilitate the organization of production and to optimize arrangements for labor force, technology to grow seedlings, power machinery, and electricity sources, the nursery location must be as near to settlements as possible.

Topography. To establish a long-term nursery, choose a flat, well-drained site or a site with a natural slope of less than 3°. However, a 3°-5° sloping site should be selected for a nursery in an area with wet soil. For the construction of a nursery in mountainous and hilly areas, select a site with a gentle slope below 5° at the foot of the mountain or in a place where the slope may be steep but the conditions are available to build horizontal terraces. When a nursery is set up on a sloping land, the southeast slope must be in the northern forest area; with the east slope, the north slope, and the northeast slope in the southern forest region; and the southeast slope or the southwest slope in the high mountain area.

Land not suitable for a nursery includes low-lying land, opaque canyons, small open spaces located in dense forests, long-term flooded swamps, river beaches below the flood line, wind outlets, slope tops, high hills, and other shady locations.

Soil. The water and nutrients that come from the soil are needed for seed germination and seedling growth, so the soil condition is an important factor when selecting a nursery site. Appropriate soil conditions for a nursery are mainly manifested in soil fertility, structure, texture, pH, and so forth. Generally, the soil of the nursery field is preferably agglomerated, with a relatively fertile sandy loam, loam, or light clay loam. The thickness of the soil layer should be >50 cm. The adaptability to soil pH value varies widely across tree species. Most coniferous tree seedlings require neutral or slightly acidic soils with a pH value of 5.0 to 7.0. The suitable soil pH value for most broad-leaved tree seedlings is 6.0 to 8.0. Salt content in the soil should be controlled below 0.1%. The heavier saline-alkaline soil is generally not selected for nursery land.

Hydrology. Water is an indispensable resource for growing seedlings in a nursery. A water source is necessary to satisfy nursery irrigation under any conditions. The water source of the nursery can be rivers, lakes, ponds, or reservoirs. If no such surface water source is available, adequate groundwater could be a concern. Nursery surface water sources are generally superior to groundwater sources, and the reason is that the temperature of the surface water is high, the water quality is soft, and contains certain nutrients, so it should be used as much as possible. Irrigation water should be fresh water, and the salt content should not exceed 0.1% to 0.15%.

Pay attention to the level of the water table when considering a site for the nursery. The appropriate groundwater level varies with soil texture. In general, the sand soil is less than 1-1.5 m; the sandy loam soil is less than 2.5 m, and the light clay loam soil is less than 4.0 m.

Insect Diseases and Bird and Animal Injury. When the number of underground pests exceeds the allowable amount prescribed by current standards (for example, local or national) or where serious bacterial infections are present (such as bacterial blight and root rot) the sites are not suitable for a nursery. If measures can be taken to control or eradicate the existing diseases and pests without affecting the seedling growth, however, these nursery sites can still be considered.

Try to avoid areas with any plants infected by disease bacteria or intermediate hosts of diseases and insect pests; also avoid trees that can attract diseases and insect pests. In general, bird habitat, rodents, and other animal-damaged land should not be used as nursery land.

In addition, the selection of a nursery needs to consider the demands of the nursery stock market, the development of the local economy, and the attitude of the local government to forestry construction. You will want to consider potential pollution to the surrounding environment when the nursery is completed and the seedling production is carried out. Pollution sources should be kept away from the site as much as possible.

2.2 Planning and Design of Forest Nursery

2.2.1 Overview of Nursery Planning and Design

The planning and design of the nursery is the overall planning and technical design of service facilities for the proposed land with seedling production potential. The purpose of nursery planning and design is to optimize a scientific and reasonable layout, make full use of the land, and arrange the investment reasonably according to the characteristics of seedlings, so as to reduce loss and waste and to grow the seedlings with diverse varieties and high quality to maximize the economic and social benefits of the nursery. A designer must first clearly and deeply understand the purpose and scale of nursery construction according to the feasibility study report of nursery construction. Conceptually, the nursery is not only the land to cultivate seedlings but also the independent management institution to cultivate, supply, or sell seedlings for afforestation. According to the usage time, a nursery is generally divided into the permanent nursery and the temporary nursery. The permanent nursery is characterized by long-time management, a large area with diverse seedling varieties, and is suitable for realizing intensive operation through mechanization and setting modern seedling production facilities. A temporary nursery generally refers to the nursery set up for the short-term afforestation tasks in a certain area, often divided into the mountain nursery and the nursery in the forest. Characteristics of the temporary nursery are that it be located close to the afforestation land, requires only short-time management, and it is made up of a small area and single seedling varieties. When the water source and soil conditions meet the basic needs of the seedlings in the temporary nursery, the existing land and facilities are used to carry out the seedling cultivation, without needing the infrastructure construction invested more funds and standardized planning and design. Therefore, the general planning and design of a nursery refers primarily to the permanent nursery.

2.2.1.1 Setting of Nursery Construction Goals

Because of the complexity of the production object, scope and conditions in agriculture and forestry, the nursery, as a place to cultivate seedlings, also has multifunctional characteristics, that is to say, a nursery cannot cultivate only the seedlings of timber forest species but also the seedlings of protective forest species, garden greening species, and fruit trees. It cannot be only a production nursery but also a scientific research nursery, even with the characteristics of tourism. Therefore, before planning and designing the nursery, the main management and service objects of the nursery must be clear, including seedling species, the grade of seedling for harvest, and its possible application areas and afforestation site conditions.

(1) Setting of Nursery Management and Service Objects

The main function of the nursery is to cultivate seedlings. Generally, according to the seedling types and service objectives, the nursery can be divided into agricultural nursery, forestry nursery, garden nursery, teaching practice and experiment nursery, and comprehensive nursery. Among them, forestry nursery can also be divided into timber forest nursery, protection forest nursery, fruit tree nursery, and special tree nursery. Because of the characteristics of intensive cultivation in agricultural production, sometimes the boundary blurs between agricultural nursery and agricultural land. Therefore, it is rare to take agricultural nursery as an independent construction project on which to carry out planning and design. Meaning, the nursery planning and design is primarily for other types of nurseries. Generally speaking, while planning the nursery, the construction company needs to make a preliminary review of the operation objectives to determine the feasibility of the project. The construction company will then make the construction features clear in the design contract and the assignment.

(2) Setting of Nursery Construction Standards

As part of the planning and design of the nursery, you need to determine the standards of the nursery. Nursery construction standards generally refer to nursery facilities, operational technology level of seedling cultivation, and management efficiency. And such standards mainly depend on the mechanization level, the degree of production specialization, the advancement and applicability of nursery facilities, the organization and management efficiency of nursery production, the quality and cost of nursery products, the market and technology development ability of nursery products, and the influence of the nursery in the region. The construction standards of the nursery can be roughly divided into the following categories:

Modern Nursery. The main characteristics of a modern nursery are as follows: Mechanization in nursery seedling production; industrialized seedling output accounting for more than 50% of the total nursery seedling output with the automation production process; all the seedlings cultivated are part of a high level of specialization; the quality of the seedlings out of the nursery meet the firstclass seedling standard in the *National Standard of the Seedling Standard of Main Afforestation Tree Species*; networking and informationization in nursery production and marketing management; the nursery has the material conditions and technical ability to develop and promote new products and exerts a wide range of influence and demonstration effect in the forest vegetation zone.

Mechanized Nursery. The main characteristics of a mechanized nursery are as follows: Mechanization is realized in nursery seedling production; the nursery has certain facility conditions for industrialized seedling, which can produce seedlings when market demand exists; the seedlings of the main tree species cultivated are in a high level of specialization; the quality of all seedlings out of the nursery has reached the above second-class seedling standard in the *National Standard of the Seedling Standard of Main Afforestation Tree Species*; seedling sales through networking and informationization; the ability to promote various new seedling technologies and to cultivate new varieties.

Specialized Nursery. The main characteristics of a specialized nursery are as follows: The specialized production machinery is adopted in some seedling production operations in the nursery according to the production needs; there can be no greenhouse seedling facilities; all species and seedlings cultivated in the nursery have a high level of specialization; the seedling technology has reached or exceeded the requirements of *National Standard in the Technical Specification for Seedling Cultivation*; the quality of seedlings out of the nursery is over the second-class seedling standard in the *National Standard of the Seedling Standard of Main Afforestation Tree Species*; seedling sales utilize networking over the internet; the nursery has the ability to promote various new seedling technology and to cultivate new varieties.

Common Nursery. The main characteristics of a common nursery areas follows: The nursery production mainly relies on handwork; generally, no greenhouse condition is available for seedling cultivation; the seedling cultivation technology meets the requirements of the *National Standard in the Technical Specification for Seedling Cultivation*; the quality of seedlings out of the nursery meets the requirements of the *National Standard in the Seedling Standard of Main Afforestation Tree Species*; selling seedlings and managing the nursery is accomplished in the traditional way.

2.2.1.2. Nursery Planning and Construction Scale

The nursery construction scale is one of the important problems that nursery planning and design has to face. The scale decisions are mainly reflected in nursery area and investment quota.

(1) Determination of Nursery Area

The determination of nursery area is based primarily on the following factors.

Development Planning of Forestry Construction and Afforestation in Local Area. The nursery area depends on its production task, which is closely related to the development of local forestry and urban greening. The annual scale of afforestation by artificial seedling planting, composition of afforestation tree species, and the possible interannual changes in the development stage and stable period of forestry construction will cause great changes in the demand for seedlings. In general, the task of afforestation areas is heavy, and the planned land area for forestry development is large, as is the planned nursery area. However, for the regions where the interannual afforestation task and artificial regeneration task change greatly, when planning the scale of the nursery area, neither the extreme upper limit nor the average value of seedling demand in individual years should be taken. Instead, 70% to 80% of the maximum seedling demand in historical years or 90% of the maximum annual seedling demand in forestry planning in this region should be taken as the basis for determining the nursery area.

Distribution Density and Transport Seedling Radius of the Nursery in Local Area. Generally, the shorter the average distance is between nurseries, the higher the distribution density of nurseries and the smaller the transport seedling radius and the seedling area are, so the smaller the nursery area is, and vice versa.

Effective Utilization Rate of Nursery Land. The proportion of nursery production land to the total area is the effective utilization rate of nursery land, which determines the land potential to cultivate seedlings in the nursery. The lower the effective utilization rate is, the larger the nursery area should be.

Technical Measures in Seedling Cultivation and Mechanization of Nursery Design. If the main seedling cultivation process is expected to be mechanized and large-scale equipment is planned, the nursery area should be large; whereas the nursery area should be small if the nursery work mainly relies on manual work.

Natural and Socioeconomic Conditions of Nursery Area. Natural conditions of the nursery area, such as topography and water source, as well as labor resources and traffic conditions will have an impact on the nursery scale.

(2) Determination of Nursery Investment Scale

Another common index reflecting the scale of nursery construction is construction investment. Nursery investment is affected by nursery area, modernization or mechanization of the nursery, economic strength of nursery construction company or investment department, decision-making basis of nursery construction, and other factors.

(3) Classification Standard of Nursery Construction Scale in China

According to the production objectives and tasks, the nursery construction projects in China are generally divided into four categories: Super-large nursery, large-scale nursery, medium-scale nursery, and small-scale nursery. Classification standards are shown in Table 2-1.

Table 2-1 Classification standard of nursery construction scale in China

Types	Area /hm^2	Total annual seedlings output /10 000 plants	Annual improved cuttings output /(10 000 individuals/cutting)	Annual container seedlings output /10 000 plants
Super-large nursery	≥100	≥5000	≥1000	≥1000
Large-scale nursery	≥60	≥2000	≥500	≥300
Medium-scale nursery	≥20	≥500	≥100	≥50
Small-scale nursery	<20	≥100		

Notes: Quoted from the *Construction Standard of Forest Seedling Engineering Project* (Trial)(2003), issued by the State Forestry Administration.

The super-large nursery is geared to the needs of batch production of high-quality seedlings for the whole country and large ecological areas and is able to macroscopically control and regulate the production. Through the introduction of advanced technology, breeding materials, and management experience at home and abroad, the production can be organized according to the requirements of industrialization, and production and management technology can be in line with international standards. The large-scale nursery produces high-quality seedlings in batches for a province (meaning an autonomous region, such as a municipality) or for key ecological engineering, with the function of regional regulation, demonstration, and experiment. Through the introduction of domestic and foreign scientific research achievements, new technologies, and new products, production can be carried out according to the requirements of scale and factory-like efficiency. The medium-scale nursery produces high-quality seedlings for a city (state) or key ecological project, and the nursery has certain regulating and demonstration functions. It can introduce domestic and foreign advanced technology, demonstrate superior management experiences, and utilize original breeding materials in an improved breeding center to organize production according to the scale requirements. The small-scale nursery is required to quantificationally produce seedlings for a place, county (city, bureau), general forestry, and garden engineering projects. Some of these nurseries adopt domestic and foreign advanced technology and rely on new and improved varieties provided by the large- and medium-scale nurseries to produce seedlings.

2.2.1.3 Nursery Planning and Design Procedure

Nursery construction is a project with a large amount of capital investment, with a one-time investment of hundreds of thousands, or even hundreds of millions, of Chinese renminbi (CNY). The state has strict requirements on the design companies and the procedures before and after planning and design. For large-scale projects, the design company should have a design qualification of class A.

A series of procedures must be completed during the planning and design of a nursery. The general procedure is as follows.

(1) Feasibility Study Report

Generally, the construction institution entrusts the design company to prepare the feasibility study report of the project, which needs to clarify the following contents: Project background and market demand forecast, construction conditions and advantages, preliminary construction scheme, preliminary investment estimation, benefit analysis, and project feasibility analysis.

(2) Bidding of Design

After the project is approved by the investor, the bidding system should be adopted to select the design company. The construction institution sends the prepared bidding document to the companies that intend to undertake the design, and the design companies submit the bidding document to the construction institution according to the requirements. The construction institution organizes relevant experts to evaluate bidding documents and determine the winning bidder.

(3) Signing the Contract and Issuing the Assignment

The construction institution issues the design assignment to the bid winner to clarify the tasks, completion period, and relevant requirements of the design institution. Both parties have to sign a contract to clarify their own rights and obligations.

(4) Preliminary Design

Preliminary design, also known as the overall planning and design of the nursery, is the core part of the whole nursery planning and design. After receiving the assignment and signing the contract, the design company organizes and carries out the preliminary design.

(5) The Approval Meeting for Preliminary Design

After the completion of the preliminary design, the construction institution organizes relevant experts to examine and approve the design. On the basis of the comprehensive evaluation on the design, any shortcomings of the preliminary design should be pointed out and suggestions for modification and improvement need to be put forward.

(6) Construction Drawing Design

After modification and improvement of the preliminary design, the construction drawing design of the individual civil engineering of the nursery can begin. The design of civil engineering includes details such as buildings, roads, bridges, greenhouses, and seedbanks in the nursery, and the design should be detailed enough to facilitate construction. As these buildings involve personal and property safety, the construction drawing design has to be conducted by institutions and individuals with design qualifications and that comply with relevant national regulations. If an agreement has been made that the construction drawing design is included in the overall planning and design of the nursery, the design institution should consider recruiting the designers of the civil engineering plan when organizing the design team at the beginning. Nursery planning and design mainly refers to the preliminary design in the above procedures and also includes the construction drawing design. The specific contents should be negotiated with the construction institution and made clear in the assignment and contract.

2.2.2 Preparation for Nursery Planning and Design

For the next steps in nursery planning and design work, the assignment design specifications for the nursery should have been received from the entrusted design institution, and those documents need to specify the location, scope, deadline for completing the task, and relevant requirements. The entrusted design institution will work with relevant institutions to form a leading organization and a planning and design team to collect text and drawing materials, and to make preparations for instruments, forms, stationery, funds, accommodation, transportation, communication, and so forth.

2.2.2.1 Investigation and Survey of the Nursery Site

In terms of the assignment design specifications issued by the construction institution, the design and planning institution organizes the technical backbone of the project to carry out on-the-spot inspection. During the inspection, according to the collected map data and accompanied by the local personnel familiar with the current situation of the planned nursery land, they check the regional boundary and the existing fixed facilities and buildings, such as houses, irrigation and drainage channels, water wells, ponds, roads, bare stones, electric piles, trees, and woodlands within the planned land area, and they record the current situation through texts and photos. At the same time, they check whether the topography and slope direction of the nursery site meet the requirements of nursery construction. In general, the terrain should be as flat as possible. If it is a slope, the natural inclination should be less than 3°; if a nursery is built in mountainous and hilly areas, it is better to be in a gentle slope of less than 5° at the foot of the mountain; if a nursery is to be built on a step land, each step surface should meet the above requirements. The slope direction of nursery land should meet the requirements for a given location, for example, the slope ought to be on the southeast side in a north forest area; on the east slope, the north slope, and the northeast slope in the south forest area; and on the southeast or the southwest slope of the semi-sunny slope in the high mountain area. During the investigation, if the selected site of the nursery does not meet the requirements it should be documented and a new nursery construction site should be determined through consultation. After the investigation of the nursery site, the technical personnel should be organized immediately for a topographic survey, which needs to include the relevant roads within and outside the nursery site. During the survey, obvious signs such as fixed facilities and buildings should be marked in the topographic map. Soil, land use type, earth volume of earth hummock, and the distribution area of diseases and insect pests should be mapped, respectively. The scale of a topographic map is 1:500-1:2000. The scale topographic map (or ichnography) drawn is the base map for division and final mapping. If the nursery construction area has recently been measured and such drawing materials are available, it should be used as much as possible.

2.2.2.2 Investigation on Natural Conditions of Nursery

The natural conditions of the nursery have an important influence on the design of nursery production technology, which also provides important basic data for nursery planning and design. The investigation of natural conditions of the nursery includes primarily the following contents.

(1) Soil Investigation of Nursery Land

The soil profile survey is adopted in the soil investigation. Multiple survey points should be set up in representative places of the nursery land, as generally, a profile in 1-5 hm^2, but not less than three profiles. A soil profile should be excavated at each point, with the profile specification of 1.5-2 m in length, 0.8 m in width, and depth to parent material horizon (the shallowest of 1.5 m). According to the soil profile, the soil type and the characteristics and properties of each soil layer are determined. The following factors should be recorded for each profile:

i. Profile location and number (using a sketch to mark the location).

ii. Altitude, slope, and groundwater level.

iii. Soil color, texture, structure, humidity, soil consistency, gravel content, plant root distribution, the morphological characteristics of the whole profile according to the layer, and determine the soil group, subclass, and soil species name.

Soil samples should be collected at each point for indoor physical and chemical properties analysis. The analysis indexes include pH value, organic matter content, total nitrogen content, available phosphorus content, available potassium content, hydrolytic acid content, total salt base amount, and hygroscopic moisture content of soil. Generally, the sandy loam soil, loam soil, or light clay loam soil with aggregate structure and rich texture are suitable for nursery land. Nursery land should be divided operation areas according to the investigation results, and the soil improvement project should be put forward as required.

(2) Investigation and Collection of Meteorological Data

Generally, meteorological data of the nursery should be collected from the local meteorological department, including the historical extreme meteorological factors in the area, all data of the past five years, and data of the current year or each month of the previous year. Specific indexes include:

Temperature. Annual, monthly, and daily average temperature, average temperature in January and July, absolute maximum and minimum daily temperature in air and on soil surface, sunshine duration and percentage of possible sunshine, the initial and final date of daily average temperature of 10 ℃ and the accumulated temperature during this period, the initial and final date of daily average temperature of 0 ℃.

Precipitation. Annual, monthly, and daily average precipitation, maximum precipitation, precipitation duration and distribution, the longest continuous precipitation days and amount, the longest continuous no-precipitation days, and annual evaporation.

Wind. Average wind speed, main wind direction, maximum wind speed, frequency, and wind days of each wind direction in each month.

Frost and Snow. Snowfall and snow cover days and the initial and final date, the maximum snow depth, the start and end date of frost, the frost-free period, fog days and the longest continuous hours of fog at one time, hail days and sandstorm, thunderstorm days, the depth of frozen soil layer, the maximum depth of frozen soil layer, and the freeze and thaw dates at the depth of 10 cm and 20 cm.

Local Microclimate. Small terrains with insufficient light, low temperature, and strong wind should not be used as nursery land.

(3) Hydrological Investigation

Water source is an indispensable environmental factor for the nursery. Nursery planning and construction has to investigate the source of water for irrigation, which requires not only sufficient quantities but also a salt content that does not exceed 0.15%. Proper depth of groundwater level should be below 1-1.5 m in sandy soil area, below 2.5 m in sandy loam soil, and below 4 m in clay loam.

Marshlands with long-term water accumulation and land below the flood line level should not be used as nursery land.

(4) Investigation of Diseases and Insect Pests

The investigation of diseases and insect pests includes the local situation of plant diseases and insect pests, such as the types, occurrence frequency, and hazard rating. A stratified survey conducted by digging pits is usually adopted to investigate diseases and insect pests in nursery land. The sample pit covers an area of 1.0 m × 1.0 m and is excavated to the depth of the parent rock. Number of sample pits: 5 soil pits are dug in areas smaller than 5 hm^2; 6-10 soil pits are dug in areas of 6-20 hm^2; 11-15 soil pits are dug in areas of 21-30 hm^2; 16-20 soil pits are dug in areas of 31-50 hm^2; 21-30 soil pits are dug in areas larger than 50 hm^2. The species, quantity, degree of damage to plants, disease history, and control methods of diseases and insect pests are investigated in the soil pits. By investigation, disease and pest control can be incorporated into the nursery planning and design.

(5) Vegetation

The local main vegetation types, mainly herbaceous species, should be investigated to understand the degree of possible weed damage to future seedling production. The survey adopts the method of quadrat survey, with the area of 2.0 m × 2.0 m, and 5-10 quadrats are set up equably per hectare.

2.2.2.3 Investigation of Social and Economic Conditions

Social and economic conditions have an important impact on the nursery operation and development, and they are components that have to be seriously considered in nursery planning and design. The investigation should include the location of the nursery, the economic conditions of neighboring institutions and villages, the social development situation, traffic conditions, power supply, water resources, and labor supply. It is also necessary to investigate the seedling production capacity and level of nurseries in the surrounding areas so as to master the market situation and corresponding production capacity of seedlings.

2.2.2.4 Related Material Collection of Equipment and Quota

Professional production equipment is needed in nursery production, especially for modern, large-scale nurseries. Therefore, in the process of nursery planning and design, you need to determine what professional equipment and/or mechanical equipment will be needed in the nursery operation, including type, model, manufacturer, price, and so forth. For some nurseries, because of limited start-up investment finances, it is impossible to purchase all the equipment, so it is necessary to investigate the possibility of renting the equipment locally.

The quota and changes in market price of construction and production costs related to local infrastructure and labor should be collected as well.

2.2.3 Contents of Nursery Planning and Design

The core work of nursery planning and design is the overall planning and design of the nursery on the basis of field investigation and relevant data collection, that is, preliminary design. Nursery planning and design should be carried out in accordance with the feasibility study report approved by the appropriate departments, including the nursery site planning and nursery engineering design.

2.2.3.1 Nursery Site Planning

According to the requirements of natural topography, production technology, functional division and external connection, the nursery site should be carefully planned in terms of the calculation of the size of the area and land use division according to local conditions. These calculations are important for production reducing consumption, beautifying the nursery, and using the land efficiently.

(1) Calculation of Nursery Area

The total area of nursery land includes production land area and any auxiliary land area. Size of the nursery area should accommodate the tasks for which the nursery is intended. In the case of a land resource shortage, it is very important to efficiently arrange the seedling land according to the seedling production task to avoid the waste of arable land resources.

Calculation of Production Land Area. Production land refers to the land directly used for seedling cultivation, which includes sowing cultivation area, asexual propagation area, transplanting cultivation area, experimental area, fruit tree seedling cultivation area, cutting orchard, and container seedling production area. And it is determined according to the seedling production task of each tree species, seedling yield per unit area, and rotation system adopted.

According to the type, quantity, specification requirements, production duration, seedling cultivation method, rotation and/or fallow crop areas, and the seedling yield per unit area of each tree species, the seedling area of a certain tree species is calculated using the following formula:

$$P = \sum_{i=1}^{n} \frac{N \times A}{E_i} \times \frac{B}{C}$$

In which:

P is the seedling area of a certain tree species (hm^2);

N is seedling quantity of the tree species needed each year (individual plant);

A is the duration of seedling cultivation (year);

E_i is the seedling yield per unit area of the tree species (individual plant);

n is tree species quantity;

B is the quantity of zones in crop rotation;

C is quantity of crop rotation zones occupied by the annual seedling cultivation of the tree species;

B/C is used only when implementing crop rotation.

For outplanted seedlings cultivated for more than two years, the production land can be calculated by using the above formula, respectively, for sowing area and transplanting area, and then adding them to determine the total area. When determining the total area of production land, the loss of seedlings in the process of seedling production, such as tending, lifting, and storage, should be considered, and the area can be increased by 3% to 5% according to the calculation results. After calculating the production land of seedlings of each tree species and other production land, these production land areas should be summarized to obtain the total production land area.

Calculation of Auxiliary Land Area. Auxiliary land refers to the land without seedlings, including road systems, drainage and irrigation systems, buildings, yard, shelter forest belt, and so forth. According to the *Design specifications of nursery garden engineering of forest* (LY/T 128—1992), auxiliary land for a large-scale nursery should not exceed 25% of the total nursery area, and auxiliary land for medium- and small-scale nurseries should not exceed 30% of the total nursery area.

Production land area plus auxiliary land area is the total area of nursery land. In the actual planning and design process, however, the total area of the nursery is often determined before the planning stages because of land acquisition availability and property rights. Circumstances are few in which the nursery area is determined completely according to the needs of seedling production. Therefore, the calculation of the nursery area must be based on the seedling cultivation methods commonly used, the location characteristics of the nursery land, the water source and soil fertility conditions, how the operation areas and other auxiliary lands are divided. After those considerations are determined, then their areas can be combined and summed.

(2) Nursery Division

To make full use of the land for facilitating production and management, you will need a comprehensive division of nursery land and a topographic map with 1∶500-1∶2000 scale based on a field survey. Then, according to the seedling characteristics, tree species characteristics, and natural conditions of nursery land, the divisions can be carried out. Nursery land division includes production lands and non-production lands.

Principles of Nursery Land Division

i. Land use planning should be convenient for nursery seedling production and management.

ii. The operation area of the nursery should be carried out based on the comprehensive consideration of various natural conditions of the nursery and the biological characteristics and ecological habits of tree species.

iii. Nursery division should be consistent with the seedling production technology and management level.

iv. The operation area should follow the north-south trend and have a proper proportion of length and width. Generally, the length of the operation area is required to be 200-300 m for the large-scale nursery or for the modern nursery with a high mechanization level and to be 50-100 m for the medium-scale nursery or for the nursery with livestock farming as the main activity. The width of the operation area is 1/2 or 1/3 of the length, and in areas with good drainage it can be wider; or with poor drainage, it can be narrower.

Production Land Division. According to the purpose and area of nursery management, the contents of the nursery production area differ greatly. But according to the methods of seedling cultivation, the forest nursery generally includes a sowing and seedling cultivation area, asexual propagation area, transplanting area, and greenhouse area. The suitable location of each production area has to be determined according to the characteristics of the types of seedling production and the conditions of nursery land (mainly terrain, soil, water source, management, for example.). Each production area should be kept as complete as possible and not be divided into separate blocks that are not adjacent to each other (Figure 2-1).

To facilitate production and management, the production area is usually divided into several operation areas based on the road, and the size of the operation areas depend on the nursery scale, topography, and mechanization, which should be 1-3 hm^2, with a square or rectangle shape. To

Figure 2-1 Planned nursery production area (Photo by Liu Yong).

facilitate calculations, the area should be the integer as much as possible. If the length is 100 m and the width is 100 m, then each operation area is 1 hm^2.

When the nursery area is large and many types of seedlings are being grown, the cultivation area of different types of seedlings should be conveniently arranged.

i. Sowing and Seedling Cultivation Area. Because the resistance of young seedlings to the external environment is weak, seedling management requires close attention. Therefore, select the land area with flat terrain, gentle slope inclination, thick soil layer, good fertility, convenient irrigation and drainage, and leeward sunny orientation for the placement of young seedlings.

ii. Asexual Propagation Area. According to the biological characteristics of each tree species, select a location that meets the requirements of cultivation technologies such as cutting and grafting, for example, choose loose soil, good fertility, and good irrigation and drainage.

iii. Transplanting Area. Independent and complete sections should be selected according to seedling cultivation specifications, growth speed, and characteristics of tree species. Transplant seedlings have developed root systems and strong resistance, so they could be set in a section with medium soil condition.

iv. Experimental Area. According to the characteristics and technological conditions of tissue culture, new species introduction, and new varieties, an area near the office site that is convenient for observation and management should be selected, in combination with the greenhouse for plant division.

v. Greenhouse Area. Greenhouse facilities in large-scale, modern nurseries belong to the nursery production area. Construction of a greenhouse requires a flat terrain, or with only a gentle slope not more than 1%. Avoid building greenhouse groups on a slope inclined to the north. If the greenhouses are built on the north slope, the design must overcome the mutual blocking of light between the greenhouses. The sun light irradiation angle is smaller in the morning and evening time on the north slope, which will prolong the shadow length of any obstacle. In the case of greenhouses, this increases the spacing between greenhouses, thus increasing the overall area of the greenhouses. For the construction of a glass greenhouse, the foundation has to be stable. Deep soil layer, high organic matter content, good drainage, and a not-too-high groundwater level are required for the greenhouse with soil cultivation. The greenhouse also requires a stable water source, electricity, and a heat source. Therefore, when planning the nursery, the greenhouse is generally arranged near the office site.

vi. Cutting Orchard. Soil conditions of the cutting orchard are the same as those needed for the asexual propagation area.

vii. Germplasm Resource Bank. Land area and soil conditions of the germplasm resource bank are the same as those for the transplanting area.

Non-production Land Division.

i. Road Network. The road network should be determined according to the topography, terrain, and convenience of seedling production. In general, a main road is set in the middle of the nursery, which connects the yard, warehouse, and machine room in the nursery; it is also connected with the external traffic trunk outside the nursery. Its width is subject to the passing vehicles, generally 3-4 m for small- and medium-scale nurseries and 5-8 m for large-scale nurseries. To guarantee the production, facilitate management and life and meet the needs of machine farming, the layout of the by-road and by-path in the nursery should be arranged. The by-road and by-path should be able to reach each operation area with a width of 2-5 m. Temporary footpath with width of 0.5-1.0 m can be set in the operation area as necessary. Around the large-scale nursery with high level of mechanization, nursery ring road can be set to be convenient for vehicles, with the width of 3-5 m.

ii. Irrigation System. The irrigation system is mainly composed of water source, water lifting, water delivery, and water distribution (Figure 2-2). The water delivery system has the greatest impact on the layout of the nursery. The system is divided into a main channel and branch channels. The function of the main channel is to directly draw water from the water source to supply water to all of the nursery land. It has a larger capacity specification with a general channel width of 1-3 m. The branch canals supply water from the main canal to a specific seedling cultivation area of the nursery. These canals have a smaller specification and a width of 0.7-2 m. The specifications of the channels vary according to the nursery irrigation area and the primary irrigation amount, and the principle is to ensure high efficiency irrigation in the dry season, but to not occupy more land. The channel gradient (the ratio of the difference in elevation between any two endpoints to the horizontal distance between them) is 0.003-0.007. The aboveground, open channel irrigation method is easy to construct, with low cost, and is widely used in China, but it has several disadvantages including the amount of space it requires, which means less land for plants, more leakage, higher management

Figure 2-2 Water delivery equipment of nursery irrigation system (Photo by Liu Yong).

cost, and inconvenient cultivation. Pipeline water delivery and sprinkling irrigation should be used in more modern nurseries: A fixed sprinkling irrigation system should be used in large-scale nurseries, while a mobile sprinkling irrigation system should be used in small- and medium-scale nurseries.

iii. Drainage System. The drainage system is mainly composed of dikes, interception ditches, and drainage ditches. The drainage ditch should be set in low-lying places, such as both sides of the road. Its specification is determined according to the local precipitation, topography, and soil conditions, so as to ensure that the ponding (also referred to as the pooling of water) can be quickly removed after a heavy water event, such as rain, and less land can be occupied. Generally, the depth of the main ditch is 0.6-1.0 m, and the width is 1-2 m; the depth of branch ditch is 0.3-0.5 m, and the width is 0.8-1.0 m. The irrigation and drainage system should be considered comprehensively and coordinated with the road network.

iv. Windbreak Forest Belt. Nurseries with hazards of big winds and moving sand (wind-sand) will need to construct a protective forest belt. It should be designed according to the protection performance of tree species in the forest belt, combined with the nursery land use and the road network (Figure 2-3).

v. Nursery-related houses, yard, warehouse, and machine room should be built in places with high terrain, poor soil conditions, convenient management operations, and convenient transportation. Those facilities in large-scale nurseries are generally located in the center of nurseries, whereas for small- and medium-scale nurseries, the staff dormitory and residence can be built in suitable places outside the nurseries (Figure 2-4).

Figure 2-3 Windbreak forest belt of the nursery in the distance (Photo by Liu Yong).

Figure 2-4 A nursery house (a), warehouse (b), machine room (c), and yard (d) (Photo by Liu Yong).

After the division is completed, the nursery plan is drawn according to the division results, and the scale of the nursery plan is generally 1:2000. The plan should show the production area and operation area of all types of seedlings, as well as the locations of roads, water wells, irrigation and drainage channels, buildings, yards, protective forests, and so forth, detailed with different colors. In addition, the plan needs to at least indicate the scale, orientation, and legend (Figure 2-5).

Figure 2-5 Nursery plan (Photo by Peng Zuodeng).

2.2.3.2 Nursery Engineering Design

A standardized nursery should be composed of several relatively independent but interrelated engineering projects, which include the production area transformation and soil improvement, water supply, drainage, roads, power supply, communication, protective forest and warehouse, machine maintenance, septic tank, meteorological station, greenhouse, and administrative and living facilities.

(1) Soil Improvement Engineering

To utilize the land effectively, the soil improvement project should be carried out for nursery land with excessive water content and with barren soil or degraded soil fertility because of continuous seedling cultivation.

A soil improvement engineering design includes determining the type of soil improvement; determining the modes and measures of soil improvement; determining the technology of soil improvement, the proportion of mechanized operation, and the modeling of equipment; and calculating the engineering quantity of soil improvement.

(2) Water Supply Engineering

The nursery must be equipped with a water supply system to ensure the irrigation of seedling production and to meet the needs of domestic water. The water supply system should make full use of local water sources, and engineering should reasonably determine the project details, its scale, and type of structures.

Water supply engineering design mainly includes water source engineering, water diversion works, and irrigation system engineering.

When it is necessary to construct a dam (sluice) diversion by using natural water sources such as river flow, the design should be in accordance with the regulations of *Code for Energy Economy Design of Hydropower Project* (NB/T 35061—2015) issued by the Ministry of Water and Power, National Energy Administration.

When it is necessary to dig amotor-pumped well for utilization of an underground water source, the motor-pumped well type, depth, and density should be designed according to regulations of *Operation Specifications for Hydrogeological Drilling of Water Supply and Shaft Sinking* (CJJ/T 13—2013) and *Technical code for tube well* (GB 50296—2014) issued by the Ministry of Housing and Urban-Rural Development.

To improve the irrigation water temperature and the need of water storage irrigation, the reservoir or water tower can be designed according to the specific needs, and its type and scale should be determined based on the pondage.

When flood irrigation is adopted, the main channel (pipe) and branch channel (pipe) should be a permanent structure; temporary irrigation channels are used primarily around the operation area. The width and depth of various channels should be determined according to the water consumption, flow rate, and geology in the seedling production area, and the structure type of main and branch channels should be determined based on the principle of using local materials that are simple and effective, for example, rubble (pebble) paving can be used generally.

The design requirements of the flood irrigation channel: First, the channel should be designed according to the needs of seedling irrigation, and its plane direction should be arranged from high to low along the terrain. Second, the channel gradient should be 0.1% to 0.4%, and the maximum gradient should not exceed 1%. Water drop structures should be built in areas where the drop is too high. Three, the side slope of the channel should be 1:1.

When sprinkling irrigation is adopted, the design should be determined according to the topography, soil, climate, hydrology, hydrogeology, and economic conditions in the irrigation area. Base the design on the principle of adapting measures to local conditions, so consider the appropriate system that will make full use of the existing water conservancy facilities, be practical, technologically advanced, economically reasonable, safe, and applicable.

A pipeline sprinkling irrigation system and unit sprinkling irrigation system for sprinkling irrigation in the nursery may be needed. The specific design of a sprinkling irrigation system should include the technical parameters of the broader irrigation system, water source analysis, hydraulic power calculation of pipeline, equipment selection, and engineering construction design. Refer to the regulations of *Technical code for sprinkler engineering* (GB/T 50085—2007). The fixed pipes for sprinkling irrigation in the nursery should be buried 50 cm below the ground surface and below the permafrost in the cold area.

(3) Drainage Engineering

To prevent water outside the nursery from invading and to drain any water pooling in the nursery, the drainage engineering should be designed according to the nursery topography, terrain, storm runoff, and geological conditions. Drainage engineering includes the intercepting ditch that prevents the external water invasion and the drainage system composed of drainage ditch network in and out of the nursery.

Drainage engineering design requirements:

Large Drainage Ditch. This section is the outlet of the drainage ditch network and directly flows into the river, lake, or public drainage system or a low-lying safe zone. The section of large drainage ditch should be determined according to the water discharge, but its bottom width and depth should not be less than 0.5 m.

Middle Drainage Ditch. This section should be set along the side of the branch road and have a bottom width of 0.3-0.5 m and a depth of 0.3-0.6 m.

Small Drainage Ditch. This section should be set beside the fork road, and its width and depth can be determined according to the actual situation.

Large and Medium Drainage Ditches. These ditches should adopt the permanent structure of rubble (pebble) paving, with the side slope of 1:1.

Drainage Ditch Network and Irrigation Channel Network. These networks should be located on one side of the road, forming ditches, channels, and roads in parallel continuous arrangement.

Intercepting Ditch Outside the Nursery. This section should be determined according to the water discharge, but its bottom width and depth should not be less than 0.5 m.

(4) Road Engineering

According to the usage, nursery roads can be divided into three types: Main road, byroad(s), and branch road(s).

Design Requirements of Main Road in the Nursery. The main road of the large-scale and super-large nursery can be designed according to the secondary standard of forest highways. Medium- and small-scale nursery can design the main road according to the tertiary standard of forest highways.

Design Requirements of Byroads. For the byroad, the subgrade width should be 3.5 m and other technical indexes should be designed according to the fourth standard of forest highways. The ring nursery road is designed according to the standard of byroads.

Design Requirements of Branch Road. The branch road refers to the tractor and pedestrian road in the operation areas, and its subgrade width should be designed as 2 m.

(5) Power Supply and Communication Engineering

The nursery power supply project should be designed according to the power source conditions including electrical load and power supply mode, based on the principle of making full use of the local power source, saving energy, and setting it up as economically as possible. Small-scale generating units can be set up where no power supply is available.

When the electrical load of the nursery is low, the transformer capacity is below 180 kW, and the environmental characteristics allow, a pole transformer station can be erected. An independent substation may need to be used for a nursery with a higher electrical load. Safety protection facilities should be set around the substation or transformer station.

Generally, the nursery communication adopts the wire communication of the overhead open wire, and the wireless communication can be adopted when the conditions are available.

(6) Protective Forest Engineering

The protective forest belt in the nursery should be designed according to the degree of windsand damage. The forest belt design should be based on good protection benefits and rational use of land, for example, taking into consideration damage prevention, nursery beautification and ecological environment improvement, and economic benefits. General regulations are as follows: In small-scale nurseries, a forest belt should be set perpendicularly to the direction of the main wind; in medium-scale nurseries, a forest belt should be set around the perimeter; in large-scale and super-large nurseries, in addition to the ring nursery forest belt, several auxiliary forest belts should be set throughout the nursery according to the protection performance of tree species, roads, and channels.

The width of the forest belt needs to be determined according to the climatic conditions, soil structure, and the specifications or performance of the protective tree species. Generally, the width of the main forest belt is 8-10 m and that of the auxiliary forest belt is 2-4 m.

The tree species with fast growth and good protective performance should be selected for the forest belt, and the mixed trees and shrubs with semi-ventilation type are suitable for the forest belt structure. Tree species with serious diseases and insect pests, and the intermediate parasitism of seedling diseases and insect pests, should be avoided. To protect the nursery land from animal and bird harm, undershrub and hedgerows with thorns and strong sprouting potential can be planted in the low layer of the forest belt.

(7) Production Facilities Supporting Engineering

The nursery should be equipped with production machine equipment, transportation equipment, and manual utensils according to the seedling cultivation task, production and management level and actual needs, and in line with the principle of "favorable to production, economy and effectiveness".

Production machine equipment of the nursery includes wheeled and walking tractors and their supporting ploughs, ploughshares, harrows, suppressors, rotary cultivators, ridge-making machines, seeders, spraying machines, bulldozers, root-cutting machines, plant lifters, filling and sowing equipment, and perhaps more given the scope and specialties of a given nursery. Nursery transportation equipment includes automobiles, walking tractors, rubber tire vehicles, dung trucks, handcarts, and so forth. Manual production tools include ploughs, harrows, pickaxes, hoes, and a variety of other tools. The supporting engineering should be designed according to the nursery production machine equipment planning. Supporting projects include warehouses for storing materials, pesticides, fertilizers, seeds, grains, oils, tools, and other production materials; and garages for parking various vehicles, as well as a seedling cellar, fertilizer yards (septic tanks), sunning ground, fencing, livestock sheds, machinery repair rooms, fire stations, and meteorological stations, as examples.

According to the needs of production tasks and scientific research project arrangement, the construction scale for the nursery greenhouse and plastic greenhouse should be reasonably determined and their type selected and designed.

(8) Nursery Management and Living Facilities Engineering

Based on being favorable to production, being convenient for management, and making life easier, the nursery administration and living facilities construction projects should be arranged in a unified and convenient way. When the nursery is constructed near cities and towns, the management and living building land should be in accordance with the national or local standards; when building nurseries in forest regions, refer to the relevant provisions of forest farm standards in the *Civil Building Grade Standard of Forestry Bureau (field)* (Trial) (LY/J 111—1987). At the same time, the characteristics of more seasonal temporary workers in nursery production employment should be considered in the design.

Public and living welfare facilities should make full use of local social services and cooperation conditions, and they should not be built separately. If necessary, building associations can be adopted.

The engineering design of the nursery should reasonably arrange the engineering projects according to the scale and technical requirements of nursery production. See Table 2-2 for specific control parameters.

2.2.3.3 Seedling Cultivation Process and Technical Design

Seedling cultivation is a systematic project consisting of a series of continuous cultivation techniques, and any improper measures will affect the final seedling yield and quality. At the same time, the cultivation technology of each kind of seedling has great flexibility and regionalism. Therefore, it makes sense to design seedling cultivation technology according to the ecological characteristics of the tree species and local conditions, to control the unfavorable conditions for seedling growth and development to the maximum extent, and to give full play to the resource advantages conducive to seedling growth so as to achieve the goals of high quality, high yield, and high efficiency with the lowest cost in the shortest time.

Because many tree species will be cultivated in the nursery, it is generally impossible to carry out detailed seedling cultivation technology and technical design for each tree species. Rather, those details can be designed separately according to the type of seedlings. Focus the design on the main processes and technology for the production of various types of seedlings in the most technologically advanced and economically reasonable manner possible. For example, the sowing seedling cultivation is required to clarify plans for land management technology, fertilization technology, seed treatment technology, seeding technology, and seedling management technology; container seedlings

Table 2-2 Technical and economic index control table of main projects design in the nursery

Engineering classification	Project name		Usage	Unit of measurement	Super large	Large scale	Medium scale	Small scale	Reference base price (in 10 000 yuan)	Remarks
Seedling cultivation production engineering	Integrated laboratory building		Tissue culture works	m²	≤1500	≤500	≤200		0.08-0.1	Frame, light steel
			Lab	m²	≤500	≤300	≤200		0.07-0.08	Brick
			Checkout room	m²	≤200	≤100	≤80	≤60	0.07-0.08	Brick
			Archives	m²	≤200	≤80	≤30	≤30	0.08-0.1	Brick
			Office	m²	≤100	≤80	≤60	≤50	0.05-0.07	Brick
	Production housing			m²	≤300	≤150	≤50	≤50	0.05-0.07	Brick
	Warehouse			m²	≤400	≤200	≤80	≤80	0.08-0.09	Brick
	Seedling cellar			Set	≤10	≤10	≤5	≤3	4-5	
	Greenhouse		Automatic control greenhouse	m²	≤5000	≤3000	≤1000		0.08-0.1	With equipment
			Solar greenhouse	m²	≤8000	≤5000	Determine as needed		0.04-0.05	
	Plastic greenhouse			m²	≤8000	≤6000	Determine as needed		0.003-0.005	
	Seedling hardening field			m²	≤20 000	≤10 000	≤6000	≤1000	0.005-0.008	
	Shed			m²	≤10 000	≤7000	≤5000	≤3000	0.008-0.009	
	Soil improvement			hm²		Determine as needed			2.5-3	Not more than 2/3 of the total area
Production equipment	Automatic filling and sowing production line			Set	1	1			30-50	
	Simple filling and sowing production line			Set		1	1	1	10-15	
	Greenhouse equipment			Set In 10 000 yuan						
	Tissue culture equipment			In 10 000 yuan	≤150	≤100	≤50			
	Storage equipment			Set	2	1	1	1	10-20	
	Inspection equipment			Set	2	1	1	1	10-15	
	Seedling machines and tools equipment			Set	5	3	2	1	20-30	
Production equipment	Irrigation and drainage equipment		Mobile sprinkling irrigation system	hm²		Determine as needed			2.4-2.6	
			Fixed sprinkling irrigation system	hm²		Determine as needed			3-3.5	
			Trickle irrigation system	hm²		Determine as needed			1.5-2	
			Irrigation and drainage channel	m		Determine as needed			0.005-0.006	

(Continues)

Engineering classification	Project name	Usage	Unit of measurement	Super large	Large scale	Medium scale	Small scale	Reference base price (in 10 000 yuan)	Remarks
Production equipment	Production vehicle	Agricultural vehicle	Set	≤4	≤3	≤2	1	8-10	
		Tractors	Set	≤4	≤3	1	1	8-10	
		Forklift	Set	≤5	≤3				
		Motorbike	Set	≤2	1				
Ancillary engineering		Power supply system	In 10 000 yuan	≤200	≤100	≤60	≤40	10-30	
		Water and heating system	In 10 000 yuan	≤150	≤100	≤60	≤40		
		Motor-pumped well	Set	4	3	2	1	20-35	With supporting equipment such as water lift
		Water tower	Set	2	1	1	1	10-15	Brick
		Impounding reservoir (can be used for warming water)	Set	≤8	≤5	≤3	≤2	4-5	Cement
		Communication equipment	Set	2	2	1	1	1-2	
		Information equipment	Set	2	2	1	1	3-4	
		Meteorological equipment	Set	2	1	1	1	5-10	
		Propaganda and education equipment	Set	1	1			20-30	
		Office equipment	Set	2	1	1	1	5-8	
		Firefighting equipment	Set	5	2	1		5-8	
	Road	Artery outside the nursery	km	≤5	≤4	≤3	≤1	30-50	
		Artery in the nursery	km		Determine as needed			8-10	
		By-path	km		Determine as needed				
	Enclosure	General enclosure	m	≤2000	≤1500	≤1000	≤1000	0.01	Brick wall fence
		Biological enclosure	m	≤2000	≤1500	≤1000	≤1000	0.005	
		Field greening	In 10 000 yuan	≤100	≤40	≤20			

Source: Quoted from the *Standard for Forest Seedling Engineering Project Construction* (Trial), in 2003, issued by the State Forestry Administration.

Notes: Set refers to a set of equipment that may be include many parts.

need a design for technical processes such as container types, medium ratio and filling, seeding, environmental control, seedling packaging, and transportation, for example.

The quality of seedlings cultivated in the nursery should meet the requirements of *Tree seeding quality grading of major species for afforestation* (GB 6000—1999), and the seedling cultivation technology should be designed in accordance with the requirements of the *National Standard Specifications for Seedling Cultivation Technology*.

2.2.3.4 Organization and Operating Management

Organization and operating management system plays an important role in the nursery construction. When determining the organizational setup, management system, and personnel quota, one should focus on cultivating high-quality and high-yield seedlings, aim at creating good economic benefits, be guided by the market, carry out diversified operations, and abandon the old management modes of administration and public institutions. Instead establish the management mode and operation mechanism of modern enterprises according to the socialist market economy law, and implement independent operation, independent accounting, and self-financing. The state, province (autonomous region, municipality directly under the central government), prefecture (city), and county-level seedling management organizations should be responsible for the management of forest seedling projects. According to the regulations in China, the staffing of a seedling production base should be determined according to the standard of one person per hm^2.

2.2.3.5 Estimating of Investment Budget and Seedling Cost

(1) Investment Budget Estimate

After the above parts are completed, the investment budget estimate needs to be made. The project includes two parts: One-time investment and circulating fund. One-time investment in nursery construction refers to the investment in various basic construction and mechanical equipment during the establishment of the nursery, such as nursery land measurement, land leveling, soil improvement, buildings, roads, drainage and irrigation system, protection forest, greening engineering, electric power facilities, mechanical equipment, and so forth, which should be counted in the form of a detailed list and itemized summary sheet. According to the actual situation of local seed and seedling production, combined with the specific conditions of the nursery, the circulating fund should be estimated by an itemized, detailed estimation method.

In the total investment of a nursery engineering construction project, the investment for the seedling production portion should not be less than 60%, and that for auxiliary and supporting projects should not be more than 30%.

(2) Seedling Cost Estimate

Seedling cost includes direct cost and indirect cost. Direct cost refers to the production cost directly used for seedlings, such as seed cost, seedling cost, labor cost, material cost, and so on; indirect cost refers to the cost not directly used for seedling production, such as depreciation of one-time investment in nursery construction, annual expenditure budget of the nursery, and so forth. The one-time investment in nursery construction is depreciated by an annual average, in which the civil engineering part is depreciated by the average of 30 years of comprehensive depreciation period; the equipment part is depreciated by the average of 12 years of comprehensive depreciation period; and the intangible and deferred assets are amortized by the average of 10 years. The annual expenditure budget of the nursery includes the salaries of cadres and managers, office expenses, annual maintenance expenses of buildings, and welfare expenses of workers.

When calculating the cost of seedlings, the total indirect cost will be allocated to the relevant seedlings in direct proportion to the area occupied by each kind of seedlings, or the total cost of seedling cultivation. Finally, the direct and indirect production costs of all types of seedlings are added to obtain the total cost of seedlings.

2.2.3.6 Construction Period and Annual Capital Arrangement

The annual construction plan and capital arrangements should closely follow the construction period requirements of the construction institution.

2.2.3.7 Benefit Evaluation

The benefit evaluation of a seedling production base project is mainly reflected in two aspects: Economic benefit and social benefit, with emphasis on an input-output comparison.

2.2.4 Nursery Planning and Design Outcomes

The final outcomes of nursery planning and design generally include two parts: overall planning and design instructions and nursery planning and design drawings.

2.2.4.1 Compilation of Nursery Planning and Design Instruction

Design instruction is the text material of nursery planning and design, which along with design drawings are two indispensable parts of nursery planning and design. The contents that cannot be expressed on the drawings must be explained in the instructions, which is generally divided into two parts: Overview and design.

(1) Overview

This part describes the current situation of management conditions and natural conditions of the nursery in this area; makes the analysis; points out the favorable and unfavorable factors for seedling cultivation work along with the corresponding solutions to the unfavorable factors; and puts forward the guiding ideology, design basis, and design principle of nursery planning and design.

Management Conditions. Location of the nursery and the economy, production, and labor conditions of the local residents. Traffic conditions in the nursery. Power and mechanization conditions. Surrounding environmental conditions (such as a natural barrier, natural water source, and so forth).

Natural Conditions. Climatic conditions; soil conditions; plant diseases, insect pests, and vegetation conditions.

(2) Design

Itemized Overall Planning and Design.

i. Nursery construction objectives and product scheme.

ii. The nursery land division, calculation of partition area, and general layout design instructions, including the size of operation area, the configuration of each seedling cultivation area, the design of road system, the design of drainage and irrigation system, and the design of protective forest belt and fence.

iii. Capital construction scheme and equipment selection.

iv. Design of seedling cultivation technology.

v. Safe production, energy conservation and emission reduction, environmental assessment and protection measures.

vi. Technical support scheme.

vii. Organization and operation management scheme.

viii. Investment budget for nursery construction.

ix. Cost calculation and benefit evaluation of seedlings.

Special Design Description of the Nursery. In the construction design stage, the key civil engineering of the nursery, such as greenhouse engineering, road engineering, water supply and drainage engineering, construction engineering in field area, and seedling cultivation technology of main tree species, needs to be designed separately and prepared with instructions. The instructions explain the design basis, design ideas, specific contents, technical parameters, economic and technical indexes, and investment budget estimates of each single construction project.

2.2.4.2 Nursery Planning and Design Drawing

(1) Drawing of Current Situation of the Nursery

When drawing the plan, first of all it is necessary to make the specific location and boundaries of the nursery clear and to collect the relevant natural conditions, operating conditions, meteorological data, and other relevant data. The current situation drawing generally includes topographic map, land use plan map, soil distribution map, hydrological map, disaster impact analysis map, vegetation map, and geological map. Whether various drawings are needed should be determined according to the actual situation. In the design process, the current situation drawing is generally required to be provided along with the assignment by party A (the owner) of the contract.

(2) Basic Drawings of Nursery Design

According to the determined layout and design scheme of the nursery production area, the location, layout, and specific design drawings of nursery roads, channels, ditches, forest belts, and construction areas are drawn on the topographic map with the scale of 1:500-1:2000.

The main design drawings include the following.

Overall Design Scheme Drawing. According to the planned floor area of each operation area, the plan sketch of nursery design is drawn, and the formal drawing is drawn after the formal design scheme is determined through consultation and modification. For the formal design drawing, the roads, ditches, channels, forest belts, construction areas, seedling cultivation operation areas, and so forth should be drawn in proportion according to the scale of topographic map, and the direction of the drainage and irrigation should be indicated by arrows. The overall design drawing needs to accurately indicate the legend, scale, north direction, and other contents, and each area should be numbered.

General Road Design Drawing. The location of main roads, byroads and by-paths, width of various pavement, and drainage longitudinal slope are determined, marking the pavement materials and forms of the various roads. The contour lines should be drawn with dotted lines on the drawings, and the roads of various grades need to be indicated with lines of different thicknesses; in addition, the control elevation of main sections should be indicated.

Landscape Planting Design Drawing. According to the arrangement of the overall design, the layout diagram of the administrative area, roadsides, and shelterbelt plants needs to be drawn. The names of tree species, hedges, and flowers should be marked according to the requirements of landscape design.

General Design Drawing of Irrigation and Drainage System. According to the requirements of the overall planning, the irrigation system and drainage channel should be designed. At the same time, the blowdown pipeline, heating mode, and pipelines in the administrative area need to be mapped.

Electrical Planning Map. The nursery design should draw the partition power supply facilities, power distribution mode, cable laying, lighting mode of road network, and the location of communication and broadcasting.

Nursery Architecture Design Drawing. The various buildings and garden buildings involved in the nursery should be designed separately and drawn according to the architectural design requirements.

All design drawings should indicate the drawing head, legend, scale, north direction, title block, and brief description of the design content.

Questions for Review

1. What are the types of nurseries?
2. What conditions should be considered in the selection of a nursery site?
3. How could one reflect the characteristics of the city when planning and designing the urban nursery?

References and Additional Readings

CHENG F Y, 2012. Nursery of landscape plants[M]. Beijing: China Forestry Publishing House: 35-80.

FORESTRY MINISTRY, 1987. Civil building grade standard of forestry bureau (field) (Trial): LY/T 111—1987 [S]. Beijing: Standards Press of China.

STATE FORESTRY ADMINISTRATIOn, 2003. Construction standard of forest seedling engineering project (Trial) [G]. Beijing: State Forestry Administration.

JIN T S, 1992. Nursery of trees[M]. Harbin: Heilongjiang Science and Technology Press.

MINISTRY OF HOUSING AND URBAN-RURAL DEVELOPMENT, 2013. Operation specifications for hydrogeological drilling of water supply and shaft sinking: CJJ/T 13—2013[S]. Beijing: Standards Press of China.

NATIONAL STANDARD SEEDLING STANDARD OF MAIN AFFORESTATION TREE SPECIES, 1999. Tree seeding quality grading of major species for afforestation: GB 6000—1999[S]. Beijing: Standards Press of China.

SHEN G F, ZHAI M P, 2011. Silviculture[M]. 2nd ed. Beijing: China Forestry Publishing House.

FORESTRY MINISTRY, 1992. Design specifications of nursery garden engineering of forest: LY/T 128—1992[S]. Beijing: Standards Press of China.

THE MINISTRY OF WATER AND POWER, NATIONAL ENERGY ADMINISTRATION, 2015. Code for energy economy design of hydropower project: NB/T 35061—2015[S]. Beijing: Standards Press of China.

STATE FORESTRY ADMINISTRATION, 2003. Standard for forest seedling engineering project construction (Trial) [G]. Beijing: State Forestry Administration.

STATE-OWNED FOREST FARM AND FOREST SEEDLING WORK STATION OF STATE FORESTRY ADMINISTRATION, 2014. Construction standard of forest seed and seedling project[M]. Beijing: China Forestry Publishing House.

THE MINISTRY OF CONSTRUCTION, 2007. Technical code for sprinkler engineering: GB/T 50085—2007[S]. Beijing: Standards Press of China.

THE MINISTRY OF HOUSING AND URBAN-RURAL DEVELOPMENT, 2014. Technical code for tube well: GB 50296—2014[S]. Beijing: Standards Press of China.

ZHAI M P, SHEN G F, 2016. Silviculture[M]. 3rd edition. Beijing: China Forestry Publishing House.

Chapter 3　Tree Germplasm Resources and Breeding of New Varieties

Wang Jun

Chapter Summary: Germplasm resources are the basis for maintaining plant genetic diversity and for breeding new varieties. This chapter starts from the basic concept of germplasm resources and introduces the significance and characteristics of tree germplasm resources. It describes the principles, methods, and issues that need attention in collecting, introducing, and preserving germplasm resources, especially the methods of new variety breeding and the basic procedures of new variety protection and validation. Readers will gain an in-depth understanding of the importance of tree germplasm resources and the basic methods of germplasm collection and conservation, new variety breeding, and variety conservation.

Natural tree species are rich in variation, and different variation types either have certain adaptive significance or have certain ornamental, economic, or ecological value that make up the material basis for the new tree variety breeding. All tree species have a certain natural distribution range, which is formed in the process of long-term evolution or artificial cultivation. When a tree grows in its natural distribution area, it is called an indigenous tree species. When a tree is planted outside its natural distribution area, it is known as an exotic. For afforestation tree species, indigenous tree species can be selected because of their adaptability, and exotics can also be introduced as they may form varieties that are suitable for cultivating locally after domestication. Through the collection, introduction, and evaluation of tree germplasm resources, further selection of excellent germplasm and the breeding of new varieties are necessary procedures to solve the problem of "seeds" in seed and seedling technology. Therefore, understanding and mastering tree germplasm resources and new variety breeding are necessary components of seed and seedling technology research and teaching. Through the study of this chapter, readers are expected to fully understand the importance of germplasm resources for the breeding of new tree varieties and the maintenance of species diversity. They also need to understand the characteristics of tree germplasm resources; master the basic principles, methods, and procedures of collection, introduction, and preservation of germplasm resources; comprehensively use a variety of breeding technologies to select new tree varieties; establish a correct awareness of new variety protection; and understand the relationship between new variety protection and variety approval.

3.1　Significance of Tree Germplasm Resources

Germplasm resources, also known as gene resources or genetic resources, are all genetic materials that control biological traits in a species. According to Article 92 of the *Seed Law of the People's Republic of China* (hereafter referred to as the *Seed Law*), germplasm resources are the basic materials for the selection and breeding of new plant varieties, including the propagation materials of various cultivars, wild resources, and special types, as well as kinds of genetic materials artificially created by using the above propagation materials. Therefore, germplasm resources include not only wild resources but also artificially created and cultivated resources. Tree germplasm resources refer to the woody plant resources with the functions of afforestation, landscape beautification, climate regulation, environment improvement, and pollution absorption, including trees, shrubs, and other resources. To make the current breeding work effective and to ensure the long-term and sustainable development of the breeding work, it is not only necessary to have a clear breeding goal and adopt appropriate breeding strategy and techniques, but also to collect and make reasonable use of abundant tree germplasm resources. These factors are especially critical when breeding new varieties.

First, abundant germplasm resources are the material basis for the selection and breeding of new varieties. The development of human history shows that the formation of cultivars in crops and fruit trees is the process of choosing and using natural resources according to their own needs. All kinds of genetic materials needed for tree breeding are also reserved in the vast natural resources. Their excellent genes could be pyramided through selection and hybridization methods to form new varieties with good economic performances and strong environmental adaptabilities that meet the needs of human production and people's needs. Many landscape trees used in urban greening, such as *Magnolia wufengensis* cv. Jiaohong 1 (a variety of *Magnolia*), *Ulmus pumila* cv. Jinye (a variety of elm tree), *Styphnolobium japonicum* cv. Golden Stem (a variety of the Chinese scholar tree), and *Sophora japonica* var. *japonica* f. *pendula* Hort., are selected from natural variation types.

Second, the abundant germplasm resources can powerfully ensure the sustainable breeding of trees. In the breeding process of trees, due to the breeding goal is usually focused on a few traits, the genetic basis of the population would be narrowed down continuously, which is not conducive to sustainable breeding. Therefore, we need to persistently collect and expand germplasm resources and to add excellent germplasms to the breeding population, so as to improve the level of genetic diversity and ensure the sustainability of breeding. Moreover, with the development of society and the changes of environment, human requirements for new varieties are constantly updated. For example, in recent years, people have more in-depth understanding of the hazards of particulate pollutants, especially $PM_{2.5}$. The tree species with strong particulate absorption capacity, such as Chinese juniper (*Sabina chinensis*), Chinese scholar tree (*Styphnolobium japonicum*), and tree of heaven (*Ailanthus altissima*), are widely recognized in urban greening construction.

Third, the effective protection of germplasm resources is of great significance for maintaining the species diversity and ecosystem stability on the earth. Urbanization is a process in which the natural ecosystem is constantly destroyed and the human interference is constantly strengthened. Protecting and conserving tree germplasm resources can provide an *ex situ* conservation place for endangered species, serving as a "post house" for migration of animals and plants. The preservation of tree germplasm resources could even be beneficial to form a suitable habitat for the endangered species, effectively enhance forest biodiversity, and buffer the damage of urban development to the ecosystem. Additionally, the effective conservation and protection of tree germplasm resources also contribute to the adjustment of urban climate, the improvement of urban environment, the increase of livable quality, and the decrease of the negative impact of urbanization process.

3.2　Characteristics of Tree Germplasm Resources

Tree germplasm resources serve for afforestation and ecosystem maintenance, owning their unique functional status. In the long history of human social development, the cultivation and utilization of trees not only provide important means of production and livelihood but also form a unique humanistic connotation, which has been deeply integrated into human culture.

First, the target characters of tree germplasm resources and new variety breeding mainly depend on the needs of greening construction. In general, a timber forest requires trees with characteristics of fast growth, straight trunks, thin lateral branches, excellent wood quality, and strong resistance; for economic forests, the performance of characteristics such as reproductive cycle, fruiting ability, and fruit quality should be considered. Urban trees are mainly used for landscaping and landscape decoration, which require beautiful tree shape, good landscape effect, favorable for human health, strong resistance, and efficient ecological function. The absorption of and protective ability against urban pollution and the evaluation of pollen sensitization are especially important considerations at present.

Second, trees often have strong humanistic connotations. Poplar (*Populus* spp.) is one of the oldest cultivated species in China. It can be traced back to the 7th century BC, when the phrase "there is a little white poplar outside the east gate, with dense leaves" appeared in *the Book of Songs*. In the Warring States period, there were also records of poplar reproduction in *Hui Zi*. A description of the leaf characteristics of white poplar and Cathay poplar was included in *Ancient and Modern Notes*. The agricultural book of *Qi Min Yao Shu* in the 6th century AD has recorded the afforestation technology of poplar in detail. More than 1300 years ago, "the Song of Guanlong" in *the Book of Jin* described "poplar and locust trees are the border trees of Chang'an Avenue", which indicated that poplar trees had become the greening tree species of urban blocks at that time. Many tree species have fine moral meanings in Chinese traditional culture. For example, Yulan magnolia (*Magnolia denudata*), Chinese flowering apple (*Malus spectabilis*), tree peony (*Paeonia suffruticosa*), and Sweet osmanthus flower (*Osmanthus fragrans*) in *Yutang Fugui* represent good luck and happiness, while pine (*Pinus* spp.) and cypress (*Cupressus* spp.) represent longevity. In western countries, many tree species are also endowed with religious or mythological connotation. For example, firs (*Abies* spp.), which are beautiful and evergreen, are usually used as the Christmas tree, and oaks (*Quercus* spp.) are the first tree species created in ancient Greek mythology and they represent glory, strength, and indomitable spirit because of the characteristics of longevity and hard wood. These humanistic attributes have also become an important guide for the selection of tree germplasm resources.

Third, when selecting tree germplasm resources, we should pay attention to the utilization of indigenous tree resources. Indigenous tree species are the selection result of natural evolution and long-term cultivation, with good adaptability to local, natural environmental conditions generally. More than 8000 species of woody plants have been found in China, including more than 2000 species of arbors and more than 6000 species of shrubs. Approximately 1000 species of trees have a history of artificial cultivation. Abundant indigenous tree resources provide us with a favorable basis for further selection of afforestation and landscape trees. However, most of the indigenous tree resources lack sufficient research on biological and forestry characteristics. To ensure the effective usage of germplasm resources, much work is needed to strengthen the basic research of indigenous tree resources.

Fourth, exotic tree species supply important tree germplasm resources. Because of the phylogeny of plants and the historical changes of natural distribution areas, many plants are able to grow outside their natural distribution areas, with good character performance, which improves the local forestry productivity. Moreover, with the aggravation of industrial pollution and soil erosion in modern times, some indigenous tree species are difficult to fully adapt to the adverse environmental changes, while some exotic tree species may have stronger adaptability to the current habitat. For example, black locust (*Robinia pseudoacacia*), native to North America, has stronger adaptability in most areas of northern China; jacaranda (*Jacaranda mimosifolia*), native to Brazil and Argentina, has better adaptability in South Africa, Australia, and southern China. In addition, the introduction of exotic tree species is an important means to broaden the genetic diversity of trees.

3.3 Collection and Conservation of Tree Germplasm Resources

To protect the genetic diversity of trees and effectively carry out genetic improvement, we need to systematically collect, conserve, and comprehensively evaluate the tree germplasm resources. Collection of tree germplasm resources should consider the valuable indigenous tree species and recognize the importance of introducing the excellent tree resources from other regions or abroad. Collected germplasm resources should be conserved, evaluated, shared, and exchanged to give full play to its benefits. In recent years, China has established a national forest germplasm resources platform, which effectively integrates the national tree germplasm resource banks from different regions and different species and is of great significance for promoting the rational utilization of tree germplasm resources and improving the effect of tree genetic improvement.

3.3.1 Collection of Indigenous Tree Germplasm Resources

Indigenous tree species have experienced long-term natural selection or artificial cultivation history under local natural conditions, with adaptability to local environmental conditions. They can be directly put into production or slightly improved to give full play, without causing adaptive disasters. Moreover, in the natural distribution area, indigenous tree germplasm resources are rich and readily adaptable to being genetically improved. Therefore, in the genetic improvement of trees, we should pay special attention to the collection and utilization of indigenous germplasm resources. For example, indigenous tree species such as Chinese white poplar (*Populus tomentosa*), Japanese pagoda tree (*Sophora japonica*), and weeping willow (*Salix babylonica*) play an important role in afforestation and greening construction in China.

Before collecting indigenous germplasm resources, an appropriate survey should be organized to systematically investigate the basic situation, distribution areas, and special types of germplasm resources so as to provide a basis for the collection work. The investigation includes regional economic and natural conditions, general plant information, resource stock, biological and morphological characteristics, and economic characters of resources, especially the characteristics of some special types of germplasm materials.

The collection of germplasm resources should be carried out in a range of environmental gradients in the whole distribution area of species, covering all representative sites in the distribution area of species as far as possible. Among them, the provenance in the central distribution area of plants has complex gene composition, a variety of excellent economic traits, and a wide range of adaptive variation, with rich genetic diversity; however, the provenances in the marginal distribution area have experienced natural selection under extreme environmental conditions. The strong selection pressure has an impact on the gene frequency in the population, and as a result, germplasm materials with potential value may appear that have strong adaptability to extreme environmental conditions. Therefore, attention should be paid to both central and marginal provenances when collecting and preserving germplasm resources.

Attention should also be paid to collecting germplasm resources at different variation sources. A large number of studies show that sources of intraspecific genetic variation for trees should include geographical provenance variation, stand variation within provenance, individual variation within stand, and variation in different parts of the individual. Among them, geographical provenance variation and individual variation within stand are the most important. Long-term studies on cold tolerance of loblolly pine (*Pinus taeda*) have shown that geographical provenance variation and individual variation within a stand accounted for 70% and 30% of the variation components, respectively. Therefore, germplasm resources with different levels of genetic variation should be collected. In addition, when collecting tree germplasm resources, we should pay attention to the collection of cultivated and wild types of the same tree species, and the special types with excellent characteristics, unique landscape effect, strong stress resistance, and special ecological functions, including valuable ancient trees and large trees. The ancient trees and large trees have been preserved through long-term environmental selection, and consequently they often have good adaptability and character performances and can be utilized directly after propagation.

During the collection of germplasm resources, some specific work should be completed, such as measurement of basic characteristics of selected materials, location by GPS in the collection area, and the collection chart drawing. After the field work, the name, source, natural conditions of origin, biological characteristics, and economic value of the collected germplasm materials should be recorded and archived in time to provide complete basic information for preservation and utilization.

3.3.2 Introduction of Tree Germplasm Resources

The process of artificially introducing a tree species from the original distribution area to a new cultivation area is called "introduction of exotics". This process is the migration of tree germplasm resources in the range of its utilization, and it is an important form of germplasm resources utilization, with the characteristics of simple operation and quick effectiveness. Especially in the urban greening construction, the introduction of exotic tree germplasm resources with high ornamental value, stable ecological function, and strong adaptability to meet the needs of tree species diversity in landscape greening has achieved remarkable effects. China has a long history of tree introduction. As early as 401 AD, sycamore (*Platanus* spp.), had been introduced to chang'an. Introduced tree species, such as Himalayan cedar (*Cedrus deodara*), magnolia (*Magnolia grandiflora*), and boxwood (*Buxus megistophylla*), have also been known for a long time and have been widely cultivated in urban greening. Introduced velvet ash (*Fraxinus velutina*) in Tianjin, with a strong salt and alkali resistance, was designated as the city tree of Tianjin in 1984.

3.3.2.1 Factors to Be Considered in the Introduction of Exotics

Given the differences in the genetic characteristics of the species and the environmental conditions in the natural distribution areas, the following factors should be fully considered before introducing exotics to ensure the effectiveness of the introduction.

(1) Performance of Exotic Tree Species in Origin

To select the suitable introduced tree species, first determine the purpose of introducing exotics according to the needs of the introduction sites. For the introduction of urban greening tree species, we should fully consider whether the tree shape is beautiful, whether it is pollution resistant, whether it has dust retention capacity, and whether it has allergens, as well as considering its growth characteristics, adaptability, and disease and insect resistance. Although the economic characteristics of trees are closely related to environmental conditions, generally speaking, it is difficult for trees with poor performance in the origin to have good genetic performance after introduction to new areas. A large number of introduction practices show that the economic characteristics of exotic species in the new area are similar to those in the original area. For example, some species of Cupressaceae show that the trunk texture is not straight in the origin and is still distorted after being introduced into China; and Chinese chestnut (*Castanea mollissima*), which has the characteristic of chestnut blight resistance, has become an important germplasm resource for local chestnut blight resistance breeding after being introduced into the United States.

(2) Genetic Characteristics of Exotic Tree Species

Before introduction of exotics, we should have a deep understanding of the genetic characteristics of the target species, including the adaptability of the tree species and the characteristics of intraspecific genetic variation. The adaptability of different tree species varies greatly. For example, *Sequoia sempervirens* (the sole living species of the genus *Sequoia*) originated in California and southern Oregon along the Pacific Coast of the United States at an altitude of less than 1000 m and with climate features that are warm in winter and cool in summer, and rainy in winter. Because of weak adaptability, the introduction of *Sequoia sempervirens* almost failed in different site conditions. However, maidenhair tree (*Ginkgo biloba*) of China has been introduced in Japan, the eastern region and central region of the United States, and some areas of Europe for more than 100 years, because of its strong adaptability. In addition, for species with a wide natural distribution, there are abundant intraspecific genetic variations, therefore, when introducing, we should pay attention to the introduction of different geographical races and ecotypes in order to make full use of ecological adaptability within the species and to ensure the introduction effect. Thus, in the introduction of trees, we have generally recognized the differences in growth, adaptability, and economic characteristics of different provenances and have attached great importance to the collection and testing of provenances.

(3) Similarity Level of Main Ecological Conditions between the Origin and the Introduction Sites

When conducting long-distance introduction, we need to study in detail the similarity of the main ecological conditions between the origin sites of exotic tree species and the introduction sites. In general, the closer the ecological conditions are, the more likely the introduction is to succeed. Heinrich Mayr, a German silviculturist, put forward the Theory of Climatic Analogues in his writings, for example, *Fremdländische Wald und Parkbäume für Europa* published in 1906 and *Waldbau Auf Naturgesetzlicher Grundlage* published in 1909. He thought that whether the introduction of woody plants was successful or not depended on whether the climate conditions of the origin and the introduction sites were similar. The theory of climate analogues can avoid the blindness of the introduction and has a certain guiding role for the successful introduction. However, while emphasizing the restriction of climate on the growth of trees, the theory ignores the comprehensive effects of other environmental factors such as temperature, photoperiod, precipitation, and soil conditions and underestimates the change ability of genetic potential and variability of tree species to environmental adaptation and human ability to domesticate trees, thus limiting the scope of introduction. some species grow well even when the environmental conditions do not match the original exactly. For example, black locust has abundant rainfall in its origin areas of North America, with an annual precipitation of 1016 to 1524 mm, an average temperature in January of 1.7 to 7.2 ℃, and an average temperature in July of 21.0 to 26.7 ℃. *Robinia pseudoacacia* introduced to the western region of China, where there is a rainfall of 400 to 500 mm and the lowest and the highest temperatures break through those of the origin, shows strong adaptability and can grow regularly. Thus, the introduction of exotics between regions with differing climatic conditions may also be successful, or even better than the growth in the origin. In general, Mediterranean climate tree species are difficult to adapt to continental climate conditions, but not vice versa; tree species from areas with drastic climate change are readily accepted in areas with moderate climate change, but not vice versa; tree species from high latitude and high altitude areas are not easy to introduce to areas with low latitude and low altitude, and vice versa; introduction of exotics from areas with high latitude and low altitude to low latitude and high altitude is easy to succeed; tree species suitable to acid soil should not be introduced to alkaline soil areas, and vice versa.

(4) Historical and Ecological Conditions of Tree Species

The adaptability of tree species is related not only to the ecological conditions in the modern distribution area but also to the ecological conditions in the history of phylogeny. Therefore, the introduction of exotic tree species may still be successful even if ecological conditions differ significantly between the origin and the introduction sites. In 1953, the Soviet botanist Кульсясов put forward the ecological history analysis method of plant introduction and domestication based on the experimental analysis of more than 3000 tree species. He believed that the modern distribution area of some plants was forced upon them during glacial events in geological history, and so the species may not be in the most suitable distribution area. When the plants are introduced to other areas, they might display better growth and development. In the long evolutionary process, many trees have experienced complex historical and ecological conditions, and selection pressure prompted them to produce a wide range of adaptive variation, which was inherited by their offspring. When the actual ecological conditions of the introduction site are similar to the historical ecological conditions of the introduced tree species, especially when they are the same or similar to the adaptive historical ecological conditions that the tree species has experienced, the introduction may succeed even if the actual ecological conditions of the two sites are quite different. For example, dawn redwood (*Metasequoia glyptostroboides*), known as "living fossil plants", was widely distributed all over the world before the Quaternary glacial period. The species had wide genetic adaptability, but almost all of them became extinct during the geological changes of the glacial period, and only a small number of remains have been found in East Sichuan, West Hubei, and Longshan County of Hunan Province. After being rediscovered in 1946, it was introduced to all parts of the world and cultivated widely.

3.3.2.2 Introduction Procedure

The introduction procedure includes the whole process from the selection of exotic tree species, material collection, seed and stock quarantine, registration, introduction test to the exploration of domestication measures, to becoming a local cultivated tree species. Introduction of exotic species is a systematic and complex process.

(1) Selection of Exotic Tree Species

The selection of introduced tree species should fully consider the actual needs of the afforestation industry. Tree species or varieties with good greening effect, beautiful tree shape, high ornamental value, and strong adaptability could be selected. At the same time, we should consider the possibility of introduction, analyze the ecological adaptability of the introduced objects, and select the introduced materials from the areas with similar ecological conditions. Before the introduction, conduct a full demonstration and scientific risk assessment to analyze the dependence and competition relationships between the introduced species and the local original species, to fully assess its impact on the local environment, and to pay special attention to the potential hazard of exotic tree species becoming an invasive species.

(2) Collection of Introduction Materials

Introduction materials mainly include seeds or asexual propagation materials, which can be collected by organizing and sending personnel to the origin sites and/or by entrusting institutions or personnel already located in the origin sites. It is also feasible to make commercial purchases through formal channels. Collection of the introduced materials should strictly comply with the laws or regulations on plant protection of the origin area or country of origin. To ensure the quality of seeds, full mature seeds should be collected from fine and healthy plants without pests and diseases, should be processed as necessary after collection, and should be promptly transported to the introduction sites for preservation and sowing. As for the asexual propagation materials, such as cuttings and scions, sealing them with wax and then packing them in a waterproof manner should be carried out immediately after collection to ensure the survival rate of vegetative propagation.

(3) Seedling Quarantine and Registration

The materials being introduced may carry pathogens or insects that are not found in the introduction site. If the introduction site lacks natural enemies of potential pathogens or insects that may cause diseases and insect damage, it could lead to introduction failure and also cause losses to the local tree species forest resources. Numerous examples of ecological disasters have been caused by improper introduction. For example, the fall webworm moth (*Hyphantria cunea*), which originated in the United States, has been transmitted to Europe, Japan, and North Korea successively. In 1979, it was transmitted to Liaodong Peninsula in China, and now it has spread to the "Three North" areas of China, seriously threatening the landscaping and forestry production everywhere and causing huge economic losses. Therefore, it is necessary to strictly implement the national quarantine regulations on animals and plants and to report the introduction materials according to the regulations and procedures. Exotic species can be imported only after passing the quarantine.

For introduction materials, detailed registration, numbering, and recording should be carried out for reference and filing. Registered items include the species of trees and names of varieties, types, sources, and quantities of breeding materials; collection dates and processing treatment measures; transportation method; date of receiving the introduced materials and preservation measures, as well as the characteristics of various traits of the introduced objects and their performance in the country of origin.

(4) Introduction Tests

Preliminary Test. The preliminary test refers to a preliminary observation and analysis of the introduced tree species on the performance of the target economic characters and ecological adaptability, along with the techniques of seed treatment, seedling, and cultivation, to preliminarily screen the promising tree species (varieties or provenances). The preliminary test is generally carried out in the isolated testing area to prevent any losses caused by the invasion of exotic pests. To avoid the influence of improper cultivation and management measures at the initial stage of introduction on the character performance, the observation period of the preliminary test should be extended appropriately, so as to ensure the true performance of the introduced tree species.

Materials with good performance in the target characteristic and adaptability can be further expanded to the evaluative test.

Evaluative Test. This test refers to a comparative test and a regional test on the introduced tree species (varieties or provenances) with fine growth performance and potential production value after the selection based on the preliminary test. The goals are to further understand the genetic variation characteristics of each introduced material and their interaction with the environmental conditions of the introduction site; to compare and analyze its adaptability in the new environment; to study the main pests and diseases, prevention measures, and cultivation techniques; to determine the suitable conditions and scope of the introduced materials; and to select the species (varieties or provenances) with development prospects under different site conditions. The comparative test and regional test can be carried out at the same time in multiple trial points. A range of site conditions and climate conditions can be selected and tested according to the standard experiment design to provide a reliable basis for the comprehensive evaluation of the introduced materials.

Productive Test. For the introduced materials that have passed the evaluation test successfully, the production test has to be carried out before the large-scale promotion and application. According to the technical measures allowed in the production, the comprehensive performance of the selected materials by the evaluation test should be verified through the production test of a certain area. The comprehensive performance of the selected materials under the complex conditions of large-scale cultivation can be reflected through the productive test, so as to solve the problems that may not have appeared in the small-scale evaluation test.

Identification and Promotion. Materials that have reached the introduction target through productive testing can be applied foridentification and promotion. The identification content focuses primarily on the utilization value in production, including economic benefits, suitable range and conditions, and key cultivation measures. After the identification, the introduced species (varieties or provenances) can be promoted in the designated area by the forestry department and the production institution. The achievements of introduction should be appraised and popularized in time, and the principle of no popularization without appraisal should be insisted upon. In order to speed up the introduction process, the promising trees can be selectively propagated at a certain scale while testing and observing to properly shorten the introduction and promotion time.

3.3.2.3 Criteria for Successful Introduction

To evaluate whether the introduction of a tree species or variety is successful, we need to make a comprehensive evaluation from the aspects of adaptability, introduction benefit, and propagation ability after introduction and cultivation, including: Adapting to the natural or cultivated environment conditions of the introduction site, and being able to grow normally without special cultivation and protection measures; maintaining the original economic value and ornamental value; being able to propagate with the inherent propagation mode and maintaining its excellent characters; no obvious or fatal diseases and insect pests; forming sexual or asexual varieties for popularization and application; no adverse ecological consequences such as ecological invasion and pathogen transmission.

3.3.2.4 Issues to Note in the Introduction Process

The introduction process needs to be very scientific and systematic. To ensure the introduction effect and benefit and to make effective use of the introduction resources, the following issues should be paid attention to in the introduction work.

(1) Insist on the Principle of "Active and Prudent" instead of Rush

Although the introduction work has the advantages of less investment, quick effect, and simple operation, the introduction process needs to consistent with the objective laws of tree development. On the one hand, we should actively use the existing excellent variety resources in the outside area or abroad; on the other hand, we should advocate the scientific introduction and prevent blind introduction and overgeneralization, to avoid losses. Therefore, it is very important to demonstrate the feasibility of the introduction, adhere to the principle of "a small amount of trial introduction, multi-point test, long-term test, and gradual promotion", and strictly follow the introduction procedure.

(2) Combine with the Introduction of Supporting Technologies

An effective guarantee for productivity improvement is to match improved varieties with excellent technologies. Based on the various genetic backgrounds of introduced tree species, the supporting propagation and cultivation techniques may differ; therefore, when carrying out the introduction work, we should pay attention to the supporting propagation techniques and cultivation measures to ensure the success of the introduction. In addition, we should implement supporting techniques adapted to the site and ecological conditions of the introduction site through transformation and absorption, so as to reduce the dependence on the technology of the origin site and to accelerate the process of introduction and popularization. Attention should also be paid to the introduction of trees, undergrowth shrubs, and even soil microorganisms accompanying the introduced tree species. In 1974, *Pinus caribaea* was introduced from Honduras to Guangdong Province of China. However, many seedlings died in that year. After investigation, researchers found that the deaths were related to mycorrhizal deficiency, and the survival rate of seedlings increased significantly after inoculation of the mycorrhizal fungi.

(3) Combination of Introduction and Breeding

Combining the introduction and domestication of trees with breeding techniques can fully utilize the introduced germplasm resources, continuously exert the value of the introduced materials, and maximize the benefit of the introduction. Therefore, in the introduction work, we should pay attention to the collection of a variety of geographical provenances, families, and excellent clones to enrich the genetic diversity and to conduct breeding studies on the basis of character evaluation. The introduced magnolia, *Liriodendron tulipifera*, was crossed with *Liriodendron chinense* by breeders of Nanjing Forestry University, and the *Liriodendron* hybrids have been produced with significant heterosis characterized by strong adaptability, vigorous growth, a straight trunk, and a beautiful crown, showing huge values in afforestation.

(4) Prevent Invasion of Exotic Pests

Introduction brings great benefits, but it also faces serious risks, such as the invasion of exotic pests, including exotic pathogens, pests, and other hazards; the threat of exotic species to the survival of native trees; and even damage to the ecosystem and human health. Crofton weed (*Eupatorium adenophora*), which originated from Mexico, has been introduced as an ornamental plant around the world since the 19th century. Because of its strong reproductive capacity, it has occupied the living space of other plants, making it difficult for other plants to grow, seriously damaging the biodiversity, and becoming a global invasive species. Staghorn sumac (*Rhus typhina*), which originated in Europe and America, was introduced to China in 1959, and because of its good adaptability and strong reproductive capacity, it has gradually occupied the living space of many green trees. At present, there have been some disputes about the planting of *Rhus typhina*.

(5) Pay Attention to the Conservation of Germplasm Resources of Exotic Tree Species

The introduction process is actually the transfer and utilization of plant germplasm resources. An important part of the introduction strategy is to preserve the germplasm resources of exotic tree species, which are the basis of long-term breeding and improvement and the premise of sustainable management. While preserving and evaluating germplasm resources, continuous introduction and improvement should be carried out to enrich and supplement resources. The introduction of European olive trees (*Olea europaea*) in China began in the early 20th century, and later, Zhou Enlai, premier of China, received more than 10,000 seedings of five varieties during his visit to Albania in 1964; during 1978 to 1987, more than 170 germplasm resources were successively introduced supporting by the introduction projects of the Food and Agriculture Organization of the United Nations (FAO), which provide important materials for the selection and cultivation studies of *Olea europaea* in China.

3.3.3 Construction of Germplasm Resource Nurseries

Germplasm resources collected and introduced must be properly preserved so as not to be lost. At present, the main ways to preserve tree germplasm resources are conservation *in situ*, conservation *ex situ*, and facility preservation. Conservation *in situ* does not change the way in which the natural habitat of the protected object is preserved, and its main purpose is to prevent further losses usually caused by human activities. Nature reserves, national forest parks, wetlands and fine natural forests all have the function of conservation *in situ*. Conservation *ex situ* is to change the conservation form of the protected object's habitat, and to build a new plant population under suitable conditions by collecting seeds, spikes and other propagation materials of the protected object. Facility preservation is a way to eliminate environmental pressure and damage and effectively preserve germplasm materials under artificial control.

A germplasm resource nursery is a specialized nursery and facility for centralized preservation and evaluation of collected and introduced tree germplasm resources. Germplasm resource nurseries prevent the loss of species germplasm resources, maintain genetic diversity, and systematically evaluate the genetic characteristics of germplasm resources to provide raw materials for continuous breeding research. Germplasm resource nurseries belong to public resources and should be open to use by the public.

(1) Site Selection

The germplasm resource nursery should be constructed according to the characteristics of the introduced and collected tree germplasm resources. It should be reasonably arranged, setting preservation points in the suitable climate zone and ecological area. To meet the needs of forestry development, the germplasm resource nursery should be constructed in an area with flat terrain, deep layers of fertile soil, sandy loam or loam with good drainage, and irrigation suitable for the soil condition. The nursery site should have a clear land ownership and a long-term, stable protection and management basis.

(2) Scale and Layout

The scale of the germplasm resource nursery should be aimed at fully preserving the ecological and genetic stability and diversity of tree species populations. Generally, such as nursery is required to preserve at least 50 families in each provenance, more than 50 individual trees in each family, and more than 10 clones. Additionally, it should maintain not less than 100 individuals for precious and rare tree species, not less than 50 individuals for endangered species, and not less than 100 individuals for introduced trees.

In general, the collection and propagation area (Figure 3-1) and the preservation area of germplasm resources should be included in a germplasm resource nursery. If necessary, a germplasm isolation area should be established to prevent the potential harm of exotic tree species. The main functions of the collection and propagation area are to propagate the collected germplasm resources, such as seeds and scions; to carry out the observation of seedling characteristics, such as seed emergence rate and seedling growth; and to conduct research on seedling cultivation technology, so as to lay a foundation for further scale propagation of varieties. The germplasm

Figure 3-1　Collection and propagation area of poplar germplasm resources in Sangganhe Poplar High-Yield Forest Experimental Bureau of Shanxi Province, China (Photo by Wang Jun).

preservation area is the core area of the germplasm resource nursery. In the preservation area, the collected germplasm resources are planted according to the appropriate density and managed by the conventional cultivation method, so as to facilitate the long-term observation and evaluation of the genetic performance of the germplasm resources. The content of genetic evaluation includes analysis of population genetic structure, geographic variation mode, and population genetic diversity, as well as the observation of production characters, landscape characters, growth characters, and adaptability in order to screen the germplasm with high economic value or improvement potential for subsequent production or breeding.

The construction of the germplasm resource nursery should be equipped with ideal production infrastructure, sound management norms and systems, reasonable staffing, including professional and technical personnel suitable for carrying out the collection and preservation of forest germplasm resources. The resource nursery also should cooperate with scientific research and teaching institutions for a long time to form the technical support and to carry out scientific education reasonably and appropriately.

(3) Facility Preservation Bank

The facility preservation bank uses a low-temperature storage room, liquid nitrogen storage tank, and other facilities to preserve plant seeds, pollen, and *in vitro* propagation materials, which can preserve thousands of germplasm materials in a relatively small space and conserve a lot of human, material, and land resources. The facility preservation bank is a powerful supplement to the germplasm preservation nursery. So far, The Southwest China Germplasm Bank of Wild Species and the Millennium Seed Bank, Royal Botanic Gardens, Kew (Figure 3-2) are the only two wildlife germplasm resource preservation facilities in the world established in accordance with international standards.

With the continuous attention paid to the safety of the forest seed industry in China, the national forest germplasm resource facility preservation bank has begun to be built. A high-level facility preservation main bank and its auxiliary facilities will be constructed in Beijing and six facility preservation sub-banks will be constructed in other places, which are of great significance to achieving the long-term and safe preservation of forest germplasm resources in China, to preventing the loss of forest germplasm resources, and to promoting the progress of science and technology of forest seed industry and the maintenance of biodiversity.

(4) Information Management System of Germplasm Resource

A complete information management system for germplasm resources should be developed and established to store all information obtained in all links of resource management, including source, distribution, cultivation status, morphological and biological characteristics, economic and ecological value, collection or introduction location and quantity, preservation location and quantity, survival situation, character observation and evaluation, promotion and utilization process, contact person information, and so forth. Establishment of the information management system is convenient not only for the storage and exchange of information but also for the deep processing, analysis, and utilization of data. In recent years, National Forestry and Grassland Administration

(a)

(b)

**Figure 3-2　Millennium Seed Bank of the Royal Botanic Gardens, Kew
(Photos by Dr. Han Biao, Shandong Forest Germplasm Resource Center).**

(NFGA) has established a national forest germplasm resource platform. The forest germplasm resource conservation centers (banks) at all levels can integrate the data of germplasm resources in this platform, effectively promoting the material sharing and information exchange of germplasm resources.

3.4 Breeding and Protection of New Tree Varieties

With the acceleration of urbanization and social development, the demand for new tree varieties in forestry production and urban construction is also expanding continuously. For the collected and introduced tree germplasm resources, after proper preservation and comprehensive evaluation, the potential materials can be directly selected according to breeding objectives. New variation types can also be created through breeding technology, which is the process of germplasm resource innovation. The ultimate goal of tree germplasm innovation is to select and breed new varieties, and popularize their utilization in urban greening and forestry production, so as to promote social development.

3.4.1 Basic Methods of Germplasm Resource Innovation

Although the breeding methods based on the development of modern biotechnology, such as genetic engineering, are constantly updated, the most commonly used methods of new tree variety breeding are still selective breeding, hybrid breeding, and polyploidy breeding.

3.4.1.1 Selective Breeding

Selective breeding refers to a breeding method of obtaining excellent types or varieties through the comparison, identification, and reproduction of groups or individuals with excellent ornamental or economic characters from existing natural populations according to certain standards and objectives. Selective breeding can directly use the natural multilevel genetic variation types to select the provenances, families, and clones with excellent target characters. It has the advantages of short breeding cycle, quick production of varieties, low cost, and produces a remarkable improvement effect.

(1) Basis of Selective Breeding

The life cycle of trees is long, and most of them are in wild and semi-wild states and are cross pollinated. Long-term sexual reproduction of trees leads to abundant intraspecific genetic variation. For the species with a wide distribution area, climate and soil conditions in the distribution area varies, so geographical provenance variation becomes obvious because of the effect of natural selection. In addition, the variations are derived from different levels including stands, inter-individuals, and parts of individuals. The existence of variations provides possibility for selection. For selective breeding of trees, geographic provenance variation and inter-individual variation within the stand are the most important sources of variation.

Bud mutation is often used in the breeding process of economic and ornamental tree species. It belongs to the utilization of genetic variation in level of different parts of the same individual and is produced by somatic mutation. The colored leaf tree *Populus euramericana* cv. Zhonghua Hongye (a variety of poplar) is a bud mutation variety selected from the mutant branches of *Populus euramericana* cv. Zhonglin 2025. *Sophora japonica* cv. Pendula (a variety of the Chinese scholar tree), which is often used in urban greening, is also a bud mutation variety of locust tree.

(2) Basic Methods of Selective Breeding

According to the source of genetic variation of tree species, selective breeding usually includes provenance selection, plus tree selection and bud mutation selection. Provenance selection refers to the process of selecting excellent provenances with high productivity and fine stability for the region of afforestation. After collecting materials from different provenances, systematic provenance trials are conducted to provide scientific basis for provenance selection and to determine the variety supply range of different provenances. Based on the results of provenance trials, we can produce better selection effect to select excellent individuals within excellent provenances.

Plus tree selection refers to the process of selecting plus trees according to the certain criteria under suitable stand conditions based on the purpose of tree improvement. The plus tree selection should be carried out in the origin stand, with the clear origin of the stand and the appropriate tree age (optimal for the middle-aged stand), and with the appropriate canopy density and stand structure. Selection in the stand that has been negative selection or upper thinning should be prohibited.

Bud mutation selection refers to the process of finding and fixing budmutation materials that meet the breeding objectives. To avoid the occurrence of chimeras, vegetative propagation methods, such as grafting and cuttage, are usually used to fix the traits of bud mutation. Bud mutation selection may sometimes require multiple fixations to stabilize the traits.

3.4.1.2 Hybrid Breeding

Hybrid breeding refers to the process of artificial sexual hybridization of different genotypes to obtain hybrids and to then make further selection and identification so as to obtain excellent varieties. Hybrid breeding technology, with advantages of simple operation that is easily mastered and low cost, is one of the most important methods in the current breeding work and the main way to produce new tree varieties. Its essence is the utilization of heterosis. Heterosis refers to the phenomenon that the hybrids derived from cross breeding of two parents with different genotypes are superior to the parents in terms of growth, vitality, fecundity, stress resistance, yield, and quality.

(1) Cross Modes

Cross modes refer to the parental amount and sequence participating in hybridization in hybrid breeding. In hybrid breeding, to gather excellent characters and obtain expected genetic gains, a variety of hybrid modes can be chosen, commonly including single cross, multiple cross, backcross, and even mixed pollination cross of multiple male parents.

i. Single cross refers to a simple one-time cross between two parents. For example, variety *Populus tomentosa* × *P. bolleana* was bred by crossing *Populus tomentosa* with *Populus bolleana*. Single cross includes orthogonal cross and reciprocal cross. Because of the influence of cytoplasmic inheritance and other factors, there may be differences in the results between orthogonal and reciprocal crosses.

ii. Multiple cross means that more than two parents are crossed twice or more, which can integrate the good characters of many parents and also enrich the genetic basis of hybrids, so as to select the excellent varieties. According to the genetic composition of the parents used in the second crossing, it can be divided into a three-way cross and a double cross. The three-way cross is to cross the hybrids of two parents with the third parent. The *Populus* × ' Nanlin' was produced by crossing the hybrid of *P. hopeiensis* and *P. tomentosa* with *P. adenopoda*. Double crossing means that four species are first matched into two single cross combinations, and then two single cross hybrids are crossed.

iii. Backcross refers to the way in which a hybrid produced by two parents crosses with one of the parents. Among them, the parent used for backcross is called the recurrent parent. According to the principle of genetics, the backcross process causes the nuclear genome of the non-recurrent parent to be gradually replaced by the recurrent parent. Therefore, by backcross, the excellent characters of the recurrent parent can be restored or strengthened in the offspring, and the incompatibility of distant hybridization can be overcome to some extent.

iv. Pollen mix cross refers to pollination of each female parent with a mixture of pollen from a number of male parents. The cross mode is easy to operate, rich in variation of offspring, convenient for selection of offspring, and even has the possibility to integrate the genetic materials of multiple male parents in a hybrid. The *Populus popularis* was produced through pollinating *P. simonii* with mixed pollen of *P. pyramidalis* and *Salix matsudana* by the Chinese Academy of Forestry. It was proved that the *P. popularis* contains genetic materials of *Salix matsudana* by isozyme and immunochemistry determination.

(2) Parental Selection and Matching

Parental selection is a process to select excellent types as parents from breeding resources according to breeding objectives. Parental matching refers to matching the parent couples that can form heterosis from the selected parent types. Correct selecting and matching of the parents is the key to ensure the successful hybridization.

When selecting parents, some basic principles need to be followed. First, select elite parents from a large number of collected germplasm resources. Second, select germplasms with prominent target characters as parents. Third, pay attention to the utilization of local varieties with better adaptability. In addition, the parents require excellent cross compatibility and fertility.

For the parental matching, the advantages and disadvantages between two parents should be complementary, taking the germplasm with fine comprehensive characters and more advantages as the female parent. To produce a progeny with large genetic segregation for selection, the provenances with different geographical origins and ecological types or the germplasms with distant genetic relationships are usually selected as parents. When the genetic relationship between parents is far, such as interspecific or intergeneric hybridization, it is known as distant hybridization. The variety *Populus simonii* × *P. euphratica* cv. Xiaohuyang-1 is the product of distant hybridization between *Populus simonii* and *Populus euphratica*, which has inherited good characteristics from both of the parents, such as the easy propagation and cold resistance of *Populus simonii* and the drought resistance and salt tolerance of *Populus euphratica*. In addition, the heritability and combining ability of parents should be considered in the parental matching process.

(3) Hybrid Methods

According to the seed maturity and nutrition supply, the methods of crossing on trees or indoor water culturing of cutting branches can be selected for hybridization. Generally speaking, for tree species with short seed maturity and small-sized seed, such as *Populus*, *Salix*, and *Ulmus*, the floral branches can be cut off before flowering and then be cultured using water in the greenhouse to conduct controlled pollination; whereas for the tree species of Rosaceae, Pinaceae, Cupressaceae, the method of crossing on the trees is adopted, because the seed maturity is long and water culturing of the cutting branches cannot meet their nutrition requirements.

(4) Operation Steps of Artificial Hybridization

Before carrying out artificial hybridization, the flowering and pollination habits of parents should be investigated, and the flower structure and inserted mode, flowering and seed maturity phenology, and flowering period consistency should be clarified, so as to lay a foundation for the hybridization operation. For parents with non-synchronous flowering periods, effective methods of pollen storage should be studied to ensure that the pollen collected in advance can maintain viability during pollination.

In the process of artificial hybridization, the plants or floral branches with strong growth and normal flower development should be selected as the hybrid stock plants. The perfect flowers should be emasculated before the petals open and the stamens are mature, and then they should be bagged and isolated in time to prevent self-pollination and foreign pollen pollutions. The emasculation operation should be careful and thorough, without damaging the pistil or puncturing the anther. For diclinous flower tree species, the female flowers should be bagged with pollination bags to isolate foreign pollen before female flowers bloom or female cones break through bud scales. Observing exudates on stigmas indicates that the stigmas probably have pollen receptivity and could be pollinated. Because of the difference of flower blooming time in different parts of a plant, it can be pollinated once a day for 2 to 3 days, to ensure sufficient pollination. After each pollination, bagging and isolation should be carried out immediately. When pollinating, be sure to mark the cross combination, pollination date, and other pertinent information. When you observe the stigma wilting, bags can be removed to ensure normal development of immature embryo. Mature seeds should be collected and processed in time and then stored or sown after recording each collection batch information in detail.

(5) Progeny Testing and Elite Hybrid Selection

Obtaining hybrids is just the starting point of hybrid breeding. The progenies need further testing and selection before new variety determination can be recognized. The progeny testing mainly includes a seedling trial, which is a comparison trial of primarily selected clones or families. Because of the long life cycle of trees, some target characteristics would be exhibited when the plant matures, so the period of progeny testing and elite hybrid selection is often long. To speed up the breeding process of new hybrid varieties, increasing attention has been paid to the indirect selection methods, such as early-late character correlation, physiological and biochemical character correlation, and marker-assisted selection, which improve the breeding efficiency.

3.4.1.3 Polyploidy Breeding

Polyploid refers to the individual whose somatic cell nucleus contains three or more sets of chromosomes. The process of breeding new polyploidy tree varieties through certain approaches is called polyploidy breeding, including polyploid induction, identification, propagation, testing, and variety certification. In fact, some important crops and fruits, such as common wheat (*Triticum aestivum*), dwarf banana (*Musa nana*), and the seedless common grape (*Vitis vinifera*), are polyploids, and many ornamental crabapple polyploid varieties are used for urban landscape purposes. In recent years, excellent triploid male poplar varieties, such as *Populus* cv. Beilinxiongzhu 1 (Figure 3-3) and *Populus* cv. Beilinxiongzhu 2, were bred by Beijing Forestry University. They both have the characteristics of rapid growth, beautiful tree shape, excellent woody properties, and strong resistance, especially without fluffy catkins, and they are particularly suitable for urban and rural greening and timber forest construction.

(1) Significance of Polyploidy Breeding

Because of the increase of chromosome sets in the cell nucleus of polyploid plants, the cells often have the characteristics of gigantism, resulting in a gigantic plant even though the cell quantity remains constant. The leaves on long shoots of triploid *Populus tomentosa* at seedling stage are giant-sized, with the leaf width of 53 cm. In 1935, a giant triploid *Populus tremula* was first discovered by Nilsson-Ehle in Sweden, which was 11%, 10%, and 36% higher in tree height, diameter at breast height (DBH), and volume growth, respectively, than the same-age diploid plant under the same site conditions. The volume growth of the triploid hybrids of *Populus tremula* × *P. tremuloides* was 1 to 2 times faster than that of local aspen. The single leaf area of tetraploid *Robinia pseudoacacia* is 1.91 times that of common diploid *Robinia pseudoacacia*; the leaf thickness is increased by 0.68 times; the fresh weight of the compound leaf is increased by 2.27 times, and the dry weight is increased by 2.13 times. Therefore, the growth process of plants can be effectively improved and the biomass can be quickly yielded through polyploidy breeding.

Genome and gene dosage effects in polyploids result in strengthened physiological and biochemical processes, exuberant metabolism, and increasing content of secondary metabolites. Therefore, polyploidy breeding is an effective way

Figure 3-3 New triploid poplar variety *Populus* cv. Beilinxiongzhu 1 (Photo by Wang Jun).

to improve the secondary metabolites, increase the utilization value of economic trees, and reduce the production cost. Researchers have reported that the rubber yield from tetraploid *Hevea brasiliensis* is 34% higher than that of the diploid parent; the paint yield from triploid *Toxicodendron vernicifluum* is 1 to 2 times higher than that of the diploid; the leaf yield of new variety 'Lucha 1' of triploid *Morus alba* is 18.21% higher than that of the diploid variety, with fine leaf quality, and the cocoon yield increased by 6.5%.

Polyploids often show high sterility. According to genetic principles, the chromosome pairing of polyploid plants is loose and irregular during meiosis, which leads to the disorder of chromosome segregation and the incomplete or unbalanced chromosome number of gametes. Therefore, the gametes produced by polyploid plants are often highly abortive. Many seedless fruits are cultivated based on this principle. A new triploid *Eriobotrya japonica* (loquat) variety 'Huayu Wuhe 1' was selected by Southwest University, China, based on the characteristics of high abortion of polyploid plants. Its fruit is larger than that of the common variety, and it has no seed nucleus, with great value for popularization.

Polyploid plants often also have the characteristics of strong ability to survive and thrive along with environmental adaptability, with advantages in dealing with biotic and abiotic stresses such as disease and insect damage, drought, and cold. As one of the driving forces of species formation and evolution, the occurrence of polyploidy in nature is often accompanied by dramatic environmental changes. Most polyploid plants also appear in areas with severe climatic and environmental changes, such as high altitude, high latitude, arctic, and desert, indicating that the adaptability of polyploid plants to adverse natural conditions is better than diploid plants. Therefore, the polyploidy breeding has good application potential in plant resistance breeding. For example, the triploid hybrid aspen variety 'Astria' is resistant to drought and barren areas, with strong rust disease resistance ability, and the tetraploid cedar *Cryptomeria fortunei* shows strong cold resistance.

(2) Pathways of Polyploid Induction

The pathways of plant polyploid induction include sexualpolyploidization and unsexual polyploidization. Sexual polyploidization refers to the process of polyploid formation by hybridization of unreduced ($2n$) gametes. For unsexual polyploidization, polyploids are induced by somatic chromosome doubling, endosperm culture, or protoplast fusion, but do not involve the sexual process of $2n$ gametes.

Because of the abnormal meiosis, many plants can spontaneously produce $2n$ gametes, so sexual polyploidization can be achieved by using natural $2n$ gamete hybridization. Investigation of 70 genotypes of *Begonia* spp. showed that 10 of them could produce $2n$ pollen. Also, $2n$ pollen plants were found in the Chinese native species *Populus tomentosa* population, which could produce $2n$ pollen at the highest ratio of 21.9%. Zhu, Lin, and Kang (1995) pollinated *Populus alba* × *P. glandulosa* with the natural $2n$ pollen of *P. tomentosa* and finally obtained 27 artificial triploids with excellent growth and woody properties, which have been widely promoted. Production of natural $2n$ pollen is influenced by both genetic and environmental factors, however, with the characteristics of instability, and the ratio of $2n$ pollen is mostly lower, which reduces its utilization effect.

The frequency of $2n$ gametes can be increased by artificially induced gamete chromosome doubling to improve the effect of polyploidy induction. Colchicine and high temperature are commonly used as mutagens for gamete chromosome doubling. As a model tree, many studies for artificial induction of $2n$ gametes in poplar have been conducted. Researchers of Beijing Forestry University achieved the stable induction effect of more than 80% $2n$ pollen by colchicine treatment on male flower bud of white poplar. They stimulated the germination of $2n$ pollen by applying a certain dose of ^{60}Co-γ ray treatment, increasing the competitiveness of $2n$ pollen in the process of fertilization and significantly improving the triploid induction rate. Moreover, the competition of haploid gametes can be avoided by using $2n$ female gamete hybridization, and 100% triploids can be produced after fertilization between $2n$ female gametes and normal male gametes. Therefore, more and more attention has been paid to the artificial induction of $2n$ female gametes. Breeders in Beijing Forestry University have developed a series of technical methods to induce $2n$ female gametes by megaspore or embryo sac chromosome doubling of poplar, which has achieved a triploid induction rate of more than 60% and laid a foundation for realizing the poplar polyploid breeding strategy of "large population, strong selection".

When fertile plants have different ploidy levels in a tree species, producing polyploids by using fertile polyploidy gametes formed by polyploid plants is also an important procedure. In some areas of northern Europe and the United States, even the tetraploid female poplar plants were directly planted in the superior *Populus* stands for open pollination. Seedlings were cultivated after collecting seeds from the tetraploid female poplar every year to obtain the triploid offspring. Production of polyploid progeny by crossing with different ploidy has also been reported in mulberry (*Morus* spp.), birch (*Betula* spp.), black locust (*Robinia pseudoacacia*), Chinese kiwifruit (*Actinidia chinensis*), and many other plants.

Somatic chromosome doubling occurs widely in nature, caused by stimulation from drastic environmental changes. Somatic chromosome doubling can be artificially induced by physical or chemical methods, such as mechanical damage, high and low temperatures, radiation, and/or chemical reagents to treat plant seeds, apical bud, callus, and other meristematic cells. Because of the asynchrony of cell division, however, it is difficult to double all cell chromosomes, usually resulting in production of mixoploids. As a single cell, a zygote can form a pure tetraploid after chromosome doubling but not mixoploid, so chromosome doubling of zygotes is the best choice for artificial induction of tetraploid plants.

In general, double fertilization of angiosperms produces both diploid zygotes and triploid endosperm cells. Therefore, triploid plants can also be obtained through endosperm culture. In 1973, Indian scholar P. S. Srivastava first obtained triploid regeneration plants from mature endosperm culture of *Putranjiva roxburghii* (an evergreen tree). Since then, the research of endosperm culture has made rapid progress. The endosperm culture of trees concentrated in non-timber species include mulberry, kiwifruit, boxthorn (*Lycium barbarum*), Chinese date, also known as buckthorn (*Ziziphus jujuba*), Chinese persimmon (*Diospyros kaki*), and others. However, many studies have also found that endosperm culture may lead to chromosomal aberrations.

Cell fusion, including somatic fusion and gamete-somatic fusion, is a new way to overcome the barriers of distant hybridization and create polyploids. The technique is based on protoplast isolation and regeneration, and it fuses the protoplasts by chemical or physical treatment. At present, regeneration plants have been obtained in nearly 100 kinds of cases of intraspecific, interspecific, and intergeneric protoplast fusion. Cell fusion has become an important means of distant-related germplasm innovation in *Citrus*.

(3) Identification of Polyploids

As the chromosome number increases from the process of polyploidy, the cell volume and physiological-biochemical characteristics of plants also change. Based on these characteristics, many methods have been developed to identify ploidy levels of plants, such as chromosome counting, flow cytometric detection, morphology identification, and molecular marker-assisted identification.

Chromosome Counting Method. Directly counting chromosomes is the most reliable method for polyploid identification (Figure 3-4). The cells used for detection can be somatic cells, mostly with meristematic zones such as stem tip, root tip, and callus; they can also be pollen mother cells. If the chromosome number in the somatic cells of the candidate material increases by a certain multiple, it can be determined that the material is a polyploid plant. However, chromosome counting method requires skilled operation, and when facing a large number of candidate materials, chromosome slide preparation is time-consuming and laborious work.

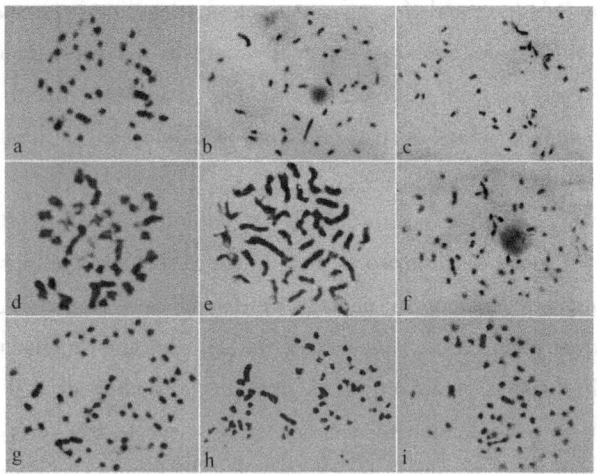

Figure 3-4 Chromosome morphology and number of different poplars (Kang Xiangyang and Wang Jun, 2010). Chromosome morphology and number ($2n = 2X = 38$) of *Populus tomentosa* (a, b), *Populus alba* (c), *Populus bolleana* (d), *Populus hopeiensis* (e). Chromosome number ($2n = 3X = 57$) of natural triploids of *Populus tomentosa* clones B381, B382, B383, and B385 (f)-(i).

Flow Cytometric Detection Method. This method uses the flow cytometer to estimate the DNA content of the plant nucleus cell, and then compares it with that of the control sample with a known ploidy level, so as to determine the ploidy levels of the candidate plants. The detection speed is fast and reliability is high, but the flow cytometry equipment is expensive.

Morphology Identification Method. This approach compares the size of leaf, flower, and other organs or the cell size of stoma and pollen by using the characteristics of giant polyploid plants, which can preliminarily determine whether the candidate plants are polyploids. Morphological identification method can be used to make a preliminary selection for a large number of induced materials, but it is still necessary to further determine the ploidy levels by chromosome counting or flow cytometric analysis.

Molecular Marker-assisted Identification Method. By using simple sequence repeats (SSR) and other molecular marker techniques, we can screen the codominant and single copy marker loci in parents and combine that with the capillary fluorescence electrophoresis analysis to determinate the ploidy levels. According to the linkage group of the marker loci, we can judge the ploidy levels of the candidate plants, and even accurately detect the number and structural variation of the single chromosome. This method has been applied to ploidy identification and chromosome aberration analysis of poplar (*Populus* spp.), willow (*Salix* spp.), and carnation (*Dianthus caryophyllus*).

3.4.2 Variety Protection and Approval

New plant varieties refer to the 'plant varieties' that have been artificially bred or developed from wild plants, with novelty, distinctness, uniformity, and stability characteristics, and with appropriate names. Variety protection and approval is not only the continuation of breeding work but also an important link before new varieties are put into production or face the market. The protection and approval of new tree varieties protects the rights and interests of breeders and also standardizes the production and management of varieties.

3.4.2.1 Variety Protection

Variety protection, also known as plant breeders' rights (PBR) or plant variety rights (PVR), is an exclusive right granted to the breeder of new plant varieties and is a form of intellectual property, which is protected by the "Regulations on the Protection of New Plant Varieties of the People's Republic of China" (hereafter "Regulations on the Protection of New Plant Varieties"). The role of new plant variety protection is to protect the rights and interests of breeders and to motivate the enthusiasm of breeders. It is a form of intellectual property protection and has the same legal characteristics as a patent right, copyright, and trademark right. Without the permission of the variety owner, any institution or individual cannot produce or sell the propagation materials of the authorized variety for commercial purposes, and they cannot reuse the propagation materials of the authorized variety for commercial purposes to produce the propagation materials of another variety.

Internationally, in order to protect the rights of breeders, the "International Convention for the Protection of New Plant Varieties" was passed on December 2, 1961, and Union Internationale pour la Protection des Obtentions Végétales (UPOV) was established at the same time. To have a common foundation in the examination and testing of new plant varieties in all countries, UPOV formulated a set of detailed test principles related to the distinctness, uniformity, and stability (the DUS test) of new plant varieties. China officially joined UPOV on April 23, 1999. At present, China is continuously formulating and improving the DUS test guidelines of new plant varieties for important species.

According to Article 13 of "Regulations on the Protection of New Varieties", the plant material applying for the plant variety right should belong to the genera or species listed in the protection list of the national plant variety. From 1999 to 2021, the National Forestry and Grassland Administration issued seven batches of protection lists of national new tree varieties successively, including 284 genera and species (the specific list can be found on the website of the New Plant Variety Protection Office of National Forestry and Grassland Administration http://www.cnpvp.net). The application and approval of varietal rights must be carried out in accordance with certain procedures (Figure 3-5), and the specific implementation is in the charge of the New Plant Variety Protection Office of National Forestry and Grassland Administration. The term of protection of the variety right should be 20 years for lianas, trees, fruit trees, and ornamental trees and 15 years for other plants since the date of authorization. Once authorized, the variety right will be protected by the examination and approval authority according to the law. Within the period of obtaining variety right, if the propagation materials of the authorized varieties are produced or sold for commercial purposes without the permission of the variety owner, the variety owner or the interested party may request the agricultural and forestry administrative department of the people's government at or above the provincial level to handle the matter in accordance with their respective functions and powers, or they can directly file a lawsuit with the people's court.

3.4.2.2 Variety Approval

Variety approval refers to the legal process by which the authoritative specialized agency reviews the newly bred varieties and determines whether they can be popularized and the suitable

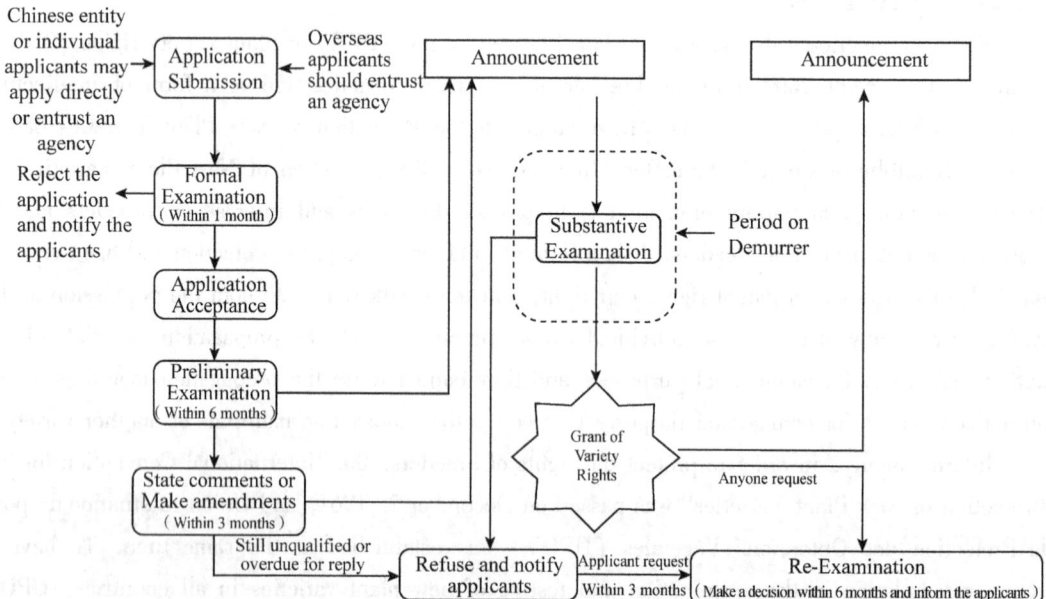

Figure 3-5 Approval process of plant variety rights (Modified according to the approval process of the Plant Variety Protection Office of National Forestry and Grassland Administration).

growth regions of promotion. According to Articles 15 and 23 of the *Seed Law*, the main forest tree varieties should be examined and approved by the variety approval committee set up by the national or provincial agricultural and forestry authorities before they are popularized and applied. Those that have not been approved should not be popularized and/or sold as improved varieties, but those that need to be used in production should be recognized by the forest tree variety approval committee. The forest propagation materials and planting materials that have passed the approval and that have significantly better yield, adaptability, resistance, and other traits than the current primary planting materials in certain growth regions can be called improved forest varieties and can be issued with an improved variety certificate. The implementation of a variety approval system is conducive to variety management and to giving full play to the role of improved varieties, so as to use land resources more fully and economically.

Note that variety approval and variety protection are stipulated by two separate national laws and regulations. If the plant materials have obtained the plant variety rights without passing the variety approval, they still cannot be produced and popularized; if the varieties will enter the production and management link, they must pass the variety approval, which has nothing to do with whether they can obtain the plant variety protection right.

Questions for Review

1. Why should we collect, preserve, and utilize the tree germplasm resources?
2. What are the main characteristics of tree germplasm resources?
3. How do we systematically collect tree germplasm resources? How do we deal with the relationship between central and marginal areas?
4. What factors should be taken into consideration when introducing exotic tree species?
5. What is the procedure of tree introduction? How do we judge whether the introduction is successful?
6. How do we preserve the collected tree germplasm resources? Try to analyze the advantages and disadvantages of different conservation methods.
7. What is hybrid breeding? What are the steps of artificial hybrid breeding?
8. Try to use the knowledge of genetics to elaborate the basic principles of parental selection and matching in hybrid breeding.
9. Why is polyploidy breeding widely used in tree breeding programs?
10. What are the basic ways of plant polyploid induction? How to identify the ploidy levels of candidate plants?
11. How do we correctly understand the relationship between variety protection and variety approval?

References and Additional Readings

CHEN C J, WANG Z H, DU J X, et al., 2012. Breeding of a new triploid mulberry variety 'Lucha 1' [J]. Science of Sericulture, 38(2): 337-342.

CHEN X Y, SHEN X H, 2005. Forest tree breeding[M]. Beijing: Higher Education Press.

CHEN X Z, LUO Z R, 2004. Dodecaploid plantlets regenerated from endosperm culture of 'Luotiantianshi' persimmon[J]. Acta Horticulturae Sinica, 31(5): 589-592.

CHENG F Y, 2012. Nursery of landscape plants[M]. Beijing: China Forestry Publishing House.

CHENG X J, 2006. Scientific research and development prospect of Chinese red-leaf poplar (*Populus* 'Zhonghua

Hongye')[J]. Modern Landscape Architecture, 5: 31-32.

DEWITTE A, EECKHAUT T, VAN HUYLENBROECK J, et al., 2009. Occurrence of viable unreduced pollen in a *Begonia* collection[J]. Euphytica, 168: 81-94.

DHAWAN O P, LAVANIA U C, 1996. Enhancing the productivity of secondary metabolites via induced polyploidy: A review[J]. Euphytica, 87: 81-89.

DUMROESE R K, LUNA T, LANDIS T D, 2009. Nursery manual for native plants: A guide for tribal nurseries [M]. Washington, DC: U. S. Department of Agriculture, Forest Service.

GUO W W, DENG X X, 1998. Somatic hyhrid plantlets regeneration between Citrus and its wild relative, Murraya paniculata, via protolast electrofusion[J]. Plant Cell Reports, 18: 297-300.

KANG X Y, 2002. Cell heredity and triploid breeding of *Populus tomentosa*[M]. Beijing: China Environmental Science Press.

KANG X Y, WANG J, 2010. Study on polyploidy induction technology of poplar[M]. Beijing: Science Press.

KUSER J E, 2000. Handbook of urban and community forestry in the northeast[M]. New York: Springer Science + Business Meida, LLC.

LI J Y, 2010. Urban forestry[M]. Beijing: Higher Education Press.

LI Y, JIANG J Z, 2006. Research progress of feed tetraploid black locust[J]. Pratacultural Science, 23(1): 41-46.

RAO J Y, LIU Y F, HUANG H W, 2012. Analysis of ploidy segregation and genetic variation of progenies of different interploidy crosses in *Actinidia chinensis*[J]. Acta Horticulturae Sinica, 39(8): 1447-1456.

SHEN H L, 2009. Seedling cultivation[M]. Beijing: China Forestry Publishing House.

SRIVASTAVA P S, 1973. Formation of triploid plantlets in endosperm cultures of *Putranjiva roxburghii* [J]. Z Pflanzenphysiol, 69: 270-273.

STATE TECHNOLOGY SUPERVISION BUREAU, 1993. Conservation principle and method of germplasm resources of forest trees: GB/T 14072—1993[S]. Beijing: China Standard Press.

THOMAS T D, BHATNAGAR A K, BHOJWANI S S, 2000. Production of triploid plants of mulberry (*Morus alba* L.) by endosperm culture[J]. Plant Cell Reports, 19: 395-399.

THOMAS T D, CHATURVEDI R, 2008. Endosperm culture: A novel method for triploid plant production[J]. Plant Cell Tissue Organ Culture, 93: 1-14.

WANG J, LI D L, KANG X Y, 2012. Induction of unreduced megaspores with high temperature during megasporogenesis in *Populus*[J]. Annals of Forest Science, 69(1): 59-67.

WANG Y L, XIE X M, HAN B, et al., 2013. Problems and thoughts in the protection of forest germplasm resource facilities[J]. Shandong Forestry Science and Technology, 3: 107-109.

XU W Y, 1988. Poplar[M]. Harbin: Heilongjiang People's Publishing House.

ZHAI M P, SHEN G F, 2016. Silviculture[M]. 3rd edition. Beijing: China Forestry Publishing House.

ZHU Z T, LIN H B, KANG X Y, 1995. Studies of allotriploid breeding of *Populus tomentosa* B301 clones[J]. Scientia Silvae Sinicae, 31(6): 499-505.

Chapter 4 Production of Improved Propagation Material of Forest Trees

Guo Sujuan

Chapter Summary: This chapter introduces the concept of seed, selection of provenance and division of seed, construction of a production base of superior tree varieties, rules of tree fruiting, and collection of and processing of seed and asexual propagation materials used in forest construction and afforestation. The purpose of this chapter is for readers to understand the significance of improved seed and provenance selection, as well as the production knowledge and technology of obtaining improved seed.

In botany, seed refers to the reproductive organ developed from the ovule, and forest tree seed is a general term for planting materials in forestry production. The seed in forestry production includes not only true seed but also fruit of similar seeds, as well as the asexual reproduction organs such as roots, stems, leaves, and buds that can be used to propagate offspring. The fruit itself can be sown directly and can cultivate seedlings to support afforestation. Generally speaking, any organ or part of its vegetative body that can be used as seeding material in forestry production is called seed, no matter what part it has developed from and no matter if it is simple or complex in morphological structure. In the *Seed Law of the People's Republic of China*, promulgated and implemented in 2000, seeds refer to planting materials or propagation materials of crops and trees, including seeds, fruits, roots, stems, seedlings, buds, leaves, and other vegetative organs, and even plant tissues, cells, organelles, and artificial seeds.

Tree seeds are the material basis of forest construction and greening, and the carrier of forest genetic genes that promote forest generation reproduction. The quality and quantity of tree seeds are directly related to the quality of the forest and the success or failure of forestry construction. Improved seed is an important basic guarantee for afforestation, it should have excellent genetic quality and sowing quality. Genetic quality is the basis, sowing quality is the guarantee; only when good conditions are present in both can it be called improved seed. Only improved seeds can provide excellent genetic guarantee for successful forest establishment.

4.1 Provenance Selection and Seed Regionalization

4.1.1 The Concept of Improved Propagation Material of Forest Trees

The national standard known as *the Standard of Examination and Approval of Improved Varieties of Forest Tree* defines "improved varieties" as follows: "The improved variety of forest trees is the propagation material with production value selected manually by strict test and identification that is proved superior to the local main planted tree species or cultivars in suitable areas." According to article 74 of the *Seed Law*, "improved varieties" are defined as follows: "The approved forest tree seeds have significantly better yield, adaptability, and resistance in a certain area than the current main planting propagation materials and planting materials." An improved variety in forestry production must be a seed with excellent genetic quality and sowing quality. The excellent genetic quality manifests primarily in the forest and greenbelt trees constructed with this seed, which have the characteristics of good ornamental, strong resistance to stress, and various functions, while the good seeding quality is reflected in the indicators such as physical characteristics of the seeds and the germination ability that meets or exceeds the national standard. Genetic quality is the basis, and sowing quality is the guarantee. Only in both cases can it be called an improved variety.

4.1.2 Provenance Selection of Improved Varieties

Improved variety is an important guarantee for the success of forest construction and urban greening, but it is not certain to succeed in afforestation. Both forest construction and urban greening must follow the principle of "matching species to the site", which includes not only matching species to the site but also matching provenance to the site. Only in this way can the adaptability and stability of trees in forests and greenbelts be guaranteed. Provenances of improved varieties are of great significance in forest construction and urban greening.

Provenance is the geographic source from which seeds or other reproductive materials are obtained. Because the same tree species have been in specific natural environments for a long time, they are bound to form genetic characteristics and geographical variation to adapt to the local conditions. If the difference between the conditions of forestland and provenance is too great, the phenomenon of poor growth or even death of trees will occur. There have been many painful lessons in this regard in China. For example, after the successful introduction of Xinjiang walnut (*Juglans regia*) in 1958 in Beijing, more than 10 provinces (autonomous regions) competed for the introduction of Xinjiang walnut with the annual introduced seed quantity of 100 000 kg. As a result, many plantings failed because of unsuitable climatic conditions, causing huge losses to forestry production. A large number of studies have proved that using improved seeds of suitable provenance for afforestation can ensure the safety of afforestation species and increase the yield by more than 10%. A study on a 28-year-old provenance test forest for Scots pine (*Pinus massoniana*) found highly significant differences among provenances, which were mainly manifested in growth characteristics and strongly controlled by genetic factors. Among the selected 12 excellent provenances and 33 excellent individual trees of *Pinus massoniana*, the average genetic gain of volume was 34.75%, and the average genetic gain of volume of excellent individual trees was 141.48%.

Iin addition, with the development of introduction and exchange and flower trade, pests and other harmful organisms are no longer blocked by natural barriers, and some dangerous diseases, insects, weeds, or harmful organisms will invade, resulting in their widespread disaster without the restriction of natural enemies. For example, Dutch elm disease was found in Holland and Belgium in 1981, and it then spread to European and American countries with the elm trees. In the 1930s, it caused destructive disasters to elm in Europe and America.

At the end of the 19th century, some countries in northern Europe, such as Sweden, failed at afforestation with pine seeds produced in Germany. Almost all pine seeds from southern France and Hungary died in Germany. We must recognize that when introducing exotic seeds for afforestation, improper selection of provenances will lead to failure of afforestation. The correct selection of suitable provenances can enhance the adaptability and resistance of forests, thus improving the stability and economic benefits of trees in forests and greenbelts.

4.1.3 The Way to Select Suitable Provenance

Suitable provenances should be found for afforestation areas. One important method is to carry out a provenance experiment, which is to collect the seeds from different provenances and use them for cultivation experiments, observe their growth and development, and select the most suitable provenance. The geographical variation law of the main tree species for afforestation can be studied, and the variation pattern can be mastered so as to accurately select and cultivate the characters needed for plantation cultivation.

Provenance research in China began in the 1950s. After the 1970s, a large-scale provenance experiment and seed regionalization were carried out all over the country, involving more than 10 tree species. Findings showed that the growth characteristics and adaptability of the same tree species vary significantly based on their geographical origins. Taking the Gmelin larch (*Larix gmelinii*) as an example, seven provenances (Tahe, Mangui, Genhe, Xinlin, Sanzhan, Wuyiling, and Hebei) growing in the same environment (Harbin) for 26 years, with tree diameter at the breast height (DBH), sapwood width, sapwood area, heartwood radius, average sapwood growth rate and so forth as parameters, have significant differences between provenances, among which the average value of growth characteristic parameters of the southernmost provenance Hebei is the highest, while the average value of Sanzhan is the lowest. In this way, the suitable provenance was selected for planting *Larix gmelinii* in Harbin.

Considering the China Armand pine (*Pinus armandii*) as an example, the productivity of different provenances differs greatly. In Tengchong, Yunnan Province, the 10-year-old tree height growth of Yunnan-Guizhou provenance is 4.7 times higher than that of Qinba Mountain, 2.3 to 2.6 times higher than that of Pingba and Dejiang provenances in Guizhou, 1.4 times higher than that of Anhua in Hunan, and 1.2 times higher than that of Yichang in the north. The basic law of provenance geographic variation is that significant differences occur among provenances in most of the characteristics. From the central production area to the marginal production area, most of the characters show a north-south gradual change trend. The central production area grows quickly, but the adaptability is not as good as the non-central production area. Only by combining the results of the provenance experiments can we better guide the rational use of seed in production.

Generally speaking, the results of a provenance test serve production through the following ways: One is to provide the basis for making seed regionalization; the other is to provide the scientific basis for the reasonable allocation and use of seeds by the user; a third is to provide the best provenance data for the user to establish the improved variety base.

4.1.4 Forest Seed Regionalization

Seed regionalization is based on the supply scope of seeds produced in each region of a tree species, given the ecological conditions, genetic characteristics, and administrative circles.

To avoid the heavy losses caused by the unclear provenance and the blind allocation and use of seeds, the forestry-developed countries implement the legal control of provenance and carry out the work of seed regionalization.

The general practice of provenance control in Germany is to divide provenance areas according to the altitude, soil, geographic location, and natural conditions of the growing area of a specific tree species. On the basis of provenance tests, the results identified by experts are determined in legal form and announced to the society through a special provenance regionalization manual, which all forestry production activities should strictly follow. The United States has demarcated the national forest seed area and has stipulated the seed collection range. It not only considers the difference of latitude but also the influence of altitude on different provenances. Every province in Canada has its own forest seed regionalization and corresponding provenance regionalization map. In practice, the work of seed collection and afforestation is carried out in strict accordance with the regionalization. Each seed area has specific seed collectors, who collect seeds regularly, attach labels, record collection details for the archives, and send the seeds to the provincial seed center for processing, storage, and seedling raising. Cultivated seedlings must be returned to their seed collection area for afforestation.

The provenance experiment was started in the 1950s in China. According to the results of the experiment, the seed areas were divided according to the climatic characteristics and growth characteristics of the main afforestation species in the various distribution areas. Seed region is a unit of provenance with similar ecological conditions and tree genetic characteristics, and it is also a unit of region for afforestation. The seed subregion is a subunit that can be divided for better control of seed usage in a seed region; that is, the ecological conditions and genetic characteristics of trees in the same seed subregion have similar characteristics. Therefore, when using seeds for afforestation, priority should be given to the allocation of seeds in the seed subarea where the afforestation site is located. If the seeds fail to meet the afforestation requirements, they should be allocated within the seed region.

4.1.5 Principle of Seed Allocation of Non-regionalization Tree Species

Seed regionalization provides a useful guide for afforestation species in production. However, *Seed Zones of Chinese Forest Trees* (GB/T 8822.1 ~ 13—1988) contains only 13 tree species. Because more than 13 tree species are needed for afforestation in China, and because most of the 13 tree species are limited to conifers, the regionalization is far from meeting the needs of production, nor does it guide usage for the broad-leaved species provenance in the construction of coniferous and broad-leaved mixed forests. Therefore, the selection of provenance of other tree species should follow these basic principles.

First, the local provenance is most suitable for the local climate and soil conditions, so the local seeds should be used as much as possible; in other words, seeds are collected locally, and seedlings are raised and afforested locally.

Second, when transferring seeds from other areas, the areas lacking guidance on seed regions should try to select seeds produced in regions with the same or similar climate and soil conditions as their own local conditions.

Third, the general rule of the seed transportation distance is that the transportation ranges from north to south and from west to east are larger than that in the opposite directions. For example, for the seeds of *Pinus massoniana* in China, the transfer latitude from north to south should not exceed 3° and that from south to north should not exceed 2°. Considering longitude, the transfer range from regions with poor climate conditions to regions with better climate conditions should not exceed 16°.

Fourth, the change of topography and altitude has great impact on the climate. When seeds are transferred vertically, generally the altitude should not exceed 300 to 500 m.

Keep in mind that the adaptability of tree species differs among species, and the range of seed transfer will probably not be the same. In production, to match species to the site and match provenances to the site, the most important thing is to strengthen the provenance test and to select seeds of the best provenance for afforestation in a given region. When the scale and time of provenance test are not enough to determine the optimum provenance, similar site conditions to the local should be selected. For the specific methods of provenance test, please refer to books and other sources that focus on forest tree breeding.

4.2 Production Base of Improved Tree Species

According to provenance testing and seed regionalization, following the principle of seed allocation basically solves the problem of geographical provenance of afforestation species. But in actual production, even if the provenance is selected correctly, who produces the seeds can still present a problem. For example, some seeds are collected by the masses from the mountains, and some are produced by specialized seed companies in improved seed bases. How should afforestation institutions choose seeds? In this regard, China has specially formulated the *Seed Law*. According to the *Seed Law*, the production and operation of commercial seeds of major trees shall be subject to the licensing system. Institutions and individuals wishing to engage in the production and operation of seeds shall apply to the local competent forestry administrative department, and the production and operation of seeds can be carried out only after obtaining the license of production and operation of seeds upon approval. At the same time, the National Afforestation Technical Regulations issued by the State Forestry Administration clearly stipulates "actively promoting the improved varieties with suitable provenance, and giving priority to the seeds produced by the improved provenance and the improved seed base".

The cycle of forest cultivation is long. Once inferior seeds are used for afforestation, it will not only affect the survival, growth, and timber of trees but also cause serious and irreparable losses. In terms of seed production, in the past, because it was the general population who collected seeds, the organization and management were not strict, and the low trees were collected but the tall trees not, and the small trees were collected but the big trees not. The improved mother trees in the stand are less fertile and difficult to harvest, and they are more likely to be abandoned by the seed collectors. The direct consequence of this vicious cycle is that the tree species degenerate, senescence ahead of time, form small old trees, and seriously affect the usage and value of the trees. Therefore, the key to ensuring the quality of seeds is to establish a special production base of improved seeds and realize the specialization of seed production.

At present, three types of improved forest seed bases are recognized in our country: Seed production stand, seed orchard, and cutting orchard. The so-called improved forest seed base refers to the place specially engaged in the production of improved seed, which is established in accordance with the requirements of the relevant provisions of the state on the construction of the seed production stand, seed orchard, and cutting orchard.

The construction of an improved forest seed base in China starts from the seed production stand. Through research on the method of delimiting or building a seed production stand by means of selecting an excellent forest stand, technical support provides for the establishment of a seed collection base of Korean pine (*Pinus koraiensis*), Chinese pine (*Pinus tabuliformis*), Scotch pine (*Pinus sylvestris*), *Pinus massoniana*, exotic pines, Yunnan pine (*Pinus yunnanensis*), black locust (*Robinia pseudoacacia*), and other tree species in production. Since the 1960s and 1970s, research on the superior tree selection of primary tree species and the construction technology of seed orchards have been started. Tens of thousands of superior trees have been discovered, which provide valuable materials for the construction of primary seed orchards and cutting orchards, as well as the supporting technology from grafting, planting for early harvest, and high yield.

4.2.1 Seed Production Stands

A seed production stand is a kind of forest stand that is built on the basis of good natural forest or good artificial forest with definite provenance, according to the construction standard of the seed production stand, through selective thinning that reserves the superior and removes the inferior, and with the goal of producing forest seeds with better genetic quality. Generally, using the seeds produced by the seed production stand for afforestation obtains a genetic gain of 3% to 7%. Because of its simple construction technology, low cost, fast production, and higher seed yield and quality than the general stand, the seed production stand is still regarded as one of the main forms of improved seed production in some areas. Good-quality improved seeds can be obtained in a short period of time by selecting the right seed production stand and managing them scientifically and reasonably (Figure 4-1).

Figure 4-1 Seed production stand of larch (Photo by Liu Yong).

(1) Conditions for Selection of Seed Production Stands

As far as possible, the climatic conditions of the seed production stand are the same or similar to those of the afforestation area. Choose a location close to the planting area; convenient transportation access, gentle terrain, sufficient light, leeward facing, and sunlight are required, so as to be convenient for forest tree fructification and production and management. The soil layer of the seed production stand should be deep and fertile, with an area of at least 2 to 3 hm^2; there should be no inferior forest of the same species nearby.

To facilitate management of the stand, choose the middle age and strong seed production stand with vigorous growth and good seed-bearing capacity. The suitable age for the establishment of the seed production stand varies with the law of seed setting, the origin of the stand, the growth environment, and the development status. Generally, the age of fast-growing tree species, artificial forest, and a forest stand with an active growth environment and good development is younger than that of a slow-growing tree species and a natural forest, respectively. The suitable age of a stand of Chinese fir (*Cunninghamia lanceolata*), *Pinus massoniana*, Chinese red pine (*Pinus elliottii*), loblolly pine (*Pinus taeda*), and other artificial forest is 10 to 15 years old; the age range of natural forest can be wider, such as *Pinus tabuliformis* with 20 to 50 years and *Pinus koraiensis* with 120 to 200 years. In addition, the best stand for the proposed seed production stand is a pure forest of the same age, with the growth and development of the stand in good condition, and with a canopy density of generally 0.5 to 0.7.

(2) Construction of Seed Production Stands

After selecting the seed collecting stand, we should select the excellent seed tree according to the national standard of building the seed production stand. To improve the genetic quality of seed trees, we need to improve the light, water, nutrient, and sanitation conditions; promote the growth and development of seed trees; and the selected seed production stand must be thinned. The principle of thinning is to "remove the inferior and retain the superior". The seed trees should be evenly distributed as far as possible, but the inferior trees should definitely be cut down, even if small open spaces are created. For dioecious tree species, attention should also be paid to the proportion and distribution of dioecious plants. Thinning intensity directly affects the growth and development of seed trees as well as the yield and quality of seeds. Basic requirements for determining the thinning intensity are to ensure the normal growth and development of the reserved seed tree, to ensure that the seed tree has a reasonable and sufficient nutrition space, and to facilitate the seed production and quality improvement of the seed tree. Make sure the trees do not cover each other, and they should not form glades and gaps in the forest, to ensure that the seed tree will not be damaged by wind, snow, and sun. To avoid the drastic change of forest environment, thinning can be divided into two to three events, gradually reaching the number of planned seed trees. In terms of canopy density, it should be kept at 0.5 to 0.6 after thinning, and the final canopy density should be 0.4 to 0.6 for most tree species.

(3) Management of Seed Production Stands

After thinning, with the canopy density decreased, the woodland has become more exposed and weeds will begin to grow, so it is necessary to weed at appropriate times. At the same time, we need to strengthen the management of water and fertilizer, tree management, pest control, and so forth.

Water and Fertilizer Management. During the flowering and fruiting period of trees, more nutrients are consumed, sometimes affecting the seed yield of the next year. Reasonable fertilization and irrigation according to soil moisture, fertility, tree development period, and so on, can avoid alternate years of good yield and bad yield, and can effectively improve the yield and quality of the seed production stand. The effect of fertilization combined with irrigation is better, and organic fertilizer should be combined with inorganic fertilizer. When the field water capacity is at less than 65% during the growth period, irrigation should be applied. Irrigation should be carried out before flower bud break, young fruit formation in early spring, and soil freezing in late autumn; it is also necessary to irrigate after each fertilization.

Tree Management. According to the research data at home and abroad, pruning the seed tree properly, that is, cutting off the dead branches at the bottom of the seed tree and the branches that hinder the development of the seed tree canopy, can improve the structure of the seed tree canopy and the light conditions. Pruning will also effectively improve the nutritional status of the seed tree, promote the seed tree to blossom and bear fruit, and improve the seed yield and quality.

Pest Control. To protect the good growth of the seed tree forest and reduce or eliminate the losses caused by diseases and insect pests, we must adhere to the principle of "prevention first, early control, and rapid elimination". We need to carry out regular investigation and research to find out the degree of damage, distribution, and occurrence of diseases and insect pests in the seed tree forest, so as to make predictions and apply preventions in advance.

4.2.2 Seed Orchards

Seed orchards are a kind of special artificial forest that are made up of superior clones or families according to the design requirements. It is managed intensively, with the goal of producing seeds with good genetic quality and sowing quality.

Extensive research and production practice has proved that the genetic gain of tree growth can be greatly increased by using seeds produced by a seed orchard—generally improved 15% to 40%. Therefore, a seed orchard is an important way for improved seed production in forestry-advanced countries in the world (Figure 4-2).

Figure 4-2 Seed orchard of Ponderosa pine (*Pinus ponderosa*) (Photo by Liu Yong).

(1) Seed Orchard Types

According to the propagation method of seed trees, seed orchards can be divided into clonal seed orchards and seedling seed orchards. A clonal seed orchard is established by asexual propagation with superior trees or individuals as materials. It has the advantages of maintaining the original excellent quality of superior trees, clear clonal source, early flowering and bearing, relative dwarfing of tree shape, and convenience for intensive management. A seed orchard of seedlings is established by free pollination seeds collected from superior trees or excellent clones, or by seedlings cultivated from controlled pollination seeds. The orchard is characterized by easy propagation, less investment, and is suitable for species with difficulty in clone propagation; or the tree species with early flowering and fruiting and short rotation period, the progeny test can be combined with seed production. The phenomenon of self-intersection is less serious than that in a clonal seed orchard. The disadvantage is late flowering and fruiting, unstable characters in a superior tree, and variations easily occur.

According to the selection and identification of the breeding materials, they are divided into the first-generation seed orchard, the first-generation seed orchard without the inferior, the first-generation improved seed orchard, and the advanced seed orchard. The advanced seed orchard is formed through multigenerational selection and cultivation, which can improve the frequency of fine genes in the forest population and can combine the fine genotypes that will better meet people's needs. With the increase of generations in the seed orchard, the improvement effect will gradually improve, so the advanced seed orchards are the future development direction. Some seed orchards of southern pine in the United States have reached the second and third generations. At present, main timber species in China, such as *Cunninghamia lanceolata*, *Pinus massoniana*, *Pinus elliottii*, *Pinus taeda*, pond cypress (*Taxodium ascendens*), China Armand pine (*Pinus armandii*), *Pinus yunnanensis*, *Pinus koraiensis*, *Pinus sylvestris*, Chinese spruce dragon (*Picea asperata*), and five major larches, have completed the construction of the first-generation seed orchards, most of which have set up a large area of genetic testing forest, and they are now in the transition period to higher generation seed orchards.

(2) Construction and Management of Seed Orchards

Please refer to *Tree Breeding Science* (Chen and Shen, 2021) and other resources for details.

4.2.3 Cutting Orchard

The cutting orchard is an improved variety base for producing branches, scions, and root segments with good genetic quality by using superior trees or clones as materials. Two main functions are important for a cutting orchard: One is to provide cuttings or roots directly for afforestation; the second is to provide clonal propagation materials for further expansion and propagation, which can be used to establish seed orchards, propagation orchards, or to cultivate clonal propagation seedlings.

The advantages of a cutting orchard include high yield and low-cost scions; the genetic quality of cuttings can be guaranteed; and convenient operation and management, easy pest control, and safe operation. The cutting orchard is generally set in the nursery, which makes it convenient to arrange in terms of labor and timeliness of picking the cuttings. The nursery location can avoid long-distance transportation and storage of cuttings, which not only improves the survival rate of cuttings but also saves labor.

4.2.3.1 Types of Cutting Orchards

Two kinds of cutting orchards are in use: Primary cutting orchards and advanced cutting orchards. The primary cutting orchard is established with materials collected from the never-tested superior trees. Its task is only to provide the branches, scions, and root segments needed for the establishment of a generation of clonal seed orchard, clonal determination, and resource conservation. The advanced cutting orchard is established by tested excellent clones, artificial crossbreeding, and selection of target trees or vegetative propagation materials collected from excellent varieties. Its purpose is to provide branches, scions, and root segments for the establishment of a generation of improved clone seed orchard or the promotion of excellent clones and varieties.

4.2.3.2 Construction Method of Cutting Orchard

The cutting orchard should be set in a place with suitable climate, fertile soil, flat terrain, convenient drainage and irrigation, and convenient transportation, and within the nursery when possible. For example, if a cutting orchard is set up in the mountainous area, the slope inclination should not be too steep, and the selected slope direction should not be too sunny nor affected by cold wind in winter.

In terms of configuration, arbor forest type is usually adopted for the cutting orchard for the purpose of providing scions, with the planting spacing of 4 to 6 m; shrub type is usually adopted for the cutting orchard for the purpose of providing branches and root segments, with the planting spacing of 0.5 to 1.5 m. The regeneration period is generally 3 to 5 years. Consider the shrub-type cutting orchard that mainly provides poplar cuttings as an example: The scion-picking plants have no obvious main trunk. Generally, 1-year-old cuttings or seedlings are planted according to the specified planting spacing. Before sprouting in the second year, the seedling stem is cut out at 10 cm away from the soil surface, and when the new branches are 10 cm high, 3 to 5 thick and strong branches with uniform distribution are selected and retained, and the rest are removed. When entering the dormant period or during the next spring, in addition to collecting cuttings, the stem can be cut out again, and the remaining stump increased by 5 cm compared to the previous year. Also, 3 to 5 winter buds should be retained for each remaining stump. This whole procedure should be repeated every year; after 3 to 5 years, the cutting orchard should be regenerated.

In practice, because of the limited number of cuttings directly provided by superior trees, it is often necessary to establish the cutting orchard for superior trees first, and then the clone propagation orchard is established after that.

The cutting orchard does not need to be isolated, but it is necessary to prevent the varieties from mixing, and the location should facilitate convenient operation and management. It can be divided by varieties or clones, and the same variety should be planted in a plot.

4.2.3.3 Management of Cutting Orchards

The management of cutting orchards includes deep plowing, fertilization, weeding, drainage, irrigation, and pest control.

(1) Timely Irrigation

A drainage and irrigation ditch should be dug in the cutting orchard, which can also be used for flood control and flood drainage. Irrigation should be done at appropriate times, for example, immediately after cutting or planting in the first year, and 8 to 10 times a year. For the first time, enough water should be used to fully saturate the soil; for the second to third times, less water can be used; for the sixth to eighth times, the water should fully saturate the soil, and for the later period of seedling growth, the water should be stopped. After the second year, according to the growth condition of scion-picking plants, local climate, and soil conditions, the number of times of irrigation can be reduced accordingly and the quantity of irrigation water increased. Generally, 6 to 8 times of irrigation can be conducted every year. At the same time, planting green manure is one of the ways to not only overcome the overgrowth of weeds but also increase the source of fertilizer, which should be paid attention to.

(2) Intertillage and Weeding

Intertillage and weeding can improve soil, eliminate weeds, promote seedling and cutting growth and root development. In the first year of cutting or planting, for the purpose of weeding and soil moisture conservation, shallow tillage and weeding was adopted, with a depth of 3 to 5 cm; after the middle and last 10 days of June, the root system has been lignified. Rainfall and irrigation can cause soil hardening, and weeds readily grow. Therefore, deep plowing, at a depth of 6 to 8 cm, and frequent weeding are required. In the whole year, intertillage and weeding should be carried out 6 to 9 times, being careful that all the weeds are removed. Shallow soil-loosening should be done without weeds; the soil shall be loosened after the rain, and intertillage should be carried out for irrigation, so that the soil in the nursery is loose and without weeds. After the second year, weeding can be deepened according to the depth of the roots, and the times can be reduced appropriately, probably 5 to 6 times in the whole year.

(3) Reasonable Fertilization

A large number of cuttings are picked every year in the cutting orchard, which has consumed nutrients and reduced the soil fertility. To ensure that the cutting orchard can provide a large number of high-quality cuttings, especially to improve the rooting rate of shoots, we need to improve the nutritional conditions of seedlings through reasonable applications of chemical fertilizer and farmyard manure according to the soil capacity of the cutting orchard.

4.2.3.4 Rejuvenating the Cutting Orchard

The yield and quality of cuttings (scions) are affected by cutting over many years, the aging of trees, the decline of growth, and the growth of rotten stumps in the cutting orchard. To prevent the degradation of the seed trees in the cutting orchard, it is necessary to rejuvenate the seed trees. The number of years until rejuvenation and regeneration are needed vary with species; for example, cuttings of a fast-growing species such as poplars—*Populus popularis*, *Populus xiaohei*, and *Populus beijingensis*—can be picked continuously for only 5 to 6 years. Another example, it is necessary to cut the stem at the end of autumn and the beginning of winter every year for *Populus tomentosa* so that new branches can germinate again from the root to form the root stump and then produce cuttings (scions). In general, after 5 to 6 years, we need to dig roots and replant, or select new nursery land and establish a new cutting orchard.

No matter which kind of improved seed production base is established, technical archives should be established that include surveys and construct text, drawings, tables, and other materials, and post-construction management measures should be tended to.

4.3 Tree Fruiting Law, Seed Collection, and Processing

4.3.1 Tree Fruiting Law

After the trees begin to bear fruit, their annual bearing capacity differs greatly. Among them, most shrub species bloom and bear fruit every year with little difference in annual bearing capacity, whereas arbor species have more bearing capacity in some years (productive years), and little or no bearing capacity in other years (non-productive years).

Different tree species exhibit great differences in the abundance and/or poor harvest phenomenon of fruit bearing. For example, the seed yield of poplar, willow, and eucalyptus is relatively stable year by year, whereas seed yield of Manchurian ash (*Fraxinus mandshurica*), Amur cork tree (*Phellodendron amurense*), and the oaks (*Quercus* spp.) is relatively obvious in productive and non-productive years; coniferous trees in alpine regions, such as *Pinus sylvestris* and *Pinus koraiensis*, often fail to bear fruit.

The main reason for the high or low yield of tree fruiting is the lack of nutrition. Most of the products of photosynthesis are consumed by the development of fruits and seeds in the abundant years, and the nutrients cannot be transported to the roots normally, thus inhibiting the metabolism and absorption function of the roots. That inhibition results in malnutrition in the critical period of flower bud differentiation, and then the next year will experience low fruit yields.

The phenomenon of the high or low yield of tree fruiting is also related to the growing environment of seed trees. If the climate is mild and the soil is fertile, the abundant and poor harvest phenomenon is not obvious. For example, for European spruce in good growing conditions, the fruiting cycle is 2 years; in less productive places the fruiting cycle takes 7 years. Severe weather may have a greater effect on fruiting than does the growing environment.

Practice has proved that the abundant and poor harvest phenomenon of trees is not unchangeable. As long as we create high nutritional conditions for the growth of trees, strengthen the tending management, prune scientifically, apply fertilizer reasonably, and eliminate natural disasters (as much as possible), we can achieve a bumper harvest every year.

4.3.2 Seed Collection

Seed collection is an important part of forest seed production, even though it is seasonal work. Whether the work is scientific and timely directly affects the quality and yield of seeds, as well as the development of the seed industry. To obtain a large number of improved seeds, we need to correctly select the seed tree for seed collection, be familiar with the tree fruiting law, predict the seed yield, master the characteristics of seed maturity and abscission habits, do well in all preparations before seed collection, make a feasible seed collection plan, select appropriate seed collection methods and tools, and do a good job with seed registration.

4.3.2.1 Selection of Seed Collecting Stands

According to the provisions of the *Seed Law*, seed collection stands include seed orchards, seed production stands, general seed collection forests and temporary seed collection forests, and group and scattered excellent seed trees. The best way to collect seeds is to collect them from the seed orchard or the seed production stands in the area where the local climate and soil conditions are close to the afforestation site. Given that the area of the improved tree variety base is small, and the seed yield is not enough to meet the requirements of afforestation, an excellent stand of natural forest or artificial forest can be selected, and even the scattered trees meeting the standard of excellent seed trees can be selected for seed collection.

4.3.2.2 Prediction of Seed Yield

To make a scientific plan for seed collection and to provide a scientific basis for seed collection preparation, seed storage, allocation, and management, we need to predict and forecast the yield of seeds and fruits. After many years of research and summarizing the experience of prediction and forecast of forest seed production, the forestry science and technology workers in China put forward some simple and practical prediction methods with high precision for *Cunninghamia lanceolata*, *Pinus tabuliformis*, and *Pinus sylvestris*, such as visual method, evaluation by sample plot, and use of a mean sample tree. The prediction method introduced from abroad is also studied according to the actual situation of our country, and the adjustment measures are put forward. For example, the research on estimation of cones in a visible hemihedral crown found that this method is difficult to implement in dense stands. The basic idea of applying this method in Sweden, the country inventing this method, is to use the observed value as the relative index of a certain region's fruit yield, which is used for macro decision-making, and it is not required to estimate the absolute yield. Therefore, it can be assumed that, through years of observation and verification, in areas with higher stand density, it is possible to select some single trees that are easily observed, such as marginal trees or sparse forest trees, and to take their average hemihedral crown cone number as the index of fruit bearing in this area. This approach avoids the difficulty of estimating when density is high (Fangyuan Yu, et al., 1992).

At present, a whole set of prediction systems of forest fruit production is gradually being established in production, including: The prediction of forest fruit production is carried out in the near ripening period of fruit; the prediction method can select and use a visual method, evaluation by sample plot, a method of mean sample tree, standard branch method, visible hemihedral crown cone estimation method, and so on; the prediction results are filled in according to tree species, collection area, and type of forest. Results are reported below, step by step.

(1) Visual Method

Visual method is also known as phenology. In this method, the seed yield grade is estimated by observing the flowering and fruiting of seed trees. According to the data over the years, the seed yield can be calculated, that is, the seed setting of the seed production stand is observed in the flowering stage, seed formation stage, and seed maturity stage.

In China, the four grades of flowering and fruiting—abundance, good, normal, and bad—were evaluated as follows:

i. In an abundance year, much blooming and fruit bearing occurs, which is 80% or higher of the highest number of flowering and fruiting in the past years.

ii. A good year, more blooming and fruit bearing occurs than in a normal year, accounting for 60% to 80% of the highest number of flowering and fruiting over the years.

iii. In a normal year, medium flowering and fruiting occurs, accounting for 30% to 60% of the highest amount of flowering and fruiting over the years.

iv. In a bad year, less flowering and fruiting occurs, accounting for less than 30% of the highest amount of flowering and fruiting over the years.

For specific observations, a group of 3 to 5 persons with practical experience with this method should be organized to follow the pre-determined survey route, randomly set points, evaluate the grade, and finally summarize the situation of each point to comprehensively evaluate the flowering and fruiting grade of the entire forest stand.

This method requires observers to be accurate in visual inspection and skilled in flowering and setting, otherwise subjective differences will occur. To check the results of the visual inspection, the method of mean sample tree or standard branch method can be used.

(2) Evaluation by Sample Plot

Evaluation by sample plot is also known as the measurement method. Within the seed collection stand, several representative standard plots are set up. Each standard plot should contain 30 to 50 trees. All fruits are harvested and weighed. The standard plot area is measured to estimate the total stand fruit quantities. The seed harvest of the year is estimated with reference to the previous year's harvest and seed yield.

(3) Standard Branch Method

In the seed collection stand, 10 to 15 trees are randomly selected. Among the 3 layers of upper, middle, and lower portions of the dark and sunny sides of each crown, branches with a length of about 1 m are randomly selected as standard branches to calculate the number of flowers and seeds on branches and the number of branches with the average length of 1 m. Then comparisons can be made to the number of flowers and fruits of the standard branches in good, common, and poor years in the history of the tree species, and based on that, an evaluation can be made regarding the seed level and seed harvest.

(4) Method of Mean Sample Tree

This method calculates the yield based on a linear relationship between the diameter of the seed tree and the number of seed. Mean sample tree refers to trees that are medium in height and in diameter. The main method is to select representative plots and to set standard plots in the seed collection forest stands. Each standard plot should have 150 to 200 trees. We would then measure the area of the standard plots; survey each tree; determine its DBH, tree height, and crown width; and calculate the average. Of the mean sample trees, 5 to 10 were selected in the standard plot and all the fruits were harvested. The average single-plant seed yield was calculated, so as to calculate the standard plot seed yield and the total stand seed yield and actual harvest. Fruit quantities of the whole forest multiplied by the seeding rate of the tree species is the seed yield of the whole forest.

Because all the fruits cannot be cleaned when collecting standing trees, the calculated total forest seed yield can be multiplied by 70% to 80% according to the seed collection technology and the growth of the trees, which is the actual collected amount.

(5) Visible Hemihedral Crown Estimation Method

In the sampling stand, more than 50 sample trees are randomly selected, and standing at a point similar to the height of the tree, the number of fruits of the visible hemihedral crown of each sample tree is counted and the average value is calculated. This value is substituted into the dependent equation of visible hemihedral crown fruit number and whole crown fruit number to obtain the average fruit number of each sample tree, which is then multiplied by the total number of forest trees to determine the fruit number of the whole forest. According to the recovery ratio and seed yield over the years, the seed harvest yield can be estimated.

When using this method, we need to first establish the correlation equation between the visible hemihedral crown fruit yield and the full crown fruit yield of the tree species.

4.3.2.3 Seed Maturity

Seed maturation is a process in which the fertilized egg cells develop into a complete embryo with radicle, germ, cotyledon, and hypocotyl.

(1) Process of Seed Ripening

Generally, seed ripening includes two processes: Physiological maturity and morphological maturity. Individual tree species have physiological after-ripening phenomenon.

Physiological Maturity. When the nutrients in the seeds accumulate to a certain extent and the embryo has the ability of germination, physiological maturity is achieved. At this time, the seed has high water content, with the internal nutrients still in a soluble state, and a non-dense seed coat. The seed is not full, has weak resistance, and the seeds are difficult to store and easy to lose their germination ability.

Morphological Maturity. When the biochemical changes inside the seeds are basically over, the accumulation of nutrients has stopped, and the outside of the seeds shows the characteristics of maturity, morphological maturity is achieved. At this time, the water content of seeds has decreased, the activity of enzymes has decreased, and the nutrients have turned into insoluble fat, protein, starch, and so on. The seed coat is hard, dense, resistant to damage, and acceptable for storage. The respiratory function was weak, and seeds began to enter into dormancy. Seeds have germination ability. In appearance, the seeds are full and hard, with specific color and smell. Therefore, the morphological maturity is often used as the symbol of the seed collecting period in production.

Physiological After-ripening. Most of the species enter morphological maturity after physiological maturity. But a few tree species, such as *Ginkgo biloba*, have mature characteristics in morphology, but the embryo is not fully developed. It requires a period of time to have germination ability, which is called physiological after-ripening.

(2) Seed Mature Period

Species have seed mature periods that differ from one another. For example, poplar (*Populus* spp.), willow (*Salix* spp.), and elm (*Ulmus* spp.) mature in spring; mulberry (*Morus* spp.) trees mature in early summer; tree-of-heaven (*Ailanthus altissima*) and black locust (*Robinia pseudoacacia*) mature in late summer; sawtooth oak (*Quercus acutissima*) and Oriental arborvitae (*Platycladus orientalis*) mature in early autumn; pines, such as *Pinus tabuliformis* and white-barked pine (*Pinus bungeana*), birch, Asian hazel (*Corylus heterophylla*), and ginkgo (*Ginkgo biloba*) mature in autumn; *Cunninghamia lanceolata*, *Pinus massoniana*, and tea-oil camellia (*Camellia oleifera*) mature in late autumn.

In addition to the internal factors of the tree species, the maturity period of the seeds is influenced by the region, year, weather, soil, tree canopy, and human activities.

The mature period of seeds of the same tree species varies in different regions. For example, the seed maturity time of Chinese cottonwood (*Populus simonii*) is in the middle of June in the south of Heilongjiang, the first 10 days of June in the north of Liaoning, the last 10 days of May in the south of Liaoning, and the first 10 days of May in Beijing. Elms do the same. For example, the mature period of elms in Beijing is around the first 10 days of May, and in Heilongjiang it is as late as the last 10 days of May to the middle of June.

Given the terrain and environmental conditions, the maturity period of the same tree species and the same area can also differ. For example, the mature period is earlier in sunny or low-altitude areas, and later in shady or high-altitude areas. In different years, because of different weather conditions, the seed maturity also varies. Generally, in the years with high temperature and little precipitation, the seeds mature earlier, and in the years with rainy, wet, and cold, they mature later. Soil conditions also affect the maturity period of trees; for example, seeds of trees growing on sandy soil and sandy loam mature earlier than those growing on clayey and moist soil. For marginal trees of the same tree species, seeds of isolated trees mature earlier than those in the dense forest. On the same tree species, the seeds in the upper part of the crown and the sunny side mature earlier than those in the lower part and the shady side.

Human business activities will also advance or postpone the mature period; for example, reasonable fertilization and water supply, and improving the light conditions, can advance the mature period.

(3) Characteristics of Mature Seeds

When the seeds of each tree species reach true maturity, they show different characteristics. Differences are reflected in changes of color, smell, and peel surface, which can be used to determine the maturity period of seed (Figure 4-3).

Cones. Cone scales are dry, sclerotic, micro-cracked, and discolored. For example, seeds of *Cunninghamia lanceolata*, larch, and *Pinus massoniana* change from green to yellow green or yellow brown, and the cone scales are slightly cracked; seeds of *Pinus tabuliformis* and *Picea asperata* change to brown, and the apex of the cone scales is contrary-flexure.

Dry Fruits. The pericarp changes from green to yellow, brown to black. The pericarp is dry,

Figure 4-3 Mature fruits of golden rain tree (*Koelreuteria paniculata*) (a) and Oriental arborvitae (*Platycladus orientalis*) (b) (Photo by Liu Yong).

compact, and hard. Among them, the pericarp of capsular fruit and pods is cracked along the suture due to dryness, as seen in *Robinia pseudoacacia*, Persian silk tree (*Albizia julibrissin*), and Chinese mahogany (*Toona sinensis*). *Paulownia* pericarps change from cyan to russet, brown, reddish brown, and the pericarp is tight and hardened; white frost appears on the pericarp of Chinese honey locust (*Gleditsia sinensis*) and other tree species. Among nut fruits, the cupule of *Quercus* trees is grayish brown, and the peel is light brown to sepia, with glossiness; the key fruits of *Fraxinus mandshurica* and other trees are yellow brown; the pericarp of Chinese tallow (*Sapium sebiferum*), *Toona sinensis*, and dragontree (*Paulownia fortunei*) turns from green into black-brown, cracks, dries, and hardens.

Fleshy Fruits. The pericarp softens and the color changes with the species; for example, camphor tree (*Cinnamomum camphora*), Zhennan (*Phoebe zhennan*), glossy privet (*Ligustrum lucidum*), and *Phellodendron amurense* change from green to purple black; *Sabina chinensis* is purple; *Ginkgo biloba* and Hong Kong kumquat (*Fortunella hindsii*) are yellow; some berries have white frost on the pericarp. The fleshy fruit is mostly green, and after maturing, the fruit becomes soft, fragrant, sweet, brightly colored, without sour and astringent taste.

(4) The Judgment Method of Seed Maturity

Many methods can determine the mature seed period.

i. The most commonly used method to judge the maturity of seed is according to the color change of cones or fruits (Figure 4-3).

ii. Incision Method. According to the development of embryo and endosperm, it can be observed by the naked eye or by X-ray without incision.

iii. Specific Gravity Method. It is suitable for cones and can be operated in the field, which is simple and easy to conduct. The mixture of water (specific volume 1.0), linseed oil (specific volume 0.93), and kerosene (specific volume 0.8) is made into a mixed liquor according to a certain proportion, and then the cone is put in. The mature cones float; otherwise they sink.

iv. Biochemical Method. This method refers to the determination and analysis of a decrease in sugar content and crude fat content to determine the maturity of seed.

Judgment methods i, ii, and iii are quickly made. The biochemical method is slower, but it is the scientific and accurate method to ensure the seed yield and quality.

4.3.2.4 Seed Scattering

(1) Seed Scattering Period

Seeds of most tree species are scattered after maturing because of the abscission layer produced by carpopodium. The scattering period of seed varies with species. For some tree species, such as poplar, willow, birch, and elm, seeds scatter immediately after maturing; for *Pinus tabuliformis*, *Platycladus orientalis*, *Quercus*, silkworm mulberry (*Morus alba*), and smoketree (*Cotinus coggygria*), seeds scatter after maturing for a short time; for *Pinus sylvestris*, *Pinus massoniana*, London planetree (*Platanus acerifolia*), *Ailanthus altissima*, chinaberry (*Melia azedarach*), *Robinia pseudoacacia*, false indigo bush (*Amorpha fruticosa*), Chinese ash (*Fraxinus chinensis*), and box elder (*Acer negundo*), seeds will scatter after seed ripening for a long time.

The scattering period of seeds is related to the genetic characteristics of the tree species, and it is also affected by external environmental factors, such as air temperature, light, rainfall, relative humidity of air, wind, and soil moisture. With high temperature, dry air, and high wind speed, the seeds lose water quickly and fall off early; otherwise, they fall off later in the season.

The time of seed abscission is closely related to seed quality. In general, the quality of seeds dropped in the early and middle stage is good, and the quantity of seeds is greater during the early and middle stage, but the quality of seeds dropped in the later stage is poor. However, the seeds of *Quercus variabilis*, which were the first to fall off, were not well developed and of poor quality.

(2) Seed Scattering Characteristics

The characteristics of seed scattering are various; for example, the whole cones of *Pinus koraiensis* fall off after ripening; when *Cunninghamia lanceolata*, *Larix gmelinii*, *Pinus massoniana*, and *Platycladus orientalis* are mature, fruit scales open and seeds scatter; the cone scales and seeds of the tree species such as *Pseudolarix amabilis*, Deodar cedar (*Cedrus deodara*), and Faber's fir (*Abies fabri*) fall at the same time. The tree species with capsule and pods usually have dehiscence of fruit and abscission of seeds; while poplar and willow seed fly apart with flocculent materials, and the split fruit cluster gradually falls off; for oak, sweet oak, beech (*Castanopsis*), fleshy fruit, and key fruit, the whole fruits are often shed.

4.3.2.5 Principle of Determining Seed Collection Period

The suitable seed collecting period is an important guarantee for obtaining seed yield and quality. The collection of seeds must be carried out after the seeds are mature. If the collection time is too early, the quality of seeds will be affected; if too late, the small seeds cannot be collected after falling off and flying away. Therefore, it is necessary to establish the seed collecting period correctly.

The seed collection period shall be determined according to the time, characteristics, and fruit size of seed ripening and abscission. The following principles shall be followed during seed collection.

Small seeds falling off or flying with the wind immediately after ripening, such as poplar, willow, elm, birch, *Paulownia fortunei*, *Cunninghamia lanceolata*, *Abies fabri*, *Pinus tabuliformis*, *Larix gmelinii*, schima (*Schima superba*), and casuarina (*Casuarina equisetifolia*), are not easy to collect after falling off and should be collected immediately before falling off after ripening. Large seeds, such as *Quercus*, Chinese chestnut (*Castanea mollissima*), *Juglans regia*, tung tree (*Vernicia fordii*), and Chinese tanbark-oak (*Castanopis sclerophylla*), which fall off immediately after ripening, should be collected from the ground or on the standing tree soon after the fruit falls off. If they are not collected promptly after falling to the ground, they will suffer from insect and animal hazards and the impact of soil temperature and humidity, resulting in low quality.

For some tree species, such as *Cinnamomum camphora*, *Phoebe zhennan*, and *Ligustrum lucidum*, the seeds have a long abscission period, but given the bright color of the mature fruit, birds will be attracted and eat the fruits that stay on the tree for a long time. Fruits should be collected from the tree without delay after the seeds reach morphological maturity.

If the seeds do not fall off for a long time after maturity, such as *Pinus sylvestris*, *Pinus massoniana*, Chinese lime (*Tilia tuan*), *Fraxinus mandshurica*, *Acer*, *Melia azedarach*, *Robinia pseudoacacia*, and *Amorpha fruticosa*, requirements for the seed collection period are not strict, and they can be collected in a less busy season. However, we still need to collect seeds as soon as possible after they reach morphological maturity, so as to avoid the damage caused by insects and birds when seeds hang too long on the trees and result in the decrease of seed quality and yield.

Some long-term dormant seeds, such as Chinese hawthorn (*Crataegus pinnatifida*) and *Tilia tuan*, can be collected before morphological maturity but after physiological maturity and be immediately seeded or laminated after collection to shorten dormancy period and improve seed germination rate.

Seed maturity is often subject to weather conditions. In the years with unusual weather conditions, there will be great changes in the maturity period. The phenological process of that year must be carefully observed to determine the seed collection period scientifically and reasonably.

4.3.2.6 Organization and Implementation of Seed Harvest

(1) Preparation before Seed Collection

Organizational Readiness. The organization and preparation work should be done well before seed collection. First, check the seed collecting forest; determine the location, area, and date of seed collection; and estimate the actual possible harvest. Then formulate the seed collection plan, organize the seed collection personnel to learn the seed collection technology, and carry out the educational training of safe production and protection of seed trees. Seed trees must be strictly protected (monitored by government organizations, such as the Seed and Seedling Station at national and province level); it is not allowed to cut trees or cut large branches for seed collection, so as to prevent looting.

Material Preparation. Before seed collection, material preparation shall be made according to the size, distance, terrain, dispersion, seed collection method, traffic conditions, possible harvest quantity, and so forth. Before seed collection, the following should be prepared well: Seed collection, climbing the tree, metering, transportation and other machines and tools, packaging supplies, labor protection supplies, temporary storage site, sunning ground, and warehouse.

"Seed Management Regulations" stipulate that when seeds are collected in a forest seed production base, where the operation manager shall organize to collect and plunder the green trees, damage the mother trees, and collect the seeds in a poor forest, the competent forestry department shall order to stop the collection, compensate for the losses, confiscate the seeds, and may impose a fine.

(2) Seed Collecting Method

The method of seed collection shall be determined according to the scattered way, fruit size, and tree height after seed ripening, as follows.

Seed Collection on Trees. This method is suitable for small seeds that are easily scattered by wind, such as poplar, willow, birch (*Betula*), swamp mahogany (*Eucalyptus robusta*), *Pinus massoniana*, *Larix gmelinii*, *Pinus sylvestris*, *Cunninghamia lanceolata*, *Platycladus orientalis*, and *Schima superba*. Generally, it is used for trees with a low trunk or those which can be climbed for collecting with tools. Commonly used seed picking tools include seed collecting fork, picking knife, seed collecting hook, tree trimmer, and seed collecting comb. The tools used for tree climbing include rope slings, foot pedal, tree climbing ring, folding ladder, and the elevator can also be used for tree climbing. In recent years, the Hunan Academy of Forestry Science has developed the equipment for climbing trees and collecting seeds, which has solved the technical problem of difficult-to-pick fine seeds from tall trees.

Switzerland, the United States, Mexico, and other countries widely use the expansion platform with lifting equipment for seed collecting. Germany, Switzerland, and other countries use helicopters and hydrogen balloons to collect seeds.

Seed Collection on the Ground. When the seeds are mature, the large seeds, such as *Juglans regia*, *Castanea mollissima*, *Vernicia fordii*, and *Camellia oleifera*, which fall off directly or need to be dropped, can be collected from the ground. Before the seeds fall off, the small twigs and other plant debris on the ground should be removed. For small and medium seeds, which are difficult to collect after they scatter, nylon net can be laid around the seed tree. Then shake the seed tree to make the seeds fall onto the net. In the United States, tree growers use a special net drawing-in machine, which can remove debris in the process of retracting the net, and purer seeds can be obtained.

In addition, it can also combine the cutting operation to pick the seed from the fallen wood. The seed tree seeds at the edge of ponds and reservoirs fall on the water surface, which can be collected on the water surface before they sink.

Mechanical Seed Collection. For the trees with tall trunks and a single fruit and other trees that have difficult-to-collect seeds, such as *Pinus koraiensis*, *Cunninghamia lanceolata*, *Pinus massoniana*, dawn redwood (*Metasequoia glyptostroboides*), *Platycladus orientalis*, *Cinnamomum camphora*, *Phellodendron amurense*, and *Juglans mandshurica*, the fruits are shaken down by mechanical power, and seeds are collected with seed collecting net or canvas. As early as in the

1980s, China successfully developed a vibrating seed collector machine for *Cunninghamia lanceolata*.

In recent years, vibrating seed collectors and vacuum sweepers have been adopted in many countries. After shaking the cones, a vacuum sweeper, similar to a road sweeper, is used to collect the cones, of which the efficiency is significantly higher than that of manual seed collection.

(3) Precautions during Seed Collection

Safety belt, safety rope, and safety helmet should be fastened well when picking seeds directly from trees. The best time for seed collection is a sunny day without wind, in which the seeds are easy to dry, easy to prepare, and the equipment is safe to operate. Seeds collected during rainy days are prone to mildew. The fruit of some tree species can crack when the air is too dry, in which case seeds can be collected in the morning when dew can help prevent seeds from scattering. Working on trees during weather with winds above level 4 is forbidden by national rules.

The seed collecting operator under the tree should pay attention to the falling of tools or broken branches and pay attention to the location of pedestrians. In addition, pay attention to the protection of stock plant resources, especially the scenic area of the park, to prevent damage to valuable trees and to avoid damage to the branches and scions. We must take care of the provenance.

(4) Seed Registration

In the process of seed collection, in order to distinguish provenance, prevent mixing, reasonably use seeds, and ensure seed quality, the seeds used or purchased on site must be registered. They should be registered in batches and packed separately. Number and label the seed packaging containers inside and outside.

4.3.3 Processing

Seed processing is a general term for the technical measures of fruit and seed threshing, seed cleaning, drying, and seed grading after seed collection. The purpose of processing is to obtain high-quality seeds that are pure and suitable for storage, transportation, and sowing.

Because of the many kinds of tree species, the method of seed processing must be determined according to the structure and characteristics of fruit and seed. To facilitate the production and processing, the seeds of different tree species are usually divided into three categories: Cones, dried fruits, and flesh fruits. Similar processing methods can be used for similar seeds.

4.3.3.1 Threshing

Threshing is one of the most important steps in the process of seed processing. Threshing is the separation of seeds from fruits. Therefore, the first step of threshing is to dry and crack the pericarp. In the process of drying and threshing, the following principles should be followed: Seeds with high water content are dried in the shade, whereas seeds with low water content are dried in the sun. Water content of seeds refers to the safe water content, that is, the minimum water content necessary for a seed to maintain its life activities. The safe water content of seeds varies with species (Table 4-1), with most having between 5% and 12%. More than 20% is high water content.

Table 4-1 Safe water content of seeds of main tree species in North China

Tree species	Safe water content/%	Tree species	Safe water content/%
Populus, Salix	5-6	*Gleditsia sinensis*	5-6
Robinia pseudoacacia	7-8	*Betula*	7-8
Ulmus pumila	7-8	*Acer truncatum*	9-10
Larix principis-rupprechtii	8	*Acer negundo*	10
Pinus tabuliformis	7-9	*Eucommia ulmoides*	10
Ailanthus altissima	9	*Fraxinus chinensis*	9-11
Sabina chinensis	8-10	*Tilia tuan*	10-12
Platycladus orientalis	8-10	*Quercus acutissima*	25-30
Ginkgo biloba	20-25	*Quercus variabilis*	25-30
Crataegus pinnatifida	20-25	*Castanea mollissima*	40-45

(1) Threshing of Cones

Under natural conditions, the mature cones gradually lose water, and then the cone scales craze and the seeds come out. Therefore, the key to seed selection from cones is to dry and crack the cone scales. Two threshing methods of cones are possible: Natural drying and artificial heating drying.

Natural Drying Method. When no artificial drying room is available and the climate is warmer, the cones can be exposed to the sunlight or air-dried in a dry and ventilated place in the shade until the seeds come out. For example, the cones of *Pinus tabuliformis*, *Platycladus orientalis*, *Cunninghamia lanceolata*, *Larix gmelinii*, and *Picea asperata* can be exposed to the sun for 3 to 10 days. After the cone scale cracks, the seeds can be detached. Cones of *Pinus massoniana* and *Pinus sylvestris* are cracked slowly and the threshing is not thorough. The stack retting (i.e., soaking) method can be used for *Pinus massoniana*, and the stack retting time can be shortened by 7 to 10 days by using 2% to 3% limewater or plant ash water to drench the cones. Cones of *Pinus sylvestris* var. *mongolica* can open early by sun exposure after being immersed in water for 1 to 2 days. Cones of *Pinus koraiensis* and *Pinus armandii* were dried in the sun or in the shade for a few days, and when the cone scales were dehydrated, they were threshed by using a wooden stick.

Artificial Heat Drying. This method dries the cones by putting them in a drying room or other warm room. Generally, the drying room is equipped with a heating element, and the temperature and humidity can be controlled. The IHT hot air cone dryer developed by Changtu Forestry Machinery Factory of Liaoning Province in the 1980s is effective; in 2004, a new type of cone drying equipment developed by Harbin Forestry Machinery Research Institute of State Forestry Administration passed the installation and acceptance phases. Using this equipment, which is automatically controlled by an industrial computer, can open 99% of the cones, with a 98% seed

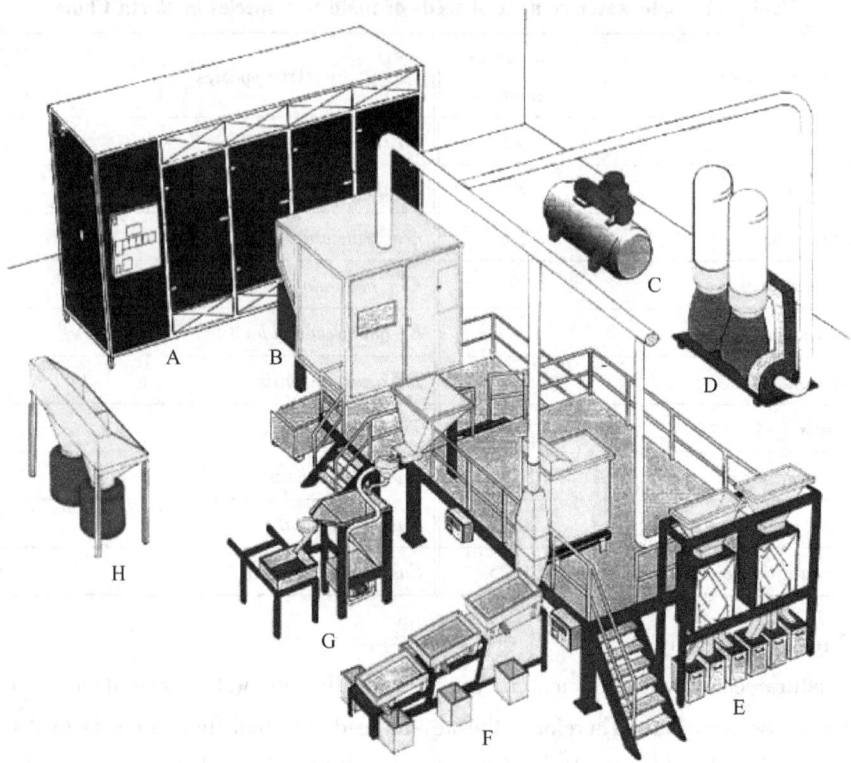

Figure 4-4 Cone/seed processing device of BCC Company in Sweden.
A. Cone/seed drying oven; B. Removal/threshing equipment; C. Air compressor; D. Filter unit; E. Gravity separator; F. Seed cleaning and seed classifier; G. Water separator; and H. Cone filling station.

survival rate. Foreign countries, such as the United States, Japan, the former Soviet Union, Sweden, and other countries have designed artificial drying rooms with high production efficiency (Figure 4-4). The cone dryer can thresh, clean seed, dry, and grade at one time.

The temperature for the artificial heat drying of cones should not be too high, because excessive heat will reduce the germination rate of the seeds. The suitable temperature for larch and spruce is 40 to 45 ℃; for *Cunninghamia lanceolata*, Japanese cedar (*Cryptomeria fortunei*), *Pinus sylvestris*, and *Pinus elliottii*, it is generally not more than 50 ℃; for European pine, it is 54 ℃. Cones with high water content should be pre-dried at 20 to 25 ℃, and then gradually heated in the drying room to avoid sudden high temperature, which will cause the loss of seed viability.

(2) Threshing of Dried Fruits

Dried fruits that crack after ripening are called dehiscent fruit, such as capsules and pods; those that do not crack after ripening are called indehiscent fruits, such as samaras and nuts. Based on fruit structure and water content, the methods of threshing differ.

Capsule. Poplar, willow, and other similar trees that have high water content and small seeds, and large capsules with high water content, such as *Camellia oleifera* and *Vernicia fordii*, are threshed in the shade. The capsules of *Toona sinensis*, *Schima superba*, and *Sapium sebiferum* can be peeled out after drying. The small capsules of some tree species, such as *Eucalyptus robusta* and *Paulownia fortunei*, should be taken back to the nursery and dried indoors for threshing.

Pods. Seeds of *Robinia pseudoacacia*, *Gleditsia sinensis*, *Albizia julibrissin*, and *Acacia* usually have low moisture content and a hard, dense seed coat. They are dried and threshed by a sunning dry method.

Samaras. Seeds from Chinese wingnut (*Pterocarya stenoptera*), maple, *Ailanthus altissima*, *Fraxinus chinensis*, and others, generally do not need to remove the alas (wings) when processing; they only need to properly be dried and then the debris is removed. Seeds of *Ulmus pumila* and hardy rubber tree (*Eucommia ulmoides*) have low water content, but excessive water loss can affect the germination rate. They should be processed by drying in the shade.

Nuts. Large nuts, such as *Quercus*, *Castanopsis sclerophylla*, and *Castanea mollissima*, cannot be exposed to the sun because of their high-water content, so they are processed by drying in the shade. Small nuts, such as *Betula* spp. and Japanese alder (*Alnus japonica*), can be spread thinly (3 to 4 cm thick) to dry and then lightly beaten or wrapped in a linen bag for kneading the seeds. The small nuts of *Platanus* are dried in the sun after harvest, smashing the fruit ball (a formation of many small nuts), depilated, and threshed.

(3) Threshing of Fleshy Fruits

Fleshy fruits include berries, drupes, collective fruit, and seeds with an aril (which is an extra seed covering, typically colored and hairy or fleshy). Its pericarp is mostly fleshy and easy to ferment and rot, and it should be processed after collection. Processing includes softening pulp, breaking pulp, extracting seeds with water, drying, and cleaning seeds. For example, for *Ginkgo biloba*, *Morus*, sea-buckthorn (*Hippophae rhamnoides*), Siberian apricot (*Armeniaca sibirica*), *Melia azedarach*, and *Sabina chinensis*, the pericarp can be softened by stack retting, crushed with a stick, or kneaded on a sieve, then washed and rinsed with running water to remove the pulp and separate out the wet seeds that were dried in the shade.

4.3.3.2 Seed Cleaning

Seed cleaning removes the scale, peel, seed coat, fruit stalk, branch and leaf fragments, empty seeds, waste seeds, and soil that are mingled with the seeds. Its purpose is to improve seed purity. According to the size and specific gravity of seeds and inclusions, air separation, water separation, screening, or grain separation are generally used.

(1) Air Separation

Air separation is suitable for medium and small seeds. Because of the different weight of full seeds and inclusions, they are separated by wind with tools such as winnowers, dustpans, and so forth.

(2) Screening

Given the differing sizes of seeds and inclusions, sieves with different bore diameter were used to remove inclusions. During screening, the physical function of sifter rotation can also be used to separate the seeds of empty and half-empty grains.

(3) Water Separation

The seed cleaning method with different specific gravity of seed and inclusion was used. For example, for *Ginkgo biloba*, *Platycladus orientalis*, *Quercus*, prickly ash (*Zanthoxylum bungeanum*), and other tree species, seeds can be immersed in the water during water separation. After a little stirring, the seeds will be full and will sink, and the empty and mothed seeds will float upward, which are then easy to separate.

(4) Grain Separation

Grain separation selects seeds with large size, plump, normal color, and no diseases and insect pests. This method is suitable for cleaning large seeds, such as *Juglans regia*, *Castanea mollissima*, *Vernicia fordii*, and *Camellia oleifera*.

4.3.3.3 Seed Drying

After cleaning, seeds should be promptly dried. The degree of drying is generally based on the safe moisture content. Only the seeds meeting the standard can be stored or transported safely. If the seeds are sown immediately after harvest, drying is not necessary.

Natural drying method is mainly used for seed drying. According to the water content of seeds, the method of drying in the sun or drying in the shade can be used. No matter which method is adopted, the seeds should be thinned and turned frequently, so that the moisture of seeds will diminish as soon as possible and the quality of seeds can be guaranteed.

In forestry-developed countries, the whole process of seed preparation has been mechanized, and the commonly used machines include a seed and cone drying box, seed threshing production line, seed cleaning and grading machine, seed wing removal machine, water separator, and gravity separator (Figure 4-4).

4.3.3.4 Seed Grading

To achieve seed separation, the use of sieves with different apertures to separate large and small seeds; the adoption of wind separation to classify the seeds with different weights, or the use of seed dielectric separation technology are conducive to improving seed quality. If the seeds after classification are used for sowing, the emergence is neat, with even growth, then management will be more convenient.

The grading standard must refer to the *National Forest Seed Quality Standard*. According to the quality indexes of seed purity, germination rate (or viability and excellent degree), and water content, the seed quality of 115 main afforestation tree species in China is divided into three grades.

After years of research, the forestry workers in China have basically found out the fruiting rules of the main afforestation species in China, determined the corresponding seed maturity and collection period, and put forward scientific methods for seed collection and processing. The national standard of the People's Republic of China, *Classification of forest tree seed quality* (GB 7908—1999), which was jointly completed by Nanjing Forestry University, Beijing Forestry University, and other teaching, scientific research, and production institutions, has listed important characteristics, methods, and data (such as seed maturation characteristics, mature period, collecting period, collecting methods, processing methods, seed percentage, and thousand seed weight) of the main tree species in China. The publication is a scientific summary of the achievements in this field for many years and provides a scientific basis and practical method for guiding the collection and processing of forest seeds in China. This resource is of great significance to the production of forest seeds in China.

4.4 Collection and Processing of Asexual Propagation Materials

Asexual propagation materials include cuttings for cutting and scions for grafting propagation.

4.4.1 Cuttings Collection and Processing

4.4.1.1 Cuttings Collection

When collecting cuttings of fruit trees, select from the trees with pure varieties, good quality, strong growth, and no diseases and pests. Cuttings should not be collected on young trees with poor performance, serious diseases and pests, and no fruit, so as to prevent the deterioration of the quality of seedlings and lead to degradation. Plants with large flowers, rich colors, and a long viewing period should be selected as the seed trees for collecting the cuttings of ornamental flowers and shrubs. The first place for collecting cuttings of timber forest and green trees should be the cutting orchard. If it is necessary to collect from the big trees, collect from the sprouts at the base of the trunk or the root sprouts near the base of the trunk. The collection time of cuttings differs according to species and propagation methods.

In general, for softwood cuttings, the young and semi-lignified branches are cut during the most vigorous growth period of trees. Most of the trees were collected from May to July, such as sweet osmanthus (*Osmamthus fragrans*) in the middle of May, and *Ginkgo biloba*, *Vitis vinifera*, and common camellia (*Camellia japonica*) in the middle of June. If the shoots are collected too early, they may lose water and wilt; if the shoots are collected too late, they will be lignified, which means the auxin content will be reduced and the inhibitory substances will be increased. These collections will not be conducive to rooting. For flowering plants, such as roses, cuttings should be carried out after flowering. The appropriate collection time of day is in the morning and in the evening. At this time, the water content of the cuttings is high, the air humidity is high, the temperature is low, and the cuttings are easy to preserve. It is forbidden to collect the cuttings at noon.

In hardwood cutting, the collection from deciduous trees is generally carried out after the trees grow slowly or stop growing at the end of autumn, and before the sprouting of the next spring. Cuttings should be taken from fully lignified branches. Evergreen trees are collected before sprouting in spring; if they are collected too early, the accumulated nutrients in the trees are less, which leads to a decrease of the quality of cuttings; or, because the increment of branches is less, the available cuttings are less, which reduces the propagation coefficient. If they are collected too late, the buds on the branches will expand, which will consume nutrients and is not conducive to rooting; or, the degree of lignification of branches is too high, which is not conducive to survival of cuttings.

Cuttings can be collected in combination with summer and winter pruning of the seed tree, and usually the upper and middle branches of the seed tree should be collected. In summer, the shoots are vigorous in growth, high in photosynthesis efficiency, strong in nutrition and metabolism, all of which are conducive to rooting. The dormant branches in winter have been fully lignified, with full buds and rich nutrition storage, which is also conducive to rooting.

4.4.1.2 Cuttings Processing

In the hardwood cutting, the 1-year-old mean (middle size) branch, which is fully mature in the sun, moderate in internode, normal in color, full in bud eye, and free of pests and diseases in the upper and middle of the crown of the seed tree, is taken as the cutting. The too-thick branches with excessive growth and the thin and weak branches are not suitable for the cutting. After cutting off the branches that are too thin at the tip and have no obvious buds at the base, cut the selected branch into 3 to 4 buds (approximately 15 cm) and bundle them into groups of 50 to 100. To avoid later confusion, keep the polarity direction consistent, align the lower cut, and mark the variety name and collection location.

In softwood cutting, it is better to select the branches with moderate maturity, full axillary buds, normal leaf development, and no pests on the seed tree. If the branches are too tender, they rot easily, and if they are too old, they will take root slowly. Cuttings are generally picked and cut at the same time without storage. After collection, immediately put cuttings into a container with a small amount of water, so that the base of the cuttings soaks in water. Cuttings can absorb water to supplement the water lost by transpiration, which will prevent them from wilting. If the shoot cuttings are collected from other places, half of each leaf can be cut to reduce the water transpiration loss. Wet a towel in some ice to cool it down and then use it to wrap layer by layer. The base of the branch is wrapped with moss. Immediately open the package after it has been transported to the destination and soak the base of the cuttings in water.

Bundle the collected cuttings by tree species and varieties, and label them with varieties, collection location. and collection time.

4.4.2 Scion Collection and Processing

4.4.2.1 Scion Collection

To collect scions, select excellent and pure plants with good quality, high ornamental value or economic value, strong growth, and no diseases and insect pests as the female parent. From the middle and upper part of the periphery of the female parent, it is best to select the 1- to 2-year-old branches that have had sufficient sunlight and full development as the scion. Generally, 1-year-old branches with a short internode, robust growth, full development, full bud, no pests and diseases, and uniform thickness are better; however, for some trees, 2-year-old or older branches can also achieve higher grafting survival rate, even better than 1-year-old branches, such as the common fig (*Ficus carica*) and European olive (*Olea europaea*), as long as the branch tissues are perfect and strong. The scion of coniferous evergreen trees should have a 2-year-old branch, which has high grafting survival rate and fast growth.

Scions should be collected and stored properly during the dormancy period (after defoliation and before sprouting in the next spring) for spring grafting. If the reproduction amount is small, scions can be picked and grafted at the same time. When evergreen trees, herbs, and pulpy plants are grafted during the growing season, the scions should be picked and grafted at the same time.

The best scion for budding is to pick and graft at the same time. Leaves of the scion collected should be cut off immediately and a section of petiole should be reserved. Scions can be collected after defoliation for scion grafting, no later than 2 to 3 weeks before germination.

4.4.2.2 Scion Processing

The scions for grafting in spring are usually collected back in the dormancy period in combination with winter pruning. Every 100 scions are bundled together, with labels indicating the species or varieties, scion collecting date, and quantity, and they are then stored at a suitable low temperature. For the scions with bleeding phenomenon, meaning high content of gum and tannin, a wax sealing method is used for storage. For example, for *Juglans regia*, *Castanea mollissima*, Japanese persimmon (*Diospyros kaki*), and other plant scions, this method has good storage effect.

Wax sealing method: After the branches are collected, cut them into 8 to 13 cm long cuttings (at least 3 complete and full buds on one scion). Dissolve paraffin by a water bath method; that is, put the paraffin in a container, and then heat the container in a water bath or a water pot. When the wax liquid reaches 85 ℃ to 90 ℃, the two ends of the scion are quickly dipped in the wax liquid, leaving the surface of the scion with a thin wax film but with no bubble in the middle. Then pack a certain number of scions in a plastic bag, seal well, and store at a low temperature of 0 ℃ to 5 ℃ for later use.

The scions of succulent plants, herbaceous plants, and some plants grafted in the growing season should be picked and grafted at the same time. The leaves and the top of the branches with insufficient growth should be removed and then scions should be promptly wrapped with wet cloth. if the scion cannot be grafted right away, the lower part of the scion can be immersed in water and placed in a cool place. Change the water once or twice a day for a short-period storage of 4 to 5 days.

Questions for Review

1. What is the significance of provenance selection in forest construction? What are the ways to choose the suitable provenance?

2. What is the significance of seed regionalization? How do you allocate seeds according to seed regionalization?

3. What are the general processes of seed ripening? What are the principles to be followed in determining the seed collection period?

4. What are the procedures of seed processing? Which types of seed are suitable for drying in the shade and which for drying in the sun?

5. What are the methods of seed cleaning? What are the methods of drying forest seeds? Why do we conduct seed grading?

6. What is the basic principle of seed allocation?

7. How do we establish a seed production stand?

8. How would you briefly describe the process and characteristics of seed ripening?

9. What are the principles of determining a seed collecting period?

10. What are the principles of seed drying and threshing?

References and Additional Readings

CHEN H, ZHANG M, LU K, et al., 2017. Effect of pruning intensity on the mother trees of adult seed orchard of *Pinus massoniana*[J]. Journal of Fujian Forestry Science and Technology, 44(1): 38-42.

CHEN S, ZHAO Z, WANG H, et al., 2017. Study on fruit maturity and seed modulation of *Polygonatum multiflorum* in different fruit picking periods[J]. Seed, 36(9): 30-34.

CHENG F Y, 2012. Nursery of landscape plants[M]. Beijing: China Forestry Publishing House.

GUO H, LIU Y, WU C, et al., 2017. Research progress of seed dormancy and storage in foreign countries[J]. Journal of Northwest Forestry University, 32(4): 133-138.

LE D, LI Y, PENG F, et al., 2018. Construction and quality evaluation of thin shell pecan ear picking nursery[J]. Journal of Nanjing Forestry University (Natural Sciences), 42(2): 134-140.

LI S, ZHOU J, WANG F, et al., 2000. Study on provenance selection of *Pinus tabuliformis* in Gansu Province[J]. Scientia Silvae Sinicae (5): 40-46.

SHEN H, 2009. Seedling cultivation[M]. Beijing: China Forestry Publishing House.

SUN S, 2002. Forest seedling technology[M]. Beijing: Jindun Press.

WANG W, GUO S, 2011. Effects of ultra-dry storage and rewetting on the germination and physiological characteristics of *Pinus bungeana* seeds[J]. Seed, 30(10): 19-24.

ZHAI M P, JIA L, GUO S, 2001. Silviculture[M]. Beijing: China Central Radio and Television University Press.

ZHAI M P, SHEN G F, et al., 2016. Silviculture[M]. 3rd edition. Beijing: China Forestry Publishing House.

ZHAO Q, WANG A, WANG Y, et al., 2016. Comparison of seedling growth and photosynthetic characteristics of 26 provenances of *Pinus tabuliformis*[J]. Journal of Northeast Forestry University, 44(11): 19-23.

ZHU Y, 2018. Collection and treatment of larch seeds in North China[J]. Modern Horticulture(15): 99-100.

CHEN X Y, SHEN X H, 2021. Tree breeding science[M]. Beijing: Higher Education Press.

Chapter 5　Propagative Materials Storage and Seed Quality Evaluation for Tree Species

Fu Xiangxiang

Chapter Summary: This chapter introduces the physiological changes of tree seeds during storage, the factors affecting seed storage tolerance, and expounds the storage methods applicable to different types of seeds. The types and characteristics of propagation materials used for vegetative propagation, physiological changes during storage, and storage methods are also introduced. The function and content of tree seed quality assessment, the basis and principle of inspection, the procedure of inspection, and the methods of relevant indexes are systematically discussed. Through the study of this chapter, readers can have a systematic understanding of the storage characteristics, storage methods, and quality assessment of tree non-sexual propagation materials (seeds) and vegetative propagation materials (vegetative organs) and have significant guidance for production practice.

5.1　Seed Storage Physiology and Storage Method

Seeds need to be stored for a long or short period until climatic or other factors permit sowing or planting. The objectives of seed storage are to reduce deterioration of the seed, maintain germination capacity and vigor to the greatest extent, and ensure seed sowing value under suitable storage conditions, which is controlled by facilities and advanced storage technology.

Storage duration of seed depends on tree species and storage purposes. The storage conditions and seed longevity of various types of seeds (normal type, intermediate type, and recalcitrant type) are completely different. Seeds for germplasm conservation need to be stored for a long time, whereas the seeds that will be sown very soon after collection in practice only need to be stored temporarily. For coated seeds and artificial seeds, conditions for storage are much stricter than that for normal seeds. Therefore, to improve seed longevity after storage, understanding the characteristics of seeds during storage could provide the basis for determining suitable storage conditions and scientific management strategy.

5.1.1　Life Activities and Metabolism Changes during Seed Storage

5.1.1.1　Effect of Respiration on Seed Storage

Seeds are living organisms, but respiration is basically in a static state within dry or dormant seeds. The respiration of seeds during storage affects their safety and longevity.

(1) Respiratory Characteristics of Seeds during Storage

Seed respiration is a process, in which a series of redox reactions happens on reserved substances in living tissues with the participation of enzymes and oxygen (O_2), resulting in the release of carbon dioxide (CO_2) and water (H_2O), while producing energy at the same time. Respiration in seeds is the main manifestation of life activities during the storage process.

Two types of respiration exist in a seed during storage: Aerobic respiration and anaerobic respiration. The definition of aerobic respiration is that storage substances are decomposed completely into CO_2 and H_2O along with abundant energy, with the existence of adequate oxygen. Chemically, the process looks like this:

$$C_6H_{12}O_6 + 6O_2 \longrightarrow 6CO_2 \uparrow + 6H_2O + \text{Energy} \ (2870.22 \text{ kJ})$$

With anaerobic respiration, the decomposition of storage matter occurs insufficiently, producing incomplete oxidation products (e. g., alcohols, aldehydes, acids, and others) and releasing less energy as well (see chemical formula below) under anoxia. Alcohols, aldehydes, and acids produced by anaerobic respiration are toxic to embryo cells.

$$C_6H_{12}O_6 \longrightarrow 2C_3H_4O_3 + 4H \longrightarrow 2C_3H_6O_3 + 75.31 \text{ kJ}$$

The character of seed respiration varies with environmental conditions, seed types, and seed quality. Since seeds of normal and intermediate type are usually stored under environmental conditions that are dry, low temperature, and airtight, hypoxia respiration with low respiratory intensity is dominant. And yet, aerobic respiration with high intensity is a priority. Commonly, two types of respiration in seeds co-exist during storage. For a seed pile stored under the condition of ventilation, aerobic respiration is the main mode, although anoxic respiration may occur at the bottom of the seed pile. In cases of poor ventilation, anoxic respiration becomes dominant. For a seed pile with high moisture, temperature within the pile will rise gradually because of strong respiration. With poor ventilation, a long duration of anaerobic respiration will accumulate excessive toxic respiration products, such as ethanol, which could inhibit normal respiratory metabolism and/or poison the seed embryo, resulting in death.

(2) Effect of Respiration on Seed Storage

Respiration has two-sided effects on seeds. On the one hand, it can promote after-ripening of seeds; on the other hand, excessive respiration can bring about negative impacts such as the following.

Too much respiration consumes a mass of storage nutrients, which shortens seed longevity. The water vapor produced by respiration gathers in the seed pile, resulting in the "sweating" phenomenon, which in turn increases the seed moisture. As O_2 is essential for seed respiration, anaerobic respiration tends to occur in the inner portion of a seed pile when insufficient O_2 is available for respiration, which could be a result from over-consumed oxygen during high intensity respiration and poor ventilation. As described above, anaerobic respiration produces toxic substances, which can not be absorbed by seed itself. Integrated, these adverse factors speed up seed deterioration. Heat generated by respiration raises the temperature by scattering through the seed pile, which further strengthens seed respiration. In short, excessive respiratory activity in the seed pile not only increases seed moisture and temperature but also enhances the respiratory intensity. As a result, it also promotes the activities of pests and microorganisms in the seed pile, aggravating the feeding consumption and harm to seeds. In turn, the mass of heat and water vapor released by strong activities of pests and microorganisms reinforces the respiratory intensity of seeds.

5.1.1.2 Metabolism during Seed Storage

In the process of storage, the phenomenon of aging occurs irreversibly within seeds. Generally, seed aging refers to a natural senescence process. During aging, a series of changes presents first at the biochemical level, then at the physiological level, and finally affects the morphology and structure of the cells.

(1) Changes in Reserve Substances

The main components stored in seed include starch, protein, and fat. During storage, starch, protein, and fat could be the substrates for respiratory metabolism. Among them, fat more readily hydrolyzes and oxidizes in comparison to sugar and protein. In addition, a seed containing a high content of oil is prone to deterioration. Seeds containing more protein and starch, which is more stable and durable, will have greater longevity than the fatty seeds.

(2) Changes at Physiological and Biochemical Levels

When seed deteriorates, the internal membrane system is damaged due to peroxidation, inducing the increment of permeability, as well as a massive accumulation of various toxic substances, such as lipid hydroperoxides (ROOH) and malondialdehyde (MDA). Also, a synthesized system of phytohormones, such as gibberellin, cytokinin, and ethylene, contribute to gradual seed decline. Finally, as the metabolism of storage substances slows down, the synthesis ability of RNA and protein decreases as well.

5.1.2 Factors Affecting Seed Storage

Although seed deterioration is inevitable, reasonable storage can prolong seed longevity efficiently. Normal-type seeds can be stored for a long time in an environment with low temperature and humidity. Being sensitive to dehydration and low temperature, recalcitrant seeds with suitable moisture content stored at constant temperature could maintain vigor for a short duration. Obviously, storage longevity of seeds mainly depends on their own characteristics and storage conditions.

5.1.2.1 Seed Characteristics

The intrinsic characteristics of seed include genetic character and seed status in itself. Controlled by genetic factors, the life span of seeds varies greatly, some of which can last for thousands of years, such as the Oriental plane tree (*Nelumbo nucifera*) and the bitter blue lupin (*Lupinus micranthus*), while some hold for only a few days, such as poplars (*Populus* spp.) and willows (*Salix* spp.). Heredity governs seed longevity mainly by controlling seed type, coat structure, and type of storage substance. Seeds can be divided into three types according to their longevity.

(1) Short-life Seed

Generally, the life span of short-lived seeds is less than 3 years, most of which are recalcitrant seeds and intermediate seeds, such as poplar (*Populus*), willow (*Salix*), hopea (*Hopea hainanensis*), Chinese chestnut (*Castanea mollissima*), and camellia (*Camellia sinensis*). They manifest low resistance to low temperature and moisture content, or are rich in fat substance, or possess the thin and brittle testa.

(2) Middle-life Seed

These seeds are also known as regular-life seed, with a life span from 3 to 15 years. The majority of normal-type seed falls into middle-life seed, such as cypress and tulip trees (*Liriodendron* spp.).

(3) Long-life Seed

The longevity of long-lived seeds is usually more than 15 years. All hard seeds are assigned to this type. Generally, seeds in this type possess attributes such as lower oil content, tight and tough testa, or impermeable covering; for instance, locust trees *Robinia pseudoacacia* and *Gleditsia sinensis*.

The state of the seed itself refers to the maturity, integrity, and planting quality. Owing to higher moisture content, more water-soluble substances, stronger respiration, as well as more susceptibility to diseases, immature seeds have lower viability and shorter longevity. Mature seeds contain more stable storage substances, which could support respiratory consumption for a longer period so as to maintain higher vitality and longevity. Moreover, intact and full seeds show higher viability and longer life, with more storage materials accumulated inside to help resist deterioration; conversely, those with mechanical damage, pest and microorganism erosion, enhanced respiration, and storage materials susceptible to consumption, lead to lower vitality and shorter life.

The sowing quality of seeds includes seed purity, moisture, germination ability (vitality, plumpness), health status, which are affected by seed maturity, integrity, and method of processing and treatment. Among sowing quality, seed moisture plays the dominant role in how well seeds survive storage.

During the storage of seeds, excessive moisture easily triggers excessive respiration, which is followed by production of increased heat and vapor, and consequently heats the seed pile. Furthermore, strong respiration overconsumes the oxygen, resulting in hypoxia respiration to produce a large amount of ethanol, which poisons the embryo cells and lowers seed vitality. In addition, seeds with high moisture are readily infected and harmed by bacteria, molds, and pests, thus reducing the seed germinability. Overall, the water content of seeds exerts significant effects on storage life, which is supported by numerous studies in the literature. However, reasonable moisture for storage depends on the type of seeds: Lower moisture provides suitable storage for normal-type seeds; maintaining higher moisture is conducive to storage for recalcitrant seeds.

On the whole, the higher the sowing quality of seed, the longer the seed life; on the contrary, lower sowing quality reduces the storability of seed.

5.1.2.2 Storage Conditions

The environmental factors affecting seed storage life include air relative humidity (RH), temperature of storage, gas, light, microorganisms, and pests.

(1) Air Relative Humidity (RH)

Fluctuation of seed moisture content during storage depends on the relative humidity (RH) of the air. When RH is higher than the equilibrium moisture content (MC) of seed, seed MC will increase by absorbing water from the air around the seed, accelerating the metabolism inside the seed and accelerating the downward trend of seed vigor and shortening the life of seeds. Alternatively, when RH is lower than equilibrium MC of seed, the seed tends to be dried gradually as water escapes from seeds into the air. Metabolism and deterioration inside the seed slows down, which benefits seed storage. Therefore, for high quality seed storage, keep the air dry in the seed storage environment.

Maintaining air RH for storage is usually determined based on each individual situation. For normal-type seeds with lower MC, it is suitable to store them at an RH of approximately 30% for medium- and long-term preservation. For short-term storage, namely temporary storage such as overwintering, it is better to store seeds at air RH balanced with the safe MC of seeds, which is about 60% to 70%. Considering the safe MC of seeds and the actual situation, air RH for storage should generally be lower than 65%.

(2) Storage Temperature

The temperature of bulk seed fluctuates with surrounding air temperature. The rising temperature of bulk seed could accelerate its respiration, along with promotion of microorganism activity and decreasing vigor of the seeds, which can lead to reduction of seed longevity. By comparison, low temperature for seed storage favors retarding metabolism and inhibiting microorganism activity, which improves seed storability.

In general, seed stored at low temperature, such as 0 ℃, -10 ℃, -18 ℃, or lower, is suggested for medium- or long-term preservation. For temporary storage, seeds can be stored at 15 ℃ or room temperature.

(3) Ventilation

In addition to O_2 and CO_2, water vapor and heat are also in the air. When seeds are stored in ventilation conditions, the oxygen for respiration is sufficient. Vapor and heat produced by seed respiration increase MC and temperature of seeds, intensify seed metabolism, all of which could shorten seed longevity. Therefore, the normal- and intermediate-type seeds stored in airtight conditions is beneficial to maintaining seed vigor. Airtight storage will block the spread of oxygen for seed respiration, and block the exchange of water vapor and heat, thus inhibiting seed activity and reducing material consumption, which maintains seed vigor for the long-term. Given that the seeds with high moisture respirate vigorously and produce large quantities of CO_2, toxic substances, water vapor, and heat, proper ventilation ensures seeds can carry out aerobic respiration, reduces the temperature and humidity in the warehouse, and decelerates the deterioration process of seeds.

(4) Other Factors

Besides the above factors, light-proof storage should be considered. In light conditions, seed deterioration will accelerate, which is harmful to seed storage. In addition, keeping the warehouse clean can reduce infection from pests and microorganisms. As mentioned earlier, the amount of vapor and heat released by the propagation and activity of pests and microorganisms worsens storage conditions, directly or indirectly harming the seeds. However, maintaining dry, low temperature and airtight conditions in the warehouse can effectively inhibit the activities of pest and microorganisms.

To sum up, the factors that affect seed storability interact with one another. Among them, seeds' MC is the dominant factor affecting storability. Respiration intensity of seeds with low MC is usually much lower than that with high MC, even at a high temperature. Similarly, the respiration intensity of seed with high MC at low temperature is much lower than at a high temperature. Therefore, a combination of a dry environment with a low temperature is usually more effective for seed storage. The air RH and temperature for storage are not only affected by the respiratory intensity of seeds but also by the ventilation conditions. More important, these factors also influence the activities of pests and microorganisms, and vice versa. Therefore, the optimum regime could be drawn up to improve long-term seed storability by integrated analysis of seed characteristics and storage conditions.

5.1.3 Methods for Seed Storage

Seed characteristics and storage purposes determine storage method. For seeds for production, making a storage strategy first depends on economic benefit, with subsequent consideration for climate conditions in the seed storage environment, storage duration, and seed value. For seeds intended for germplasm preservation, the performance of storage facilities, storage duration, and storage effect are the priorities for consideration. Various regimes for seed storage are described as follows.

5.1.3.1 Dry Storage

Dry storage is defined as seeds that are stored in dry surroundings in which a certain low temperature and RH should be ensured. Normal- and intermediate-type seeds are suitable for dry storage. According to storage span and specific measure, dry storage can be divided into more detailed methods as described below.

(1) Ordinary Dry Storage

Seeds with an acceptable MC are deposited into containers or packages, which are placed in a sterilized, dry, and ventilated room. It is suitable to store the majority of plant seeds for a short term. Generally, this type of dry storage lacks equipment to control environmental factors, so the temperature and air RH in the open room cannot be maintained in a suitable state for the long term. Further, the temperature and MC of bulk seed exposed to open circumstances will fluctuate with that of storage surroundings.

Because it is simple and economical, ordinary dry storage is a preferred method to store large-scale quantities of seeds intended for production. In general, the vigor of seed stored for 1 to 2 years can be maintained. When the storage period exceeds 3 years, however, seed vigor may obviously decline.

To maintain seed vigor, seeds should be cleaned, graded, and dried prior to storage. In addition, the storage warehouse should be cleaned and disinfected. When entering the warehouse seeds should be registered and filed, and then inspected and tested regularly. Also, routine management for the storage warehouse, such as ventilation, heat dissipation, and so forth should be done well.

(2) Low-temperature Dry Storage

Seeds suitable for dry storage can be stored at a low temperature (below 0℃) for medium and long terms. This method is applicable to all seeds suitable for dry storage. Low-temperature dry storage method requires equipment to control the temperature and air RH of the warehouse. Generally, the range of temperature for low-temperature dry storage is 0 to 5 ℃ or −20 to −10 ℃. Considering the higher investment and operation costs, low-temperature storage is adopted only for small-scale seed storage or germplasm resources preservation, especially in tropical areas.

(3) Dry Storage in a Sealed Package

Prior to storage, seeds' MC should be lowered to a standard, recommended amount for storage; after packaged with an airtight container or packing material, seeds could be stored under regular warehouse conditions. At certain temperatures, seeds stored in a sealed package maintain their viability for a long time, extend their life span, and also facilitate exchange and transport.

Beneficial effects result from sealed storage: It prevents oxygen supply from external air humidity to impact seed MC, leading to lower respiration in seed. Moreover, the sealed conditions depress the growth and reproduction of aerobic microorganisms, thus prolonging the longevity of seeds. Owing to good effect and low cost, sealed storage is the preferred recommendation in regions where rainfall and air RH are relative higher than elsewhere.

The sealed seeds cannot be stored at high temperature conditions; if they are, seeds will die with accelerated speed. At high temperature, the respiration of seeds will speed up, resulting in serious hypoxia in the sealed container, followed by anaerobic respiration, which produces the environment for toxic substances to accelerate the deterioration of seeds. In addition, the occurrence of anaerobic diseases, such as those from fungi, will be promoted at high temperatures; it becomes more serious when the MC of seeds is high. Therefore, the good effects of sealed storage can be guaranteed only at lower temperatures.

(4) Ultra-dry Storage

Ultra-dry storage refers to ultra-dried seeds with an MC less than 5% that are sealed and stored at or below room temperature. Usually, ultra-dry storage is used for conserving germplasm resources and breeding materials. And, it is suitable to store the majority of normal-type seed. Tolerance of desiccation depends on seed type: Fatty-seed endures greater desiccation, which is fit for ultra-dry storage, while the desiccation tolerance of starch and protein seeds varies greatly. Tolerance to desiccation for these types of seed should be further verified.

The key technologies for ultra-dry storage include the following aspects.

Achievement of Ultra-low MC. It is difficult to lower seed MC to less than 5% by regular drying technology; the vigor of seeds will be reduced or even lost when dried at a high temperature. At present, the common drying techniques include freeze vacuum drying, silica gel drying, and desiccant drying at room temperature. These drying methods not only effectively reduce seed MC to less than 5% but also maintain seed viability.

Pretreatment for Germination of Ultra-dry Seeds. Ultra-dry seeds are easy to damage during imbibition. Their germinability will be lower if pretreated by soaking directly in water. Based on the principle of seed "infiltration" and "repair," ultra-dry seeds primed by polyethylene glycol (PEG) or treated with imbibition-drying will balance seed MC and could effectively prevent imbibition damage, so that high-vigor seedlings can be achieved.

(5) Ultra-low Temperature Storage

Storing dry seeds at ultra-low temperature ($-196℃$) makes their metabolism inactive so as to achieve long-term storage.

Liquid nitrogen is usually used as the refrigerant for ultra-low temperature storage. Technically, it provides the possibility of "indefinite" storage of seeds. But problems might occur during the process of cooling and reheating, so experimentation in those stages of storage is necessary. No mechanical equipment, such as an air conditioner, is needed, but liquid nitrogen and tanks for containing the liquid nitrogen are needed for ultra-low temperature storage. In spite of simple equipment, the cost of liquid nitrogen is expensive. Therefore, ultra-dry storage is mainly limited to preservation of precious and rare germplasm that needs long-term storage.

Techniques for ultra-low temperature storage are listed as follows.

Determination of Seed MC. Only seeds with suitable MC can survive in liquid nitrogen.

Freezing and Thawing Techniques. Proper speed of freezing and heating is crucial for seed survival.

Selection of Packing Material. Packing material isolates the seeds from the liquid nitrogen, which prevents seeds from freeze injury.

Addition of Cryoprotectant. Common cryoprotectants include DMSO, glycerin, and PEG.

Germinative Method after Thawing. Germination conditions for thawed seeds should be optimized.

Ultra-low temperature storage provides the possibility for long-term conservation of germplasm resources, but the related technology needs further improvement.

5.1.3.2 Wet Storage

Seeds stored at humid, low temperature, with adequate ventilation is called wet storage. Seeds suitable for wet storage include dormant seeds and recalcitrant seeds. Dormancy of seeds can be gradually released after wet storage, and germination will then be uniform and quick. Thus, seeds with deep dormancy characteristics, such as most of the seeds of the maple (*Acer* spp.), should be stored in moist surroundings before sowing. Being sensitive to dehydration and low temperature, recalcitrant seeds such as ginkgo (*Ginkgo biloba*), oaks (*Quercus* spp.), chestnuts (*Castanea* spp.), Xinjiang walnut (*Juglans regia*), and camellias (*Camellia oleifera*) should be stored in humid conditions, which could provide the proper water content for maintaining seed viability.

The basic requirements for wet storage are that seeds should be embedded in moist and loose media; ventilation is necessary to guarantee adequate oxygen for active respiration for seeds with high MC and to decrease heat produced by respiration from the seed bank; and seeds should be disinfected before storage to prevent infection from microorganisms and fungi. Also, storage at a suitable low temperature can inhibit the activity of mold and maintain seed dormancy. Overall, since high respiratory intensity during storage results in short storage duration, wet storage is usually adopted for temporary (overwintering) storage.

Methods for wet storage are outdoor burial and indoor stacking.

(1) Outdoor Burial

A pit for seed storage can be prepared on a shady slope with good drainage and loose soil. In principle, store the seeds in the layers below the frozen soil and above the water level. The mixture of seeds and wet sand with water content of 60% (1:3, v/v) is set into the pit, which is then covered with sand and straw to maintain ventilation. Note that the thin layer of seed can disperse the heat produced by respiration and guarantee adequate ventilation.

Massive amounts of seed stored by outdoor burial is too much to monitor thoroughly. This storage method is preferred for winter in the northern areas that have low temperature and dry atmosphere. Seeds stored in this manner in southern areas with rain, high temperatures, and heavy and sticky soil with poor drainage tend to germinate too soon or rot. Thus, it is important to monitor seeds during storage.

(2) Indoor Storage

Layered seeds and wet sands, or a mixture of seeds with sand (snow) (1:3, v/v) can be stored indoors, in a basement, or in a shady shelter that is dry and has ventilation. To facilitate inspection and ventilation, the mixture can be stacked into ridges with a footpath. When seed quantity is small, the mixture of seeds and sand can also be placed in an aeriferous container located in a ventilated basement.

5.2 Storage Physiology and Storage Method of Vegetative Propagules

Forest is composed of various tree species. Among widely planted species, the majority of seedlings cultivated for afforestation can be obtained through sexual reproduction, but for some ornamental cultivars, or those with a special shape, cultivated material can be nurtured only through vegetative propagation.

Asexual propagation, also known as vegetative propagation, cultivates saplings using vegetative propagules (such as stem, branch, leaf, bud, and root). The characteristics of vegetative propagation are able to deliver fine traits from female parent to offspring and allow the plants to blossom and bear fruit earlier than when working with seeds. This propagation method is especially suitable for cultivars with performance goals of high ornamental value, rapid growth, and good bearing (economic forest), but with a long cycle before bearing fruit. The vegetative propagation seedlings are extensively applied in practice for their rapid growth and uniformity, as well as for needing only simple management. For example, the process works well with clonal plantlets of *Eucalyptus* spp., hybrid offspring of Chinese tulip trees (*Liriodendron chinense*) propagated by tissue culture, cutting seedlings of red tip photinia (*Photinia* × *fraseri*), grafting seedlings of ornamental Chinese flowering crabapple, and cultivars of dogwoods (*Cornus* spp.).

Large-scale asexual propagation includes cutting, grafting, and tissue culture. Usually, the propagules for vegetative propagation are excised organs. In practice, since vegetative propagules are collected from cutting orchards or growing fields of elite adult individuals, which are far from the nursery, or the collected propagules cannot be immediately used for propagation because of unfavorable factors, such as climate and other conditions, it may be necessary to store the material provisionally to maintain vigor in the propagules for use at a later time.

In nature, seeds evolved protective features to adapt to the changes of seasons and climate in the process of evolution, so seed vitality could last for a long time under favorable conditions. Unlike seeds, vegetative organs that have been separated from maternal plants lack a corresponding protective structure; their vigor decreases rapidly under regular conditions, leading to the failure of vegetative reproduction. According to the characteristics of asexual propagation, vegetative propagules generally refer to the organs in a dormant state (hardwood for cutting or grafting in spring) before sap fluxion, or at vigorous growth stages (greenwood for cutting or for grafting in autumn, as well as leaves, juvenile stems, or buds for tissue culture). Because dormant propagules contain abundant macromolecular nutrients, low growth hormone, and low enzyme activity, they are easy to store and storage duration could extend for a long period. However, when reserve nutrients consist largely of soluble micromolecule substances, high content of growth hormones is present, and high enzyme activity occurs in thriving vegetative organs, vegetative material can be stored for only a short time.

5.2.1 Types of Asexual Propagules

With the development of asexual propagation technology, the processes have developed from conventional vegetative propagation technology, such as cutting and grafting, to tissue culture for factory production of plants and artificial seeds. All these are widely used as asexual propagation materials for plant production.

(1) Common Propagules

Common propagules refer to cuttings for cutting propagation and scions for grafting propagation, including vegetative organs at the state of dormancy and growth, that is, stems (shoots), buds, roots, and leaves.

(2) Micro-vegetative Propagules

A micro-vegetative propagule is defined as the periodic biological tissues (meristem) produced during tissue culture, including protoplast, cell, callus, organ, embryo, and embryoid. Embryoids are a mass of tissue produced during tissue culture processes that resemble embryos. Within the embryoid is the meristem, a type of tissue that can develop into different plant parts, such as a leaf, root, and stem. Above mentioned propagules could survive when stored only at ultra-low temperature instead of normal conditions.

Artificial seed is produced by embedding meristem in the capsule, which is also a micro-propagule. The structure of artificial seed consists of three parts, an artificial embryo, artificial endosperm, and artificial seed coat, and they are analogous to that of a natural seed. An artificial seed embryo contains two types of meristem: Somatic embryo and non-somatic embryo. A somatic embryo is produced by tissue culture, and the function is the same as a zygote embryo in natural seeds; non-somatic embryos refer to vegetative propagules (differentiated tissue) without growth polarity, such as shoot tips, nodal segments, hairy roots, and calli, as well as protocorm and protocorm-like bodies (PLB) with growth polarity. Artificial endosperm could provide nutrients and auxin for embryoid metabolism, growth, and development. Since the artificial coat acts as the layer for protecting the inner embryo, it should possess the characteristics of air and water permeability, fixed shape, and resistance to mechanical shock. To date, artificial seeds have been widely used in medicinal plants, ornamental plants, fruit trees, and tree species with economic benefits.

5.2.2 Storage Physiology of Vegetative Propagules

During the storage process, the physiological indices of vegetative propagules, including water content, storage nutrients, and inhibitory substances, change significantly, which could affect their survival. Generally, the size of propagules has a great influence on its physiological changes: The larger the size is, the more nutrients and water it contains, the better the storability; in comparison, the smaller the size is, the less nutrients and water it contains, the weaker the storability.

(1) Moisture Content of Propagules

Research has proved that water content in propagation tissues positively affects survival. For instance, the rooting rate of cuttings from mulberry (*Morus alba*) and poplar (*Populus russkii*) lowered sharply when the water content of cuttings was less than 50%. To slow down the loss of water, propagules are usually sealed using plastic or paraffin before storage.

(2) Relative Conductivity

Relative conductivity is also an important physiological index to measure the activity of reproductive material. With the extension of storage duration and the decrease of water content, it will increase obviously. Among tested factors affecting rooting rate, the relative electrical conductivity of stored cuttings from *Populus russkii* is the most pivotal physiological index, followed in importance by proline and moisture content of tissue.

(3) Proline

Proline is one type of amino acid. Normally proline content rises sharply in the early stage of water loss, then tends to increase slowly with the continued loss of water. However, its content will decline when water loss is serious. Massive quantities of proline will accumulate in vegetative propagules at the beginning of storage. Subsequently, it will decrease with declining activity of propagules.

(4) Reserve Substances

In the process of storage, in the main reserve of nutrients stored in propagules, macromolecular substances are degraded into micromolecular soluble nutrients to provide the energy for metabolism. During storage, the macromolecular substances show a downward trend, whereas the soluble micromolecular nutrients present a unimodal fluctuant pattern, showing an upward prior to a downward trend along with the extension of the storage period. The degradation of macromolecular substances induces the rise of soluble nutrients. Among micromolecular substances, increasing soluble sugar acts as the role of osmotic adjustment when water loss happens, which is beneficial for storage of propagules.

(5) Growth Regulators

Growth regulators primarily refer to five categories of phytohormones. Among them, some secondary metabolites, such as jasmonic acid, salicylic acid, brassinolide, and other steroids, as well as tannins and phenolic acids that accumulate in mature tissues could play the roles of regulation. Among regulators, abscisic acid, tannin, and phenolic acid often have negative effects on vegetative reproduction and the survival of reproduction.

Documented in scientific literature is that the content of inhibitory substances in propagules stored at low temperature could be lowered, and the concentration of growth-promoting substances could be increased so as to improve the survival of vegetative propagation. The storage duration at low temperature depends on tree species or variety.

5.2.3 Preservation Methods of Asexual Propagules

Preservation of asexual propagules involves storing the propagules under controlled conditions, which can ensure that propagules maintain the minimum consumption and maximum vitality.

5.2.3.1 Preservation Methods Suitable to Regular Asexual Propagules

(1) Dormant Asexual Propagules

Dormant propagation materials could be preserved at a low temperature for hibernation until being cut prior to sap flowing or being grafted after the sap is flowing. As dormant scions (buds) are in the state of dormancy and consume less nutrients, the storage at low temperature could last a longer period. In production, these propagules are usually stored for hibernation. In detail, after being wrapped with plastic film or sealed with wax, dormant material could be stored in an environment kept at 0 ℃ to 5 ℃, or in a cellar with a low temperature, or buried in wet sand to overwinter. Regular checking (once every 1 to 2 weeks) should be executed during storage. In general, such storage could prolong dormancy for 1 to 2 months.

(2) Growing Asexual Propagules

The asexual propagation materials in a growing period, such as buds and leaves, should be used for production as soon as they are harvested. Otherwise, it should be stored appropriately. One to two leaves could be kept in propagules for temporary storage, to avoid water loss and maintain their viability. The following methods can be selected for divergent demands.

Stored in Wet Sand. Put the cuttings into wet sand, and the suitable burial depth is for the first bud at the base to be parallel to the sand surface; keep the sand damp.

Stored in Water. Put the cuttings into the water, and the suitable submersed depth is for the first bud at the base to be parallel to the water surface. Water should be refreshed once a day.

Store at Low Temperature. Store the wrapped cuttings with moisturizing material at 0℃ to

4℃.

Store at Constant Temperature. Store the wrapped cuttings with moisturizing material at room temperature, in which the relative humidity of air is above 80%.

Publications have recorded that the duration of above storage methods generally last 5 to 7 days. However, the detailed storage duration relies on tree species (variety), types, and maturity of propagules.

5.2.3.2 Preservation of Micropropagules

As micropropagules, such as protoplasts, cells, calluses, organs, and embryos (embryoids), are of small size and contain few nutrients, the effective storage method is to store at an ultra-low temperature. At ultra-low temperature, the meristem is basically in the state of "life suspension", which greatly reduces or stops the deterioration related to metabolism, so as to extend their longevity.

Similarly, artificial seeds readily germinate and lose water under constant temperature, so they need to be preserved under special conditions. Documented storage methods involve low temperature storage, drying storage, inhibition storage, liquid paraffin storage, and integrated multimethod storage. Among them, drying storage and low temperature storage are accepted widely, and are the best choices of storage research at present.

(1) Dry Storage

After maturity, the embryo of normal-type seed enters a static stage, which improves its storability. Desiccation of the embryo is the only way for seed development and maturity to occur. Drying treatment can activate germinative genes and store redundant mRNA for further germination. Therefore, desiccation is the most effective way to prolong the storage life of a somatic embryo. During the drying process, we should pay more attention to improve the desiccation tolerance of the embryo, and pretreatment with abscisic acid (ABA) can improve the desiccation tolerance of artificial seeds.

(2) Low Temperature Storage

Low temperatures can effectively inhibit the respiration of an artificial embryo and make it enter a dormant state. Artificial seeds can be stored for 1 to 2 months at 4 ℃, and for longer terms at ultra-low temperatures (lower than −80 ℃). At ultra-low temperatures, the metabolism and life activities of living cells in artificial seeds almost stop completely, so that mutation in genetic characteristics and the potential of morphogenesis rarely happen. The main method of ultra-low temperature storage for artificial seeds is pre-culture drying, which is to say, the artificial seeds will be dried as pretreatment prior to storage at an ultra-low temperature.

As an economical, non-toxic, and stable liquid substance, liquid paraffin is often used as the agent to store plant calluses.

5.3 Quality Evaluation of Tree Seeds

Seed is the most important material of production for urban forests. The seedlings applied to construction of an urban forest are mainly raised from seeds. The quality of seeds is affected not only by seed sources (such as the stand or maternal tree for seed collection) but also by the process of seed production, including seed-collecting season and method as well as techniques involving seed processing, storage, and transportation. To satisfy the needs for seed quality management, the *Seed Law of the People's Republic of China* (2021, revised edition) was issued and implemented in 2022. It emphasizes the status and role of forest seeds, and more important, on the regulations for supervision and management of forest seed quality. On this basis, the state and local administrative measures for management of seed quality and local standards are being developed. The quality of tree seeds not only will determine the success or failure of raising seedlings and the quality of seedlings, which could affect the landscape of forest, but also will affect the adaptability of seedlings to the environment of urban forests. Therefore, evaluation for tree seed quality will ensure high-quality seeds to meet the needs of multi-tree species, multi-objectives, and multi-specifications for construction of the urban forest.

5.3.1 Overview for Evaluation of Tree Seed Quality

5.3.1.1 Effect of Seed Quality Evaluation

The purpose of seed quality evaluation is to ensure that qualified seeds are used for construction of urban forests. The role of seed quality testing occurs throughout the process of seed production, processing, storage, sales, and use (Figure 5-1).

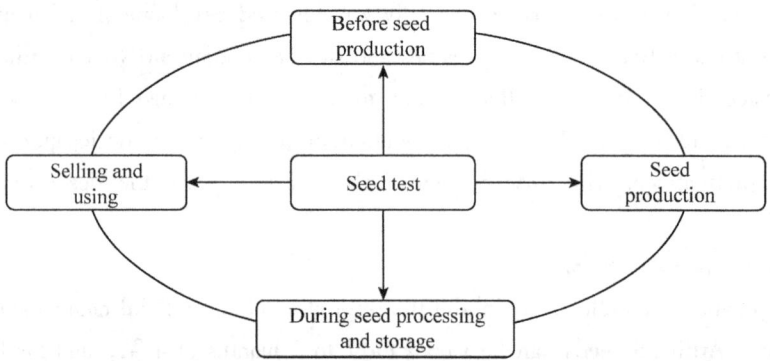

Figure 5-1 System of evaluations for tree seed quality.

First, seed quality assessment runs throughout the process of seed production. Collecting season is determined by monitoring seed quality during seed maturity, for differences in ripening time depends on type, provenance, site condition of the stand for seed collection, as well as annual climate.

Second, seed quality indicators, such as seed germination, purity, moisture content, and purity could be affected by seed processing. Appropriate procedures and machinery parameters for seed processing could be determined based on the evaluation of seed quality.

Third, storage condition is specific for seed type and affects seed longevity. Seed quality testing during storage under various storage conditions could optimize storage conditions for specific seed types.

Finally, seed quality evaluation is not only the basis of grading seed quality but also the basis of pricing when the seeds are available for sale and use. Moreover, evaluation can prevent the fake and inferior seeds from being traded, which can ensure sound development of seed and seedling markets. In practice, the sowing quantity could be decided according to seed quality as well, which provides the basis for making the production plan and management.

5.3.1.2 Contents of Seed Quality Evaluation

(1) Definition of Seed Quality

According to the latest revised edition of the *Seed Law*, "seed" is defined as the materials of crops or trees for planting or multiplying, including seed and fruit, as well as root, stem, seedling, bud, leaf, and so forth. In production, seed and fruit are the most common types of "seed", which are the main objects for seed quality evaluation.

In general, seed quality of forest trees contains two aspects. One is the internal quality (variety quality or genetic quality), which refers to the seed quality traits determined by genetic factors, such as production performance, adaptability, resistance, nutrition, and processing quality. Variety quality consists of seed authenticity and uniformity. Authenticity is assigned as the true identity of cultivated varieties, while uniformity is designated as variety purity. Another aspect is sowing quality, which mainly refers to the quality of seeds, including seed purity, germination, vitality, vigor, moisture content, thousand seed weight, and health status.

(2) Content of Seed Quality Evaluation

Seed quality evaluation includes two aspects: Variety quality/genetic quality refers to the authenticity and uniformity of varieties; sowing quality refers to three basic indicators, seed purity, germination percentage, and water content, must meet the minimum requirements of effective regulations (i. e., seed quality standard, contract agreement, and seed label).

5.3.1.3 Basis and Principle for Seed Quality Evaluation

(1) Basis of Seed Quality Evaluation

Seed quality evaluation is determined primarily by seed quality testing. The judgement of seed quality is carried out by comparing the test results with the target requirements (or values), then drawing a conclusion. Evaluation of seed quality is based on the corresponding seed quality criteria. Generally, the system of seed quality standards is divided into four levels:

National Standards. This standard refers to the seed quality criteria and technical regulations related to seed production and management issued by the state, such as *Rules for Forest Tree Seed Testing* (GB/T 2772—1999), *Classification of forest tree seed quality* (GB 7908—1999), and *Tree Seed Storage* (GB/T 10016—1988).

Industry Standards. It refers to the seed quality evaluation regulations and relevant seed quality standards issued by the national administrative departments (forestry and construction) as needed, such as *Woody Plant Seed Pretreatment for Germinating* (LY/T 1880—2010), *Embryo excision test for woody plants* (LY/T 1881—2010), and *Woody seedling for green space and landscape* (CJ/T 24—2018).

Local Standards. Local governments at all levels issue the standards relevant to seed quality to strengthen the management of seed quality and promote the development of the local seed industry.

Enterprise Standards. Standards are set internally based on seed types in production and sales within the enterprise. Specification requirements from enterprise standards should be higher than that from national, industry, and local standards, when stated contents are the same among them.

When determining seed quality, national or industry standards are executed first, followed by local standards, and finally by enterprise standards.

(2) Principles of Seed Quality Evaluation

Variety quality evaluation (genetic quality evaluation) depends on its authenticity and uniformity. Evaluations for authenticity and purity of varieties are based on field and laboratory tests. For quality purity, the purity of a seed lot is decided by the lower result, if the result tested in the field differs from that in the laboratory. Further, if the purity tested in the field fails to meet the minimum requirements of the national standard, the qualified seed lot can be used as propagation material only after removing the impurities. Also, if the purity tested in the laboratory is lower than the minimum standard of the national classification standard, it cannot be further used for raising seedlings.

Sowing quality evaluation is based on indicators, such as seed purity, germination percentage (soundness, viability), 1000-seed weight, water content, and health. In detail, the sowing quality of a seed lot is classified according to the above tested indexes.

In brief, a seed lot with high purity, clarity, as well as high germination percentage (vitality, excellent degree), moderate water content, full seeds, healthy and free of disease and insect infection could be ranked as an elite seed lot. Therefore, it is necessary to strictly evaluate seed quality throughout seed production, purchase, processing, storage, and sales to ensure high-quality seeds for production.

5.3.1.4 Classification and Procedure of Seed Quality Evaluation

(1) Classification

Classification evaluations could be performed to meet a variety of needs.

First, according to the function, seed quality evaluation can be divided into:

Internal Inspection, Also Known as Self-inspection. Evaluation of seed quality is executed by the institutions of seed production, trading, or user. They would then designate the grade of the seeds to support further work, such as seed collection, processing, pricing, as well as making the production plan.

Supervision Inspection. Seed quality management department authorizes seed testing organizations to test seed quality within its jurisdiction, so as to supervise and manage the seed quality.

Arbitration Inspection. The arbitration institution, authoritative institution, or trade parties adopt arbitration procedures and methods to test the seed quality, then submit the arbitration results.

Although the evaluation aims differ in the above three categories, the common concerns focus on controlling and ensuring seed quality.

Second, according to testing place sites, seed quality evaluation can be divided into:

Field Test. In the process of seed production, according to the performances of plants growing in the field, the purity of the cultivar could be measured, and the heterologous individuals, weeds, diseases, and insect infections could also be investigated.

Laboratory Test. After seed collection, sampling and inspecting are carried out in the process of seed processing, storage, and trading. Laboratory tests may include seed authenticity, variety purity, clarity, germination percentage (viability), 1000-seed weight, moisture content, and seed health status.

Plot Planting Test. The authenticity and uniformity of seeds sowed in a field plot are identified by their characteristics performed during a growing season, using the standard variety as the control.

Field test and plot planting test aim at identifying the authenticity and purity of varieties, as well as evaluating the quality of varieties. Tests in the laboratory mainly involve evaluation of sowing quality.

(2) Testing Procedure

Seed quality evaluation must be carried out according to prescribed procedures, ensuring the scientificity, fairness, and reliability of evaluation (Figure 5-2). When evaluating bulk seed, it should be divided into several seed lots first, then the submitted sample is obtained by a sampling procedure, sequenced as: Primary sample, composite sample, and submitted sample. The submitted sample should be registered after the inspection agency receives it and prior to testing. The first step of the testing procedure is to test seed purity; then, germination percentage, viability, soundness, 1000-seed weight, and so forth can be tested for pure seeds obtained from purity analysis. Seed samples for testing moisture content and health state should be sampled separately. The samples should remain sealed and be tested as soon as possible.

5.3.2 Progress of Tree Seed Quality Testing

5.3.2.1 Overview of International Seed Quality Testing

(1) Origin and Development

Seed testing originated in Europe. From the middle of the 18th century to the middle of the 19th century, with the development of the seed trade, many cases involving the sale of fake and inferior seeds occurred throughout the European countries. Seed inspection was born at that moment to maintain the development of an honest seed trade. In 1869, a German scientist, Friedrich Nobbe, established the first laboratory in the world for seed testing in Salander. The laboratory carried out testing work such as seed authenticity, clarity, and germination percentage. In 1876, he compiled and published the *Handbook of Seed Science*. Therefore, Dr. Nobbe is recognized as the founder of seed science and seed testing science.

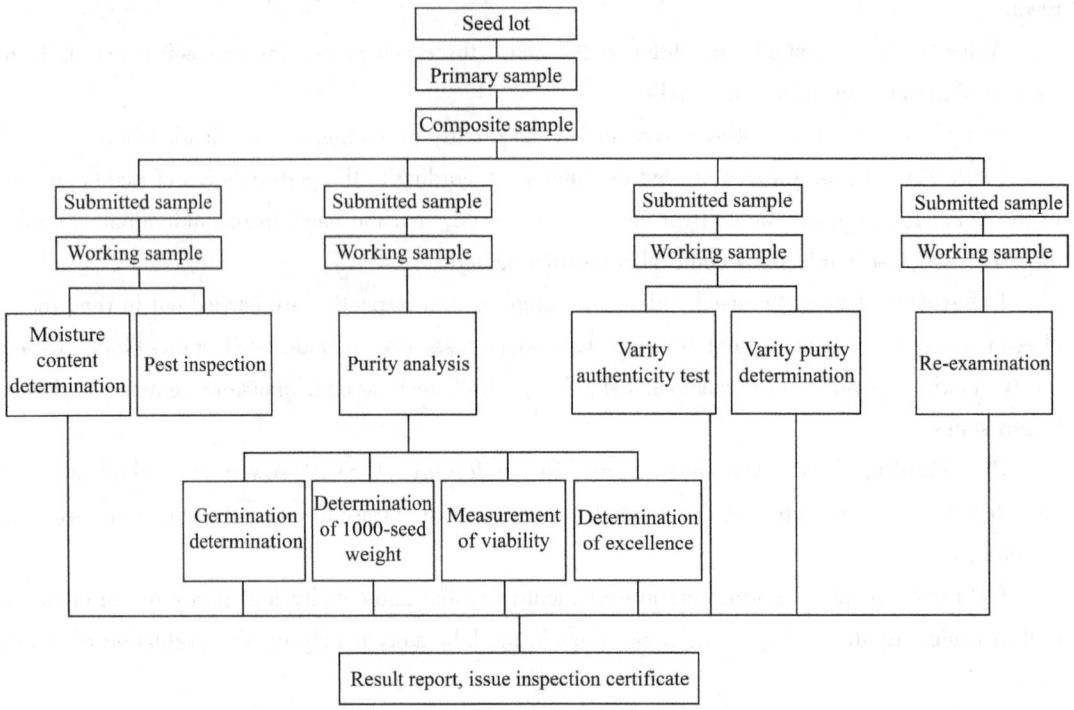

Figure 5-2　Evaluation procedure for forest seed quality.

After that first laboratory for seed testing, more laboratories were established in Denmark, Austria, Netherlands, Belgium, and Italy in succession. In 1875, the first European meeting of seed inspection was held in Austria. The main points of seed inspection and the basic principles of seed quality control were discussed in this meeting. In 1876, the United States established the first agricultural research station in North America responsible for seed inspection; then, the United States issued standard regulations for seed inspection in 1897. In 1890 and 1892, Nordic countries held meetings in Denmark and Sweden, respectively, to formulate and review regulations for seed inspection. At the beginning of 20th century, many countries in Asia and other continents also set up seed inspection stations to carry out seed inspection.

With the development of the international seed trade, seed testing technology needs to be standardized to strengthen international seed trade. Therefore, international joint testing of seeds has been put on the agenda. In 1906, the first international seed testing conference was held in Hamburg, Germany; in 1908, the United States and Canada established the Association of Official Seed Analysts (AOSA); in 1921, the first conference of the European Seed Treatment Assurance (ESTA) was established and held in France; in 1924, the fourth international seed testing meeting was held in Cambridge, United Kingdom, and the International Seed Testing Association (ISTA) was founded officially. Since its establishment, ISTA has successively held many world conferences around the world and has developed and revised the International Rules for Seed Testing many times.

Meanwhile, in 1885, a German named E. O. Harz edited *Agricultural Seed Science*; in 1922,

German Wittmach also compiled *Agricultural Seed Science*; in 1932, Wan taro Kondo from Japan published *Agroforestry Seed Science*; in 1944, R. H. Porter summarized the achievements of American seed test and compiled *Agricultural and Horticultural Seed Quality Test*; in 1958, Filsova from the Soviet Union summarized seed inspection technology in his country and compiled *Seed Testing and Research Methods* and *Methods of Seed Quality Determination*.

(2) International Seed Testing Association

International Seed Testing Association (ISTA) is a worldwide intergovernmental association, which consists of national official seed testing laboratories (stations) and technology experts. It is composed of seed scientists and members of seed testing stations from various countries in the world. ISTA is a nonprofit organization funded by Member States, which includes 83 member states/distinct economies and encompasses more than 130 member laboratories, 46 individual members, and 43 association members as of the end of 2023.

The primary goal of ISTA is to develop, adopt, and issue standard procedures for seed sampling and inspection and to promote a wide consensus on standard procedures in international trade. Second, ISTA plays important roles in actively pushing forward research work in the areas of seed science and technology, including sampling, testing, storage, processing, and popularization, as well as in encouraging seed certification of varieties (cultivars), holding worldwide seed conferences, and carrying out training.

ISTA has 20 technical committees, which are responsible for standardizing and legalizing seed sampling and quality testing methods with advanced scientific knowledge. Among them, the Forest Tree and Shrub Seed Committee (FTS) is the only technical committee focusing on forest seed detection, with members primarily from the United States, Canada, and Malaysia. FTS emphasizes the inspection of forest seed and related research work, such as seed storage, seed pretreatment (stratification), and seed diseases; in addition, it develops the methods suitable for forest seed detection and introduces them into the ISTA regulations, so as to apply to the international trade of forest seed.

Association of Official Seed Analysts (AOSA) was founded in 1908; it aims to coordinate the consistency of seed inspection methods, results, and reports among seed inspection stations (laboratories) from various countries. AOSA members include federal, state (province), and university seed laboratories in the United States and Canada, and membership is also extended to joint laboratories (government agencies and research institutes outside the Member States) and honorary members (people who have made outstanding contributions to AOSA or the industry).

The functions of AOSA include three aspects: Establishing AOSA regulations for seed testing; improving and revising seed testing regulations and procedures; and promoting and assisting state and federal governments to legislate on seed.

5.3.2.2 Progress of Tree Seed Quality Inspection in China

Before 1949, no specific testing institution for forest seeds was located in China, and the related work was accomplished by the food department and commodity inspection agency. In the early 1950s, some institutions of scientific research, teaching, and production launched seed testing research. In the middle of the 1950s, provincial institutions for forest seed testing were successively established. The Technical Regulations for Forest Seed Quality Testing was issued in 1956, then Trial Measures for Forest Seed Management was formulated in 1978 by the State Forestry Administration of the People's Republic of China. Since 1978, many provinces set up forest seed and seedling management stations and seed testing laboratories. In 1982, two national testing centers for forest seed quality were established; subsequently, another six national testing institutions, distributed in eight provinces (autonomous regions), were successively set up. The main tasks of the national testing centers are to undertake the national testing for the quality of forest seed and seedlings, to evaluate and supervise the quality of seeds and seedlings used in national major forestry projects, and to provide the basis for macro-control of national forestry projects. Furthermore, based on the quality evaluation, the market operation of forest seed and seedling is supervised and regulated.

5.3.2.3 Seed Quality Standards

(1) International Standard of Tree Seed Quality

Given the regional distribution of most tree species, few international seed quality standards have been established for forest seeds. Currently, the only existing standards include the *Arbor and Shrub Seed Manual* published by ISTA, *Standard Specification for Seed Starter Mix* (1994) published in the United States, *Nursery Stock* (1905) published in Britain, and *Nursery Stock* (1990) published in France.

(2) Quality Standard of Forest Seed in China

In China, *Forest Seed Testing Methods* (GB 2772—1981) and *Forest Seed* (GB 7908—1987) were formulated since 1978, on the basis of Rules for Forest Tree Seed Testing (ISTA, 1976), Seed Testing and Research Methods, and Seed Quality Determination Methods (Filsova, 1958), combined with current practices in forest seed production.

After the 1990s, with the development of a seed industry under the new emphasis on forestry, "Forest Seed Testing Methods" and "Forest Seed" were revised and renamed as *Rules for Forest Tree Seed Testing* (GB 2772—1999) and *Classification of Forest Tree Seed Quality* (GB 7908—1999) and were officially issued and implemented in 2000. In addition, the main national standards related to forest seeds include *Forest Tree Seed Storage* (GB 10016—1988), *Forest Tree Seed Collecting Technology* (GB/T 16619—1996), and industry standards such as the *Woody Plant Seed Pretreatment for Germinating* (LY/T 1880—2010). At the same time, each province (autonomous region, municipality directly under the central government) has formulated and promulgated a series of local standards of seed quality in combination with their own regional characteristics, in which tree species unlisted in the national standards were supplemented. The promulgation and implementation of these regulations and quality standards meet the development of Chinese forest seed industry and play positive roles in normalizing industry operations and market trade and in protecting the safety of forest production.

5.3.2.4 Certificate of Seed Quality Evaluation Issuance

(1) The International Certificate of Seed Inspection

The following certificates can be issued after the application procedure is completed and approved by the government.

Orange International Seed Lot Certificate. Both sampling from the lot and testing of the sample are carried out under the responsibility of an accredited laboratory.

Green International Seed Lot Certificate. Sampling from the lot and testing of the sample are carried out under the responsibility of two accredited laboratories from different countries.

Blue International Seed Sample Certificate. The accredited laboratory is responsible only for testing the sample as submitted. It is not responsible for the relationship between the sample and any seed lot from which it may have been derived.

The principles for filling in the three certificates are the same, but the filled contents in them differ from one another. The results reported on a Blue International Seed Sample Certificate refer strictly to the sample at the time of receipt; sampling from the lot is not under the responsibility of an accredited laboratory. In international seed trade, only an Orange or Green International Seed Lot Certificate is considered an effective quality certificate.

(2) China Seed Inspection Certificate

According to the actual situation of seed quality evaluation in China and the requirements of *Rules for Forest Tree Seed Testing* (GB 2772—1999), the certificate of forest seed quality evaluation must be issued by an authoritative inspection institution with inspection qualification. Among the various sampling units, two types of forest seed quality evaluation certificates are used: Seed lot quality inspection certificate and seed sample quality inspection certificate.

Seed Lot Quality Inspection Certificate. Samples submitted for inspection should be obtained by the authorized inspection agency or by sampling under its supervision, and the quality inspection certificate will be issued by the authorized inspection agency after an inspection has been completed.

Seed Sample Quality Inspection Certificate. Samples submitted for inspection could be taken by an unauthorized inspection agency, but the certificate would be issued by an authorized inspection agency after inspection. The inspection agency is responsible only for the samples submitted for inspection and is not responsible for the representativeness of the samples submitted for inspection.

5.3.3 Principle and Method of Inspection for Seed Sowing Quality

Inspection of seed sowing quality mainly focuses on the comprehensive evaluation on seed purity, germination ability, 1000-seed weight, and moisture content, so as to determine the seed quality grade. In addition, the evaluation method based on seed viability and vigor is being gradually extended.

5.3.3.1 Sampling

Sampling is the first step for seed quality testing. Scientific sampling is necessary to obtain a submitted sample and a working sample for further testing. Sampling needs to obtain a representative sample of a size suitable for tests, in which the probability of a constituent being present is determined only by its level of occurrence in the seed lot. Thus, the basic principle of sampling is that the collected sample should be representative; that is, the constituents in the submitted sample should be consistent with that in seed lot, so as to reflect the proportion of these components.

(1) Seed Physical Characteristics Affecting Sampling

The representativeness of a sample is affected by many factors. An experienced sampler is important, and the seed physical characteristics are important factors that will have an impact on the representativeness of the sample. For example:

Seed Scattering. A cone with a certain slope is often formed when granular objects fall from above onto the flat surface. This characteristic, which determines the angle of beveled inclination, is called scattering; while the angle of beveled inclination is called the natural beveled angle. Seed is also a kind of scattered object, so various seeds display specific scattering and natural beveled angle determined by scattering. The natural beveled angle of seed varies depending on seed shape, coat structure, and moisture content of seeds. The more spherical the seed shape, the smoother the seed coat, and the smaller the natural beveled angle of the seed. In the same seed lot, the seed coat becomes rough with rising moisture content, leading to an increase of friction between seeds, so that the natural beveled angle of seeds becomes bigger. In addition, the natural beveled angle often increases if the seed batch becomes hot and moldy.

Automatic Grading of Seeds. A lot contains various seeds, such as full, empty, intact, and damaged seeds and all kinds of mixture materials. Because of different specific gravity and scattering, the natural beveled angles formed by various components also differ. Therefore, when the seed lot is moved, redistribution of each of the components in the seed batch often happens, resulting in the phenomenon of automatic seed classification. For example, when pouring seeds from above into the storage unit, the fullest and heaviest seeds always fall in the center of the seed batch, whereas empty and shriveled seeds, as well as flighty mixtures, likely gather along the walls. In comparison, when seeds flow from the lower part of the storage unit, seeds in the central part flow down first, then the seeds dispersed on the wall descend. Therefore, the phenomenon of automatic grading is more serious when seeds flow down from a larger storage bank.

The automatic grading of seeds results in heterogeneity among various parts of the seed batch, which brings many difficulties to the sampling work. Therefore, it is necessary to pay full attention to this problem when sampling. Avoiding the negative effects from automatic grading could ensure a fully representative sample.

(2) Division of Seed Batches

Seed lot refers to a batch of seeds collected from the same tree species that have the following conditions: Seeds were collected within a county during the same harvest period; processed, modulated, and stored with the same method; mixed thoroughly to ensure all components distributed evenly and randomly in the seed lot; and the lot size fit the specified quantity.

For the seed lot exceeding a specified quantity, it is necessary to divide it into a sub-lot based on seed weight. According to the upper limitation of seed weight, the regulation for dividing a sublot is described as follows: 10 000 kg for extra-size seeds, such as walnut (*Juglans regia*) and Chinese chestnut (*Castanea mollissima*); less than 5000 kg for large-size seeds, such as *Camellia oleifera* and Siberian apricot (*Armeniaca sibirica*); 3500 kg for medium-size seeds, such as Korean pine (*Pinus koraiensis*), China Armand pine (*P. armandii*), and elaeagnus (*Elaeagnus angustifolia*); 1000 kg for small-size seeds, such as Chinese pine (*Pinus tabuliformis*), Gmelin larch (*Larix gmelinii*), Chinese fir (*Cunninghamia lanceolata*), and black locust (*Robinia pseudoacacia*); and 250 kg for extra-small-size seeds, such as swamp mahogany (*Eucalyptus robusta*), white mulberry (*Morus alba*), dragontree (*Paulownia fortunei*), and casuarina (*Casuarina equisetifolia*). A seed lot in excess of 5% prescribed quantity must be subdivided into more seed lots, each of which is not larger than the prescribed quantity.

(3) Sampling

The primary sample is obtained from the seed batch, and the composite sample is obtained after mixing the primary sample, and the submitted sample is taken from the mixed sample.

Primary Sample. A seed batch is filled in multiple containers, and the containers to be sampled must be selected at random. Primary samples must be drawn from the top, middle, and bottom of the selected containers by using various sampling apparatus or by hand (suitable for seeds that are hard to flow, e. g., with shell or wing).

Composite Sample. All the primary samples are combined and mixed thoroughly to form the composite samples. The weight of the composite samples shall not be less than 10 times that of the submitted samples.

Submitted Sample. Submitted samples must be obtained by reducing the composite sample to an appropriate size using various dividers. The required size of submitted samples are as follows:

In general, the working sample weights for purity analysis are calculated to contain at least 2500 pure seeds, and the submitted sample weights for all tests should be at least two to three times the weight for purity analysis. The weight of large-size seeds should be at least 1000 g, and the quantity of the extra-large-size seeds should be no less than 500 seeds. The required weight of a sample for seed health testing is half of the samples used for the determination of purity. The minimum weight for the determination of moisture content is 50 g, and 100 g of large-size seed needs to be sliced. Variety authenticity identification samples are provided according to the requirements of the determination method.

For plant variety and cultivar identification, the sample should be submitted depending on the testing method. Submitted samples must be packed in proper containers: The seeds with wings should be packed in nondeformable containers such as wooden boxes. The submitted samples for moisture testing and samples that have been dried to a low moisture content must be packed in moisture-proof containers, which contain as little air as possible. The submitted seeds with high moisture content (stored in the wet condition) should be packed in breathable containers under higher humidity during transit. The submitted samples for the health test shall be packed in glass or plastic bottles.

Working Sample. Representative seeds that are separated from the submitted sample form the working sample, which is for the determination of a certain quality index.

The above submitted sample and working sample are obtained by sample reduction method.

The common methods for sample division are as follows:

Mechanical Divider Method. The samples are obtained using apparatus dividers according to the specified procedure.

Quartering (manual halting, sample quartering; diagonal method). Quartering is also called "diagonal method" and "cross differentiation method", and it is used to obtain a sample by using a sampling plate. After preliminary mixing, the composite sample or submitted sample is poured evenly and shaped into a square. The square is then divided into four parts along two diagonal lines with the dividing plate. Next remove the seeds in the relative triangle area, and the remaining seeds are fully mixed. The sample is successively halved in this way until the required size is obtained.

Spoon Sampling. This method is for obtaining a working sample from submitted samples. After preliminary mixing, pour the submitted sample over the tray and shape into a square. With the sampling spoon, take small portions of seed from not less than five random places. Sufficient portions of seed are taken to constitute a subsample of the required size.

5.3.3.2 Determination of Direct Evaluation Indexes

The indexes for direct evaluation of seed quality include seed purity, germination percentage, and moisture content.

(1) Purity Analysis

The purity is the percentage by weight of all component parts (pure seed, inert matter, and other seeds). It is an important index of sowing quality and one of the main factors to consider for seed grading. The purity of seeds could affect the storage duration of seeds because inert matter is usually the pyrogen in the seed pile. Also, the purity affects sowing quantity in the field, which can further affect the homogeneity and uniformity of seedlings. To reduce the hole rate (referring to the empty chambers after sowing that can occur with low quality seed) and improve the seedling survival rate, seed purity is required to be more than 98% when used on the automatic seeding line. The key to purity analysis is to distinguish pure seeds, other seeds, and inert matter. Pure seeds include intact seed units, seeds either with or without seed wing, seeds either with or without seed coat, and multiple seed units. Other plant seeds must include seed units of any other plant species other than that of pure seeds in taxonomy. Inert matter includes seeds that are deficient in germination ability, damaged severely, seeds without covering, as well as all other matter and structures not defined as pure seed.

One whole working sample, two half working samples, and two whole working samples all can be used for seed purity analysis.

(2) Germination Test

Germination rate is the most direct and reliable index for seed quality evaluation. Under field conditions, the results are unstable and cannot be repeated, as the changing environment could result in an inconsistent condition of germination. Therefore, laboratory methods have been adopted, in which the conditions have been standardized to enable the seed to germinate uniformly and quickly. Test results are accurate and reliable.

Germination of a seed is the emergence and development of the seedling to a specific stage, under laboratory conditions. The aspect of its essential structures indicates whether it is able to develop further into a qualified seedling under normal conditions in the field. The germination percentage of a seed lot is the proportion by number of seeds that have produced seedlings classified as normal seedlings under the specified conditions and within the specified time period.

Germination Conditions. Germination testing is to determine the germination ability of seeds under the specified conditions and within the specified period. The specified conditions include growing media, germination temperature, moisture, aeration, and light.

The growing medium can be sterilized paper, pure sand, and soil with a pH range of 6.0 to 7.5. Paper growing media could be filter paper, gauze, or absorbent cotton, which is commonly suitable for germination of small- and medium-sized seeds. Sand growing media can be fine sand, perlite, or vermiculite with a size of 0.05 to 0.8 mm, which is suitable for germination of large-size seeds, recalcitrant seeds, or dormant seeds. Loam with better physical and chemical properties should be selected as the soil growing media, which is mainly used for seeds with diseases when germinating on the paper and sand media.

Temperature for the germination test is divided into two types: Constant temperature and alternating temperatures. The constant temperature of 25 ℃ or 30 ℃ is fit for germination of the majority of tree seeds. The alternating temperatures of 20 to 30 ℃, simulating the temperature change of day and night alternation, is more suitable for some species. The daily alternating temperatures set the lower temperature to maintain 20 ℃ for 16 hours and the higher temperature of 30 ℃ for 8 hours. In general, studies support that the alternative or lower temperature for germination is in favor of dormant seeds.

Water is the essential condition for seed germination. The water content for sand growing media should be 60% to 80% of its saturated water content. For cotton (paper) growing media, sufficient water remains in the media after draining off excess water. For soil growing media, moisture is enough until the soil is broken by hand with a gentle pressure. It should be maintained at 90% to 95% of relative humidity of the air during germination.

Aeration is needed for seed germination. Seeds of most of species can germinate either in light or in darkness. In certain cases, light must promote seed germination of minority species.

Seed Pretreatment. Published literature revealed that nearly one-third of forest seeds are dormant, so seed dormancy must be relieved prior to germination. Baskin and Baskin (2004) divided seed dormancy into five types: Physical dormancy, morphological dormancy, physiological dormancy, morphophysiological dormancy, and comprehensive dormancy. The corresponding methods to release dormancy were also proposed depending on the dormant type. For example, most of the tree seeds in Aceraceae are classified into physiological dormancy, which is released by stratification. For seeds with physical dormancy, physical methods are adopted to soften the seed coat, so as to improve the permeability of the seed coat.

Methods of Germination. Generally, germination of large-sized, medium-sized, and small-sized seeds can be tested by using standard test procedures. For tiny seeds, however, for which purity analysis is difficult to carry out, germination testing by weighted replicates is allowed.

i. Standard Method for Germination Test. Pure seeds fraction of a purity analysis is divided into 4 parts by the quartering method, and 25 seeds are randomly selected from each part to form 100 seeds as a replicate. A total of four 100-seed groups are taken as four replicates. Also, 4 replicates can be extracted by seed counter. Germination testing for sampled seeds could be carried out under suitable conditions.

ii. Testing Seeds by Weighted Replicates. Four replicates of seeds drawn for the germination test have a prescribed weight in each replicate that ranges from 0.1 g to 0.25 g. The test result is reported as the number of normal seedlings produced by the weight of seed material examined, that is, plants/g. This method is suitable for tiny seeds, such as most *Eucalyptus*, (birch) *Betula*, and alder (*Alnus*), for which it is difficult to separate pure seeds by regular methods.

Observation Records and Seedling Evaluation. The germination status should be regularly observed and recorded during testing. When the essential seedling structures sufficiently develop, the evaluation for seedlings could be carried out according to the following principles.

i. Normal Seedling. It shows the potential for continued development into qualified plants when grown in good quality soil and under favorable conditions of moisture, temperature, and light. Normal seedlings include intact seedlings, seedlings with slight defects, and seedlings with secondary infection.

ii. Abnormal Seedling. It does not show the potential to develop into a qualified plant when grown in good quality soil and under favorable conditions of moisture, temperature, and light. Abnormal seedlings include damaged seedlings, deformed or unbalanced seedlings, and decayed seedlings.

The specific requirements for judging normal seedlings or abnormal seedlings vary depending on the tree species. The *Seedling Evaluation Manual* published by ISTA can be used as the reference for seedling evaluation.

Seeds that have not germinated by the end of the test period should be cut and identified one by one. Ungerminated seeds include hard seeds, fresh seeds, dead seeds, astringent seeds, embryo-less seeds, and insect-damaged seeds. If fresh seeds or hard seeds are prevalent, the reason of non-germination should be analyzed by seed viability test.

The result of the Germination Test is Calculated at the End of the Germination Test Period.

i. Germination Percentage. Which is expressed as the percentage by number of normal seedlings in the total tested seeds at the prescribed conditions and duration. The percentage is calculated as follows:

$$\text{Germination percentage } (\%) = n/N \times 100$$

Where N is the number of total tested seeds; n refers to the number of normal seedlings at the prescribed conditions and duration. Percentages are rounded to the nearest whole number.

ii. Absolute Germination Percentage. The index refers to the percentage of full seeds in the tested seeds. It is calculated according to the formula:

$$\text{Absolute germination percentage } (\%) = \frac{n}{(N-a)} \times 100$$

Where n is the number of normal seedlings at the prescribed conditions and test duration; N is the number of whole tested seeds; a is the number of empty and astringent seeds in the tested seeds.

iii. Germination Energy. Which is expressed as the percentage of the number of normal seedlings to the total tested seeds when daily germinating percentage reached its peak. Germination energy reflects the degree of germinating speed and uniformity. Comparing two seed lots with the same germination percentage, the higher the germination energy of the seed lot, the higher quality it is.

iv. Mean Germination Time (MGT). Which is defined as the average germination time required for the tested seeds. It is usually expressed with days for majority seeds, or with hours for seeds with fast germinating speed. The calculation formula is as follows:

$$MGT = \frac{\sum d \cdot n}{\sum n}$$

Where d is the duration (d) from bedded to normal seedlings formation; n is the number of normal seedlings at prescribed conditions and test duration. MGT is an index measuring the speed of germination. A seed lot with shorter MGT means it is germinating quickly and uniformly, reflecting its strong vitality.

v. Germination index (GI). Which is a comprehensive indicator reflecting germination speed and percentage for a seed lot, which is calculated by the following formula:

$$GI = \sum \frac{n}{d}$$

Where the meanings of n and d are the same as above.

Comparing two seed lots with the same germination percentage, the one with the higher germination index, the better the seed quality. To a certain extent, GI indicates seed vigor.

(3) Determination of Moisture Content (MC)

Seed moisture content refers to the percentage of weight loss of the original sample when it is dried. MC is an important factor affecting the safety of seed storage, and one of the main indexes for seed grading as well. Therefore, it is necessary to monitor the MC of seeds, which will be stored, sold, and further used.

Water in seeds exists in two forms: free and bound state. Free water, also known as unbound water, exists in the intercellular space and displays the features of ordinary water. It could flow through the intercellular space freely and escape easily from seeds influenced by temperature and humidity. Bound water is also called adsorbed water or combined water, which is adsorbed by hydrophilic colloids, such as starch and protein in the seeds. Bound water is difficult to evaporate from the seeds, except at a higher temperature for a long time. In addition, a third state of water in seeds is called a compounded state, also known as tissue water. These compounds contain hydrogen (H) and oxygen (O) elements, which could be converted into water potentially, such as in sugars.

This form of water readily decomposes and carbonizes, resulting in weight loss and high moisture content under high temperature for a long time.

The method of seed moisture content determination should ensure that the free and bound water in the seed are rolatilize, and ensure to avoid the loss of oxidation, decomposition, or other volatile substances in seeds.

Two methods can determine the moisture content: Standard and rapid determination.

Standard Method. The principle of standard measurement is determined by the loss of weight after drying, including a low-constant-temperature method, a high-constant-temperature method, and a predrying method.

The low-constant-temperature method refers to seeds that are dried in an oven at 103 ℃ ± 2 ℃ for 17 hours ± 1 hours. The free water and bound water are heated to escape from seeds. The moisture content can be obtained according to the loss in weight. This method is suitable for all tree seeds, and the relative humidity of the surrounding air must be less than 70% when testing.

The high-constant-temperature method means seeds are dried in an oven at 130 to 133 ℃ for 1 to 4 hours. Only the free water is evaporated from the sample. Also, the moisture content can be obtained based on the loss in weight. No requirement is prescribed for relative humidity of the air in the laboratory.

The predrying method is suitable for a sample with high initial moisture content (>17%). In detail, after being predried in an oven at 70 ℃ for 2 to 5 hours, the sample is reweighed to determine the loss in weight. The moisture content is calculated until it is below 17%; then, the seeds are cut into small pieces, and a final moisture content is determined by the low-constant-temperature method or the high-constant-temperature method.

Some tree seeds contain unsaturated fatty acids and volatile substances, such as aromatic oil, which are easy to evaporate under high temperature, so that measured moisture content could be higher than actual moisture content. Alternatively, low-constant-temperature method or reduced-pressure method is suitable for this kind of seeds. For the reduced-pressure method, a lower temperature for drying is suitable when samples contain substances with low melting points or thermal instability. In this situation, water hardly volatilizes. For the seeds with a high content of soluble sugar, such as the immature seeds, soluble sugar tends to form a grid structure when drying, which affects water diffusion from the inner seed. For this type of seed, a vacuum drying oven is preferred to determine the moisture content.

Rapid Determination. This technique rapidly determines the moisture content by using electronic instruments (such as capacitance moisture tester or resistive moisture tester) and an infrared moisture tester. According to the principles of measurement, the available equipment can be divided into resistive moisture meters, capacitance moisture meters, infrared moisture meters and microwave moisture meters. At present, this rapid method is applied primarily for determining the moisture content of crop seeds. As the accuracy in determination results is lower when used in tree seeds, these tools are not widely applied across woody species.

In the formal inspection report and quality label, the moisture content of a sample should be determined by a standard method. As a general estimate, during seed acquisition, transportation, drying, and processing, for example, the moisture content could be attained by rapid determination.

All above determining methods are destructive. For plants, non-destructive rapid determination is more valuable. With the development of technology, rapid and non-destructive determination methods, such as visible light and near-infrared spectroscopy, have attracted more attention and have been attempted to determinate moisture content in crop seed. However, extensive application should be postponed until the process is supported by improved technology.

5.3.3.3 Determination of Indirect Evaluation Indexes

As a direct evaluation index for tree seeds, the germination percentage is often difficult to test quickly because of dormancy or a long germination duration. Therefore, seed quality must be evaluated by indirect measurement. Indirect evaluation indexes typically include seed soundness (cutting test), 1000-seed weight, seed viability, and seed vigor.

(1) Seed Soundness (Cutting Test)

This indicator requires a close inspection for the external and internal status of seeds. For tree seeds without a suitable technique to test their viability at present, or when purchasing seeds at the production site, the seed quality can be identified according to seed soundness. Commonly, three methods are used to determine seed soundness: Cutting, extrusion, and oil pressing.

(2) Weight Determination

Weight of seed is one of the most important indexes to evaluate seed quality, which is usually expressed as 1000-seed weight (TSW). The TSW is the weight of 1000 pure seeds from the submitted sample under air-dry conditions (measured in grams). Weight is related to both seed size and seed plumpness. The higher the TSW, the better the seed lot quality.

The method for determining TSW includes 100-seed weight, 1000-seed weight, and whole-seed fraction weight. Among them, 100-seed weight is the most common. In detail, 8 replicates, 100-seed in each replicate, are randomly selected from pure seeds by hand or by seed counter, then weighed, respectively. Finally, the average weight, standard deviation, and coefficient of variation are calculated based on the weight of 8 replicas:

Variance: $S^2 = \dfrac{n(\sum x^2) - (\sum x)^2}{n(n-1)}$

Standard deviation: $S = \sqrt{S^2}$

Coefficient of variation: $Var = \dfrac{S}{\bar{x}} \times 100\%$

Where x is the weight of each replica, g; n is the number of replicas; S is the standard deviation; $\bar{x}$ is the average weight of 100-seeds, g.

The coefficient of variation should be less than 6.0 for seed lots with great differences in seed size, whereas it should be less than 4.0 for general seed lots. The seed weight could be calculated according to 8 replicas only when coefficient of variation meets above rules. Otherwise, another 8 replicas should be determined, and the average and standard deviation will be calculated based on the weight of 16 replicas. Replicas whose difference from average is more than two times the standard deviation could be abandoned, and 1000-seed weight could be calculated based on the remaining replicas (1000-seed weight = $\bar{x} \times 10$).

(3) Seed Viability Test

Seed viability assesses germination potential by physical or chemical methods. It is the rapid way to determine the germinative ability in seeds with dormancy or to evaluate fresh seeds at the end of a germination test. The popular methods include staining method, excised embryo test, and X-ray test.

Staining Method. The common staining processes are topographical tetrazolium testing and indigo staining.

i. Topographical Tetrazolium Staining. A colorless solution of 2, 3, 4-triphenyl tetrazolium chloride (or bromide), TTC (TTB) or TZ can be used as an indicator to reveal the reduction processes that take place within living cells. During the determination process, the colorless TTC solution is imbibed by the seed. Within the seed tissues it interacts with the reduction process of living cells and accepts hydrogen from the dehydrogenase. By hydrogenation of 2, 3, 5-trifenyl tetrazolium chloride, a red, stable, and non-diffusible substance, triphenyl formazan, is produced in living cells. This interaction makes it possible to distinguish the red-colored living parts from colorless dead cells in the seed.

Tetrazolium staining is an enzymatic reaction, which is not only affected by enzyme activity but also by substrate concentration, reaction temperature, pH, and more. TTC with a concentration profile of 0.1% to 1%, pH of 6.5 to 7.5 is a common recommendation. During a suitable range of temperature, staining rate increases with the rise of temperature, indicating that the reaction rate doubles every 10 ℃ increment. For example, the sample needs to stain for 4 hours at 20 ℃, but only 2 hours at 30 ℃. The maximum staining temperature should be lower than 45 ℃.

Each seed is examined and evaluated as viable or non-viable on the basis of the tetrazolium staining patterns and the tissue revealed. Viable seeds stain completely and non-viable seeds are colorless completely; partly stained seeds will also appear. In partly stained seeds, large or small flaccid areas become apparent in various parts of the seeds; the position and size of these areas (irrelevant to shade of color) are the basis of evaluating seed viability (Figure 5-3).

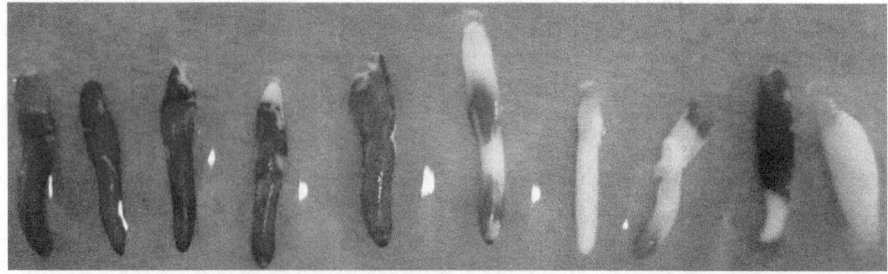

Figure 5-3 Determination of embryo viability of jack tree (*Sinojackia xylocarpa*) by tetrazolium staining (Provided by Shen Yongbao).
From left to right, the first 5 embryos are viable and the last 5 are non-viable.

Seed viability is not equal to germination percentage completely. The viability measured by the TTC method is slightly higher than the actual germination percentage. Therefore, seed viability can assess the potential germinability of seeds, but it cannot replace the germination test. A regression equation, which is constructed based on the correlation between viability and germination percentage, can be predictive of the actual germination percentage.

Owing to its consistent reliability and fast reaction speed, tetrazolium staining has been used extensively to determinate seed viability of woody plants, especial dormant seeds.

ii. Indigo Carmine Stain. Indigo, also known as indigo carmine with the formula of $C_{18}H_8O_2N_2(SO_5)_2Na_2$, is a water-soluble blue powder. The dead tissue stained in blue could be defined as non-viable, since it can absorb the indigo and be stained blue. Accordingly, stained location and size are the basis of evaluating seed viability.

Indigo solution of 0.05% to 0.1% is prepared with distilled water as it is needed, and it should be stored in darkness. The advantages of the indigo carmine test manifest in the quick staining and the testing accuracy. Indigo staining is a physical diffusion process, which means the damaged living tissue may also be stained, more or less, resulting in deviations in interpreting the viability evaluation.

How do the two staining methods compare? Seed viability evaluated by tetrazolium staining is closer to actual germination percentage, whereas indigo carmine staining is quite different from the germination percentage. Liu et al. (2012) compared the effects of the two staining methods on evaluating seed quality of sandalwood and concluded that tetrazolium testing was more reliable.

After staining, the viability of seed is determined one by one according to staining location and proportion in seed. From this identification, seed can be classified as viable or non-viable. The standards of tetrazolium and indigo carmine staining for some tree species are listed in the Rules for Forest Tree Seed Testing.

Excised Embryo Test for Viability. Embryos are excised and incubated under prescribed conditions for 5 to 14 days. Viable embryos either remain firm and fresh or show evidence of growth and differentiation, for example, water-swelling of seed, expansion and greening of cotyledons, radicle and lateral root formation, and the epicotyl and the first leaf formation. Non-viable embryos show signs of decay (Figure 5-4).

The excised embryo test has been used to determine seed viability of woody plants, especially for seeds that germinate slowly or show deep dormancy, such as seeds of Chinese tallow (*Sapium sebiferum*). *International Seed Testing Regulations* and *Excised Seed Embryo Culture of Woody Plants* list incubation conditions for excised embryo in some tree species, such as *Acer* spp., *Pinus* spp., mountain-ash (*Sorbus* spp.), and linden trees (*Tilia* spp.).

X-ray Test. Seed is placed between photosensitive material (film or paper) and an X-ray with a wavelength of 0.1 to 1 nm. Various types of seed tissue absorb the X-rays to varying extents, depending on their thickness and/or density. The sensitive photographic emulsion is excited to varying degrees, depending on the amount of radiation that passes through the seed, thus creating an X-radiograph. The darkest region in the image corresponds to the parts in the seed where X-ray pass

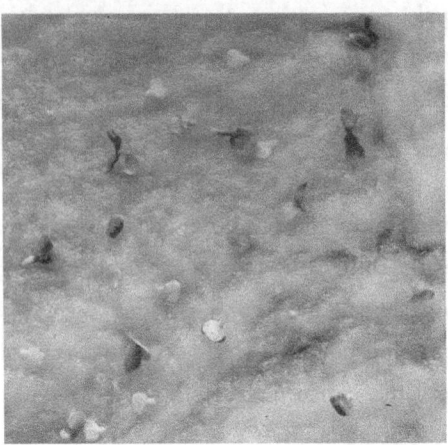

Figure 5-4　Excised embryo culture of *Sapium sebiferum* seeds (Provided by Li Shuxian).

through most easily, while the brighter area is the part with higher density in the seed. X-radiography could provide a quick, non-destructive method of differentiating between filled, empty, insect-damaged, and physically damaged seed from the visible morphological characteristics evidence on an X-radiograph.

At present, the application of an X-ray test tends to develop digital imaging technology. The instrument is a Faxitron UltroFocus X-ray imaging system, which has also been widely applied on inspecting seeds of forest trees. In the past, X-radiograph has been imaged on photographic paper directly by using an Hy-35 agricultural X-ray machine in China. For comparison, digital imaging is simple, fast, and very effective; photographic paper imaging is very effective, but the procedure is tedious and has been gradually replaced by digital imaging.

According to the differences of photography method and evaluation regulations, X-radiography is divided into direct X-radiography and contrast X-radiography.

ⅰ. Direct X-radiography. Unprocessed seed is projected on fluorescent screen, X-ray film, or photographic paper or is formed directly with digital imaging. This technique can detect not only the status of seed development and fullness but also mechanical damage and insect damage. Thus, it is the basis for judging accurately the vitality of fresh seeds. Through image interpretation, seeds can be identified with an X-radiography and divided into categories of full seeds, empty seeds, insect-damaged seeds, and mechanical damaged seeds (Figure 5-5).

ⅱ. Contrast X-radiography (XC). Seeds should be pretreated with contrast agents before being imaged. Dead tissue is infiltrated by contrast agents because of the loss of selective permeability. The contrast agents in dead tissue can strongly absorb the X-rays, showing the contrast of density on the radiograph, which is the basis for determining seed viability. Depending on the selectivity and affinity of contrast agents to various tissues, the living, damaged, and dead tissues within the pretreated seed are more likely to be distinguished on the X-radiograph.

Popular liquid contrast agents are $BaCl_2$, $AgNO_3$, NaI, and KI solutions. After being treated with these contrast agents, the dead or mechanical damaged seed will absorb these agents. The

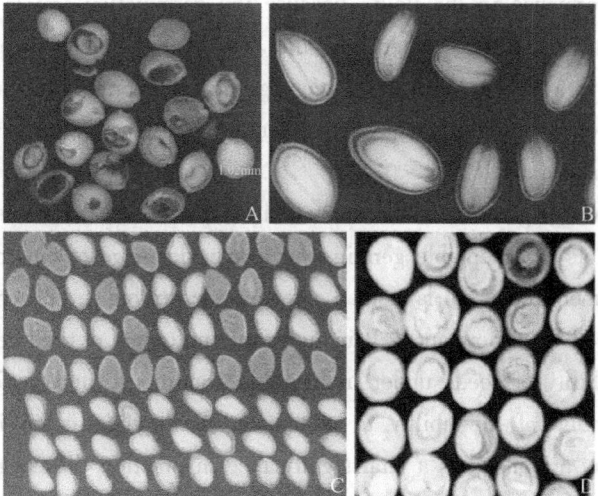

Figure 5-5 Direct X-radiography of forest tree seeds (Photos by Fu Xiangxiang).
(A) Eastern redbud (*Cercis canadensis*) and (B) *Pinus tabuliformis* are digital images; (C) Chinese red pine (*Pinus elliottii*) and D) linden (*Tilia miqueliana*) are photographic paper images.

heavy metal elements in the solution absorb X-ray strongly and create a clear image on the radiograph. By comparison, no contrast agents are left in living tissue, so X-ray readily passes through and leaves the film black. Note that the contrast agents mentioned above are usually toxic and will injure the tested seeds.

Water can also be used as a contrast agent since it will strongly absorb X-rays. Water as the contrast agent is called the IDX (I is culture, D is drying, X is X-ray) method. Both living and dead seeds can absorb water, but the speed of water loss from the living tissue differs from that of the dead tissue during drying. In the process of drying, water loss from the living tissue is slow, resulting in higher moisture content; conversely, lower moisture content in the dead tissue results from fast water loss. As lower water content absorbs fewer X-rays, the dead or injured tissues are imaged clearly on the radiograph. By comparison, greater amounts of water in the living

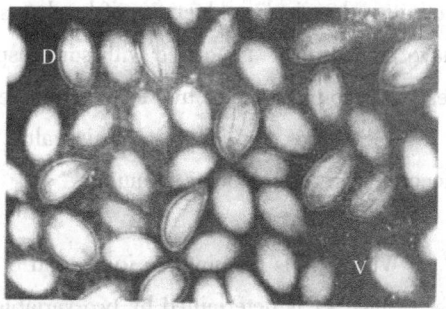

Figure 5-6 Determination of seed viability of Yunnan pine (*Pinus yunnanensis*) by IDX test (Cited from Shen at al., 1998). Note that D is the non-vital seed and V is the vital one.

tissue absorbs a greater amount of X-rays, so fewer rays pass through the image, creating a bright and homogeneous radiograph showing blurred details in seeds (Figure 5-6). IDX method is a non-destructive test. The tested seeds can also be used for control germination test. Therefore, IDX test is widely applied to test seed viability of tree species.

(4) Seed Vigor Testing

A seed germination test reflects only the germination ability in the laboratory, which may deviate from the performance in the field, because the germination test is carried out under the optimum environmental conditions rather than the complicated and changeable conditions in the field. It could result in a high germination percentage in the laboratory but low results in the field, which happens more often under adverse environmental conditions. Therefore, a seed vigor test is very meaningful and necessary.

Definition. A seed vigor test is the sum of those properties that determine the activity and performance of seed lots of acceptable germination in a wide range of environments. Seed vigor is not a simple measurement index but is instead a concept describing several characteristics associated with the aspects of seed lot performance. It includes the uniformity of seed germination and seedling growth, emergence ability of seeds under unfavorable conditions, and performance after storage, particularly the retention of the ability to germinate. Seed vigor is an important indicator for evaluating seed quality and also has a vital significance in practice. The vigorous seed lots can be screened before sowing; in addition, the parameters of seed processing, storage, and treatment could be optimized by vigor testing, so as to ensure and improve seed quality. Moreover, vigor testing could help breeders develop new varieties with stress resistance.

Determination Method. A seed vigor test is either a direct or an indirect analytical procedure to evaluate the vigor of seeds. Direct tests, such as a low-temperature-treatment test, a Hiltner test, and others, reproduce environmental stresses or other conditions in the laboratory. The emergence percentage and/or growth rate, as well as toughness of seedlings, are recorded. Indirect tests measure biochemical, physiological, and physical characteristics of the seed, for example, enzyme activity, electrical conductivity of leachates, respiratory intensity, and accelerated aging (AA), that have proved to be associated with some aspect of seed vigor in the laboratory. The main evaluation tests commonly used for tree seed vigor are as follows.

i. Seedling Growth Test. This value is expressed by vigor index (VI), which is calculated according to the following formula:

$$VI = GI \times S$$

Where GI is germination index and S is seedling biomass (g) or average root length (cm).

Since VI is determined by two variables, seed germination rate and seedling growth, it is more informative than only a single germination index.

ii. Conductivity Test. Which is a commonly used method of physiological and biochemical determination and is the only test listed in the current regulations of ISTA. The basic principle is that the extent of electrolyte leakage (such as amino acids, organic sugars, and other ions) from seed tissue is affected by the ability of cell membrane reconstruction and repair at the early stage of seed imbibition. The faster the repair of the membrane integrity, the less the electrolytes leak. Seed lots with high vigor could reconstruct the membrane rapidly and repair damage to the maximum extent, whereas seed lots with low vigor display the poor ability of damage repair. Therefore, the conductivity of seed lots with high vigor is lower than those with low vigor. Furthermore, the conductivity is negatively correlated with the rate of emergence in the field.

iii. Accelerated Aging Test. Which is used to predict storage potential of seed lots. The theoretical basis is that seed aging will be accelerated under high temperatures (40-50 ℃) along with high relative humidity (100%), and the degree of seed deterioration under such conditions for a few days is equivalent to the aging that would occur over months or years in an actual field environment. High-vigor seed lots subjected to aging treatment can still germinate normally, whereas low-vigor seed lots will germinate more abnormal seedlings that will even die. Research has shown that ultra-dry seeds could significantly improve anti-aging ability.

(5) Relationship between Seed Viability and Seed Vigor

Seed vigor and seed viability are easily confused; they are related but are also different. Seed viability refers to the existence or absence of seed life, namely viability. Seed vigor is a comprehensive concept, which not only involves the existence or absence of life but also reflects the ability of seedlings to survive under heterogeneous environmental conditions. In the deterioration process, the difference between seed viability and seed vigor is significant. As presented in Figure 5-7, at the beginning of seed deterioration (point A), little change in seed viability but the initial descent in seed vigor is occurring; with the development of deterioration to a certain extent (point B), an obvious divergence shows initial downturn in seed viability along with an abrupt decrease in seed vigor. At point C, when serious deterioration happens, the seed viability still maintains 50% or so but the seed vigor is very low and has lost its useful value.

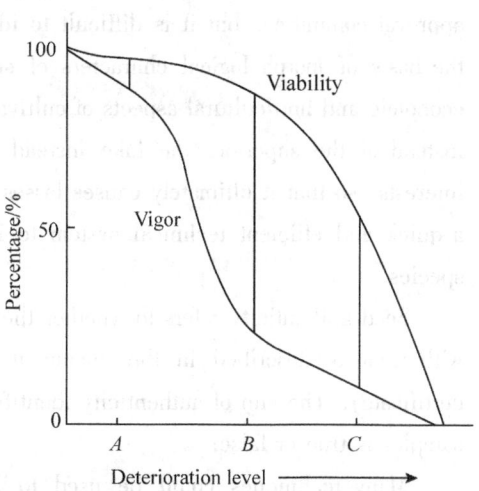

Figure 5-7 Relationship between seed viability and seed vigor during the process of deterioration (Cited from Delouche et al., 1965).

5.3.3.4 Coated Seed Testing

Coated seed refers to a single unit in which the seed is covered evenly and completely by a specific coating agent according to a certain ratio of "agent/seed" and forms a superficial film. Seeds for coating treatment should be the finest seeds with elite quality of commerce and planting that have been strictly selected and processed. Coating material is made of insecticides, fungicides, microelements, growth regulators, and so on by special processing technologies. Coating seeds is beneficial to precision sowing, which will save approximately 3% of sowing quantity; moreover, it will greatly improve seedling survival.

Coated seeds are divided into 5 categories: Coated seeds, heavy pelletized seeds, fast-growing pelletized seeds, flat pelletized seeds, and fast split pelletized seeds.

Quality evaluation for coated seeds is similar to that of regular seeds, but differences occur in a few details. According to the regulations of ISTA, when a purity analysis for coated seeds is carried out, the working sample should be separated into pure pellets, unpelleted seeds, and inert matter. Germination tests on coated seeds must be made with pellets from the pure pellet fraction of a purity test. Because coating materials of pelleted seeds may affect water absorption and germination, the duration of a germination test can be appropriately prolonged.

5.3.4 Authenticity Identification of Tree Seeds

In the past 10 years, remarkable achievements have been made in breeding new varieties and cultivars of forest trees, including tree species for timber, non-timber, and horticulture. The large number of varieties and cultivars has been certified or recognized by the national or provincial variety approval committee, but it is difficult to identify and characterize varieties and cultivars solely on the basis of morphological characters of seeds or seedlings. Meanwhile, for the high values in economic and horticultural aspects of cultivated species (varieties), the phenomenon of "the inferior instead of the superior, the fake instead of the true" happens commonly, driven by economic interests, so that it ultimately causes losses to forest production. Therefore, it is urgent to establish a quick and efficient technical system to identify the authenticity of seeds and seedlings of tree species.

Seed authenticity refers to whether the varieties, species, or genera of a seed lot is in accord with what is described in the document (the label, variety certificate, or quality inspection certificate). The aim of authenticity identification is to clarify whether the identity of seed/seedling samples is true or false.

Many techniques could be used to identify the authenticity of seeds and seedlings. For example, they could be divided into morphological identification, physical and chemical identification, physiological and biochemical identification, cytological identification, and/or molecular identification.

Currently, the identification of species/varieties of tree species depends primarily on morphological identification, assisted by physiological and biochemical identification (mainly isozyme and electrophoretogram of storage proteins in seeds). The development of molecular technology brings about a revolutionary breakthrough for identification of varieties. Besides its specificity, simple operation, fast procedure, and accurate result, molecular identification is not affected by seasons and environmental conditions; it has become the most advanced identification technology for varieties identification.

5.3.4.1 Morphological Identification

Morphological characteristics of seeds vary with tree species (varieties), which are manifested in seed shape, size, smoothness of seed coat, and appendage(s) of seed. The selected characters for identification are divided into main traits, fine traits, special traits, and variable traits according to their significance and stability. Among them, the stability of variable traits is too poor to be applied in identification. Although the advantages of morphological identification are simple and fast, it is only suitable for seeds with large size and abundant morphological variation.

5.3.4.2 Biochemical Identification

Biochemical identification mainly refers to electrophoresis identification, which uses electrophoresis technology to analyze the components of isozymes and proteins in species (varieties).

The technology then determines the differences in biochemical indicators to further distinguish between species (varieties).

At present, isozymes and proteins are the main objects for electrophoretic analysis. Genetically, the component difference of protein or enzyme is controlled by genetic differences from species (varieties). In essence, the differences in enzyme and protein result from genetic differences, that is, the differences of species (varieties). Thus, advanced electrophoresis technology could analyze accurately the differences of proteins or isoenzymes stored in seeds, and then for that, identify species (varieties).

In detail, identification from electrophoretic analysis depends on the number, position, width, and color shade of electrophoretic bands.

Number of Bands. The number of bands varies across different species.

Band Position (Rf value). Despite the same number of bands in different species/varieties, the positions of specific bands are divergent.

Color Shade of Band. Owing to the dosage effect of a gene, the color shade of a band varies depending on its species/varieties.

Band Color. The electrophoretic bands display colors that differ after staining based on various components. For example, after staining, the isoenzyme band of α-amylase presents as transparent white, β-amylase shows as pink, R-amylase shows as light blue, while Q-amylase displays as red or brown.

The electrophoretic technique has been widely used in identification of species/varieties of trees; for instance, various species in the larch (*Larix*) trees were identified based on the PI difference of protein electrophoretic analysis; also, Chinese date (*Ziziphus jujuba*) varieties could be distinguished by electrophoretic bands that have a combination of peroxidase, lipase, and amylase. However, electrophoretic identification has some limitations because of its less locus, band pattern variability resulting from various development stages and heterogeneous environment conditions.

5.3.4.3 Cytological Identification

Cytological identification is the basis for the differences of karyotype and chromosomal bands depending on species (varieties). It mainly includes the number, size of chromosome, karyotype, band-type, and meiosis behavior. Band-type analysis includes C-band, G-band, Q-band, R-band, and others. For example, the triploid of *Populus tomentosa* was found out by karyotype analysis, while species differences in *Pinus* were revealed by fluorescence banding.

Although cytological identification of species (varieties) is simple, challenges remain in determining differences in the karyotype and band type across species (varieties), as well as sectioning techniques. For species with a large number but a small size of chromosomes, karyotype analysis is difficult to distinguish chromosomes one by one, which restricts greatly their extensive application.

5.3.4.4 Molecular Identification

A molecular marker is an ideal genetic marker developed after the morphological marker, cytological marker, and biochemical marker. The molecular marker is based on the polymorphism of biological macromolecules, especially the genetic material nucleic acid.

Compared with stated methods, DNA molecular markers possess outstanding advantages as follows:

It takes the DNA genetic material as a study object, which is not affected by tissue type, development stage, and so on. All DNA extracted from any part of a plant is suitable for early identification of species/varieties at the stage of seeds seedling.

A sample for molecular analysis has nothing to do with environmental conditions. Observed variations derive only from DNA sequence difference of the allele gene. The stability of analysis is beneficial to reveal genetic differences among varieties and to exclude phenotypic variation caused by heterogeneous environment.

Numerous loci distribute throughout the entire genome.

Much variation in alleles can be found because of high polymorphism.

Homozygous and heterozygous genotypes could be identified by using co-dominant markers.

For authenticity identification of species/varieties, the standards of molecular marker selection lies on its simple, reliable, stable, and repeatable; most important, accurate results should be obtained as fast as possible.

Comprehensive review of publications reveals that efficient molecular markers to identify species (varieties) authenticity include random amplified length polymorphism (RAPD), amplified fragment length polymorphism (AFLP), simple sequence repeat (ISSR), microsatellite (SSR), among others. In terms of reliability and simplicity, SSR markers are the first preference, followed by ISSR, RAPD, and AFLP markers. In fact, the more efficient transition procedure from random PCR to specific PCR is that specific fragments can be screened by using RAPD or ISSR first, then the sample be cloned, sequenced, and then transformed into sequence amplification characteristic region (SCAR) markers. Besides the characteristics of molecular markers, selection should consider the traits of plant materials and the associated research background.

Questions for Review

1. What are the factors that affect the storage life of tree seeds? How do they interact and restrict each other?

2. What are the storage methods for tree seeds? Please illustrate the suitable seed types for different storage methods.

3. What physiological index changes in asexual propagation materials during storage reflect the vigor of propagation materials? Describe the change trend of these physiological indexes during storage.

4. What are the types of asexual propagation materials? Describe the suitable storage method.

5. As an important type of asexual propagation materials, what are the characteristics of artificial seeds compared with natural seeds?

6. What is the process of tree seed quality testing?

7. What is the difference and connection between seed viability and seed vigor?

8. What is the principle of using molecular markers to identify the authenticity of tree seeds? What are the advantages and disadvantages compared with other identification methods?

9. What are the physiological activities of seeds during storage?

10. What are the factors that affect seed vitality?

11. What is the best way to store seeds according to their safe moisture content?

References and Additional Readings

ADMINISTRATION OF QUALITY SUPERVISION OF THE PEOPLE'S REPUBLIC OF CHINA, 1999. Rules for forest tree seed testing: GB 2772—1999[S]. Beijing: Standards Press of China.

BASKIN J M, BASKIN C C, 2004. A classification system for seed dormancy[J]. Seed Science Research, 14: 1-16.

CHEN S Y, CHOU S H, TSAI C C, et al., 2015. Effects of moist cold stratification on germination, plant growth regulators, metabolites and embryo ultrastructure in seeds of *Acer morrisonense* (Sapindaceae)[J]. Plant Physiology and Biochemistry, 94: 165-173.

ADMINISTRATION OF QUALITY SUPERVISION OF THE PEOPLE'S REPUBLIC OF CHINA, 1999. Classification of forest tree seed quality : GB 7908—1999[S]. Beijing: Standards Press of China.

DELOUCHE J C, CALDWELL W P, 1965. Seed vigor and vigor tests[J]. Proceedings of the Association of Official Seed Analysts, 50: 124-129.

STATE FORESTRY ADMINISTRATION, 2010. Embryo excision test for woody plants: LY/T 1881—2010[S]. Beijing: Standards Press of China.

GOODMAN R C, JACOBS D F, KARRFALT R P, 2006. Assessing viability of northern red oak acorns with X-rays: application for seed managers[J]. Native Plants Journal, 7(3): 279-283.

INTERNATIONAL SEED TESTING ASSOCIATION (ISTA). Various years. Seedling evaluation manual [S]. Switzerland: ISTA. Accessed at htttps: //www. seedtest. org.

INTERNATIONAL SEED TESTING ASSOCIATION (ISTA). 2017. International rules for seed testing [S]. Switzerland: ISTA. Accessed at htttps: //www. seedtest. org.

JIN H. 2006. Seed Biology[M]. Beijing: Higher Education Press.

LI H, CHENG P, ZHENG Z, et al., 2011. Effects of low temperature storage on the physiological indexes and survival rate of Russian poplar[J]. Journal of Northeast Forestry University, 39(3): 12-14.

LIU X, XU D, YANG Z, et al., 2012. Fast determination method of sandalwood seed viability[J]. Journal of Nanjing Forestry University (Natural Sciences), 36(4): 67-70.

MINISTRY OF HOUSING AND URBAN-RURAL DEVELOPMENT, 2018. Woody seedling for green space and landscape: CJ/T 24—2018[S]. Accessed at https: //www. zixin. com. cn/doc/15672. html.

ROBERTS E H, 1983. Loss of seed viability during storage[A]// Thompson R J, 1983. Advances in research and technology of seeds[C]. Wageningen: Pudoc.

STATE FORESTRY ADMINISTRATION, 1999. Rules for forest tree seed testing: GB 2772—1999[S]. Beijing: Standards Press of China.

SEED LAW OF THE PEOPLE'S REPUBLIC OF CHINA, 2021. Revised edition. Accessed at https: //baike. baidu. com/item/中华人民共和国种子法/4471386.

SHEN Y, JIN T, 1998. Determination of seed viability of *Pinus yunnanensis* by X-ray water contrast method[J]. Scientia Silvae Sinicae, 34(2): 111-114.

STATE OWNED FOREST FARM AND FOREST SEED AND SEEDLING WORK STATION OF STATE FORESTRY ADMINISTRATION, 2001. Chinese woody plant seeds[M]. Beijing: China Forestry Publishing House.

ADMINISTRATION OF QUALITY SUPERVISION OF THE PEOPLE'S REPUBLIC OF CHINA, 1988. Tree seed storage: GB/T 10016—1988[S]. Accessed at http: //m. foodcta. com/spbz/detail27554. html.

STATE FOREST ADIMINSTRATION, 2010. State woody plant seed pretreatment for germinating: LY/T 1880—2010

[S]. Beijing: Standards Press of China.

ZHAI M P, SHEN G F, 2016. Silviculture[M]. 3rd edition. Beijing: China Forestry Publishing House.

ZHENG G H, 2004. Seed physiology research[M]. Beijing: Science Press.

ZHENG Z, ZHANG M, YANG Y, 2014. Ten winter scion collection and storage technologies of fruit mulberry[J]. Shaanxi Journal of Agricultural Sciences, 60(4): 121, 128.

Chapter 6 Seedling Growth and Physiology

Wang Weiwei, Shen Hailong

Chapter Summary: This chapter briefly discusses the types of seedlings, the growth period and characteristics of seedlings, and the relationship between abiotic and biotic environmental factors and seedling cultivation. With the goal of fully understanding and knowing seedling type, seedling growth, and development period and characteristics, we should deeply grasp the relationship between various abiotic environmental factors and seedling growth and development, and understand the influence of diseases, pests, and weeds on seedling cultivation in combination with previous knowledge. Mycorrhizae, rhizobia, and artificial substrates are closely related to the development of modern seedling technology, and sufficient attention should be paid to its development status. Principles of plant physiology and forest ecology are needed to comprehensively consider the relationship between seedling growth and development indicators and biological and non-biological environmental factors, as well as the coordination and interaction between various indicators and factors. Attention must be given to the role of various ecological factors, the characteristics of dominant factors and limiting factors, the compensation role between ecological factors, and the irreplaceability among ecological factors, so as to consciously use these principles in the practice of seedling cultivation to improve the level of nursery stock cultivation technology.

6.1 Seedling Types, Characteristics, and Seedling Age

6.1.1 Seedling Types and Characteristics

Nursery stock is a type of afforestation material with intact roots and stems that is propagated from forest seeds. According to various classifications, seedlings can be divided into many types (Shen, 2009), as outlined below.

According to the propagation materials, seedlings can be divided into seedlings and vegetative propagation seedlings. Seedlings are directly propagated from seeds, whereas vegetative propagation seedlings are propagated from roots, branches, leaves, and other vegetative organs and tissues. According to the type of organs, tissues, and technical methods, vegetative propagation seedlings can be further divided into cuttings, root cuttings, layering, burying, root sucker, leaf cuttings, grafted seedlings, and so forth. Cutting seedlings are obtained by cutting a part of a branch and rooting it in soil (substrate). Root cutting seedlings are cut from a part of tree roots and are then rooted in soil (substrate). Layering seedlings are achieved by burying a part of growing branches in soil (substrate) or wrapping them by using soil (substrate), which are then cut off from the mother tree after rooting. Burying seedlings are cultivated by burying a whole tree branch horizontally in the soil (substrate) and allowing it to make roots. Root seedlings refer to trees with strong sprouting ability, for which breaking soil or digging ditches near the roots and causing mechanical damage to the roots promotes a large number of sprouts to produce more roots. Leaf cutting seedlings can be propagated by rooting in soil (substrate) with broad-leaved tree leaves or a part of leaves and the needles of conifers. Grafted seedlings are obtained by cutting a part of the branch, or by cutting only the bud as scion and connecting it to the trunk and stump-root of the same or a different species, so that the two can heal into one seedling.

According to the method of seedling cultivation, they can be divided into bareroot seedlings and container seedlings. Bareroot seedlings are cultivated in the field without the root lump formed by substrate and root system, making the root system unprotected when it is out of a nursery. Container seedlings are cultivated in seedling containers, with a root lump formed by substrate and root system when it is out of a nursery.

According to the time of seedling cultivation, we can have 1-year-old seedlings and perennial seedlings. 1-year-old seedlings are obtained by sexual or vegetative propagation methods, such as sowing or cutting, among which the 1-year-old seedlings obtained by sowing are called sowing seedlings. Perennial seedlings refer to the seedlings cultivated in the nursery for more than 2 years, which are divided into reserved bed seedlings and transplanted seedlings according to the transplanting process used. Reserved bed seedlings refer to the seedlings that are produced without transplanting, also known as reserved nursery seedlings; transplanting seedlings refer to the seedlings that are re-cultivated after one or several transplantings, also known as bed changing seedlings. The large seedlings used in urban forest construction may be seedlings that were transplanted and cultivated many times.

According to whether the nursery environment is controlled manually, plants can be divided into test-tube plantlets, greenhouse seedlings, and field seedlings. Test-tube plantlets are cultivated with artificial medium in a laboratory test tube (including other containers), a process that is also known as tissue culture seedlings or micropropagation seedlings. Greenhouse seedlings are cultivated in a greenhouse; field seedlings are cultivated in an open nursery, and most bareroot seedlings are field seedlings.

According to the size of seedlings, they can be divided into standard seedlings and large seedlings. Standard seedlings are commonly used in current production, for which conifers range from 1 to 4 years of age, and broad-leaved trees are mostly 1-year-olds. Large seedlings are cultivated in a nursery for many years. In many cases, urban forest construction uses large seedlings, and landscaping large seedlings are mostly shaped in the cultivation process, so they are also called shaped seedlings or stereotyped seedlings.

According to the seedling cultivation substrate, seedlings can be grown with or without soil. Conventional seedlings are cultivated with soil. Soilless seedlings usually refer to hydroponic culture, that is, the method of directly cultivating seedlings with a nutrient solution. Seedlings grown in artificial medium also belong to soilless seedling methods.

According to the quality of seedlings, they can be categorized into substandard seedlings, qualified seedlings, target seedlings, and optimal seedlings. Substandard seedlings refer to seedlings whose specifications and vitality fail to meet the requirements of the nursery technical regulations or standards and cannot be used for afforestation. Qualified seedlings meet the requirements and can be used for afforestation. Target seedlings are cultivated in large quantities in a nursery, and their physiology, morphology, and genetic characteristics can adapt to the afforestation site conditions. Optimal seedlings are those that can meet the lowest overall cost of afforestation while also meeting the requirements of afforestation survival rate and early growth.

6.1.2 Seedling Age

The age of the seedlings is expressed in Arabic numerals (that is, 1, 2, 3, and so on) as a unit of seedling age that undergoes an annual growth cycle. The first number represents the number of years that a seedling has grown in the initial nursery after it has been formed by seeds or asexual reproduction materials; the second number represents the number of years it has grown on the transplanted ground since the first transplanting; the third number represents the number of years it has grown on the transplanted ground after the second transplanting, and so on. The numbers are spaced by short horizontal lines (known as hyphens or dashes), and the sum of the numbers is the age of the seedlings. Examples are as follows:

1-0 denotes a seedling that has not yet been transplanted, that is, 1-year-old seedlings.

2-0 means a seedling that has not been transplanted in 2 years, that is, reserved bed seedlings.

1-1 means a 2-year-old seedling that has been transplanted once and cultivated for 1 year after transplantation.

1-1-1 means a 3-year-old seedling that has been transplanted twice and cultivated for 1 year after each transplantation.

0.5-0 indicates that seedlings have completed about half of an annual growth cycle.

0.3-0.7 means a 1-year-old seedling that has been transplanted once, 3/10 of an annual growth cycle for pre-transplant culture and 7/10 after transplantation.

1_1-0 denotes cutting seedlings (root cuttings or grafted seedlings) with a 1-year-old stem and a 1-year-old root that has not been transplanted.

1_2-0 denotes cutting seedlings (root cuttings or grafted seedlings) with a 1-year-old stem and a 2-year-old root and it has not yet been transplanted.

1_2-1 refers to the cuttings (rooting or grafting) as transplanted seedlings with 2-year-old stems and 3-year-old roots; they have been transplanted once and cultivated for 1 year after transplantation.

Note that in *Tree seeding quality grading of major species for afforestation* in GB 6000—1999, parentheses bracket the lower foot, that is, 1(2)-0, 1(2)-1, etc., which has the same meaning: The root age of cuttings, root cuttings or grafted seedlings in the original nursery (before transplantation).

At present, no unified expression method is in place for container seedling age. In North America, "container type + volume" or "container type + container diameter depth + transplantation time" is generally used to represent container seedling age. The former, such as "styro 4", indicates a foam-integrated block container made by Styrofoam Company, which is approximately 4 cubic inches (65 cm^3) in volume; the latter, such as "PSB 313 B 1+0", denotes the 1-year-old container seedlings that have been cultivated in a foam-integrated block container produced by Styrofoam with a diameter of 3 cm, a depth of 13 cm, and planted under B site conditions. China can refer to the above methods, but given the complexity of the container seedling situation in our country, it cannot fully correspond with foreign countries. Our own container seedling age expression method should be established in the future.

6.2 Seedling Growth Rhythm

The type of seedling growth and the rhythm of growth and development are closely related to the normal selection of seedling cultivation measures (Shen, 2009). In the actual seedling production, appropriate seedling technical measures must be taken according to the growth types and growth periods.

6.2.1 General Characteristics of Seedling Height, Diameter, and Root Growth

(1) Seedling Height Growth

According to the length of growth period, it can be divided into two types: Spring growth type and full-term growth type.

Spring growth type. It is also called *prophase growth type*. The height growth period of seedlings and the extension of side branches are very short, only 1 to 2 months in the north and 1 to 3 months in the south, and only 1 growth per growing season. Generally, height growth ends during May to June. The spring growth type seedlings show the characteristics of short-term high-growth from 2 years of life; that is, the seedlings enter the fast-growing period after a very short growth period in spring; the fast-growing period lasts for a short time, and the height growth stops soon after the fast-growing period. After the main leaf growth, leaf area expansion, the new, young ends gradually become woody, and the winter buds grow. Given that spring growth seedlings develop quickly, nutrient accumulation in the previous year is very important.

Full-term growth type. It refers to the tree species whose seedling height growth lasts for a long time across the entire growth season. The growth period of northern tree species is 3 to 6 months and that of southern tree species is 6 to 8 months, some more than 9 months. The height growth of the full-term growth type seedlings continues throughout the growth season, and the growth of leaves and the woody new branches is carried out at the same time, with full woody state reached in autumn. In the annual growth cycle, there are generally 1 to 2 suspended periods of height growth.

The essence of the two growth types is limited growth (prophase growth type) and infinite growth (full-term growth type). The actual performance of the two growth types is affected by both the tree species and the existence of buds. The amount of growth depends on the size of the embryo, the amount of energy accumulated in the seed, and whether the environmental conditions for germination growth are positive. The prophase growth type seedlings usually form specific terminal buds at the end of the first growing season, whereas trees of full-term growth type seedlings do not. For the second and subsequent growing seasons, most tree species show their specific height growth types. Prophase growth type species, such as pine, generally show determinate growth controlled by terminal buds, whereas the full-term growth type species (such as poplar, willow, and larch, for example) are affected only by environmental conditions.

(2) Seedling Diameter Growth

The diameter growth peak and height growth peak of seedlings are staggered. Diameter growth also has a growth pause. The diameter growth peaks in summer and autumn, after the height growth peaks. In autumn, diameter growth stops later than height growth, which is a common rule of many tree species. In spring, the top buds of seedlings more than 2 years old germinate first, producing plant growth regulator, which is transported down through the cambium and stimulates the growth of cambium. Therefore, the small growth peak of diameter appears first, and then the first fast-growing peak of height growth appears.

(3) Root Growth

Root growth has several peaks during the year and is staggered with the peak of height growth. The growth peak of roots in summer and autumn occurs after the height growth peak. The stop period of root growth is later than that of height growth. Root growth peak is close to or at the same time as the peak of diameter growth. The root growth amount is highest in summer, second in spring, and least in autumn. In addition to needing temperature, soil moisture, and nutrients for growth, some tree species also require aeration conditions, for example, larch and *Pinus* seedlings are sensitive to soil aeration conditions.

6.2.2 Annual Growth Phase of Sowing Seedlings

The sowing seedlings refer to the 1-year-old seedlings, which have a range of growth and development characteristics with different requirements for environmental conditions and management in different periods from sowing and entering the dormancy period. Generally, growth can be divided into four phases: Emergence phase, seedling phase, fast-growing phase, and seedling hardening phase. The 1-year-old seedlings across a variety of tree species have their own growth and development pattern. It is necessary to carry out targeted management for different seedlings in the process of seedling cultivation.

(1) Emergence Phase

The emergence phase is from sowing to seedling emergence, the emergence of true leaves in the aboveground part (for coniferous trees the shell falls off or the needles just unfold), and before the emergence of lateral roots in the underground part. This growth period varies by tree species, germination accelerating methods, soil conditions, meteorological conditions, sowing methods, and sowing seasons. It takes 10 to 20 days for general tree species and 40 to 50 days for the tree species germinating slowly. During the emergence phase, seeds grow and develop into seedlings, the cotyledons of broad-leaved trees appear (the true leaves of the tree species whose cotyledons remain in soil are not unfolded), and the cotyledons of conifer trees emerge with the seed coat not shed and without the acrospires. The underground part grows faster without lateral roots, while the aboveground part grows more slowly. Seedlings grow by the nutrients stored in the seeds and do not have their own ability to produce nutrients, so the seedlings are weak in resistance.

(2) Seedling Phase

Seedling phase refers to the period from the first true leaf appearing on the ground, with lateral roots appearing in the underground part, to the beginning of height growth of seedlings. This period's duration is generally 3 to 8 weeks, varying by tree species. In this phase, true leaves appear in the aboveground part, lateral roots appear in the underground part, photosynthesis begins to produce nutrients, and the number of leaves increase and the leaf area gradually expands. In the early stage, the height of seedlings grows slowly, but root growth is fast and multiple lateral roots develop. Given that the main absorbing roots grow more than 10 cm in length in the later stage, and the growth speed of aboveground parts changes from slow to fast, the size of individual seedlings increases significantly, which requires more water and nutrients.

(3) Fast-growing Phase

Fast-growing phase is the most vigorous period of seedling growth, which is the period between the acceleration and the deceleration of seedling height growth under normal conditions. The duration of this period varies with species and environmental conditions. For spring-sowing trees in Beijing, the fast-growing phase of seedlings is approximately 3 months from mid-May to mid-August. Fast-growing phase is the key period of seedling growth. In this phase, the biomass of seedlings increases rapidly and reaches the maximum, and the number of leaves and the size of a single leaf increased. The height growth, root collar diameter growth, and root growth of the seedlings reach more than 60% of the annual growth, forming developed roots and vegetative organs. In this phase, fast-growing trees grow lateral branches, and the growth range of seedling roots is larger.

(4) Hardening Phase of Seedlings

The hardening phase of seedlings refers to the period when the aboveground and underground parts of seedlings are fully lignified and enter into hibernation. This phase starts from the rapid decline of seedling height growth to the stop of seedling diameter and root growth. In the hardening phase, the height growth rate of the seedlings decreases rapidly and then stops forming the terminal bud, while the seedling diameter and root system continue to grow with a small growth peak and then they also stop. Water content of the seedlings decreases gradually, and the dry biomass increases gradually. The whole seedlings were completely lignified and the seedlings' resistance to low temperature and drought increases. Leaves of deciduous trees fall off, and trees enter into dormancy.

The relationship between seedling height growth, diameter growth, root growth, and bud formation in different growth phases is shown in Figure 6-1.

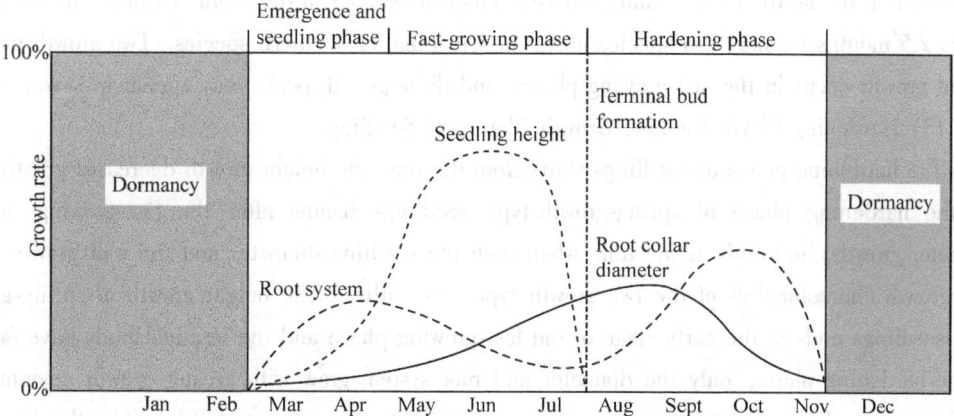

Figure 6-1 Relationship between seedling height growth, diameter growth, root growth, and bud formation across growth phases (Adapted from Landis et al., 1998).

6.2.3 Annual Growth Phase of Reserved Bed Seedlings

The annual growth of reserved bed seedlings can be divided into three phases: Early growth phase, fast-growing phase, and late growth phase. The biggest difference for this type of seedling compared with 1-year-old seedlings is that there has been no emergence phase, and it shows the characteristics of prophase growth type and full-term growth type.

(1) Early Growth Phase

The early growth phase is from the expansion of winter buds to the rapid increase of height growth. The seedling height grows slowly, but the root grows quickly. Duration of the early growth period of spring-growth-type seedlings is very short, about 2 to 3 weeks, whereas that of the full-term-growth-type seedlings lasts for 1 to 2 months.

(2) Fast-growing Phase

The fast-growing phase starts from the time the height growth of seedlings increases by a large scale and ends at the time the height growth of seedlings significantly decreases for the full-term-growth-type seedlings; or, after the peak of diameter growth for the spring-growth-type seedlings. This phase is the period in which the growth of both aboveground parts and roots accounts for the largest proportion of the annual growth. However, the height growth period between the two growth type seedlings differs greatly. For the spring-growth type, the fast-growing phase of seedling height growth ends in May or June. The duration is generally 3 to 6 weeks for northern species and 1 to 2 months for southern species. Height growth of spring-growth-type seedlings accounted for more than 90% of the whole year. Seedling height growth will stop soon after a significant decrease in the height growth rate. From then on, other events happen, such as the growth of leaves, the expansion of leaf area, the increase of leaf quantity, the gradual hardening of new shoots, and the emergence of winter buds in summer. After the height growth stops, the diameter and root system continue to grow. The vigorous growth period (peak) is approximately 1 to 2 months after the height growth stopped. For the full-term growth type, the fast-growing phase of seedlings ends from August to early

September in the north of China and from September to October in the south of China. Its duration is 1.5 to 2.5 months for northern species and 3 to 4 months for southern species. Two growth peaks of height growth occur in the fast-growing phase, and three growth peaks may appear in several cases.

(3) Hardening Phase (or Late Growth Phase) of Seedlings

The hardening phase of seedlings starts from the time the height growth decreases greatly (that is, the hardening phase of spring-growth-type seedlings begins after the fast-growing peak of diameter growth) and ends at the time when both the seedling diameter and the root growth stops. The growth characteristics of the two growth types also differ. The height growth of spring-growth-type seedlings ends at the early stage of the fast-growing phase and the terminal buds have formed. In the hardening phase, only the diameter and root system grow with greater growth amount. The height growth of the full-term-growth-type seedlings has a short growth period in the hardening phase, and then the terminal bud appears; the diameter and the root system have a small growth peak in the lignification stage, but the growth amount is not great.

6.2.4 Annual Growth Phase of Transplanted Seedlings

The annual growth of transplanted seedlings is generally divided into survival phase, early growth phase, fast-growing phase, and seedling hardening phase (late growth phase). The greatest difference with 1-year-old seedlings and reserved bed seedlings is a survival phase (recovery phase). Attention should be paid to ensure survival of seedlings, and other phases are the same as those of reserved bed seedlings.

(1) Survival Phase

The survival phase starts from the time of transplanting to the time when the aboveground parts begin to grow and the root system in the underground recovers its absorption function. The root system of seedlings has been cut off, and part of the fibrous roots that absorb water and nutrients are cut off, which reduces the ability of seedlings to absorb water and inorganic nutrients. Therefore, seedlings need this recovery phase after transplantation. Because of the increase of the planting spacing, the light condition has been improved and the vegetative area has been enlarged, allowing the uncut roots to recover their functions quickly. The cut roots form a callus on the cut surface, and many new roots germinate from the callus and its vicinity; as a result, the diameter growth of transplanted seedlings increases. The duration of the survival phase is 10 to 30 days.

(2) Early Growth Phase

Early growth phase starts at the time the aboveground parts start to grow, and the new roots of underground parts grow until the time the height growth of seedlings has increased greatly. The aboveground parts do not grow quickly until the later phase. The root system continues to grow, and new roots are produced from the callus of roots. The height growth performance of the two growth-type seedlings is the same as that of the reserved bed seedlings.

(3) Fast-growing Phase

The beginning and ending time of the fast-growing phase is the same as that of reserved bed seedlings, but the emergence time is later. The growth characteristics of aboveground and underground parts are the same as that of reserved bed seedlings. A growth-suspending phenomenon takes place for the full-term-growth-type seedlings in the fast-growing phase, while the transplanted seedlings sometimes appear later than the reserved bed seedlings.

(4) Hardening Phase of Seedlings

The start and end of the hardening phase and the growth characteristics of transplanted seedlings are similar to that of reserved bed seedlings (see above).

6.2.5 Annual Growth Phase of Cutting Seedlings

The annual growth cycle of cutting seedlings can be divided into four phases: Survival phase, seedling phase (early growth phase), fast-growing phase, and seedling hardening phase.

(1) Survival Phase

The survival phase of deciduous tree species starts from the time the cuttings are inserted into the soil to the time when the lower part of the cuttings grows roots, the upper part produces leaves, and the new seedlings can produce nutrients independently. For evergreen tree species, this phase is from the time the cuttings are inserted into the soil to the time when the cuttings produce adventitious roots. In the survival phase, no roots are on the cuttings and no leaves have grown for the deciduous species, and the main source of nutrients is stored in the cuttings themselves. Besides the original water content of cuttings, water is absorbed from the soil (matrix) through xylem ducts at the lower cut of the cuttings. The duration of the survival phase varies greatly among tree species. It takes 2 to 8 weeks for fast-rooting species; for example, 2 to 4 weeks for willows, Chinese tamarisk (*Tamarix chinensis*), and poplars (*Populus cathayana* and *Populus nigra*) and 5 to 7 weeks for Chinese white poplar (*Populus tomentosa*) and Korean boxwood (*Buxus sinica*). It takes 3 to 6 months or even 1 year for slow-rooting coniferous trees; for example, 3 to 3.5 months for dawn redwood (*Metasequoia glyptostroboides*) and 3.5 to 5 months for Himalayan cedar (*Cedrus deodara*). Softwood cuttings also root at the callus, which is faster than that of dormant cuttings, resulting in shorter duration of the survival phase. For example, it takes 3 to 6 weeks for *Metasequoia glyptostroboides* and 7 to 9 weeks for *Cedrus deodara*.

(2) Seedling Phase

The cuttings of deciduous tree species produce young stems on the aboveground parts, so it is called a seedling phase. It starts from the time when adventitious roots grow in the underground part of the cuttings and leaves sprout at the upper end, and it ends at the time when the height growth significantly increases. Although evergreen tree species have the aboveground part, they grow slowly, which is called the early growth stage. This stage starts from the time the underground part has produced adventitious roots and the aboveground part begins to grow, and it ends at the time when the height growth increases greatly. Cutting seedlings perform the growth characteristics of the two growth types in the year of striking. The duration of seedling stage or early growth stage is about 2 weeks for spring-growth type and 1 to 2 months for the full-term-growth type. During this stage, the seedlings by cutting have developed adventitious roots, which can absorb water and organic nutrients from the soil. The leaves of the aboveground parts can produce carbohydrates, therefore the roots grow quickly in the early stage, and the number and length of roots increase rapidly, while the aboveground parts grow slowly. In the later stage, the aboveground parts accelerate growth and move into the fast-growing phase.

(3) Fast-growing Phase and Hardening Phase

The start and end times and growth characteristics of cutting seedlings in the fast-growing and hardening phases are the same as those for reserved bed seedlings.

Compared with cutting seedlings, a graft union process is used for the grafted seedlings, which is equivalent to the survival stage of cutting seedlings. Other phases are basically the same as those of cutting seedlings. The annual growth process of the burying cuttings is basically the same as that of the cutting seedlings.

6.2.6 Annual Growth Phase of Container Seedlings

Container seedlings grow quickly with strong controllability under the optimized environment of artificial control in most cases, so their growth is generally divided into three basic phases.

(1) Emergence Phase (Early Growth Phase)

For the sowing seedlings, this phase starts at the time of sowing, then seed germination, until the growth of true leaves. For the cutting seedlings, this phase is from striking the cuttings into the container to the time when the cuttings take root and the stems start to grow. It can be divided into germination stage and early growth stage. This stage mainly ensures that the seeds germinate into seedlings or the cuttings take root.

(2) Fast-growing Phase

The fast-growing phase starts when the seedling height begins to grow at an exponential or faster rate until the seedling reaches a predetermined height. Spring-growth-type seedlings reach the required height when the terminal bud is formed (measures need to be taken to avoid the formation of a terminal bud when it does not reach the predetermined height). For the full-term-growth-type seedlings, the terminal bud does not form, and so seedling growth cannot end automatically. The end of this period needs to be artificially controlled by observing that the seedling has reached the desired height and by stopping the height growth promotion measures.

(3) Hardening Phase

The hardening phase starts from the formation of terminal buds or when seedlings reach a predetermined height and continues until the seedling enters into dormancy. At this phase, the energy of height growth is transferred to the thickening growth and root growth of seedlings to ensure that the diameter of seedlings also reaches the required thickness. The lateral buds have formed and the roots continue to grow, and the two physiological processes of dormancy induction and stress adaptation are completed.

6.2.7 Growth Phase of Tissue Culture Seedlings

At present, some broad-leaved trees use tissue culture micropropagation technology to cultivate seedlings, such as poplar, eucalyptus, clove, rhododendron, white birch, and other special varieties. The whole process of tree micropropagation from inoculating explants on growth medium to growing into complete plants with roots can be divided into five phases: The establishment phase of stable aseptic culture system; the proliferation, growth, and strengthening phase of stable culture system; the rooting induction of stem buds to form seedlings; the transplanting and domestication of rooting seedlings; and the cultivation of commercial seedlings. The first three phases refer to the establishment period of plantlets under conditions of complete artificial control; the fourth phase is equivalent to the survival period of transplanted seedlings or cutting seedlings, which is mainly promoted by air humidity control; and the fifth phase is similar to the cultivation of reserved bed seedlings or container seedlings, which can be divided by reference to the corresponding stages, and corresponding cultivation measures can be taken.

6.2.8 Annual Growth Phase of Large-size Seedlings

Large-size seedlings refer to seedlings that have been transplanted several times. The growth phase was divided according to the transplanted seedlings in the year of transplantation, and it was divided according to the reserved bed seedlings in other years.

6.3 Factors Affecting Seedling Growth

6.3.1 Environmental Factors Affecting Seedling Growth

The ecological factors of a nursery comprise abiotic and biotic factors. Abiotic factors include atmosphere (above ground) and soil (below ground). Atmospheric and soil factors are related to each other, and they have a comprehensive impact on the growth and development of seedlings together with biotic factors, such as disease, insects, and fungi.

The atmospheric factors affecting seedling cultivation mainly include temperature, humidity (moisture), light, and carbon dioxide. Carbon dioxide, water vapor, and light participate in photosynthesis, respiration, and transpiration. These physiological processes are related to leaf temperature and stomatal function, which in turn are related to carbon dioxide concentration, light intensity, humidity, and temperature. In addition, seedling growth is related to the amount of clouds and their shape, wind direction and speed, solar radiation, and precipitation (Chi and Zhou, 1991), but these are indirect factors, which work by having an effect on air temperature, humidity, light, and carbon dioxide concentration.

Soil factors affecting seedling cultivation include, for example, soil temperature, soil moisture, soil air, soil texture, soil structure, soil mineral nutrition, soil organic matter, soil pH, soil thermal properties, and soil toxicological properties. Seedlings absorb and utilize soil nutrients and water, of which the amount and availability are directly or indirectly affected by the above soil factors.

(1) Temperature

Temperature plays a direct role in plant metabolism, and each reaction requires a specific temperature range. Temperature also affects and controls other processes of growth, such as transpiration, respiration, and photosynthesis. The suitable temperature range can be divided into three basic points: The lowest temperature, the most suitable temperature, and the highest temperature, and seedlings grow best under the most suitable temperature. The temperate tree species begin to grow when the temperature ranges from 0 to 10 ℃, but 18 to 30 ℃ is the most suitable temperature. They will grow very slowly below 15 ℃, and the growth is limited above 30 ℃. Temperature should be correspondingly raised for tropical tree species. The normal growth and development of seedlings requires soil (substrate) temperature of 18 to 20 ℃ and approximately 25 ℃ air temperature at the stems and leaves between layers of seedlings, thus the surface temperature of the seedbed (substrate) should always be maintained at 30 to 35 ℃, but not higher than 40 to 46 ℃ or lower than −2 ℃ (which can be particularly damaging if it lasts for more than 2 hours) (Chi and Zhou, 1991).

Seed germination also has the most suitable lowest and highest temperature, which is directly related to the characteristics of tree species. Generally, tree seeds can germinate at 5 to 8 ℃, faster at 10 to 15 ℃, and most vigorously at 20 to 25 ℃. The optimum temperature for germination is 15 ℃ for sawtooth oak (*Quercus acutissima*), 20 to 25℃ for Chinese pine (*Pinus tabuliformis*) and Oriental arborvitae (*Platycladus orientalis*), 25 ℃ for Scots pine (*Pinus massoniana*) and Chinese fir (*Cunninghamia lanceolata*), and 30 ℃ for tree-of-heaven (*Ailanthus altissima*). The germination rate of *Pinus tabuliformis* at 35 ℃ (although germinating faster) is lower than that at 25 ℃, and that of *Cunninghamia lanceolata* at 36 ℃ was still as high as 86%. Korean pine (*Pinus koraiensis*) can germinate successfully only when the air temperature reaches 15 ℃ (8-13 ℃, at 5 cm underground) after accelerating germination treatment with the low temperature of 0 to 5 ℃ and the high temperature of 15 to 20 ℃ (Qi, 1992). Manchurian ash (*Fraxinus mandshurica*) can germinate after 90 to 120 days at 15 to 20 ℃ and 90 to 120 days at 3 to 5 ℃ (Zhang et al., 2007).

The air temperature and ground temperature required for rooting cuttings differ from those just given in the previous paragraph. Poplar, willow, and other deciduous broad-leaved tree species can take root at lower ground temperatures, but the most suitable ground temperature for most tree species is 15 to 20 ℃. While some evergreen broad-leaved tree species need higher ground temperature for rooting, generally 23 to 25 ℃ is ideal. Cuttings of *Populus tomentosa* can form the callus and root primordium in a wide range of temperatures (4-30 ℃), whereas cuttings of white poplar (*Populus alba*) can ensure a higher rooting rate only when the soil temperature of the cutting bed is stable at 12 ℃. The highest rooting rate is found at 20 ℃/10 ℃ in day/night temperatures for Monterey pine (*Pinus radiata*) cuttings.

The stem of a plant starts to metabolize above 0 ℃ (when the average temperature is stable at 1-2 ℃), while the root system usually begins to metabolize under a soil temperature greater than 5 ℃. Temperate tree species, such as *Pinus koraiensis*, must be vernalized for a period of time (below 0 ℃ or even −10 ℃, for more than 15 days) to release dormancy and germinate.

The optimum temperature differs depending on the stage of seedling development. A moderate and stable temperature pattern is required for a given seedling stage. Higher temperature is required in the fast-growing stage, but the temperature near the seedling layer should not exceed 30 ℃. The optimal temperature in the lignification stage should maintain for 4 to 6 weeks to promote diameter and root system growth, and a temperature slightly higher than 0 ℃ for 4 to 6 weeks (especially at night) will promote bud formation and improve cold resistance.

A temperature too high or too low is not conducive to the growth of seedlings. High temperature will lead to sunburn hazard, which readily occurs in late spring, early summer, and winter. Low temperature can result in frost, frost heaving, and so forth. Frost often occurs at the beginning or near the end of seedling growth in spring and autumn.

(2) Moisture Condition

Water condition is controlled by air humidity and soil moisture.

Air Humidity. Appropriate air humidity, which usually refers to the relative humidity, can promote the growth of seedlings. Experience has pointed out that high humidity (60%-90%) is needed in seedling emergence stage (early growth stage); moderate humidity (50%-80%) is needed in fast-growing stage, and ventilation should be better for canopy closure; and low humidity is required in lignification stage. In general, 60% to 80% air humidity can ensure normal growth of seedlings in the growing season. Different tree species are affected by humidity in varying degrees. Mongolian Scots pine (*Pinus sylvestris* var. *mongolica*) is not sensitive to air humidity, while Gmelin larch (*Larix gmelinii*), *Pinus koraiensis*, and Amur cork tree (*Phellodendron amurense*) have a strict requirement on humidity; specifically, when the air humidity is less than 39%, the leaves will stop growing, and when the air humidity is less than 20%, physiological water loss, stem and leaf wilting, or even dry death may occur (Chi and Zhou, 1991).

In vegetative propagation, humidity control is more important. High humidity (90%-100%) is generally required at the initial stage, because cuttings need to reduce transpiration to maintain swelling pressure to form roots, and grafting needs a high humidity environment to avoid water stress. In recent years, cutting practice at home and abroad has proved that the air humidity near the seedling layer is more important than that of soil (substrate). During the rooting period after cutting, the 60% to 80% soil moisture content (the ratio of the difference between wet soil weight and dry soil weight to dry soil weight) and the 90% relative air humidity of saturation, is more conducive to rooting and seedling survival, while the humidity of soil (substrate) should not be too high.

Soil Moisture. Reaching the appropriate water content is necessary for tree seed germination; for example, the relative water content of *Pinus koraiensis* for seed germination should be 50% to 60%; the relative water content (the ratio of the difference between fresh weight and dry weight of seedlings to fresh weight of seedlings) can reach more than 90% when the seedlings emerge, and after the seedlings enter the fast-growing stage, the relative water content is between 70% and 80%. The aboveground part of seedlings stops growing in autumn with a relative water content of 65% to 75% (Jin, 1985). The water supply of seeds and seedlings is mainly from soil; therefore, if the soil water supply is insufficient, the water balance of seeds and/or seedlings will be lost, which will seriously affect the germination and growth.

The effect of soil moisture and air humidity on seedling growth is sometimes more significant than that of soil mineral nutrition. Even between wilting water content and field water capacity, the effect of soil moisture on seedling growth is not equivalent. The effect of soil water deficiency on seedling height growth is more obvious than the effect on root collar diameter growth in the seedling growth period. During the fast-growing period of seedlings, leaf node spacing will shorten, the height growth of seedlings will stop, and apical buds will form when soil moisture is insufficient. Water has a great influence on seedling morphology. Under poor water conditions, the ratio of leaf thickness to leaf area of seedlings increases significantly, the leaf color becomes yellow, the reflectivity decreases, and the luster is lacking. By comparison, excessive water content also changes the normal morphology of seedlings, such as thin leaves and small root systems (Jin, 1985).

Water plays an important role in ecological environment regulation of the nursery. When the surface temperature of the seedbed is too high in summer, a small amount of irrigation can be adopted to reduce the temperature and avoid sunburn. When the seedlings are damaged by frost in spring and autumn, irrigation can save the seedlings. Appropriate irrigation is also beneficial for seedlings when they are damaged by pesticide, wind, and hail—events that hinder the growth and development of the seedlings (Jin, 1985).

(3) Soil Texture

Soil texture affects water storage, water supply, fertilizer conservation, fertilizer supply, heat conduction, temperature conductivity, and cultivability of the soil (Table 6-1), and thus plays an important role in the growth and development of seedlings.

The height of a seedbed or ridge should be lower on sandy soil, no matter whether plantings are in a ridge culture or bed culture. A large amount of fertilizer should be applied, mainly organic fertilizer, but it should be applied according to the principle of "less amount, more times" when applying chemical fertilizer. Sandy soil has a lower heat capacity, with soil temperature increasing rapidly in spring, commonly known as "hot soil", in which sowing and cutting emergence is 1 to 2 days faster than that of clay soil. For ridge culture or bed culture in clay, the height of the bed or ridge should be higher as compared to sandy soil. Clay has a high heat capacity, but the soil temperature increases slowly in spring, so it is commonly known as "cold soil". Generally, when sowing or striking cuttings, the seedling emergence is slower because of the slow temperature increase. When fertilizing clay soil, the characteristics of poor soil aeration and weak aerobic microbial activity must be considered to avoid too deep fertilization, which would delay fertilizer efficiency. In a growing season, the seedlings grow better on sandy soil in the early stage but then grow better on clay in the later stages (Jin, 1985). Loam provides an optimal soil environment for seedling cultivation. Loam has an ideal number of macropores and a considerable number of small voids, good ventilation and water permeability, strong water and fertilizer retention, good soil thermal condition, good tillage, and a longer suitable cultivation period.

Table 6-1 Ecological properties of different soil texture types

Ecological properties	Sandy	Loam	Clay
Water retention capacity	Weak	Medium	Strong
Fertilizer retention capacity	Weak	Medium	Strong
Warming speed	Fast	Medium	Slow
Soil aeration	Good	Medium	Poor
Tillage	Good	Medium	Poor
Nutrient conversion speed	Fast	Medium	Slow
Emergence difficulty	Easy	Easy	Difficult
Unfavorable delayed dormancy	No	Controlled	Yes

Source: Adapted from Jin (1985).

(4) Soil Mineral Nutrition

The nitrogen, phosphorus, potassium, calcium, magnesium, iron, sulfur, and microelements that seedlings need are all absorbed from the soil by their root system. The total number of mineral elements in the soil can generally meet the needs of seedlings, but the available nutrient content is not always sufficient; for example, the available phosphorus in the northern nursery soil is generally lacking.

Nitrate nitrogen is the main quick-acting nitrogen in the nursery soil, and it is greatly affected by soil temperature. The content of nitrate nitrogen in the nursery soil is high in summer but low in spring. The nitrogen absorbed by seedlings is mainly NH_4^+ and NO_3^-, followed by a small amount of NO_2^- and organic ammonia (protein nitrogen and humus nitrogen). The amount of nitrogen absorption is closely related to seedling growth, morphogenesis, and seedling resistance (disease resistance, cold resistance, and drought resistance, for example). Across tree species, the response to and amount of requirements differ in terms of the forms of nitrogen. In the seedling stage, the tree species requiring low nitrogen are *Pinus koraiensis*, *Fraxinus mandshurica*, Mongolian oak (*Quercus mongolica*), and Manchurian walnut (*Juglans mandshurica*), for example. Tree species requiring high nitrogen include *Larix gmelinii*, Olgan larch (*Larix olgensis*), *Phellodendron amurense*, Korean spruce (*Picea koraiensis*), and *Pinus sylvestris* var. *mongolica*, for example. And examples of those requiring a medium level of nitrogen include Chinese cottonwood (*Populus simonii*), Amur linden (*Tilia amurensis*), Japanese white birch (*Betula platyphylla*), and Manchuria fir (*Abies holophylla*) (Jin, 1985).

Phosphateform (PO_4^{3-}, HPO_4^{2-}) is the main phosphorus available for root system absorption. The amount of phosphorus uptake is closely related to the growth and development and the resistance of seedlings. The performance of seedlings with insufficient phosphorus uptake is similar to that of insufficient nitrogen uptake in many aspects, such as stunted plants, poor root development, shrunken leaves, and the leaf color presenting as almost purple. The response of tree species to phosphorus also differs across species. 1-year-old seedlings of *Larix olgensis* and *Pinus sylvestris* var. *mongolica* need more phosphorus than those of *Pinus koraiensis* and *Fraxinus mandshurica*.

Potassium is taken up through the roots from the soil in the form of K^+ and exists in inorganic form. Insufficient potassium absorption will reduce growth, affect lignification, and reduce disease resistance and cold resistance of seedlings. Sulfur, calcium, magnesium, iron, and microelements have their own independent role on seedling growth. Sodium is harmful to seedlings, but adding calcium with 5% sodium content to the soil, that is, maintaining a calcium: Sodium ratio equal to 5 : 95, can eliminate the harm of sodium ions. Calcium and hydrogen ions can eliminate high levels of poisonous effects caused by either one alone (Jin, 1985). Seedlings of different tree species are required to maintain a certain calcium and magnesium ion ratio, so as to avoid the antagonistic effects between calcium and potassium, and iron and aluminum ions.

(5) Soil Acidity and Alkalinity

The pH value of soil is the indicator of soil acidity and alkalinity, which affects the growth and development of seedlings primarily by influencing the availability of soil nutrients and soil microbial activities. If the soil pH value is too high or too low, the balance of mineral nutrients will be broken down, and some substances will dissolve, which creates an excess and causes poisoning of seedlings. When the pH value is too low, aluminum, iron, manganese, zinc, and copper in the soil may lead to toxicity to seedlings, and the hydrogen ion (H^+) itself can directly damage seedlings. When the pH value is too high, the hydroxyl ion (OH^-) will also directly damage the seedlings. In the acidic soil, excessive manganese ions will have a toxic effect, which can be relieved by calcium. The toxicity resulting from too much ammonium manganese in the soil can be alleviated by the application of calcium. The application of calcium alone to acid soil can help to absorb ammonium nitrogen.

The adaptability of seedlings to pH value varies greatly, and most tree species grow well under neutral and slightly acidic conditions. Generally, the pH should be 5.0 to 7.5 for conifers and 6.0 to 8.0 for broad-leaved trees. South (2017) found that the pH value of 4.5 to 5.0 seems to be the best when cultivating pine seedlings on sandy soil, which can effectively reduce the seedling loss caused by damping-off disease (Figure 6-2), reduce the number of nematode populations, increase the nitrogen level (reducing the need to apply nitrogen fertilizer), reduce the number of weeds in the nursery, increase the number of beneficial trichoderma (fungi), and increase the biomass of seedlings.

(6) Soil Organic Matter

Soil organic matter can provide nutrients for seedling growth and development, enhance soil water and fertilizer conservation capacity and buffering capacity, improve soil physical properties, and promote soil microbial activities. In the process of growing seedlings around the world, the key consideration of land use, land cultivation, and land protection is to increase the content of soil organic matter. It is very important to keep high soil organic matter content in the nursery for improving ground temperature, maintaining pliable soil structure, and regulating soil water and fertilizer supply abilities. The application amount of organic fertilizer should not be too much at one time. Applying 45 000 kg of peat per hectare is an ideal amount according to the soil-improving experience provided by the Lesser Khingan Mountains forest nursery. Results show that the soil is

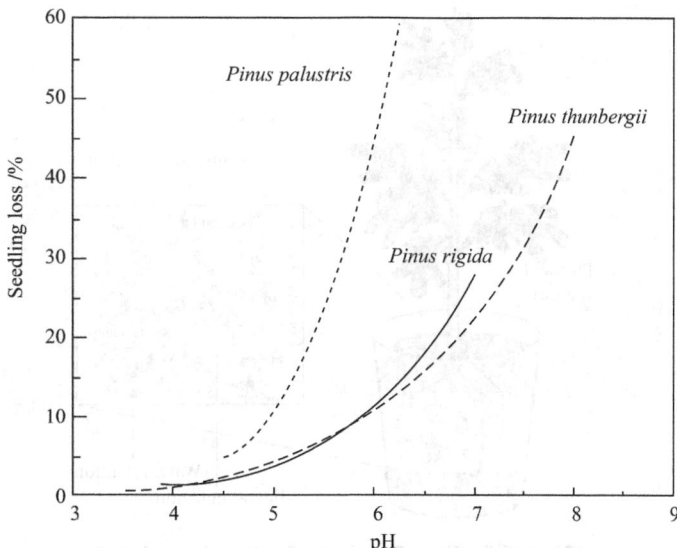

Figure 6-2   Low-level soil pH (pH 4.5-5.0) can effectively reduce the damping-off disease of pine seedlings (Adapted from South, 2017).

loose with more voids, and the small aggregates of 0.005 to 0.1 mm increase after one year's application of organic materials. The macro aggregates (0.5-1.0 mm) form after two years, and the aggregates of 1 to 2 mm appear after three years. The formation of aggregates enhances the functions of water and fertilizer conservation and soil water, fertilizer, gas, and heat regulation, resulting in favorable conditions for seedling growth.

(7) Artificial Substrate

Cultivation substrate, also known as artificial soil, pot soil, pot mixture, soil mixture, and mixed fertilizer, is a type of artificial substrate for seedling cultivation. Artificial substrate is generally used for container seedling, which requires water, air, mineral nutrition, and physical support conditions for the plants growing in the container (Figure 6-3).

The substrate should be equipped with the characteristics of slight acidity, high CEC, low self-fertility, reasonable porosity, no pests, reasonable price, sufficient and stable source, high homogeneity and volume stability, durable storage, and easy wetting (Landis, 1998). Special attention should be paid to the regulation of the substrate pH value, which is generally kept in the range of 5.5 to 6.5, but it depends on the specific tree species. For example, red pine (*Pinus resinosa*) grows best at pH 5.0 to 5.3, grows normally at 4.5 to 6.0, and grows poorly above 6.0.

(8) Light Conditions

Light condition is the most complex, changeable, and important factor affecting plant growth and development, taking into account light intensity, length of light time, and light quality.

As far as a leaf of a seedling is concerned, the optimal illumination for photosynthesis is much lower than that of full sunlight. In terms of a seedling with luxuriant branches and leaves, however, even under full sunlight many leaves will not receive enough illumination for maximum photosynthesis because they are covered by the upper and lower leaves. For the seedlings cultivated

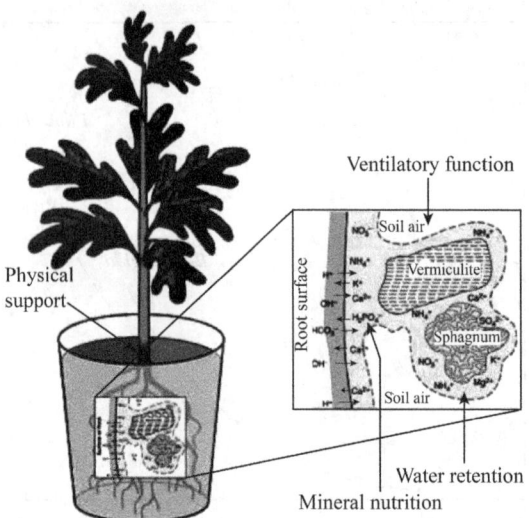

Figure 6-3 Four functions of cultivation substrate (Adapted from Dumroese et al., 2022).

in the field, the so-called optimum light intensity for photosynthesis means that in a certain period of time, the net effect of light in a certain comprehensive situation is more favorable for photosynthesis than that in another comprehensive situation. The shade species such as Yeddo spruce (*Picea jezoensis*), Khingan fir (*Abies nephrolepis*), and Manchurian fir (*Abies holophylla*) grow better under moderate shade (the needles are dark green, with a high leaf number and seedling height growth) than do those under full light. When the seedling density reaches the seedbed closing, a certain seedling height should be met. Shade-tolerant species present an arch shape, with seedlings on two sides of the bed being shorter and the ones in the center being taller. Shade-intolerant species present a V shape, with seedlings on both sides of the bed being taller and the ones in the center being shorter. The increase and decrease of light affect the transpiration of seedlings, the effectiveness of potassium fertilizer on seedlings, and the lignification degree of stems (strong light promotes the lignification of stems). When the seedlings newly emerge, sufficient sunlight can improve the resistance of seedlings to adverse situations.

Plants produce more carbohydrates under red light, whereas under blue and violet light, more protein and fat would be synthesized. UV light with shorter wavelengths than 280 nm has a strong destructive effect on seedlings, and UV light or blue violet light with slightly longer wavelengths can inhibit the elongation of seedlings, resulting in short and thick seedlings, and induce them to phototaxis. Wavelengths of 650 nm can affect the stem elongation of seedlings and seed germination; however, infrared rays can improve the temperature and transpiration of seedlings. The research of tree physiology has confirmed that far red light can promote the growth of internodes between leaves; under dark or monochromatic blue light, the germination rate of Faber's fir (*Abies fabri*) and Chinese lime (*Tilia tuan*) seeds is not as high as those under red-orange light; and the germination rate of Chinese spruce dragon (*Picea asperata*), Scotch pine (*Pinus sylvestris*), *Larix gmelinii*, and

Betula platyphylla seeds has no obvious difference under dark or monochromatic blue light and red-orange light.

The height growth time of southern tree species is not related to sunshineduration, while that of northern tree species is related. In the process of tree seed germination, temperature and light interact. Results have shown that for unfrozen European birch seed, the germination situation under long days was better than that under short days under the germination temperature of 15 ℃, but seed germination under both long days and short days was better when the temperature was 20 ℃.

Light is not always beneficial to seedlings, and strong light can cause sunburn damage. If the newly emerged seedlings from spring sowing are exposed to sunlight the morning after a frost, they often suffer serious losses, but the damage can be reduced if they are shaded in time.

At the early stage of growth (the construction period), seed germination was not affected by light intensity, but it was significantly affected by light quality and light time. For example, red light promotes seed germination of loblolly pine, whereas far-infrared light inhibits germination. Durations of 8 to 12 hours of light per day is beneficial to seed germination of most tree species, while 16 hours of light is better than 12 hours for Douglas-fir. The cotyledons of germinated coniferous seedlings begin to photosynthesize, and the development of true leaves depends on the photosynthetic products of cotyledons. Light intensity reaching 55 $\mu mol/(s \cdot m^2)$ (3000 lx) could meet the seedling growth requirements of most tree species, although that of red pine required 120 $\mu mol/(s \cdot m^2)$ (6500 lx), and that of southern pine could not be shaded. Germination needs strong light; and light, temperature, and humidity, as well as mulch, comprehensively influence germination, so attention must be given to adjust these factors to the optimal situation. During the fast-growing period, the light should be gradually adjusted to near the light saturation point. Shading may not be carried out unless it is very necessary, and it should be adjusted every day. When the seedling height reaches 80% to 90% of the requirement, the light time can be reduced to promote the formation of terminal buds. In the later growth stage (lignification stage), the first step is to stop the height growth to allow the terminal buds to fully form, to stimulate the growth of diameter and root, and to promote the gradual lignification. Therefore, reducing the light duration is the first step during this time.

Light is very important for cutting seedlings. Formation of new roots of cuttings depends on the nutrients contained in the cuttings, and on the nutrients and plant hormones formed in the process of assimilation. Results show that light can improve the ground temperature and promote rooting for hardwood cutting. For softwood cutting, the suitable light intensity is conducive to photosynthesis, making nutrients, maintaining the nutrient balance of cuttings, and promoting rooting. However, when the water cannot be replenished in time under excessive light intensity, the cuttings lose water, so it is necessary to shade properly. The number of leaves of cuttings often plays a decisive role in the rooting condition for softwood cutting.

(9) Carbon Dioxide

The basic elements that make up the seedlings, including carbon, oxygen, and hydrogen, are abundant in the ecological environment of a nursery. Carbon comes from carbon dioxide in the air and soil. Carbon dioxide content in soil is high, as much as tens of times higher than that in the air. Carbon dioxide is released through soil respiration and absorbed by seedling leaves. A close relationship exists between soil carbon dioxide emission and soil temperature. Carbon dioxide is higher in the layer of air that is found within 1 to 2 cm near the ground. The content of carbon dioxide in the air is low and uniform in the whole height of the seedling leaf layer, but higher if the seedbed has a high leaf area index and leaf closure. Carbon dioxide content is very low in the air of a greenhouse. Researchers report that the ideal carbon dioxide concentration for plant photosynthesis is approximately 1% (1000 mg/kg), and the carbon dioxide concentration is 280 to 350 mg/kg in atmospheric environment, 200 to 400 mg/kg in the nursery, and a much lower carbon dioxide concentration in greenhouse, but other closed seedling raising facilities are the limiting factors. Improving the concentration of carbon dioxide in the air can promote the formation of callus as soon as possible and increase the rooting rate of cuttings. For raising seedlings in the field, ventilation should be strengthened, and dry ice can be applied to increase the carbon dioxide concentration when it is necessary. Ventilation, dry ice application, and carbon fuel combustion are mainly used to increase carbon dioxide concentration in a greenhouse.

6.3.2 Biological Factors Affecting Seedling Growth

In addition to temperature and other environmental factors, seedling growth is affected by surrounding biological factors (weeds, diseases, insects, mycorrhizae, and other organisms). The biological environmental factors of seedlings include both beneficial and harmful elements. Beneficial organisms refer to mycorrhizal fungi and rhizobia, while harmful organisms refer to pathogens, pests, and weeds. Birds, mice, rabbits, deer, and so forth can sometimes cause damage to seedlings, but it is not common.

(1) Mycorrhizal Fungi

Plant rhizosphere microorganisms and higher plant absorbing roots often form a combination of parasitic, symbiotic, and saprophytic relationships. The combination of parasitic bacteria and roots is beneficial only to microorganisms, but they will cause varying degrees of harm to the host plants. Saprophytic form is that the rhizosphere microorganisms obtain the required nutrients from the dead plant roots. Symbiotic microorganisms do not cause damage to plant tissues, and both are mutually beneficial, so they are called beneficial microorganisms. Mycorrhizal fungi are one of the most direct and influential species in seedling cultivation. Three main types of mycorrhizae occur: Ectomycorrhiza, endomycorrhiza, and ectendomycorrhiza, along with other types, such as mixed mycorrhizae and pseudo mycorrhizae.

The effect of mycorrhizae on seedlings is to enhance the absorption of water and nutrients, especially phosphorus and nitrogen. In addition, it can enhance plant stress resistance and immunity, improve nursery soil, and produce growth hormone, so as to ultimately promote seedling growth and improve seedling quality, nursery productivity, and afforestation survival rate. Mycorrhizae can effectively expand the absorption area of host plant roots. Song et al. (2005) used arbuscular mycorrhizal (AM) fungi to artificially inoculate Ussuri poplar (*Populus ussuriensis*) seedlings, which showed that the main root length, root collar diameter, lateral root number, and root biomass of mycorrhizal seedlings were significantly different from those of control seedlings. They also showed that mycorrhizal fungi increased the root volume and total absorption area of seedlings, especially the active absorption area of seedling roots. With mycorrhizae, seedlings can absorb water through numerous slender hyphae and continuously supply themselves, which can improve soil water use efficiency especially under drought conditions. All the absorbing roots of well-developed mycorrhizal roots form mycorrhizae, which can increase the speed of absorbing and transporting water by 10 times (Hua, 1999). Mycorrhizal fungi can utilize both organic and inorganic nitrogen. In the process of nitrogen metabolism, mycorrhizal fungi can absorb nitrogen from soil and transport it to the roots of plants, which can transfer organic nitrogen into inorganic nitrogen that can be used by plants. Mycorrhizal fungi can effectively promote the absorption and utilization of phosphorus by host plants, and vesicular arbuscular (VA) mycorrhizal fungi can directly absorb phosphorus from soil to plant after transformation; ectomycorrhiza (ECM) fungi can make plants absorb phosphorus that cannot be used in the root system. Mycorrhizal fungi can absorb and store the mineral nutrients in soil, such as Zn, Cu, Mg, Fe, S, Ca, and others, and can transfer them to plants to meet their survival needs. Mycorrhizae can improve the nutrient absorption and utilization rate in seedlings by secreting a variety of enzymes, expanding the effective utilization space of soil, maintaining the active absorption rate, and reducing the critical absorption concentration.

The adaptability of mycorrhizal fungi to environmental temperature, soil pH, and the resistance to soil toxic substances is stronger than seedlings' adaptability and stress resistance. Mycorrhizae can help seedlings grow normally by enhancing their adaptability and stress resistance under adverse environmental conditions. They can effectively enhance the adaptability of seedlings to the environment. Mycorrhizae can improve the immunity of plants through biological, physical, and chemical actions, which can prevent or reduce root diseases. Mycorrhizae has a strong function of complexing metal elements, and VA bacteria can enhance the tolerance of plants to heavy metal ions, which is critical in soils that have a high content of heavy metals. Ectomycorrhiza can protect plant roots. After forming mycorrhizae, mycorrhizal fungi produce antibiotic substances, which can exclude other microorganisms in the rhizosphere. Mycorrhizal fungi can also reduce environmental pollution, and ectomycorrhizal fungi can enduringly degrade aromatic pollutants by hydroxylation.

Mycorrhizae can improve the physical and chemical properties of soil. Mycorrhizal fungi can decompose soil organic matter, accelerate soil nutrient cycling, improve soil structure, and improve the availability of soil nutrients. The enzymes produced by mycorrhizae will decompose insoluble organic matter or fixed minerals in soil into nutrients that can be absorbed and utilized by plants, so as to improve soil fertility. Mycorrhizae can also improve soil chemical properties by increasing soil organic matter content, expanding the scope of rhizosphere mucilage layer, accelerating the weathering of mineral soil, and forming a unique forest soil microhabitat. It can not only improve the soil's physical and chemical properties but also maintain the soil structure and improve the soil cultivability.

Promoting seedling growth, improving seedling quality, and increasing nursery growth potential are the comprehensive effects of the above actions. Gong et al. (2000) carried out the inoculation test of VA mycorrhizal and ECM mycorrhizal on Himalayan birch (*Betula alnoides*) seedlings, which indicated that after 180 days of ECM inoculation, the average seedling height increased by 92.98% to 106.85%, the aboveground dry weight increased by 206.43% to 554.69%, and the underground dry weight increased by 202.83% to 566.40% compared with the control treatment. Similarly, after 90 days of VA mycorrhizal fungi inoculation, the average height, aboveground dry weight, and underground dry weight of seedlings increased by 50.48% to 63.41%, 78.65% to 151.04%, and 215.25% to 311.86% respectively. Seedlings could be used in afforestation efforts after 150 to 180 days of mycorrhiza inoculation, at least 5 months ahead of the control group. Meng and Tang (2001) used ectomycorrhizal fungi to inoculate Korean aspen (*Populus davidiana*) seedlings sown in pots. Results showed that the growth rates of seedling height, root collar diameter, lateral root number, and dry matter weight of the whole plant were 38.13%, 20.27%, 70.97%, and 33.39%, respectively, after inoculating with the *Cortinarius russus* mycorrhizal fungi. Hua et al. (1995) found that the typical ectomycorrhizal fungus *Pisolithus tinctorius* (Pt) had obvious mycorrhization effects, promoting seedling growth and increasing biological yield. The mycorrhizal rate was 100%, and the yield of qualified seedlings was increased by more than 14.6%. The average seedling height, root collar diameter, dry matter weight, lateral root number, and total root length increased by 28.1% to 71.4%, 22.8% to 49.2%, 66.7% to 457.1%, 128.0% to 200.0%, and 82.4% to 101.0%, respectively.

(2) Rhizobia

The application of rhizobia in agriculture, horticulture, forage, and other aspects of research is very in-depth, but it is not often used in the seedling cultivation of woody plants. In recent years, a small amount of research has focused on *Acacia* species (commonly known as acacia, mimosa, thorntree, and wattle).

Lv et al. (2003) collected and isolated 14 strains of rhizobia from six *Acacia* forests under a variety of site conditions and inoculated them into *Acacia crassicarpa* seedlings. Compared with the control, the height growth of *A. crassicarpa* seedlings inoculated with most strains increased by 4.5% to 18.6%, root collar diameter growth by 2.5% to 46.8%, nitrogenase activity by 7.6% to 241.8%, chlorophyll content by 11.0% to 19.1%, and the nitrate reductase activity, nitrogen content, phosphorus content, potassium content, calcium content, magnesium content, and iron content in leaves increased by 3.3% to 34.4%, 7.4% to 43.8%, 9.1% to 72.7%, 8.3%, 7.3% to 41.5%, 12.5% to 25.0%, and above 10%, respectively.

Results showed thatrhizobia inoculation could significantly promote the growth of straight-stem earleaf acacia (*Acacia auriculiformis*) seedlings. After 6 months, compared with seedlings without rhizobia inoculation, the growth of plant height, root collar diameter, total biomass, rhizobia weight, nitrogenase activity, and leaf nitrogen content increased by 1.1% to 44.8%, 6.8% to 26.2%, 10.6% to 104.3%, 18.8% to 420.8%, 28.6% to 106.1%, and 0.5% to 5.3%, respectively. At the same time, rhizobia inoculation had a significant effect on the content of total nitrogen, available phosphorus, and available potassium in soil (Zhang et al., 2005).

Kang and Li (1998) showed that the height, total biomass, and root nodule biomass of *Acacia* seedlings inoculated with different rhizobia increased by 35.38% to 160.26%, 17.85% to 238.79%, and 2.4% to 102.61%, respectively, compared with the control seedlings. Seedling height, total biomass, and root nodule biomass of various *Acacia* tree species/provenance seedlings inoculated with rhizobia increased by 2.64% to 109.82%, 1.82% to 281.48%, and 64.7% to 211.15%, respectively, compared with the control seedlings. Nitrogen content and total nitrogen content of the seedlings inoculated with rhizobia were 8.58% to 77.55% and 11.64% to 262.50% higher than those of the control.

(3) Seedling Diseases

Diseases in a nursery usually refer to infectious diseases. After the seedlings are injured, they show discoloration, necrosis, decay, wilting, deformity, and other symptoms, as well as powder, molds, and bacterial ooze, for example. The main infectious diseases are damping-off of seedlings (including seed rot, damping-off and root rot, rotten leaf and damping-off) caused by pathogens in nursery; leaf diseases (for example, rust, powdery mildew, black spot, and mosaic); branch diseases (rot and canker of poplar, willow, and locust; phytophthora of black locust [*Robinia pseudoacacia*]; shoot blight of evergreen tree); and the root diseases (root cancer, nematode, sunburn, and violet root rot).

(4) Seedling Pests

Many kinds of insects are harmful to seedlings in a nursery, and they are usually divided into leaf pests, stem pests, and underground pests. Among them, the species causing the most serious damage to the seedlings and that bring great economic losses to the nursery include underground pests (such as grubs, mole crickets, *Eucryptorrhynchus chinenis*, cutworms, wireworms, and crane flies), trunk borers (such as clearwing of *Populus tomentosa*, longicorn, and *Dioryctria splendidella*), piercing-sucking pests (such as red spider, *Platycladus orientalis* aphid, and *Pseudaulacaspis pentagona*), and defoliators (such as *Semiothisa cinerearia*, *Malacosoma neustria*, eucleid, *Clostera anachoreta*, and leaf beetle).

(5) Harmful Animals in Nursery

Ants are a great threat to seedlings by sowing the tree species with particle seeds such as poplar and willow. They can steal and eat seeds, which cause missing seedlings and ridging breaks in local plots for large-scale seedling cultivation, and they can lead to complete failure for small-scale seedling operations.

Bird damage is a great enemy of sowing seedlings. Sowing of *Pinus sylvestris* var. *mongolica*, *Picea asperata*, and *Larix gmelinii* in a forest nursery in the Lesser Khingan Mountains forest region often failed because of bird damage. It takes a lot of time and effort to rid an area of birds every year in a large, state-owned nursery sowing area.

Rats, rabbits, and deer may cause economic losses in some areas or in some years because of stealing and eating seeds and biting seedlings.

(6) Harmful Plants in Nursery

Harmful plants in a nursery refer to the weeds. Weeds are plants growing in the wrong place, that is, where they interfere with intentionally planted species; the unpopular plants; and plants that are considered to be worthless. They are the plants that interfere with people's intention for land use, the plants that are not consciously cultivated by humans, the plants with no apparent application or ornamental value (Su, 1993). Weeds that can cause serious harm in the world are shown in Table 6-2.

Environment has a profound impact on the growth and development of weeds. Among the environmental factors, changes in water and heat play an extremely important role in the growth of weeds in the field. Taking Northeast China as an example, weeds have formed the types and characteristics that adapt to different water and heat conditions after long-term natural selection. The occurrence sequence and vigorous growth periods differ, but they can be divided into the following stages (Qi, 1992): The first stage, which occurs from mid-March to mid-April, a period of winter annual weeds and most perennial weeds occurrence, and a large number of weeds in early- and mid-April; the second stage is from late April to mid-May, which is an occurrence period of plenty of 1-year-old early spring weeds, and individual perennial weeds also emerge at this time; the third stage is from mid-May to early July, which is a period of a large number of late spring weeds; the fourth stage starts from early June, which is the latest occurrence time of some 1-year-old weeds, such as *Portulaca oleracea*, *Digitaria sanguinalis*, and *Capparis spinosa* seedlings. The aboveground part of perennial weeds continues to regenerate after being eradicated, however, and early spring weeds still have an emergence period. The fifth stage is from the beginning of August to mid-September, which is the re-occurrence period of winter annual weeds and perennial weeds.

Table 6-2 Most harmful weeds around the world

English name	Latin name	Distribution (Source area)
Nut grass	*Cyperus rotundus*	Wide distribution from tropics to temperate zone (Asia)
Bermuda grass	*Cynodon dactylon*	From tropics to temperate zone (Asia, Africa)
Barnyard grass	*Echinochloa crus-galli*	Around the world (Europe)
Awnless barnyard grass	*Echinochloa colonum*	Tropics, subtropics (India)
Indian goose grass	*Eleusine indica*	Around the world except the Mediterranean coast
Johnson grass	*Sorghum halepense*	Some tropics and temperate zone, frigid zone (Mediterranean Sea, the Middle East)
Water hyacinth	*Eichhornia crassipes*	Southern Hemisphere to northern latitude of 40° (South America)
Cogon grass	*Imperata cylindrica*	Southeast Asia, Africa
Common lantana	*Lantana camara*	High-temperature zones (Asia, Africa, Central and South America)
Guinea grass or buffalo grass	*Panicum maximum*	Asia, Central and South America, Africa

Source: Adapted from Su (1993).

From the perspective of the growth rate of weeds, taking early and middle of June as a boundary, the growth rate of annual weeds is slower before June and becomes faster after the middle of June, which is related to the water and heat conditions. For example, the daily growth height of *Echinochloa crusgalli* can reach 1 to 2 cm, and the coverage and fresh weight of weeds increase sharply. The growth of seedlings in this period, however, is still in the seedling stage, when young seedlings have slow growth, low growth amount, weak resistance, and high withering rate. And seedling growth time is 15 to 20 days later than the vigorous weed growth period. This "jet lag" in vigorous growth is extremely unfavorable to seedlings. Therefore, June is the key month for weeding in the nursery.

Weeds have high adaptability (low temperature and cold tolerance, drought and flood damage resistance, barren resistance, harden land, and grass wasteland) and amazing reproductive capacity (a large number of seeds, most with vegetative propagation capacity, seed germination with a large range of adaptability). Weeds also have many characteristics of seed maturity (early maturity, long maturity period, different maturity degree and germination rates) and a variety of effective dispersing methods (including the weed seeds).

Weeds are the main competitors of seedlings. They capture the nutrients and water needed by seedlings and affect the light and air circulation. Results have shown that the nutrient consumption of weeds was two to four times more than that of seedlings. Taking spruce as an example, the nutrient consumption of weeds is three times as much as that of spruce, with an average of 36.9% of nitrogen, 10.5% of phosphorus, and 19.0% of potassium. Water consumption of weeds is approximately twice that of seedlings. The water consumption of a *Chenopodium album* is two to three times more than that of millet and corn. To produce the dry matter of 1 kg, the water requirement of *Xanthium sibiricum* is 900 kg, 720 kg for *Chenopodium album*, but only 250 kg for millet and 330 kg for maize, respectively. Weeds also have a serious impact on light. Most of the weeds grow quickly, whereas the seedlings mostly have a slow-growing seedling stage. The shading effects of weeds limit the growth of seedlings due to overgrowth in the early stage, resulting in poor growth conditions and inhibition of metabolism; they reduce carbohydrate accumulation; and they reduce soil temperature (about 3 ℃ on average). Therefore, weeds affect the decomposition of soil organic matter and the activities of microorganisms, thus affecting the growth of seedlings.

In addition, many weeds are intermediate hosts of pathogens and pests, which easily accumulate and promote the occurrence and spread of diseases and pests.

Questions for Review

1. What are the types of seedlings? What are the characteristics of each type?

2. What are the growth characteristics of seedling height, diameter, and root? And what are the division methods and characteristics of growth and development stages of different types of seedlings?

3. What are the abiotic and biotic environmental factors that affect seedling growth? What do they have to do with seedling growth?

References and Additional Readings

CHI W, ZHOU W, 1991. Seedling technology in alpine region[M]. Harbin: Northeast Forestry University Press.

DUMROESE R K, LUNA T, LANDIS T D, 2022. Growing media[A]// Jacobs D F, Landis T D, Luna T. Nursery manual for native plants: A guide for tribal nurseries. Volume 1: Nursery Management Agriculture Handbook 730 [C]. Revised edition. Washington, DC: USDA Forest Service. p 77-93.

GONG M, WANG F, CHEN Y, et al., 2000. Mycorrhizal dependency and inoculant effects on the growth of *Betula alnoides* seedlings[J]. Forest Research, 13(1): 8-14.

HUA X, 1999. New Technology of mycorrhizal application[M]. Beijing: Popular Science Press.

HUA X, LUO Y, LIU G, 1995. A study on mycorrhization of pines with vegetative inoculum of *Pisolithus tinctorius* in nursery[J]. Forest Research 8(3): 258-265.

JIN T, 1985. Seedling Cultivation Technology[M]. Harbin: Heilongjiang People's Publishing House.

KANG L, LI S, 1998. Response of *Acacias* to rhizobia inoculation[J]. Forest Research, 11(4): 343-349.

LANDIS T D, TINUS R W, MCDONALD S E, et al., 1998. The container tree nursery manual. Agriculture Handbook 674[M]. Volume 1, 1990; Volume 2, 1990; Volume 6, 1998. Washington, DC: US Department of Agriculture, Forest Service.

LV C, HUANG B, WEI Y, et al. 2003. Comparative effects of different *Acacia* rhizobia inoculated on *Acacia crassicarpa* seedlings[J]. Journal of Nanjing Forestry University (Natural Sciences Edition), 27(4): 15-18.

MENG F, TANG X, 2001. Study on the promotion effects of Mycorrhizae to *Populus davidiana* seedlings [J]. Mycosystema, 20(4): 552-555.

QI M, 1992. Forest seedling science[M]. Harbin: Northeast Forestry University Press.

ADMINISTRATION OF QUALITY SUPERVISION OF THE PEOPLE'S REPUBLIC OF CHINA, 1999. Tree seeding quality grading of major species for afforestation: GB 6000—1999[S]. Beijing: Standard Press of China.

SHEN H, 2009. Seedling cultivation[M]. Beijing: China Forestry Publishing House.

SONG F, YANG G, MENG F, et al., 2005. The effects of arbuscular mycorrhizal fungi on the radicular system of *Populus ussuriensis* seedlings[J]. Journal of Nanjing Forestry University (Natural Sciences Edition), 29(6): 35-39.

SOUTH D B, 2017. Optimum pH for growing pine seedlings[J]. Tree Planters Notes, 60(2): 49-62.

SU S, 1993. Weed science[M]. Beijing: Agricultural Publishing House.

SUN S, 1985. Handbook of forest tree seed and seedlings[M]. Volume 1, Volume 2. Beijing: China Forestry Publishing House.

SUN S X, 1993. Afforestation[M]. Beijing: China Forestry Publishing House.

ZHAI M P, SHEN G F, 2016. Silviculture[M]. 3rd edition. Beijing: China Forestry Publishing House.

ZHANG H, YU Y, HUANG B, et al., 2005. Effects of rhizobia inoculation on the growth of straight-stem *Acacia auriculaeformis* seedlings and the contents of soil nutritive elements[J]. Journal of Northeast Forestry University, 33(5): 47-50.

ZHANG P, SUN H, SHEN H, 2007. Effect of temperature on germination of stratified seeds of *Fraxinus mandshurica* Rupr. [J]. Plant Physiology Communications, 43(1): 21-24.

Chapter 7 Nursery Soil Management and Seedling Protection

Wang Aifang, Liu Yong, Bai Shulan, et al.

Chapter Summary: This chapter introduces the theory and techniques of nursery soil fertility management, water management, symbiotic bacteria management, weed management, and pest and disease control. It elaborates on the protection methods for overwintering seedlings. By understanding the contents of this chapter, readers will be able to grasp the methods and ways of regulating external factors when cultivating seedlings from a broader perspective.

7.1 Soil Fertility Management

Soil fertility, that is, the quality of soil nutrient supply, is an important factor for the growth of seedlings. Soil cultivation, fertilization, and rotation are the common measures in maintaining and improving soil fertility continuously in nurseries.

7.1.1 Soil Tillage

Soil tillage, also known as land preparation, is plowing the soil by means of physical methods, and its effects are as follows.

First, tillage can improve the physical properties of soil. Plowing may loosen the soil structure, enhance water infiltration, and thus absorb more precipitation. Plowing may cut off the capillarity and reduce water evaporation, and thus improve the capacity of soil water conservation and drought resistance. Furthermore, the air permeability of loose soil is strong, which means it can not only provide more oxygen for root respiration but also remove excess carbon dioxide to promote root respiration. In loose soil, soil porosity is high, air heat capacity is low, and soil temperature is high, which is conducive to seed germination and seedling growth in early spring. In addition, loose soil is conducive to microbial activity, therefore promoting organic matter decomposition and increasing soil nutrient supply.

Second, plowing the upper and lower soil levels makes the lower soil ripen better, which is also conducive to the formation of aggregate structure in the upper soil.

Third, leveling the land can create good conditions for irrigation, sowing, and seedling emergence.

Fourth, shallow plowing can turn over and bury the weed seeds and crop residues, destroy the living environment of pests and diseases, and reduce the occurrence of weeds and pests. Moreover, if fertilizer is sprinkled on the soil surface before shallow plowing, the fertilizer can be mixed evenly.

In short, soil tillage improves the water, fertilizer, gas, and heat conditions of the soil; improves the soil fertility; and improves the growth environment of seedlings. Soil cultivation includes activities such as leveling land, shallow plowing, plowing, harrowing, compacting, intertillage, and so on.

(1) Leveling Ground

As for a new nursery, the land may be uneven so that it is not convenient to make into a seedbed for seedlings. Cultivating in an old nursery may be difficult because of the potholes made by lifting the seedlings yearly, especially after working with large seedlings. Therefore, the land should be leveled before plowing, and rocks and the stubble of grass roots should be removed so as to be ready for the next process.

(2) Shallow Plowing and Removing Stubble

Many roots remain after the seedlings are lifted in the nursery, or stubble remains after the crops or green manure crops are harvested in the crop field. During this time, the soil water loss is a significant amount. Shallow plowing and removing stubble should be done immediately after seedling lifting or crop harvesting, to a depth of 4 to 7 cm generally. When the nursery land is newly reclaimed on raw land, abandoned land, or cut-over land, the general tillage depth is 10 to 15 cm. The shallow plowing and stubble cleaner machines can use a disc harrow, nail harrow, and so forth.

(3) Plowing

Plowing is a primary tool of soil cultivation. The effect of plowing depends on the season and the depth of plowing.

Season. Depending on the climate and soil, plowing is done generally in spring and autumn. Autumn plowing can reduce pests, promote soil ripening, improve ground temperature, maintain soil moisture, and be carried out after seedling lifting or crop harvest in the northern cold area. Early plowing can eliminate weeds as early as possible, reduce the waste of soil nutrients, and obtain a longer fallow period. Through drying the upturned soil in the sun and freezing the upturned soil, dead soil can be changed into living soil, which is conducive to nutrient decomposition. Autumn plowing could especially increase the soil porosity and expand the water-holding range, therefore increasing the ability to absorb rain and snow in autumn and winter. Waterlogging transforms into spring moisture. For sandy soil, however, autumn plowing is not suitable in windy areas in either autumn or winter.

Spring plowing takes place when the previous season's plants are harvested late or not enough laborers are available. If spring plowing is necessary, it is often done immediately after soil thaw in early spring, dependent on windiness, rising temperatures, and/or major amounts of evaporation in the spring.

The specific time of plowing should be determined according to the soil water condition. When the soil water content accounts for 50% to 60% of its saturated water content, the plowing leads to the best quality, the least resistance, and the most suitable cultivation. During field observation, a handful of soil is kneaded into a ball by hand, then the ball is held 1 meter above the ground and dropped to the ground naturally. If the soil ball is broken, it is suitable for plowing. Alternatively, if the newly cultivated land has no large soil blocks, and/or the soil blocks will break with a kick, that is the best time for plowing.

Depth. The depth of plowing depends on the nursery conditions and seedling requirements. Deep plowing has a significant effect on water conservation, along with a good effect on promoting the ripening of the deep raw soil, increasing the aggregate structure of the soil, and improving the soil fertility. On the contrary, if the plowing is too shallow, it cannot achieve the above purposes. Therefore, the saying goes, "When deep plowing and fine harrowing, there is no fear of drought and flood." The suitable depth of plowing should be determined according to factors such as seedling raising method, climate condition, soil condition, and plowing season.

The depth of plowing has a great influence on the distribution of the root system of seedlings, for which deep plowing leads to the development of a deeper root system. Seedling raising methods have different requirements based on the plow depth. Generally, when sowing, the main absorption roots are distributed in the soil layer at approximately 20 cm, so the depth of cultivated land in the sowing area is 20 to 25 cm in general soil conditions, and 25 to 35 cm for cutting seedlings and transplanting seedlings in general soil conditions because of the deep distribution of their roots.

The plow depth should also consider the climate and soil conditions. For example, it should be deeper in a dry climate than in wet conditions, deeper in clay soil than in the sand soil nursery. In saline alkali land, to improve soil, restrain the rise of the salt alkali, and wash out the alkali; deep plowing down to 40 to 50 cm has good effect, but the soil cannot be turned over. Generally, autumn plowing should be deep, and spring plowing should be shallow. In brief, soil, land, and time should be considered before plowing, so as to achieve the expected effect.

The quality requirements of plowing: Ensure that no hard soil block will be formed after plowing; do not omit plowing, as the rate of omitting plowing should be less than 1%; meet the requirements of the depth of cultivated land, but it should not be too deep; and do not turn the bottom layer of soil with poor structure over to the surface layer. Commonly used farming machines and tools include the suspended three share plow, five share plow, two-wheel two share plow, and animal-power new-type walking plow.

(4) Harrowing

Harrowing is the surface soil cultivation after plowing. Its main function is to break apart the upturned sod and crust, level the ground, break up chunks of soil, and remove weeds. Harrowing time has a great influence on the effect of plowing, which should be determined according to the climate and soil conditions. Under the climate conditions of less snow in winter and a dryer and windy spring, it is necessary to harrow the land after autumn plowing to prevent moisture loss. In the low-lying saline alkali and water wetlands, it is not necessary to harrow the land immediately after the plowing, so as to promote soil ripening and to improve soil fertility through sunning the upturned soil, but it is necessary to harrow in the early spring of the next year. After spring plowing, we must harrow the land immediately, otherwise it will lose moisture and not be conducive to sowing. As the saying goes, "Dry harrowing for dry plowing, wet harrowing for wet plowing;

plowing without harrowing can make full of rough soil (clods). "

The quality requirements of harrowing include harrowing thoroughly, finely, and evenly. Commonly used harrowing machines include a tooth harrow and disc harrow.

(5) Compacting

The main function of compacting is to crush the soil block; compact the loose soil on the surface; prevent the loss of the vaporous water in the surface soil, which is conducive to water storage; and moisture conservation. Compacting time is mostly during winter in dry and windy areas, and after sowing in other areas. Compacting should not be done on sticky soil, because it will harden the soil and prevent the seedlings from emerging. In addition, compacting should be done when the soil moisture is suitable, since it will make the soil harden when the soil moisture content is high. Commonly used machines and tools include a non-handle roller, ring roller, diamond roller, wood stone roller, and cement stone roller.

(6) Intertillage

Intertillage is a loosening operation for surface soil during the growth of seedlings. Its function is to overcome soil hardening caused by irrigation and rainfall, reduce the evaporation of soil water, reduce the soil returning to salt and alkali, promote gas exchange, increase soil permeability, create suitable conditions for the activities of soil microorganisms, improve the utilization rate of effective nutrients in the soil, and eliminate weeds. On sticky soil, intertillage can prevent the soil from cracking and promote the growth of seedlings. Intertillage is generally done five to eight times per year, mostly after irrigation, rainfall, and combined with weeding.

The depth of intertillage varies with the size of seedlings, generally 2 to 4 cm deep for small seedlings, and gradually deepens to 7 to 8 cm or even 12 cm as the seedlings grow. The principle is that the roots cannot be damaged, and the seedlings cannot be damaged or hoed off. Machines and tools commonly used in the intertillage process include the cultivator, the horse-drawn pulling hoe, and a regular handheld hoe.

7.1.2 Rotation

Rotation, also known as rotation of crops, is to plant different tree species, or to rotate the planting of tree species and crops in a certain order on the same land, which is a biological measure to improve soil fertility and ensure healthy seedlings and high yields. Planting the same type of seedlings on the same land year after year is called continuous cropping.

Research has proved that continuous cropping tends to cause diseases, insect pests, and soil fertility decline; so consequently, reduction of the quality of seedlings and the yield. The specific reasons are as follows: First, some tree species have special needs and absorptive capacity to certain nutrient elements. Cultivating the same tree species on the same nursery ground for years can cause the lack of such nutrient elements and affect the growth of seedlings. Second, cultivating seedlings of the same tree species for a long time creates a suitable environment for some pathogens and pests to develop, such as damping-off and aphid infestation. Third, roots of some species seedlings can secrete acids and toxic gases, and long-term accumulation will have toxic effects on the growth of seedlings.

Rotation involves corresponding measures taken to prevent the disadvantages of continuous cropping. Rotation has a long history in China. The *Book of Sisheng* in the Han Dynasty points out that "if the harvest is not good for two years in a row, the soil should be fallow for one year." An agricultural proverb says: "The rotation of crops is like manuring", which shows the effect of rotation on increasing production. Generally speaking, the advantages of rotation lie primarily in making full use of soil nutrients; improving soil structure and soil fertility; reducing weeds; and changing the environment of pathogens and pests to the point that conditions do not support them, therefore playing the role of biological control of pests and diseases.

However, continuous cropping has good effects in some tree species, such as conifers and oaks. These species have mycorrhizae, which can help plants absorb nutrients, and continuous cropping is conducive to the propagation of mycorrhizal fungi. But continuous cropping is not suitable when serious damping-off is present in wet and cool conditions.

Rotation methods include tree species and tree species rotation, tree species and crop rotation, and tree species and green manure rotation.

(1) Rotation among a Variety of Tree Species Seedlings

For many tree species seedlings, the rotation of seedling species that are not prone to the same diseases and insect pests and that have different soil requirements can prevent the development of some diseases and insect pests. Such rotations also avoid excessive consumption of certain nutrients in the soil. Thus, to achieve a reasonable rotation among tree species, we need to understand the requirements of different seedlings for soil, water, and nutrients; the types of diseases and insect pests that may infect the species, as well as the seedling species resistance; and the mutual benefits and adverse effects of rotation among tree species. Theoretically, a rotation of legume and non-legume, deep roots and shallow roots, fertilizer-loving and poor-soil tolerance tree species, conifer and broad-leaved trees, trees and shrubs, and so forth is advantageous. Commonly used seedling rotation modes are shown in Table 7-1.

Table 7-1 Tree species seedlings suitable and unsuitable for rotation and continuous cropping

Suitable or not for rotation	Rotation tree species	Rotation effect
Tree species suitable for rotation	Mutual rotation or continuous cropping of *Larix gmelinii*, *Pinus koraiensis*, *Pinus sylvestris*, *Pinus tabuliformis*, *Pinus densiflora*, *Platycladus orientalis*, *Pinus massoniana*, *Picea asperata*, *Abies fabri*	Seedlings grow well; because these trees have mycorrhizae, they are also suitable for continuous cropping under the condition of not serious diseases and insect pests
	Mutual rotation of *Populus*, *Ulmus pumila*, *Phellodendron amurense*	Good growth of seedlings, with fewer pests and diseases
	Rotation of *Pinus tabuliformis* and broad-leaved seedlings such as *Castanea mollissima*, *Populus*, *Lagerstroemia indica*, *Amorpha fruticose*, *Albizia julibrissin*, *Acer negundo*, *Gleditsia sinensis*	Good growth of seedlings, with fewer pests and diseases

Suitable or not for rotation	Rotation tree species	Rotation effect
Tree species suitable for rotation	Rotation of *Pinus bungeana* and *Albizia julibrissin*, *Acer negundo*, *Gleditsia sinensis*	Good growth of seedlings and reduction of damping-off
	Rotation of *Cunninghamia lanceolata*, *Pinus massoniana* and *Ulmus pumila*, *Robinia pseudoacacia*	The effect is good
Tree species unsuitable for rotation	*Larix gmelinii* with *Pyrus*, *Malus pumila*, *Populus*, *Betula*, *Robinia pseudoacacia*, *Amorpha fruticosa*	Susceptible to disease
	Pinus tabuliformis, *Pinus bungeana* with *Ulmus pumila*, *Juglans regia*, *Diospyros lotu*, *Robinia pseudoacacia*	Serious damping-off occurs
	Picea asperata with *Padus racemosa*	Susceptible to rust disease
	Sabina chinensis, *Zanthoxylum bungeanum*, *Acer saccharum*, *Malus pumila*, *Pyrus*	Susceptible to rust disease

(Continues)

(2) Rotation of Tree Seedlings and Crops

After harvest, a lot of roots are left in the soil, which can increase the organic matter of the soil, improve the soil structure, and compensate for the loss of fertility caused by the large amount of nutrients taken out of the soil by the seedlings' growth. The main crops in rotation with tree seedlings include beans, sorghum, corn, and rice. For instance, the rotation of Chinese fir (*Cunninghamia lanceolata*) seedlings and rice (*Oryza sativa*) has been adopted in southern China. Cultivating *Cunninghamia lanceolata* seedlings for one or several years and then planting one season of *Oryza sativa* can effectively reduce the occurrence of diseases, underground pests, and xerophytic weeds. Note, however, that rotation of seedlings and crops must not cause diseases and insect pests; for example, planting vegetables or potatoes in the nursery land should be avoided because that combination is prone to damping-off and insect pests. Some coniferous seedlings, such as Gmelin larch (*Larix gmelinii*), Scotch pine (*Pinus sylvestris*), and dragon spruce dragon (*Picea asperata*), rotate with soybeans, which are prone to damping-off of conifer seedlings and scarab damage.

(3) Rotation of Tree Seedlings, Green Manure, and Forage Grass

Rotation of tree seedlings with green manure and herbage is beneficial to improve soil structure and fertility. Research has reported that alfalfa can fix N at 202.5 kg per hectare; clover can fix N at 150 kg per hectare, and *Astragalus sinicus* can fix N at 112.5 kg per hectare. The rotation of green manures or pastures often uses alfalfa (*Medicago sativa*), wood sorrels (*Oxalis* spp.), Chinese milkvetch (*Astragalus sinicus*), false indigo bush (*Amorpha fruticosa*), and bush clover (*Lespedeza bicolor*).

7.1.3 Fertilization

Fertilization can directly provide nutrient elements for seedling growth by chemical or biological measures, can improve the soil fertility, improve the physical and chemical properties of the soil, and create favorable environmental conditions for the growth and development of seedlings. In the process of seedling cultivation, seedlings take up a large percentage of nutrient elements from the soil and also take away much of the fertile surface soil and most of the roots when they are lifted and out of the nursery, all of which greatly reduces the soil fertility. Relying only on the physical measures of soil farming and rotation of biological measures cannot completely make up for the lack of soil nutrient elements; doing so will reduce the quality of the seedlings that are grown. Therefore, fertilization measures are very important for soil fertility management.

7.1.3.1 Seedling Demand for Nutrient Elements

A variety of nutrient elements are needed for the growth of seedlings. According to plant analysis, dry matter in plants is made up of dozens of elements. Chief among them are 16 elements: Carbon (C), hydrogen (H), oxygen (O), nitrogen (N), phosphorus (P), potassium (K), calcium (Ca), magnesium (Mg), sulfur (S), iron (Fe), manganese (Mn), zinc (Zn), copper (Cu), boron (B), chlorine (Cl), and molybdenum (Mo). Among these, carbon, hydrogen, and oxygen account for the largest proportion (96%), and they come naturally from air and water. The other 13 elements, also known as mineral elements, need to be gained from the soil. Because plants need a large amount of nitrogen, phosphorus, potassium, calcium, magnesium, and sulfur (3.5%), these are generally called macroelements. Among them, the demand for nitrogen, phosphorus, and potassium is the largest, and their soil content is generally insufficient. These three are the essentials of fertilizer. Other elements are needed in much smaller amounts (0.5% of plant tissues) and are called microelements (Table 7-2).

Table 7-2 Plant essential nutrients: Their proportions and physiological functions in plant tissues

Nutrients	Proportion in plant tissue /%	Physiological function
C	45.0	Photosynthesis
H	6.0	Photosynthesis
O	45.0	Respiration
Total of carbon, hydrogen, and oxygen	96.0	
N	1.5	Composition of amino acids and protein
P	0.2	Energy transfer
K	1.0	Osmotic adjustment
Total of nitrogen, phosphorus, and potassium	2.7	
Ca	0.5	Construction of cell wall
Mg	0.2	Enzyme activation, components of chlorophyll
S	0.1	Amino acid construction and protein synthesis
Total of calcium, magnesium, and sulfur	0.8	
Fe	0.01	Chloroplast component, RNA synthesis
Mn	0.005	Enzyme activation

Nutrients	Proportion in plant tissue /%	Physiological function
		(Continues)
Zn	0.002	Enzyme activation, chloroplast composition
Cu	0.0006	Chloroplast component, protein synthesis
B	0.002	Assimilate transport and cell growth
Cl	0.01	Maintain cell turgor
Mo	0.00001	Composition of enzyme
Total of microelements	0.5	
Total	100	

Source: Adapted from Dumroese (2009).

Each essential nutrient element plays an important role in the growth and physiology of plants. The lack of any element will cause adverse effects on the normal growth of plants, even if the proportion of the microelement in the plant is extremely low (Table 7-2). Compared with many other factors, nutrient elements play a key role in affecting the quality of seedlings. Providing this nutrition in a timely manner and at appropriate, proportionally accurate levels can increase the growth rate of seedlings many times over, can reach the quality standard of seedlings faster than that of non-fertilization, and can further promote their growth after afforestation.

7.1.3.2 Seedling Deficiency Symptoms Related to Nutrients and Their Diagnosis

When the supply of some nutrients in the soil is insufficient, the metabolism of seedlings is affected and the external morphology will then show certain symptoms. During seedling growth, leaves may lose their green color and turn into yellow, purple, white, or other colors in a phenomenon known as chlorosis. When seedlings are in nutritional distress, a diagnosis is needed for the deficiency symptom and for the remedy in order to ensure normal growth and development of seedlings.

The following methods can be used in the diagnosis of nutrient deficiency.

Morphological Diagnosis. Carefully observe the characteristics of abnormal external morphological symptoms of seedlings, and then analyze them to determine whether the seedlings are deficient in nutrients, and if so, which nutrients are lacking (Table 7-3).

Fertilization Diagnosis. Apply some quick-acting nutrients through foliage top-dressing and observe the feedback effects on the spot.

Chemical Analysis. Test nutrient content in the soil and the content of nutrients in abnormal seedlings; compare with normal seedlings. Generally, a diagnosis can be drawn by applying morphological diagnosis and fertilization diagnosis together. If an answer is not clear, chemical analysis needs to be used. Although the chemical analysis methods require more work, the diagnosis result is accurate and reliable.

In the nutritional diagnosis, we should pay attention to the following items. Note the difference between nutrient deficiency symptoms and disease and/or insect infestations. Plant diseases and insect pests usually spread from one point into the surrounding areas, but the deficiency of a certain nutrient is limited to one point or area. Note the difference between nutrient deficiency and genetic factors. The albinism of seedlings caused by genetics occurs only in a single plant, but if caused by a nutrient deficiency it occurs in spots or in many seedlings. Note the difference of symptoms between major elements and microelements. With a deficiency of nitrogen, phosphorus, and potassium, symptoms usually start from the old leaves in the lower part of the plant, whereas the deficiency symptoms of iron, boron, manganese, and other microelements usually start from the new shoots on the top of the plant.

Table 7-3 Deficiency symptoms of partial nutrient elements in seedlings

Nutrient elements deficiency	Main symptoms
N	Leaves are yellow-green and thin; stems are short, thin, and delicate; the lower old leaves are withered and fall off; the growth of shoots is stagnant
P	Symptoms first appear on the old leaves; leaves are purple or bronze; seedlings are thin and small; apical buds are underdeveloped; lateral buds are degenerate; roots are few and slender
K	Dark green leaves; slow growth; short stem; small amount of lignification
Ca	Symptoms appear first on new leaves; leaves are small and light green, with yellow leaf tip and edge; twigs are weak; roots are thick, short, and curved
Mg	Yellow tips for conifer needles and interveinal yellowish for broad-leaved leaves
Fe	New shoots show yellow, light yellow, and milky white; symptoms gradually develop downward, and the whole plant turns yellow in severe cases
Mn	Leaves show chlorosis and form small necrotic spots
Zn	Internode growth is inhibited; leaves are severely deformed
B	With withered shoots and clusters of twigs, the fruit is malformed or most of it has fallen off the plant; leaves become thicker; color of the leaves darkens; leaves are small

7.1.3.3 Fertilization Theory and Principle

A reasonable supply of soil nutrients needs to be regulated by artificial fertilization. Therefore, how to achieve reasonable fertilization is the key to bringing the benefits of fertilization into play. Fertilization theory and principles are guidelines formed on the basis of production practice and scientific research.

(1) Fertilization Theory

Law of the Minimum. Plants need 16 kinds of essential mineral nutrient elements. Deficiency or lack of any of the elements will inhibit the growth of plants. However, which element has the greatest impact on plant growth? Liebig's Law of the Minimum states that plant growth is controlled by the most deficient mineral nutrient. That is to say, even if all other elements are sufficient, the deficiency or lack of a single element becomes the limiting factor for plant growth. The most vivid example is the "Barrel Principle". The barrels for holding water are made of several wooden boards, and the amount of water held is determined by these boards together. If one of the boards is short, the water-holding capacity of the barrel is limited, and the short board becomes the "limiting factor" of the water-holding capacity of the barrel. To increase the capacity of the barrel, replace the short board or lengthen it (Figure 7-1). Fertilization is the equivalent of lengthening the "short board".

Steady-status Nutrient. This theory contains three aspects: First, all nutrients should be supplied in a timely manner and quantitatively correspond to the relative growth rate of plants, so that the growth rate and the nutritional status of plants remain stable. The optimal nutrition is to supply all nutrients at the same rate as the highest growth rate. Because plants exhibit exponential growth, the technology of exponential fertilization came into being in order to keep a stable supply rate of nutrients.

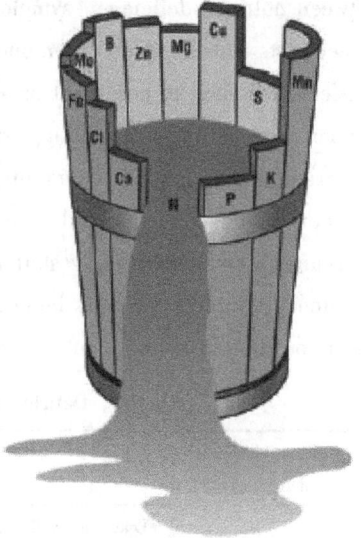

Figure 7-1 "Barrel Principle" used to explain the limiting factors of nutrient elements in plant growth (Dumroese, 2009).

Second, the appropriate concentration of nutrients in media for plant absorption and utilization is very low (< 1 μM). Ingestad and Ågren (1995) obtained the highest plant growth rate in the spray culture of plants by using a low concentration, which amounted to 1/640 and 1/35 of the concentration of traditional culture medium, based on steady-status nutrient theory. This result indicates that the concentration of nutrients in medium needed for optimum growth of plants is not high, but the flux of nutrients from the medium to the root surface is required to maintain its effectiveness and balance to flow into the root system. At the same time, low concentration can give full play to the ability of plant roots to absorb nutrients.

Third, providing all nutrients in the optimum proportion is the foundation to obtaining the maximum growth of plants. Maintaining a reasonable balance among plant nutrients can reduce or even avoid the antagonistic effect caused by the imbalance of the proportion of elements in traditional methods, so that the steady-status nutrient supply at low concentration and in accord with the growth rate of plants can play a maximum potential and greatly promote the growth of plants.

The theory of steady-status nutrients in plants plays an important role in the cultivation of afforestation seedlings. The root system of newly planted seedlings needs time to recover growth, thus their ability to uptake nutrients from the soil is poor. Therefore, the growth of seedlings at the early stage of afforestation depends largely on the transfer of nutrients stored within the plant to new roots and shoots. The initial nutrient concentration of seedlings is the key factor that may restrict the initial growth of seedlings after afforestation. The nutrition-loading technology developed by the theory of steady-status nutrients means that in the process of cultivating seedlings in nursery, according to the law of nutrient demand for the growth and development of seedlings, the amount of fertilizer increases exponentially at each application time, and the amount of nutrient provided for seedlings is synchronous with the growth of seedlings. The aim is to fix the fertilizer in the seedlings as much as possible to form a nutrient sink, which can then be utilized to promote the growth of new roots and shoot elongation after afforestation. This approach is the fundamental difference between the steady-status nutrient loading technology and the conventional fertilization technology. Seedlings cultivated by this technology have more advantages in difficult site afforestation. Since the 1990s, steady-status fertilization technology has been widely used in oak (*Quercus* spp.), spruce (*Picea* spp.), and pine (*Pinus* spp.).

(2) Principles of Fertilization

First, make clear the purpose of fertilization. Fertilizer types and application methods differ with the various purposes of fertilization. If the main purpose is to increase the nutrient elements, the readily available mineral fertilizer is the most appropriate; if the aim is to improve the soil organic matter and the soil structure in addition to increasing the nutrient elements, then organic fertilizer, green manure, and pond mud fertilizer are appropriate.

Second, fertilize according to climate conditions, which directly affect the status of nutrients in the soil and the absorption capacity of seedlings. Generally, in cold and dry conditions, fertilizer decomposes slowly and absorption capacity of the seedlings is low. In this environment, select a "hot" fertilizer (for example, horse manure or sheep manure) that decomposes easily, and apply the manure after it has fully decomposed. The amount of fertilizer in each application can be large and given with less frequency. Under the conditions of high temperature and rain, fertilizer decomposes quickly, is readily absorbed, but the nutritive value is easily leached out. In this case, a "cold" fertilizer (such as pig manure or cow manure), which decomposes slowly, should be selected. A small amount of fertilizer can be applied with more frequency. In the years with high temperatures, the first top-dressing should be applied earlier in the growing season. In the years with more rainfall, especially in nurseries that apply large amounts of fertilizer at the end of the growth season, seedlings are prone to frost damage.

Third, fertilize according to the soil conditions. Fertilization in nurseries should be carried out according to the demand of the seedlings for soil nutrients and the nutrient status of nursery soil. Apply the fertilizer that it is needed with the appropriate amount. The nutrient status of nursery land is closely related to soil types and physical and chemical properties (such as pH). Sandy soil is poor at conserving water and fertilizer, so the fertilizer should be applied in a small amount at more frequent times; a cold fertilizer such as pig manure or cow manure should be used; and the depth of fertilizer application should be deep rather than shallow. Clay soil is characterized by a strong ability to conserve water and fertilizer, which means fertilizer can be applied in a large amount but fewer times; horse manure, sheep manure, and other thermal fertilizers should be used, and the depth of fertilization should be shallow rather than deep. The characteristics of loam soil and its fertilization principles are between sand and clay soil.

An alkaline fertilizer should be used for acidic soil, and an acidic fertilizer should be used for alkaline soil. In acidic or strong acid soil, phosphorus can become fixed into iron phosphate and aluminum phosphate in the soil, which cannot be absorbed by plants, so fertilizers such as calcium magnesium phosphate, phosphate rock powder, plant ash, or lime should be applied; nitrate nitrogen is better for nitrogen fertilizer. In alkaline soil, the effect of ammonium nitrogen (such as ammonium sulfate and muriate of ammonia) is better for nitrogen fertilizer. Phosphorus can become fixed as tricalcium phosphate, which is not easily absorbed by seedlings, so water-soluble phosphate fertilizer, such as superphosphate or ammonium phosphate, should be selected.

Fourth, fertilize according to the characteristics of seedlings. Different species of tree seedlings need different amounts of nutrient elements. According to the analysis, the order of the main nutrient elements content in the dry matter of seedlings is N>Ca>K>P; the general tree species need more nitrogen, so nitrogen fertilizer is mainly used. However, legumes have rhizobia to fix nitrogen in the atmosphere, and phosphorus can promote the development of rhizobia, therefore, legumes have higher requirements for phosphorus than for nitrogen. Depending on the growth and development stages of the same tree species, the requirements of nutrient elements differ. For 1-year-old seedlings, at the seedling stage, they are sensitive to nitrogen and phosphorus; at the fast-growing stage, they have high requirements for nitrogen, phosphorus, and potassium; in the later stage of growth, top-dressing with potassium fertilizer and stopping nitrogen fertilizer application can promote the lignification of seedlings and enhance their stress resistance. With the increase of age, the amount of fertilizer needed is also increasing. The fertilizer needed for 2-year-old seedlings in seedbeds is generally two to five times higher than that of 1-year-old seedlings. The higher the density of seedlings, the more fertilizer is needed, so more fertilizer should be applied as appropriate.

Fifth, fertilize according to the nature of fertilizers. For reasonable fertilization practices, we must understand the nature of fertilizers and their effects on seedlings under different soil conditions. For example, phosphorus in phosphate rock powder is easy to dissolve and release under strong acid condition, which is suitable for the acid soil in South China but not suitable for the calcareous soil in North China. It is better to use calcium magnesium phosphate in a centralized way to prevent phosphorus from being fixed by the soil. Nitrogen fertilizer should be used in a proper centralized way, because a small amount dispersed through the soil tends to have no significant effect on the yield. The application of phosphorus and potassium fertilizer has a good effect on the soil when the nitrogen is sufficient.

7.1.3.4 Fertilizer Types and Properties

Many types of fertilizers are available, and they are generally classified into organic, inorganic, and microbial fertilizers based on their composition.

(1) Organic Fertilizer

Organic fertilizer is formed through the decomposition of organic substances, such as plant residues or human and animal manure. It not only contains a variety of nutrients, such as nitrogen, phosphorus, and potassium, but also has a long-lasting effect. That means that it continuously provides nutrients to seedlings during the whole growth process. More important, organic fertilizer can improve the physical and chemical properties of soil, promote the activities of soil microorganisms, improve the soil fertility, and have a positive effect toward improving sand and clay soil. However, organic fertilizer has some deficiencies. The quantity and proportion of various nutrients cannot fully guarantee the growth needs of all kinds of seedlings. Some nutrients, especially available nutrients, are not enough, and the proportion of nitrogen, phosphorus, and potassium may not be appropriate for the seedlings being grown. A certain amount of inorganic fertilizer must be supplemented in the rapid growth period of seedlings. Commonly used organic fertilizers in a nursery include compost, animal manure, green manure, peat, human manure, poultry manure, seabird manure, cake manure, humus, and bone meal.

(2) Inorganic Fertilizer

Inorganic fertilizer, also known as chemical fertilizer, is mainly composed of minerals, including three major elements of nitrogen, phosphorus, potassium along with microelements. The effective components of inorganic fertilizer are high, with a fast fertilizer effect and easy absorption by the seedlings. The effect on soil improvement, however, is far less than that of organic fertilizer given that the inorganic fertilizers do not contribute as broad a range of nutrients. If only inorganic fertilizer is applied year-round, soil structure would deteriorate and soil fertility would decrease. Five kinds of inorganic fertilizer are commonly used in a nursery, including nitrogen fertilizer, phosphorus fertilizer, potassium fertilizer, compound fertilizer, and microelement fertilizer.

Nitrogen Fertilizer. Commonly used nitrogen fertilizers include ammonium sulfate (physiological acid), ammonium bicarbonate (close to neutral), ammonium nitrate (neutral), urea (neutral), ammonium water (weak alkaline), calcium ammonium nitrate (weak alkaline), and lime nitrogen (alkaline). The main function is to promote the growth of plant stems and leaves, and the leaves grow thick and green.

Ammonium sulfate [$(NH_4)_2SO_4$] is a quick-acting ammonium fertilizer with nitrogen content of 20% to 21%. After it is applied to the soil, the ammonium ion is easy to be absorbed by plants or adsorbed by soil colloids, but the sulfate ions remain in the soil solution. Long-term application causes an increase in the soil acidity and hardening. It is better to use it in combination with organic fertilizer.

Ammonium nitrate (NH_4NO_3) is a quick-acting nitrogen fertilizer with nitrogen content of 33% to 35%. It contains ammonium nitrogen and nitrate nitrogen, which are readily absorbed by plants. It is not suitable to remain in soil but works well for top-dressing.

Urea [$CO(NH_2)_2$] is a quick-acting nitrogen fertilizer with nitrogen content of 44% to 46%. Long-term application has no damage to the soil, and it is suitable for top-dressing.

Ammonium bicarbonate (NH_4HCO_3) has a nitrogen content of 17%. The fertilizer is unstable, decomposes easily, and is likely to lose effectiveness. It can be used as base fertilizer and top-dressing. It should be applied deeply into the ground, covered with soil, and irrigated immediately to prevent the nitrogen from volatilization.

Phosphate Fertilizer. Common phosphate fertilizers include superphosphate (acid), calcium magnesium phosphate (slightly alkaline), phosphate rock powder (weak acid), and bone meal. Its main function is to promote the development of the plant root system and to enhance cold resistance and drought resistance of the plants.

Superphosphate is a mixture of $Ca(H_2PO_4)_2 \cdot H_2O$ and calcium sulfate. It is a water-soluble fertilizer, in which sulfate ions are easily adsorbed and fixed by the soil but they have little mobility. It cannot be fully utilized by plants in the year of application but it has a degree of after-effect.

Potash Fertilizer. Commonly used potassium fertilizers include potassium chloride (physiological acid), potassium sulfate (physiological acid), and plant ash. Its main function is to promote the robust growth of plants, with thick and hard stems, and to enhance resistance to pests and falling over (lodging).

Potassium sulfate (K_2SO_4) is a quick-acting fertilizer. After being applied into the soil, potassium ion is adsorbed by soil colloid and has relatively good mobility, so it is suitable to be used as base fertilizer.

Potassium chloride (KCl) characteristics are similar to that of potassium sulfate, but do not use potassium chloride for the chlorine-sensitive seedlings such as apple trees. Too much chlorine will have a negative effect on them.

Compound Fertilizer. This fertilizer contains two or more chemical nutrient elements. For example, diammonium phosphate contains nitrogen and phosphorus; potassium nitrate contains nitrogen and potassium; ammoniated superphosphate contains phosphorus and nitrogen. Attention should be paid to its characteristics during application; for instance, the compound fertilizer contains ammonium ions, which should not be applied on saline alkali land; a compound fertilizer containing chloride ions should not be applied to the chlorine-sensitive plants or saline alkali land; a compound fertilizer containing potassium sulfate should not be used in acidic soil. Try to avoid spreading the fertilizer on the soil surface and instead apply it deeply. Cover the soil on the surface because it is not easy to be absorbed and utilized by the plant root system when applied on the soil surface because nitrogen easily volatilizes and phosphorus is readily fixed by the soil.

Microelement Fertilizer. Iron, boron, manganese, copper, zinc, and molybdenum are not used as necessary fertilizers because of the small demand for them in seedlings. Their natural content in the soil can meet the needs of seedlings. Occasionally some soils lack microelements, however, so it may become necessary to use microelements for fertilization. The commonly used microelement fertilizers include ferrous sulfate, boric acid, manganese sulfate, copper sulfate, zinc sulfate, and ammonium molybdate. Foliage top-dressing is often applied since it would be easy to be absorbed and utilized by plants and not be fixed by soil.

Slow-release fertilizer, also known as controlled-release fertilizer, refers to the nitrogen fertilizer that can release effective nutrients slowly and provide longer benefits because of the change of chemical composition or the coating of semi-permeable or impermeable substances on the surface of fertilizer components. The most important characteristics of slow-release nitrogen fertilizer are that it can control the release rate, decompose gradually in the soil, and be absorbed and used for the crops, so that the nutrients in the fertilizer can meet the various needs of each growth stage across the whole growth period of the crops. The fertilizer effect maintains for months up to more than one year for one application. The primary fertilizers in this type include urea formaldehyde, isobutyl urea, butylene diurea, oxalyl urea, and sulfur-coated urea.

(3) Microbial Fertilizer

Microbial fertilizer is a general term for all kinds of microbial inoculum fertilizer made by using the microorganisms beneficial to the growth of seedlings in the soil, including nitrogen-fixing bacteria, rhizobia, phosphating bacteria, potassium bacteria, and other bacterial fertilizer and mycorrhizal fungal fertilizer.

Practice has proved that organic fertilizer is suitable for using as basal fertilizer, inorganic fertilizer is suitable for top-dressing, and granular phosphate fertilizer is suitable as seed fertilizer. To give full play to the fertilizer effect, a variety of fertilizers can be mixed in use or complex fertilizers can be used, such as diammonium phosphate, monoammonium phosphate, monopotassium phosphate, or compound fertilizers. Generally, organic fertilizer is mixed with inorganic fertilizer; quick-acting fertilizer is mixed with slow-acting fertilizer; and nitrogen, phosphorus, and potassium are mixed in a certain proportion. According to the experimental data, the mixture of superphosphate and organic fertilizer can improve the efficiency of phosphate fertilizer by 25% to 40% and reduce the leaching loss of nitrogen, suggesting an increasing demand for the compound fertilizer in the future.

The mixed fertilization should be made reasonably according to the nature of the fertilizers, otherwise, it will have negative effects. Please refer to Figure 7-2 to determine if various fertilizers can be used together.

7.1.3.5 Fertilization Amount

(1) Traditional Calculation Method of Fertilization Amount

The appropriate amount of fertilizer can be determined according to factors such as the amount of nutrients absorbed by seedlings (B), the nutrient content in the soil (C), and the utilization rate of fertilizer (D). If the reasonable fertilization amount is A, it can be calculated according to the following formula:

$$A = (B-C)/D$$

And yet, determining the amount of fertilizer accurately is a very complicated issue, because the amount of nutrients absorbed by seedlings, the content of nutrients in the soil, and the utilization rate of fertilizers are affected by many factors. Therefore, the calculated fertilization amount can be used only as a reference. The best fertilization amount of each tree species needs to be determined through experiments.

In China, 1-year-old seedlings are generally fertilized with 45 to 90 kg of nitrogen, 30 to 60 kg of phosphorus pentoxide, and 15 to 30 kg of potassium oxide per hectare per year, on which basis, the fertilizer amount for 2-year-old seedlings should be increased by two to five times. Then, according to the needed amount of nutrient elements per hectare and the amount of effective elements in the fertilizer, the actual fertilization amount per hectare can be roughly calculated.

Organic fertilizer plays an important role in improving soil fertility. It is used primarily as the base fertilizer, and the amount per hectare is generally 45 000-90 000 kg. The deficiency of nutrient elements in the base fertilizer should be supplemented by top-dressing. Generally, the amount of

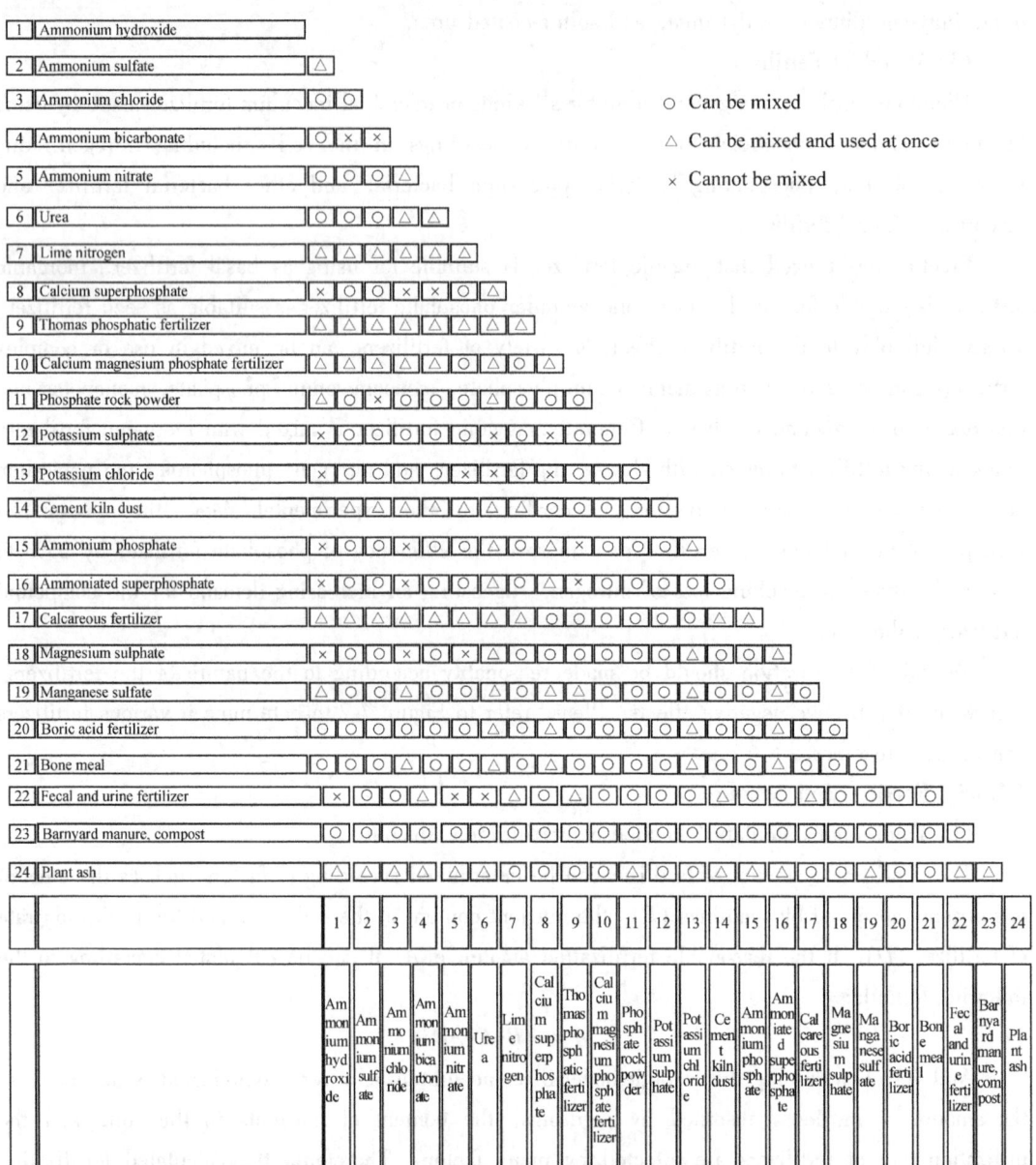

Figure 7-2 Fertilizer mixing application diagram.

soil top-dressing per hectare per each fertilization is 3750 to 5250 kg for human feces and urine, 75 to 112.5 kg for ammonium sulfate, 60 to 75 kg for urea, and about 75 kg for ammonium nitrate, ammonium chloride, and potassium chloride.

For example, calculate the amount of fertilizer applied to 1-year-old Chinese pine (*Pinus tabuliformis*) seedlings. The ratio of fertilizer elements is N : P : K = 4 : 3 : 0.5. The density of seedlings is 500 per square meter. Application amount is 160 kg of nitrogen, 120 kg of phosphorus, and 20 kg of potassium per hectare.

The compost of 90 000 kg per hectare as basal fertilizer was applied, and the equivalent nitrogen, phosphorus, and potassium elements are:

Nitrogen = 90 000 kg × 0.004 (nitrogen content) × 0.3 (utilization rate) = 108 kg

Phosphorus = 90 000 kg × 0.002 (phosphorus content) × 0.15 (utilization rate) = 27 kg

Potassium = 90 000 kg × 0.005 (potassium content) × 0.4 (utilization rate) = 180 kg

Nitrogen to be added: 160−108 kg = 52 kg

52 kg / 0.20 (nitrogen content) / 0.5 (utilization rate) = ammonium sulfate 520 kg

The amount of phosphorus and potassium to apply is calculated accordingly.

(2) Determination Method of Optimal Fertilization Amount in Steady-Status Nutrient Theory

According to the steady-status nutrient theory, increasing the nutrient reserves in seedlings is essential for the initial growth of seedlings on afforestation. To determine the optimal fertilization amount of seedlings, it is necessary to formulate multiple fertilization amounts based on experience or data. Based on the biomass at each fertilization amount, a response curve of biomass to the fertilization amount is simulated, and the sufficient fertilization amount and the optimal fertilization amount are determined by finding the inflection point based on this curve (Figure 7-3). At the point when the biomass begins to reach a maximum, the amount of fertilizer applied is the sufficient fertilization amount. As the amount of fertilizer application continues to increase and the biomass of the seedlings remains unchanged, the seedlings continue to absorb nitrogen and to increase the nitrogen concentration. When the amount of fertilizer application increases to a certain amount, the nitrogen concentration in the soil solution is too high and the stress effect appears. Then the biomass of seedlings begins to decline, and the fertilizer amount corresponding to the biomass when seedlings were about to be stressed is called the optimal fertilizer amount. Salifu and Jacobs (2006) conducted

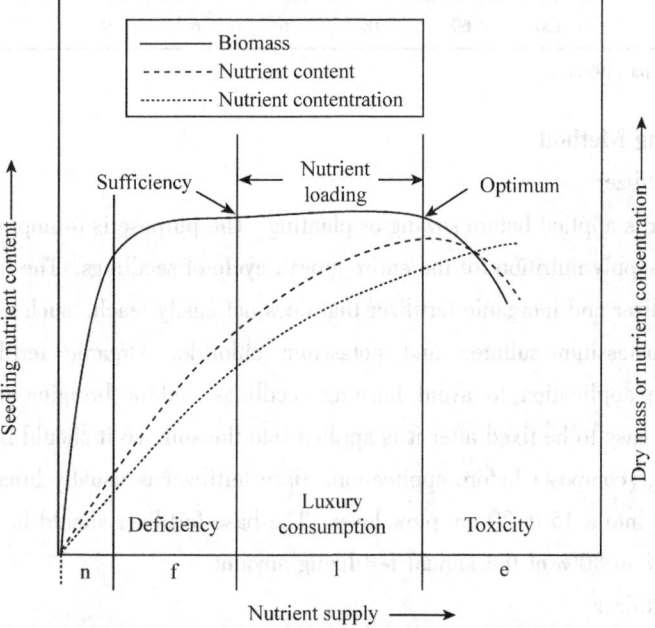

Figure 7-3 Relationship between fertilization amount and seedling growth, nutrient content and concentration in seedling (Salifu and Timmer, 2003).

exponential fertilization on red oak container seedlings and found that the seedling biomass was the highest when the fertilization amount was 25 mg/plant; when the fertilization amount exceeded 100 mg/plant, the seedling biomass began to decrease. Therefore, the sufficient fertilization amount and the optimal fertilization amount for red oak container seedlings were 25 mg/plant and 100 mg/plant, respectively (Table 7-3).

7.1.3.6 Optimum Fertilization Ratio of Nutrient Elements

The proportion among various nutrient elements in fertilization is as important as the amount of fertilization. A large number of studies show that the external nutrient concentration required for the maximum growth of various plants is not high, but the supply of nutrient elements must be balanced. Ingestad and Ågren (1995) pointed out that in general plants need the following fertilization ratio: 100N : 50P : 15K : 5Mg : 5S (weight ratio), but differences occur among tree species (Table 7-4). Each tree species needs to be tested to find the best ratio.

Table 7-4 Best proportion of nutrient elements (weight ratio) applied in several tree species

Tree species	Macroelements						Microelements
	N	P	K	Ca	Mg	S	
Paulownia tomentosa	100	75	14	7	7	9	
Paulownia elongata	100	75	18	7	8.5	9	
Populus simonii	100	70	14	7	7	9	Fe 0.7, Mn 0.4, B 0.2, Cu 0.03, Zn 0.03, Cl 0.03, Mo 0.007, Na 0.003
Robinia pseudoacacia	100	60	15	8	9	9	
Cunninghamia lanceolata	100	60	16	7	7	9	
Pinus elliottii	100	60	18	6	6	9	

Source: Zheng and Jia (1999).

7.1.3.7 Fertilizing Method

(1) Base Fertilizer

Base fertilizer is applied before sowing or planting. The purpose is to improve the soil, enhance soil fertility, and supply nutrition for the entire growth cycle of seedlings. The base fertilizer is made up of organic fertilizer and inorganic fertilizer that does not easily leach, such as ammonium sulfate, superphosphate, potassium sulfate, and potassium chloride. Organic fertilizer must be fully decomposed before application to avoid burning seedlings and/or bringing in weeds and pests. Superphosphate is easy to be fixed after it is applied into the soil, so it should be mixed with organic fertilizer for retting (compost) before application. Base fertilizer is usually broadcast on the surface of soil and plowed into a 15 to 20 cm plow layer. The base fertilizer should be sufficient, generally accounting for 70% to 80% of the annual fertilizing amount.

(2) Seed Fertilizer

Seed fertilizer is applied at the time of sowing. The main purpose is to provide the nutrient elements needed for the intensive growth of seedlings. Seed fertilizers are primarily inorganic fertilizer composed of phosphorus and refined organic fertilizer, such as human manure and cake fertilizer. Generally, granular phosphate fertilizer is applied in the sowing ditch, because of its small contact area with the soil and small fixed amount by the soil, which is conducive to root absorption. Do not use powdery phosphate fertilizer as seed fertilizer, because it is easy to burn seeds and seedlings.

(3) Top-dressing

Top-dressing is the fertilizer applied during the growth period according to the growth rhythm of seedlings. The purpose is to supplement any deficiencies in the base fertilizer and seed fertilizer. Top-dressing is made up of quick-acting inorganic fertilizer and human manure, such as urea, ammonium bicarbonate, ammonium hydroxide, potassium chloride, decomposed human manure, and superphosphate. Generally, severalfold soil is needed to mix well with the fertilizer, or water is added to dissolve and dilute the fertilizer before use. Top-dressing can be used in the soil or as a foliar application.

Soil Top-dressing. Methods for applying fertilizer to the soil include furrow application, spray (irrigation) application, and broadcast application. In furrow application, the furrow should have a depth of 6 to 10 cm between rows, be less than 10 cm away from the seedlings, and be covered immediately after the application. Spray (irrigation) application places fertilizer on the seedbed after dilution, or it is used in conjunction with irrigation to the seedling bed. Broadcast application mixes the fertilizer with a portion of fine soil and then evenly spreads it on the bed surface. These three top-dressings need irrigation after fertilization. If sprinkling irrigation is available, the fertilizer left on the stems and leaves of the seedlings can be rinsed with water to prevent burning the seedlings. Furrow application is the best in terms of fertilizer utilization rate. Using urea top-dressing as an example, the utilization rate of seedlings in furrow top-dressing is 45% in the same year, 27% for spray (irrigation) application, and 14% for broadcast application without covering soil.

Foliar Top-dressing. A solution of nutrient elements is sprayed on the stems and leaves of seedlings, and the nutrient solution is absorbed and utilized by the mesophyll cells through the cortex. Foliar top-dressing avoids the fixation of fertilizer by soil and the leaching loss in irrigation and precipitation. A smaller amount of fertilizer is needed and it has a high and speedy absorption rate, which makes it a better top-dressing method. The solution concentration for foliar top-dressing is generally 0.2% to 0.5% for urea, 7.5 to 15 kg/hm^2 for each time; 0.5% to 2.0% for superphosphate, 22.5 to 37.5 kg/hm^2 for each time; 0.3% to 1.0% for K_2SO_4, KCl, and KH_2PO_4; and 0.25% to 0.50% for other microelements. The best time for foliar top-dressing is in the morning, evening, or on cloudy days and wet-air days. In case of rainfall within two days after spraying, the fertilizer effects will likely fail, so supplementary application may be done. Slow-acting fertilizer is not suitable for foliar top-dressing, and it is easy to burn seedlings when the concentration of solution is too high. Therefore, foliar top-dressing is only a supplementary measure of nutrition and cannot completely replace soil top-dressing.

The time of top-dressing should be determined according to the growth rhythm of seedlings, especially the seasonal variation of when seedlings absorb nutrients from the soil, and through field experiments and experience. The top-dressing time of 1-year-old seedlings is usually in summer, and the quick-acting nitrogen fertilizer is applied several times to supplement a large nutrient demand during the vigorous growth period of seedlings. In some areas, P and K fertilizers are also applied as late top-dressing in the autumn to promote diameter growth and lignification of seedlings, which in turn enhance stress resistance. The time of nitrogen top-dressing in nurseries in China is generally no later than August. However, for seedlings with a long growth period and in areas with a long, frostless period or basically frostless all year-round, the time of nitrogen top-dressing can be appropriately prolonged. The timing of top-dressing differs between transplant seedlings and reserved bed seedlings. For some broad-leaved tree species (especially cutting seedlings) that have easy-to-grow roots, top-dressing soon after transplanting will have an effect, whereas for some conifers, top-dressing soon after transplanting will not work, and may even bring on adverse effects. As for the reserved bed seedlings, because they have formed a complete root system in the soil, the positive effects of top-dressing during their growth period will be obvious.

(4) Nutrient Loading

According to the theory of steady-status nutrients, nutrient loading maximizes the nutrient concentration in the seedling through top-dressing. The premise is that it will cause no harm to the growth of the seedling, will form a large nutrient sink, and will improve the afforestation effect. Exponential fertilization and fall fertilization are generally used.

Exponential Fertilization. It is a fertilizing method that adds nutrients by exponential increase in order to adapt to the relative growth rate of plants in each growth stage. A large number of comparison studies found that exponential fertilization was significantly better than the conventional fertilization in the biomass, nutrient concentration, and afforestation effect of seedlings. For example, the container seedlings of red oak were set with 8 treatments, therein, 25 C was the conventional (C) fertilization treatment, in which 1.56 mg of fertilizer per plant were applied once a week for 16 weeks, then the total amount was 25 mg/plant. 25 E was the exponential (E) fertilization treatment; with the amount of fertilization increased exponentially, the total amount of fertilization was also 25 mg/plant. Results showed that the biomass and nitrogen concentration of the latter were higher than in the former (Chen et al., 2015). At present, this method is applied primarily in container seedlings, therefore the specific fertilization methods are detailed in Chapter 9 Container Seedling Cultivation.

Fall Fertilization. It is generally the top-dressing carried out after the seedlings stop growing in height during autumn. Its purpose is not to promote the increase of seedling morphological index or biomass but is instead to increase the content of nutrient elements in the seedlings, which will improve the resistance of seedlings and promote their rapid growth during the next spring. This technology has been widely used abroad in the seedling cultivation of Norway spruce (*Picea abies*), black spruce (*Picea mariana*), loblolly pine (*Pinus taeda*), Masson pine or Chinese red pine (*Pinus elliottii*), red pine (*Pinus resinosa*), ponderosa pine (*Pinus ponderosa*), Douglas fir (*Pseudotsuga menziesii*), southern blue gum (*Eucalyptus globulus*), and oaks (*Quercus*). In China, only a few studies are available on Chinese white poplar (*Populus tomentosa*), *Larix gmelinii*, *Pinus tabuliformis*, Chinese cork oak (*Quercus variabilis*). Li et al. (2011) conducted nutrient loading on

1-year-old seedlings of Olgan larch (*Larix olgensis*), with the fertilizing amount of 60 to 90 kg/hm^2, applying the fertilizer on September 16 and October 1. Results showed that compared with the control, not only did the dry weight of roots and stems increase significantly but also the nitrogen content and cold resistance of seedlings increased significantly.

(5) Fertigation

Fertigation is a fertilization technology that combines irrigation and fertilization by adding water-soluble fertilizer along with irrigation. A fertilizer injection device, such as a bypass fertilizer tank, venturi fertilizer applicator, or injection pump, is added to the sprinkling irrigation system. A water-soluble fertilizer solution configured according to the needs of tree species can be used. When the sprinkling irrigation system is open, the fertilizer will be sprayed on the seedlings along with the irrigation water. This approach has the advantages of improving fertilizer utilization rates, saving labor, accurately controlling the amount and time of fertilizer application, rapid nutrient absorption of seedlings, and beneficial for the application of microelements. Not all nurseries will have sprinkling irrigation facilities available, however, but where possible, it can be of benefit. Shortcomings for fertigation include, for example, the fertilizer must have high solubility and salt may accumulate in the soil. Technical and management requirements are high and strict; this type of system is used primarily in greenhouse and container seedling cultivation.

7.2 Water Management

Water is the most important condition for plant growth and development. No life can survive without water. Approximately 95% of the fresh weight of plants is water. Suitable soil water condition is the basis of cultivating strong seedlings. Water not only directly participates in the growth and physiological activities of seedlings but also affects the nutrient absorption of seedlings by changing the concentration of soil solution. If watering is omitted once in the process of seedling cultivation, the seedlings may be seriously injured or even die. Therefore, water management is the most critical link in seedling production. The key to water management is to determine the water quality, method, time, and quantity of irrigation.

7.2.1 Water Quality Control

Generally, seedling growth is difficult with natural precipitation alone, so it is necessary to supplement soil water with artificial irrigation. The absorption and utilization of water by seedlings are directly related to water properties. Water quality is an important limiting factor in selecting nursery land; in can be very costly to work with poor water quality. A place with poor water quality should not be selected as nursery land. The main factors affecting water quality include salt content and impurity content.

Salt content of irrigation water is typically less than 0.2% to 0.3%. Too high salt content in water will increase the osmotic pressure of the soil solution and cause physiological drought to seedlings; damage the soil structure and reduce the permeability of soil; change the pH value and solubility of soil; and reduce the quality of seedlings.

The pH value of irrigation water is 5.5 to 6.5. Too high or too low pH will affect the availability of soil nutrients, which is not conducive to the growth of seedlings. Given that species often have different requirements for the pH value, phosphoric acid, sulfuric acid, nitric acid, and acetic acid are often used to adjust the pH value of irrigation water in practice.

Temperature of irrigation water also has a direct impact on the absorption of water and nutrients. In most circumstances, a suitable irrigation temperature is higher than 10 to 15 ℃ in spring and autumn and higher than 15℃ to 20 ℃ but less than 37℃ to 40 ℃ in summer. Water temperature can be increased by building a sun pool, artificial heating, or solar heating, for example.

Impurities in irrigation water may include sand particles, soil particles, insects, pathogen spores, and grass seeds. Such impurities can affect the irrigation system, damage irrigation equipment, and bring pests and weeds to the nursery. When dealing with fungi, bacteria, and insect eggs in the water, sodium hypochlorite or calcium hypochlorite solution can be added to the water, or pressurized chlorine gas can be injected into the irrigation system. Filters can be adopted for minimizing fine sand, weed seeds, and algae.

7.2.2 Irrigation Method

Irrigation methods include lateral irrigation, border irrigation, sprinkler irrigation, drip irrigation, and infiltration irrigation.

(1) Lateral Irrigation

Lateral irrigation, also known as ridge irrigation, is a method in which water infiltrates into the seedbed or ridge from the side and is generally used for high beds and high ridges. Advantages are that the bed surface does not easily harden, and the soil still has good ventilation performance after irrigation. Disadvantages are that compared with spray irrigation, the channels cover a large area, the irrigation quota is not easy to control, the water consumption is great, the irrigation efficiency is low, and it takes more labor.

(2) Border Irrigation

Border irrigation is also called basin irrigation. Water floods the bed surface, filling the surface and percolating downward into the soil. It is suitable for low bed and flat field operation. Advantages of flood irrigation are that it is water saving, requires less investment, and has easier operation compared to lateral irrigation. Disadvantages are that the soil structure is destroyed during irrigation, which hardens the soil; the water channels occupy more land; the irrigation efficiency is low; more labor is needed; and the irrigation amount is difficult to control.

(3) Sprinkler Irrigation

Sprinkler irrigation uses a water pump to pressurize the water, or a natural landscape drop, to transport water to the nursery through the sprinkler irrigation system. The sprinklers then spray water on the seedlings (also known as overhead irrigation). The spray-head is installed with a fixed mode and a mobile mode (Figure 7-4). The system is suitable for high bed, low bed, high ridge, low ridge, flat farming, and other operations. It has the advantages of saving water, easy control of the amount of irrigation, and preventing the soil from secondary salinization caused by excessive irrigation. Reducing the area occupied by a channel (in other irrigation systems) can improve the land utilization rate. The soil is not hardened, which prevents soil erosion; the system has high efficiency and is labor saving. Irrigation in spring has the effect of increasing ground temperature and preventing frost, and spray irrigation at high temperatures can reduce ground temperature and protect seedlings against high temperature. Irrigation is even, even if the terrain is slightly uneven. Therefore, it is an effective and widely used irrigation method. The disadvantage is that the capital investment is high, and it is restricted by wind speed. Irrigation is uneven when wind speed is greater than 3 m/s.

Figure 7-4 Fixed sprinkler irrigation system (a), mobile sprinkler irrigation system (b), and spray-head (c) (Photos by Liu Yong).

(4) Drip Irrigation

Drip irrigation conveys water to the irrigation site through a pipe and supplies water to the soil in the form of water droplets. Because it can directly transport water to the soil at the root of seedlings, it has the advantages of saving water (by 30% to 50% compared to sprinkler irrigation), high irrigation efficiency, and does not affect soil structure. However, many pipelines are needed and construction costs are high. The use of a low-pressure pipe system to slowly and evenly drop water and fertilizers that are dissolved in the water into the soil at the root of seedlings is currently the most advanced irrigation method.

(5) Infiltration Irrigation

Two types of infiltration irrigation are available: One is field infiltration irrigation for bare-root seedlings; the other is subirrigation infiltration for container seedlings.

Field infiltration irrigation introduces water into the soil by pipes, wetting the soil in the root zone of the seedlings. Nurseries can use buried pipe irrigation and/or submersible irrigation. The former irrigation makes water infiltrate into the soil through the joints or gaps in the wall of the underground pipeline; the latter raises the groundwater level to the root layer of the crop by capillary action. The main advantages of field infiltration irrigation are that it does not damage the soil structure and has high water use efficiency, which can save water by 50% to 70% compared with sprinkler irrigation. The disadvantages are large construction investments and complicated construction technology.

Subirrigation is composed of nursery rack, seepage trough, water storage tank, water pump, water delivery pipe and return pipe, and timer. Under the action of the pump, the water in the water storage tank is injected into the water seepage tank through the water delivery pipe. The container seedlings draw the required water from the bottom of the container through the capillary action of the nursery substrate, and the unused water returns to the water storage tank through the return pipe in an ongoing recycled manner (Figure 7-5). Its biggest advantages are to improve the quality of seedlings, reduce waste of water resources, and reduce pollution caused by fertilizer leaching. Research has shown that subirrigation decreased water consumption by approximately 63%; seedling height and root collar diameter with subirrigation increased by approximately 10%, and the biomass increased by more than 18.4%; nutrient content per plant increased by more than 18%; and the nutrient concentration was increased by at least 17% compared to the overhead sprinkler irrigation in *Quercus variabilis* container seedlings (Chen et al., 2015). Subirrigation is a water-saving and pollution-reducing irrigation technology that can also cultivate higher quality seedlings than sprinkler irrigation.

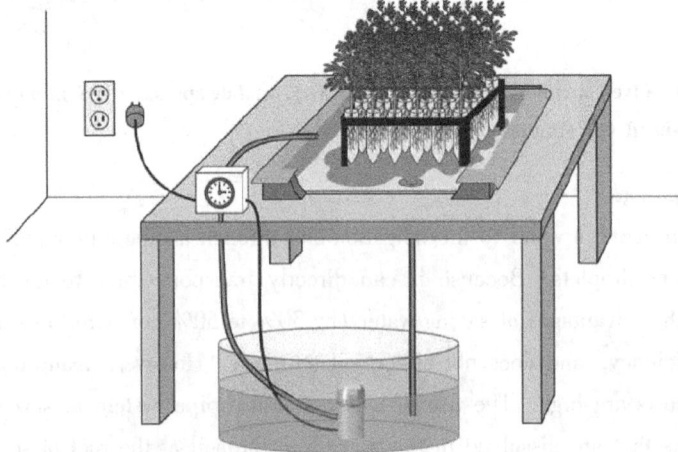

Figure 7-5 Schematic diagram of subirrigation system (Adapted from Jim Mann).

7.2.3 Irrigation Time and Amount

The growth of seedlings is limited, and even seedling death can occur, due to lack of water, but this does not mean the more water, the better. Excessive soil water will make seeds and cuttings rot, and root respiration will be limited, which often results in seedlings growing poorly or dying. In addition, excessive irrigation can cause soil secondary salinization. Reasonable irrigation decisions should take into account the biological characteristics of tree species, the growth stages of seedlings, soil conditions, weather conditions, and many other aspects.

The biological characteristics of tree species vary. Some need less water, while others need more water. For example, poplar, willow, birch, and *Larix gmelinii* need more water; therefore, more frequent irrigation is needed. Chinese ash (*Fraxinus chinensis*), Shantung maple (*Acer truncatum*), and elm tree (*Ulmus pumila*) take second place in terms of the amount of water they require. Chinese hawthorn (*Crataegus pinnatifida*), apples (*Malus* spp.), beach rose (*Rosa rugosa*), and black locust (*Robinia pseudoacacia*) need less water and are prone to yellowing when the soil moisture is too high, so they can be irrigated less. *Pinus tabuliformis*, Oriental arborvitae (*Platycladus orientalis*), and other evergreen coniferous trees like dry soil, so irrigation times can be appropriately reduced and irrigation amount can be less.

The water demand of seedlings varies according to the growth period. Irrigation times and irrigation amount should be adjusted accordingly. Before sowing, sufficient basal water should be applied to ensure that the seeds can absorb enough water to promote germination; irrigation after sowing may cause soil hardening and reduce ground temperature, therefore, irrigation should be avoided before emergence. In the seedling stage, the root system is shallow and is relatively sensitive to water demand. Although this stage needs relatively little water, seedlings benefit from more irrigation times but less irrigation amount each time. In the fast-growing stage, when seedlings are growing deep root systems and have large water demands, the irrigation times can be reduced, but the irrigation amount each time needs to be larger than in other stages. In the lignification stage, irrigation should be reduced or stopped to promote lignification and to prevent excessive growth. The wetting depth of each irrigation should reach the distribution depth of the main absorbing roots.

Weather conditions influence irrigation. If the weather is relatively dry, the soil loses water quickly, so the irrigation frequency and the irrigation amount should be greater. In sunny and windy weather, the transpiration of seedlings and soil is high, and the water consumption is greater, so the irrigation interval should be shortened.

Soil conditions also directly affect the times and amount of irrigation. For sandy soil and sandy loam with poor water-holding capacity, smaller amounts and more times of irrigation can be carried out; for clay, with strong water-holding capacity, the interval of irrigation can be extended; for bottomland and saline alkali land, the frequency of irrigation should be controlled appropriately for the specific conditions.

Soil moisture content should be checked to determine whether nursery land should be irrigated. Generally, the soil moisture content suitable for seedling growth is 60% to 80% of the field water capacity. Ground irrigation should be applied in the morning or at dusk when evaporation is low and the difference between water temperature and ground temperature is also low. If the aim is to reduce the ground temperature by sprinkling irrigation, it is suitable to irrigate at high temperatures.

7.2.4 Drainage

The drainage system of a nursery should use a ditch system to quickly drain the rainwater that cannot penetrate into the soil away from the nursery. A ditch system contributes to seedling management and protection in the rainy season. Precipitation in the rainy season in the north of China is high and concentrated. The precipitation from July to August accounts for 60% to 70% of the annual precipitation. Precipitation in the south is greater than in the north, and rainfall extends over a longer time frame. It is easy to cause excessive accumulated field water, and if the drainage does not occur in a timely manner, the field water may cause the seedlings, especially the root system of the seedlings, to suffocate and decay, to have weakened growth, to be more susceptible to pests and diseases, to have poorer quality, and even to die. Therefore, the drainage work should be arranged, and the following aspects should be paid attention to.

First, the drainage system, composed of large, medium, and small drainage ditches, should be arranged according to the terrain drop of the whole nursery during the planning, design, and construction of the nursery (see Chapter 2 for more information). Each working area must be connected to a drainage ditch, and all rainwater will finally be collected into a large drainage ditch and discharged from the nursery through the lowest part of the nursery.

Second, before rainy season is approaching, weeds and sundries in the drainage ditches should be removed to ensure smooth water removal, and all the border openings of the seedbed should be opened. After continuous rainfall or rainstorms, special personnel should check the drainage route, dredge the drainage ditches, and try to channel the water out to avoid any ponding plots.

Third, after the rainy season, the seedbed should be intertilled as quickly as possible, which will increase the soil permeability and facilitate respiration of the seedling root system.

Fourth, some fertilizers or insecticides used in the production of nursery seedlings will be discharged with water and pollute the environment. Therefore, a wastewater settling tank should be set up in large-scale nurseries, so that the drainage water, in accordance with the environmental protection standards, can be discharged through a sedimentation treatment.

7.3 Symbiotic Fungi Inoculation and Management

7.3.1 Mycorrhization Seedling Cultivation and Its Management

7.3.1.1 Concept and Significance of Mycorrhizae and Mycorrhization Seedling Cultivation

Mycorrhizae form a reciprocal, symbiotic association of the mycelium of a fungus with the roots of a vascular plant. At present, according to the morphological and anatomical characteristics of mycorrhizae, the types of mycorrhizal fungi, and the types of plants, mycorrhizae are usually classified into ectomycorrhiza, endomycorrhiza, ectendotrophic mycorrhiza, arbutoid mycorrhiza, monotropoid mycorrhiza, ericoid mycorrhiza, and orchid mycorrhizas. Among them, ectomycorrhiza and arbuscular mycorrhiza of endomycorrhiza have been studied the most at present.

(1) Ectomycorrhiza

Ectomycorrhiza is the mycorrhiza formed when mycorrhizal fungi mycelium infects non-suberification vegetative roots of host plants. After the formation of ectomycorrhiza, the mycelium of the fungus does not penetrate the interior of the cell tissue of the host root and grows only between the root cell walls of the host plant. This root system of the host plant has three main characteristics: The root shape of the host plant is usually visible to the naked eye, displaying shortening, thickening, color changes, and a variety of bifurcation characteristics (Figure 7-6), with no root cap and epidermis, root hairs degrading, and many extended hyphae on the surface of the fungal mantle (Figure 7-6); a layer of fungal mycelium is tightly interwoven to form a visible fungal mantle on the nutritive root surface of the plant; and as the fungal mycelium grows in the intercellular space of the root cortex, a grid-like structure is formed between the root cortex cells of the host plant, which is called the Hartig net (Figure 7-7).

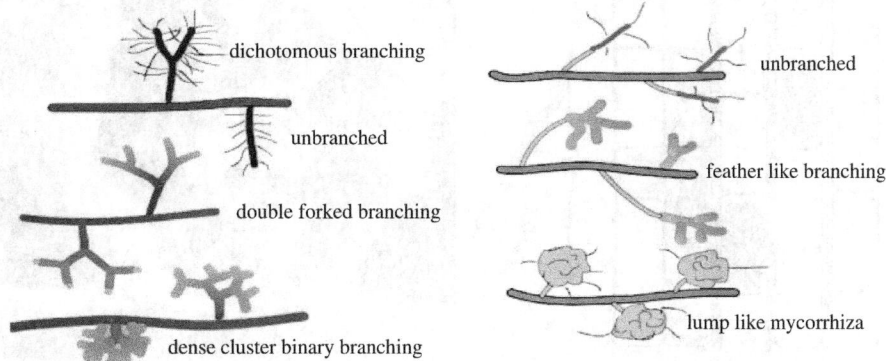

Figure 7-6 The shapes of the ectomycorrhizal bifurcation (Adapted from Brundrett et al., 1996).

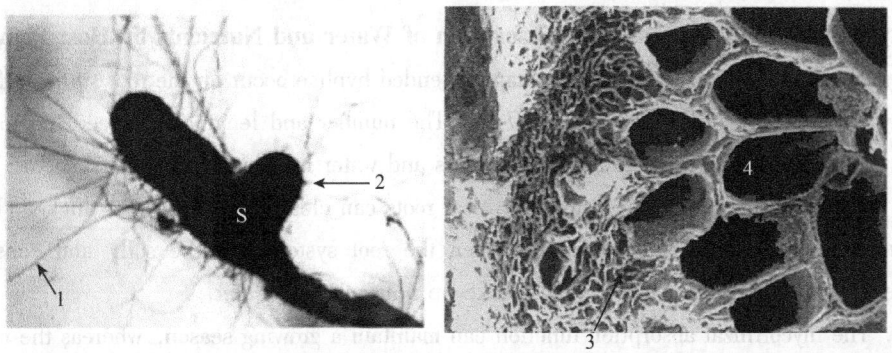

Figure 7-7 Fungal mantle of ectomycorrhiza and structure and morphology of the Hartig net
1. extended hypha; 2. fungal mantle; 3. Hartig net; 4. host cell.

(2) Arbuscular Mycorrhiza

Arbuscular mycorrhiza is a type of endomycorrhiza. Note that endomycorrhiza and endophyte are two distinct concepts. Endomycorrhiza is a symbiotic association formed by plant roots and soil fungi, which is neither a root nor a fungi but instead a symbiotic association, whereas endophytes are bacteria that exist in all organisms. A variety of endophytes are found in plants and animals, including in humans. Unlike ectomycorrhiza, after plant roots are infected with arbuscular mycorrhizal fungi, the external appearance of roots is generally difficult to distinguish with the naked eye; only under a microscope can it be seen clearly that arbuscular mycorrhizal fungi form in the root cortical cells, and endophytic hyphae occur within or between cells (Figure 7-8).

Then, mycorrhization seedling cultivation refers to the process of mycorrhizal formation (including all kinds of mycorrhizae) of seedlings by artificial inoculation of mycorrhizal fungi. The significance of mycorrhization seedling cultivation lies in the following points (only the main aspects are listed here):

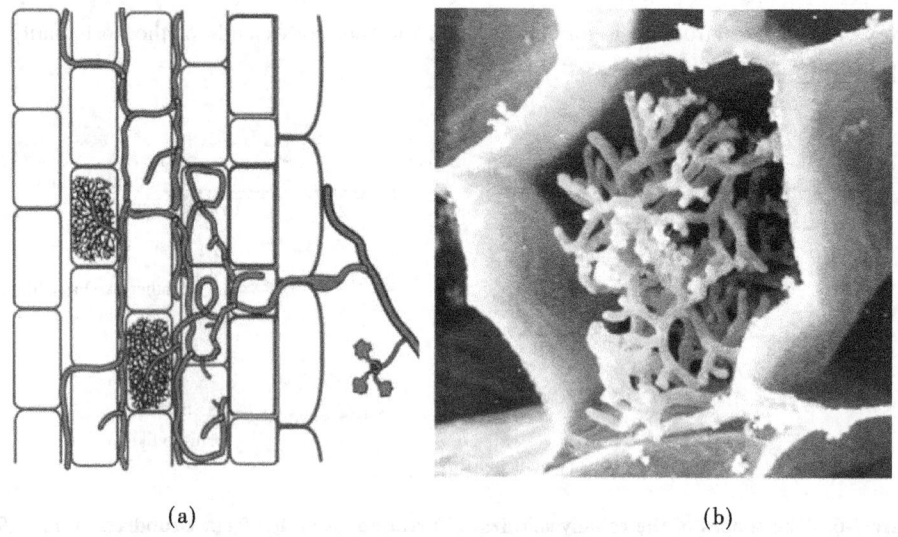

(a)   (b)

Figure 7-8 Schematic diagram in structure of arbuscular mycorrhiza (From Brundrett et al., 1996).

Mycorrhizae Can Increase the Absorption of Water and Nutrients by Host Plants

i. After the mycorrhizae formation, many extended hyphae occur on the root surface, forming a huge mycelial network in the soil (Figure 7-9). The number and length far exceed the root hairs, thereby assisting the host plant to absorb nutrients and water in a larger range.

ii. The fungal mantle on the ectomycorrhizal roots can clearly make the root thicker (Figure 7-10), which increases the contact area between the root system and the soil, and consequently increases the absorption area of the root to take up nutrients and water.

iii. The mycorrhizal absorption function can maintain a growing season, whereas the root hair's absorption function can last for only a few days. From a time perspective, mycorrhizae can help plants absorb water and various nutrients for a longer time.

iv. Extraradical mycelium of mycorrhizae is very slender and can penetrate the soil with a bulk

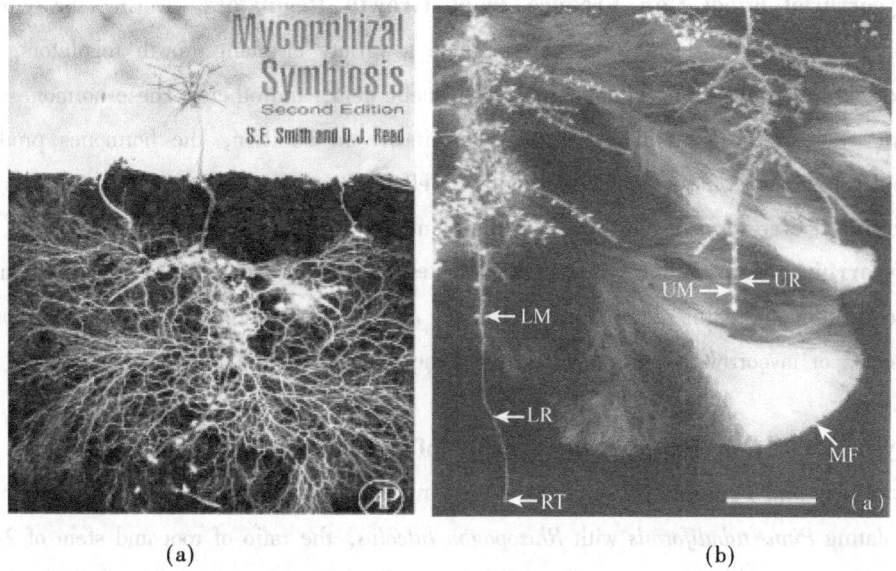

(a)　　　　　　　　　　　　　(b)

Figure 7-9　Extended mycelia of ectomycorrhiza and their large mycelial network (From Smith and Read, 1996; Brundrett et al., 1996).

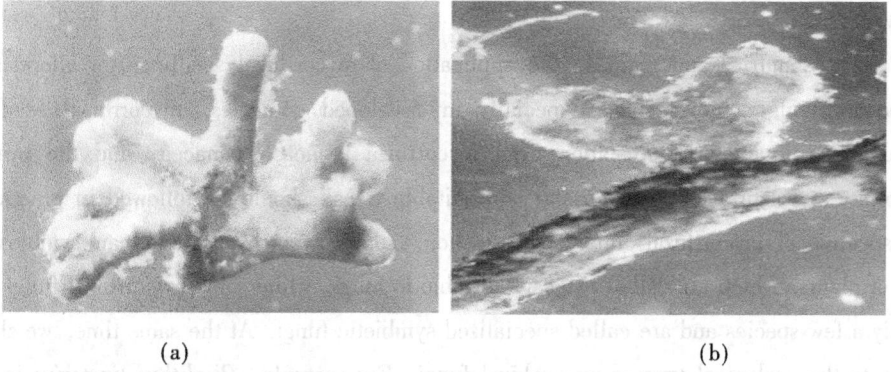

(a)　　　　　　　　　　　　　(b)

Figure 7-10　Mycorrhiza symbiosis formed by *Pinus tabuliformis* and *Suillus bovinus* (Photos by Bai Shulan).

density of 1.8 g/cm^3 to help the plant absorb the nutrients and water it needs. This type of mycelium can also extend out of the phosphorus-depleted area or to rock gaps to absorb mineral phosphorus, something the root system of plants cannot achieve. It undoubtedly expands the absorption range of the host root system.

Increase the Host Plant's Absorption of Phosphorus and Other Mineral Nutrients. In addition to promoting the absorption of phosphorus and other mineral nutrients by plants, mycorrhizal fungi can secrete organic acids and release fixed phosphate ions in the soil by chelating with iron and aluminum. Mycorrhizal fungi can also produce phosphatase, which converts insoluble phosphorus into soluble phosphorus, thereby creating favorable conditions for plants to absorb mineral phosphorus.

Mycorrhizal Fungi Can Produce Plant Growth Regulators. During the growth and symbiosis process, mycorrhizal fungi can produce a variety of plant growth regulators, such as auxin, cytokinin, gibberellin, vitamin B_1, indoleacetic acid, and others. These hormones have the same properties as those produced by the plant itself. In addition, the hormones produced by mycorrhizal fungi sometimes produce a stimulating effect as soon as the mycelium contacts the host plant, before it even forms mycorrhiza with the plant.

Mycorrhizae Can Enhance the Stress Resistance of Host Plants. Host plant stress resistance to drought, saline-alkali, pH and heavy metals, pests and diseases can be improved by the presence of mycorrhizae, thereby improving the viability of the host plants in these stressful environments.

Mycorrhizae Can Improve the Quality of Seedlings, Increase the Survival Rate of Afforestation, and Promote the Growth of Young Forests. Bai et al. (2011) have shown that by inoculating *Pinus tabuliformis* with *Rhizopogon luteolus*, the ratio of root and stem of 2-year-old *Pinus tabuliformis* seedlings is twice that of the control, the height growth is 1.5 times that of the control, and afforestation survival rate increases by 30%.

7.3.1.2 Significance and Methods for Selecting Excellent Fungal-tree Combinations in Ectomycorrhiza

In many countries and regions, the application of mycorrhizal seedlings for afforestation has achieved gratifying achievements. Especially in developed countries, mycorrhizal seedlings are required for mountainous afforestation. But mycorrhiza is not a panacea, and the principle of "appropriate mycorrhiza for suitable land and suitable trees" should be followed in practice. From the perspective of mycorrhizal fungi specialization, some mycorrhizal fungi can symbiose with a variety of plants, which are called extensive symbiotic fungi, while some mycorrhizal fungi symbiose with only a few species and are called specialized symbiotic fungi. At the same time, we should pay attention to the ecological type of mycorrhizal fungi. For example, *Pisolithus tinctorius* is a broad-spectrum mycorrhizal fungus. However, Smits reported that inoculating *Dipterocarpus* seedlings with *Pisolithus tinctorius* failed to form mycorrhizae. Researchers have also confirmed that the mycorrhiza can be formed only when *Pisolithus tinctorius* is inoculated to the root system of the original isolated species of the strain; for example, when the *Pisolithus tinctorius* is inoculated on various eucalyptus seedlings. This characteristic is because the different strains of *Pisolithus tinctorius* have formed some specific biological characteristics (i.e., ecotype) in a specific environment. Therefore, to apply mycorrhizal technology to production procedures, it is very important to separate local mycorrhizal fungi in different areas and to then carry out artificial inoculation tests on the main tree species in this area to select the best combination of mycorrhizal fungi and trees with high infection rate and strong adaptability to a specific environment. Meng et al. used four kinds of ectomycorrhizal fungus to inoculate the seedlings of Korean aspen (*Populus davidiana*) in a pot experiment. Results showed that the most obvious growth-promoting effect on *Populus davidiana* was with

Cortinarius russus, followed by *Russula foetens* and *Lactarius insulsus*. The worst response was with *Pisolithus tinctorius*, therefore, the first step in the application of mycorrhizal technology in production is the selection of an excellent combination of fungus-trees. Bai et al. (2011) have isolated a large number of indigenous mycorrhizal fungal strains on the site where the vegetation has been severely degraded for many years on the Daqing Mountain in Inner Mongolia. They selected different combinations of fungus-trees for several local arbors and shrubs, and they have successfully screened corresponding excellent fungus-trees combinations for *Pinus tabuliformis*, Mongolian Scots pine (*Pinus sylvestris* var. *mongolica*), and a variety of *Larix gmelinii* known as *Larix principis-rupprechtii*, as well as several native shrubs. This work provides an important theoretical basis and scientific reference for targeted cultivation of mycorrhizal seedlings in different regions. The specific method is as follows.

(1) Selecting Combinations of Fungus-trees in the Early Period of Seedlings

For seedling inoculation during early seedling growth, pour an appropriate amount of strains culture medium into a wide-mouth container and then sterilize it at high temperature. Afterward, according to the growth of different tree species and strains, inoculate the fungal strains into the container, and then put the sterilized or germinated plant seeds into the container; or, inoculate them simultaneously if the growth of fungal strains is synchronous with the seedlings. In this way, when the germs of seeds start to grow, their radicles can fully contact with mycorrhizal fungi. If the seedlings remain in the container for a long time (40-50 days), and at the same time, if the strains inoculated are indeed the mycorrhizal fungi of the tree species, mycorrhiza can be symbiosed on the seedling roots. Generally, mycorrhizal colonization rate of the seedlings can reach 80% to 100% after 15 to 20 days of inoculation. If they need to be transplanted again, they can continue to grow and form mycorrhiza on new substrates. This method is suitable only for inoculation of small-seed plant seedlings and needs to be carried out under sterile conditions. However, inoculation success rate of this method is extremely high, which is very suitable for small-scale tests (Figure 7-11, Figure 7-12). Based on the high infection rate of seedling inoculation, stress resistance combinations can be further selected in the seedling pot.

(a)   (b)   (c)

Figure 7-11 Schematic diagram of mycorrhizal inoculation in early period of seedlings and the eucalyptus sprout method (From Brundrett, 1996).

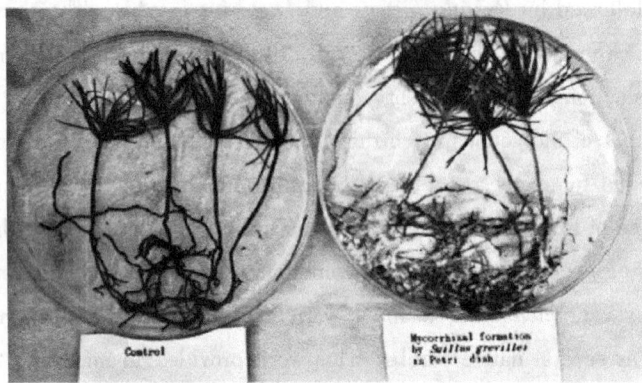

Figure 7-12 Mycorrhizal inoculation method of Chinese pine in the early period of seedlings (Photos by Bai Shulan).

(a)   (b)

Figure 7-13 Selecting drought-resistant fungus-tree combinations of Chinese pine (Photos by Bai Shulan).

(2) Selecting Combinations of Excellent Fungus-trees in Pot for Targeted Mycorrhizal Seedlings

On the basis of the selection of a mycorrhizal colonization rate in the early period of seedlings, the excellent fungus-tree combination should be selected for stress resistance or growth promoting in pot conditions to facilitate the targeted seedling cultivation of the various fungus-tree combinations. The selected tree species can be directly inoculated with mycorrhizal fungi in pot experiment to study the infection rate of the trees by the fungi, and then to study the related topics of the different cultivation directions of the fungus-tree combination, to screen for the best fungus-tree combination for various cultivation directions (Figure 7-13).

Methods: First, the seedling substrate is disinfected (high temperature and high pressure), and then the disinfected seeds are inoculated with different mycorrhizae strains under potting conditions. The inoculum can be a solid inoculum (inoculation amount 50 g/pot) or liquid inoculum (inoculation amount 5 ml/pot). After culturing for a certain period of time when mycorrhizas are symbiosed effectively, stress treatments were carried out to screen out fungus-tree combinations with resistance; or the growth amounts were measured to study their growth promotion effects, thus, to screen out the fungus-tree combinations with growth promotion effects. In the western region of

China, however, the research on the drought-resistant fungus-tree combination selection has been studied the most (Figure 7-14, Figure 7-15).

Figure 7-14 Selection of heavy metal (Zn, Cd)-resistant fungus-tree combinations of *Pinus tabuliformis* potted seedlings (aboveground growth status) (Photos by Bai Shulan).

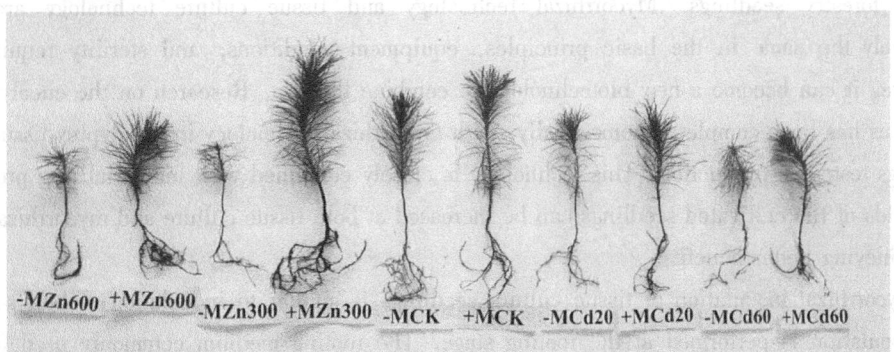

Figure 7-15 Selection of heavy metal (Zn, Cd)-resistant fungus-tree combinations of *Pinus tabuliformis* potted seedlings (root growth status) (Photos by Bai Shulan).

(3) Selection of Excellent Fungus-tree Combinations in a Nursery

From the perspective of production and application, after the selection of excellent fungus-tree combinations in pot experiment, a small-scale production trial in a nursery field is needed to test the stability of the excellent fungus-tree combinations. The method of mycorrhizal seedling cultivation in the nursery includes selection of the fungus-tree combination of seedlings and transplants. Specific methods are as follows.

Selection of the Fungus-tree Combination for Seedlings. The excellent fungus-tree combination selected under laboratory conditions could be verified in a nursery to explore its repeatability and stability. To do so, select a suitable nursery field as a test plot, sterilize the soil with a steam sterilizer, and then perform a mycorrhizal synthesis study of seedlings in the nursery field using the fungus-tree combination selected in the laboratory. After 2 to 3 months, the mycorrhizal morphology and the infection rate of mycorrhizal fungi can be examined, and then targeted cultivation studies are conducted to evaluate the characteristics of different combinations of fungus-trees in the field environment.

Selection of the Fungus-tree Combination for Transplants. Sterilize the soil by steam or by 1% to 2% ferrous sulfate. The mycorrhizal seedlings synthesized in the laboratory are planted directly into the field plots, or container seedlings from the 1-year-old seedlings are used for mycorrhizal synthesis study. The mycorrhizal morphology and the infection rate of mycorrhizal fungi are checked after 2 to 3 months, then further study a variety of target directions and evaluate the characteristics of the fungus-tree combination in the field environment.

7.3.1.3 Cultivation Method of Ectomycorrhizal Seedlings in Nursery

Through the selection of excellent fungus-tree combinations in the early period of seedling, the potted seedlings, and the nursery field, large-scale mycorrhizal seedling production of the best fungus-tree combinations can proceed.

(1) Mycorrhizal Inoculation Method for Tissue Culture Seedlings

Tissue culture technology is a relatively new biotechnology developed in forestry production in recent years, and it is also one of the important measures for industrializing the propagation of various forestry seedlings. Mycorrhizal technology and tissue culture technology are almost completely the same in the basic principles, equipment conditions, and sterility requirements; therefore, it can become a new biotechnology to combine the two. Research on the eucalyptus and pine trees has been completed domestically, with mycorrhizal technology in eucalyptus tissue culture seedlings tested in production. This technology is closely combined with industrializing propagation and yields of the cultivated seedlings can be increased at both tissue culture and mycorrhizal levels, thus achieving better benefits.

Mycorrhizal inoculation of tissue culture seedlings is similar to inoculation of bud seedlings. The inoculation is performed at the rooting stage. The rooting medium commonly used for tissue culture seedlings is not suitable for the growth of some strains, however. According to different needs, certain nutrients should be added to modify the conventional MS tissue culture medium so that it will be beneficial to the rooting of tissue culture seedlings and be suitable for growth of mycorrhizal fungi.

At present, inoculation of tissue culture seedlings is used primarily for easy-to-root broadleaved trees (especially eucalyptus and poplar). In eucalyptus, it takes approximately 20 days from transplanting budlings to seedling rooting and transplanting. Mycorrhizal inoculation can be carried out about 5 days after transplanting the budling to the bottle of tissue culture medium. Generally, 2 to 3 sets of 0.5 cm^2 plate fungus can be inoculated into each bottle. If the seedlings are left in the bottle for a long time, the seedlings can form mycorrhiza in the bottle. The time and the amount of inoculation should be determined according to the growth rate of individual strains. Generally, the inoculation of fast-growing strains can be relatively late in the growth stage, or the inoculation amount can be reduced; by contrast, the inoculation can be early or the inoculation amount can be increased (depending on the growth rate of individual strains). All work must be performed under aseptic condition, otherwise it will cause contamination that will ultimately cause a loss in seedling cultivation. This mycorrhization method does not require additional equipment, additional preparation of inoculants, nor inoculation to each sapling. It is the simplest and most effective method for mycorrhization of industrialized tissue culture seedlings. However, this method can be used only on tree species that have undergone tissue culture research and application. Because of the rapid growth of individual strains, a large number of aerial mycelia can even surround the entire tissue culture seedling, and although they will not cause seedling death, they will affect their growth. In addition to delaying inoculation and reducing the amount of inoculation, reducing the indoor temperature can control the growth rate of strains (Figure 7-16).

(2) Mechanized Inoculation of the Early Period of Seedlings

In the nursery that has achieved mechanized seeding, the fungus-tree combination selected in the laboratory and the nursery is used for mycorrhization during the sowing process. The method is to machine-sow seeds at the same time into different types of inoculum, such as solid inoculum, granule inoculum, and capsule inoculum. After the bud seedlings

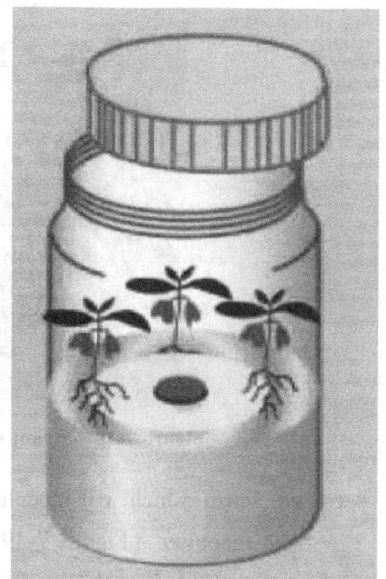

Figure 7-16 Mycorrhizal inoculation of tissue culture seedlings (From Brundrett, 1996).

grow, the roots can contact the strains of inoculum to form mycorrhiza. This method can be used only in nurseries where mechanization is implemented. Before the mechanized inoculation of the bud seedlings, the soil of the nursery bed needs to be disinfected to minimize interference by indigenous bacteria (Figure 7-17, Figure 7-18). The best disinfection method is mechanical steam sterilization. Sterilization can also be accomplished by mixing the soil with 1% ferrous sulfate. Although ferrous sulfate plays a role in sterilization, but it can also acidify the soil, which is more suitable for the growth and development of mycorrhizal fungal mycelia. Most ectomycorrhizal fungi prefer acidic soils.

(3) Seedbed Inoculation Method

Seedbed inoculation is also an intensive method to achieve seedling mycorrhization. There is no need to inoculate each seedling separately. The process can inoculate a larger number of seedlings at one time, which has a good prospect for popularization and application in production. This inoculation method is suitable for seedlings that can grow on the seedbed for a relatively long time, such as 3 to 5 months or more to obtain better results. Several types of inoculum can be applied for this inoculation method.

Figure 7-17 Steam sterilizer for domestic nurseries (Photo by Bai Shulan).

Solid Inoculum. An appropriate amount of the solid inoculum is applied to the seedbed soil, and then

Figure 7-18　Steam sterilizer for foreign nurseries (Photo by Bai Shulan).

seeds are sown, which will inoculate the seedlings.

Solid inoculum (Figure 7-19) that is typically made up of 300 g/m^2 is mixed in a large container, sprinkled on the seedbed, and then plowed into the soil before sowing (Figure 7-19). Afterward, a certain thickness of sieved moist soil is spread across the surface, the thickness of which is determined according to the size of the seed particles: For example, for larch and *Pinus sylvestris* the thickness is 0.5 to 1.0 cm, and for *Pinus tabuliformis* the thickness is 1.5 cm (Figure 7-20). The seedlings growing for a period can be infected with mycorrhiza (Figure 7-20). For the seedling management of seedbed inoculation method, moist soil is generally suitable in the emergence stage. After emergence, keep the seedbed dry; avoid excessive water and do not use chemical agents, such as herbicides. Weeding and loosening the soil can be done by hand without fertilizer.

(a)　　　　　　　　　　　　(b)　　　　　　　　　　　　(c)

Figure 7-19　Solid inoculum cultivated in plastic bags or cans (a and b) and Inoculum mix in seedbed of sowing seedlings (c) (Photos by Bai Shulan).

Mycorrhizal Fungal Fruiting Body Inoculum. Through the preliminary selection, for the determined mycorrhizal fungi of the fungus-tree combinations, fruiting bodies can be collected in the forest during the mushroom emergence period (Figure 7-21). The fruiting bodies are then pulverized to make fungus suspensions (Figure 7-22).

(a) (b) (c)

Figure 7-20 Sifting soil for seedbed after sowing (a), Inoculated Chinese pine mycorrhizal seedlings (b) and 1-year-old Chinese pine seedlings with mycorrhizae (c) (Photos by Bai Shulan).

(a) (b)

Figure 7-21 Fruiting body of *Lactarius vellereus* (a) and Fruiting body of *Suillus bovinus* (b) (Photos by Bai Shulan).

(a) (b)

Figure 7-22 Smashed fruiting bodies (a) and Configuring fruiting body fungus suspension (b) (Photos by Bai Shulan).

The fruiting body suspension is then mixed into the seedbed soil or poured on the roots of the seedlings (Figure 7-23) to achieve the purpose of inoculation. This method is very economical and effective for the use of native strains, and it saves a series of work steps in the production of artificial inoculum. In addition, the method is simple, convenient, practical, easy to master, and the mycorrhizal formation is fast.

Fungus Soil Inoculum. Nursery soil that contains mycorrhizal fungi is transferred to the seedbed to encourage the growing seedlings to form mycorrhiza. Because of limited sources of the inoculated soil, however, this method is rarely used in production.

Seedling Seedbed Inoculum. For seedlings cultivated in the nursery field, seedbed inoculation prepares the way for seedling inoculation, without the need for transplanting, which is also very effective. The inoculant can be the fungus fruiting body suspension, or pure culture liquid, or solid inoculum (Figure 7-23). After inoculation, the inoculum is covered with the soil.

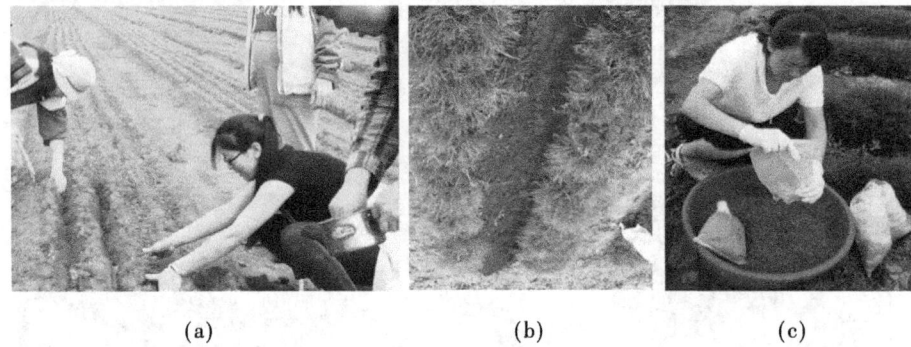

(a)　　　　　　　　　　　　(b)　　　　　　　　　　　　(c)

Figure 7-23　Inoculation of fruiting body fungus suspension at seedling stage (a). Ditching and inoculation of pure solid inoculum in seedling stage (b and c) (Photos by Bai Shulan).

(4) Transplant Inoculation Method

Transplant inoculation is the most commonly used inoculation method in China. In China's current seedling work, transplanting is required in many places and/or for certain tree species. In particular, coniferous trees rarely adopt direct seeding forestation. Transplants can form a developed root system and can significantly increase the survival rate of outplanting. Therefore, mycorrhizal inoculation can be performed at one time during the transplantation of seedlings in the nursery, which can save labor and strains, and can also achieve the purpose of batch inoculation. For example, inoculum (from smashing the fruiting bodies) can be mixed into a slurry and used for dipping roots of the seedlings (Figure 7-24), which will then be transplanted. Or, seedling can be put in a nutrient bag to achieve the purpose of inoculation. In some cases, inoculum strains are formulated into a viscous fungus suspension, and the same purpose can be achieved by soaking the seedling roots for a certain period of time (Figure 7-25). To promote the growth of new roots as soon as possible to better infect mycorrhizae, some long roots and injured roots can be cut off before dipping or soaking. The transplanting seedling inoculation method has good results.

Figure 7-24 Inoculation of seedlings with a solid inoculum (a) that was mixed with mud; dipping roots into the inoculum (b) (Photos by Bai Shulan).

Figure 7-25 Transplanting seedlings to containers after soaking their roots (Photo by Bai Shulan).

7.3.1.4 Cultivation Method of Arbuscular Mycorrhizal Seedlings

The cultivation of arbuscular mycorrhizal seedlings is similar to the cultivation of ectomycorrhizal seedlings. First of all, strains must be obtained and combined with excellent fungus-tree combinations, based on the targeted cultivated trees and their different habitats (such as drought resistance, heavy metal resistance, cold resistance, and disease resistance). Then seedbed seedling cultivation, nutritious cup seedling cultivation, and so forth can be adopted.

However, the most important difference from the ectomycorrhizal seedling cultivation is that the arbuscular mycorrhizal fungi cannot be cultured artificially, and they are not large fungi such as mushrooms, but are instead various fungal spores in the soil that cannot be observed by the naked eye. Its inoculum is mainly the rhizosphere soil of seedlings with mycorrhizae, which is the limiting factor of large-scale application of arbuscular mycorrhizal technology in production. In addition, the arbuscular mycorrhizal plants in a forest are mostly shrubs.

7.3.1.5 Problems in Mycorrhizal Application

(1) Put the Right Fungus in the Right Site

"Suitable trees for the suitable land" is advocated in afforestation, and for the application of mycorrhizal technology we should emphasize "Suitable fungus for the suitable site" and "suitable fungus for the suitable trees". Only the ideal fungus-tree combination can give full play to the beneficial effect of mycorrhizae, and different fungus-tree combinations have different ecological effects on plants. After years of research, Bai Shulan suggested that the growth-promoting effect of *Suillus bovinus* and *Suillus luteus* isolated from Daqingshan (an area in Inner Mongolia, China) was much better than that of fungus *Suillus grevillei* and *Suillus granulatus* on *Pinus tabuliformis*, whereas the drought resistance enhancement of *Rhizopogon luteolus* isolated from the Lamadong, Daqingshan, was more significant than that of other fungal species on *Pinus tabuliformis*. and Bowen believed that the utilization of phosphate rock powder in the soil by *Suillus granulatus* and *Rhizopogon luteolus* was better than that of *Boletus variegatus* and *Cenococcum geophilum*; *Suillus plorans* was suitable for being inoculated on pine or larch in cold areas at high altitudes. Obviously, in these comparisons, in addition to regional factors, differences in tree species must be considered. The effect of tree species on mycorrhizal inoculation also differs. Bai's (2011) experiment suggested that the infection rate of the collected *Suillus grevillei* isolated from Pichaigou, Daqingshan, was much faster than that of *Suillus grevillei* isolated from a Narisitai forest farm. Therefore, the effect of the ecotype of the strain on the tree species is indeed different.

For a variety of pine trees in the low altitude areas of southern China, we can consider inoculating with fungi such as *Pisolithus*, *Scleroderma*, and *Rhizopogon*, whereas for some pine trees in the north, the effect is better with fungi such as *Boletus*, *Suillus*, and *Rhizopogon*. In the north, in addition to the above fungus, some tree species can use fungi of *Lactarius*, *Laccaria*, and *Cortinarius*. At present, finding a strain that not only adapts to different areas and different species but also shows very good effects is difficult.

(2) Cooperation of Forestry Technology

No matter what kind of mycorrhizae, its application must be accompanied with appropriate production measures, and the most suitable conditions must be created for the continuous growth and mycorrhizal formation. Otherwise, mycorrhizae cannot play a good role. So, at this stage especially, pesticides and chemical fertilizers should be used as little as possible.

(3) Keep the Vitality of the Inoculum

Mycorrhizal inoculum is a biological preparation, which must be inoculated at its strongest vigor. Therefore, the inoculum must be stored in a clean environment at a low temperature (2−5 ℃), and the storage time cannot exceed 6 months.

(4) Correct and Flexible Application of Inoculation Technology

No matter which type of inoculum is used, let the inoculum contact the root as much as possible, for instance, by applying the inoculum to the root, root soaking, root dipping, root slurry, root injection, and other methods. Inoculation time is best at the seedling stage, which not only saves labor, time, and fungus inoculum, thus saving costs, but also is easy to manage and allows for vigorous growth of nutritive roots and rapid symbiosis with mycorrhizal fungus. The amount of inoculation is directly related to the formation of mycorrhiza. For young and small seedlings, the inoculation amount can be less, and for large seedlings, the inoculation amount can be increased appropriately. In general, maximum effectiveness occurs when the new nutritive root can fully contact the inoculum as it grows out.

7.3.2 Inoculation and Management of Rhizobia

Rhizobium is a type of bacteria that symbioses with legumes, forms root nodules, and fixes nitrogen in the atmosphere to supply nutrients to plants. This symbiont is called the root nodule. The main differences between a root nodule and ectomycorrhiza are as follows.

The nodule is a structure formed on the lignified root of a plant, whereas ectomycorrhiza is an association of different branching characteristics formed on the non-lignified nutritive root tip of a plant (Figure 7-26). The strains forming a root nodule are bacteria, and the strains forming mycorrhizae are soil fungi. A root nodule can fix nitrogen, and most of the plants that can form root nodules are legumes, whereas mycorrhizae cannot fix nitrogen, and the plants that can form mycorrhizae occupy more than 90% of the terrestrial plants. Mycorrhizae can help host plants absorb water and nutrients in the soil by expanding the absorption area of host roots, secreting organic acids and enzymes, and in other ways, all of which improve the growth and stress resistance of host plants. Therefore, mycorrhizae play a more prominent role in the host plants in a stressful environment.

(a)　　　　　　　　　　　　　　(b)

Figure 7-26　Root nodules on legumes (a) and Ectomycorrhiza formed on plant roots (b) (Photos by Bai Shulan).

The application of rhizobia in agriculture, horticulture, forage, and so forth, have been widely studied. Most of the plants inoculated with rhizobia in forestry are legume shrubs (although *Acacia confusa* is a tree more common in arbors). The afforestation environment in China is not ideal, especially the ecological environment is very bad in the western region of China, for example, arid soil, lack of microorganisms, and natural exposure of rhizobia to roots is very difficult after afforestation. Therefore, inoculating rhizobia in the nursery is an important research direction for afforestation quality improvement. Here we will briefly introduce the application of rhizobia in production.

(1) Strain Isolation

Because of the strong specialization of rhizobia, it is most effective to inoculate rhizobia isolated from the same tree species. The method is to select the seed tree of pre-cultured seedlings for root nodule collection, take a root nodule to the laboratory, clean the nodule, sterilize it, isolate the tissue, and purify it for many times to obtain the strain (specific method omitted here).

(2) Inoculation Method

Seed Soaking Method. The isolated rhizobia is cultured to make liquid inoculum. Seeds of the target tree species are sterilized, force germinated, and then immersed in the liquid inoculum. The soaking time is determined according to the characteristics of the seeds. Generally, the soaking time is 2 to 12 hours. Remove seeds from the liquid and allow them to dry slightly to sow.

Seed Dressing Method. Seeds are sterilized and force germinated. Then seeds are mixed evenly with the inoculum, and after drying them slightly in the shade, the seeds can be planted. Alternatively, the rhizobia inoculum can be mixed with granular fertilizer, and then planted with germinating seed at the same time.

Root Soaking Method. The seedlings are cultured first, and then the roots of the seedlings are soaked with liquid inoculum for 20 to 30 minutes before transplanting to the nursery field or to nutrition bags.

Irrigation Method. During the growth period of seedlings, a 5 to 10 cm ditch should be dug in the root soil of seedlings cultured in the seedbed. The cultivated rhizobia inoculum should then be configured as liquid and poured into the ditch, after which the ditch should be covered and leveled.

(3) Post-inoculation Management

After inoculating with rhizobia, seedlings are under a nursery's routine management.

7.4 Weed Management

Weeds are the enemy of seedling cultivation. Because weeds compete with seedlings for nutrients and water uptake in the nursery and affect light and air circulation, weeds worsen the growth and development conditions of seedlings. Furthermore, weeds inhibit the metabolism process, reduce the accumulation of beneficial substances, and therefore affect the growth and development of seedlings. Sometimes weeds lead to a large number of diseases and insect pests in the seedlings, which can cause seedling cultivation failure.

According to experimental data, the water consumption of lambsquarters (*Chenopodium album*) is three to four times more than that of *Larix gmelinii* seedlings in the same year. Therefore, weed control is not only a regular task of seedling cultivation but also an important aspect of nursery land management.

Weed control is completed by weeding operations, with the goal of eliminating weeds and ensuring the normal growth of seedlings. Commonly used weeding methods are artificial weeding, mechanical weeding, and chemical weeding. Among them, artificial weeding and mechanical weeding can be combined with intertillage, which is generally called loosening soil and weeding.

7.4.1 Loosening Soil and Weeding

Loosening soil and weeding are two different processes in seedling cultivation, but they are generally integrated, because these two measures can be completed by one operation. Loosening soil reduces compacted soil during the growth of seedlings, which belongs to the intertillage part of soil cultivation. Its main purpose is to loosen the soil, increase the permeability, and provide sufficient oxygen for root respiration. To loosen the topsoil with a hoe or machine also destroys the living environment of weeds and completes the removal of weeds. Through one operation, two measures are completed to achieve two goals.

7.4.1.1 Methods of Loosening Soil and Weeding

Loosening soil and weeding can be accomplished through two methods: Manual operation and mechanical operation. Manual operation involves pulling out weeds by hand or uprooting weeds using a hoe and other tools, which is a traditional weeding method without any side effects. Manual weeding requires high labor intensity and has low efficiency. With the increase of labor costs, the expense of manual weeding is also increasing. Generally, the labor of manual weeding accounts for 20% to 60% of the labor of the whole nursery. According to the statistics of 30 nurseries in Heilongjiang Province, the weeding cost accounts for 40% to 60% of the total labor cost of the nursery.

Mechanical operation is a method of loosening soil and weeding by using a variety of special weeding or intertillage machinery. Compared with manual operation, it is faster, more efficient, and lower in cost. It is suitable for removing weeds among large row spaces, but it cannot remove weeds among seedlings in small spaces. For example, in a large seedling area with a planting spacing of more than 1 m, a walking tractor can be used to loosen soil and weed on the seedbeds.

7.4.1.2 Time and Requirements for Loosening Soil and Weeding

Weeding must be carried out in a timely manner. The principle is "weeding early, weeding small weeds, and weeding thoroughly", Weeds in the nursery should be removed before the rainy season. Loosening soil and weeding generally occurs five to eight times a year, mostly after irrigation and rainfall.

The depth of soil loosening varies with the size of seedlings, generally 2 to 4 cm for small seedlings, and gradually deepens to 7 to 8 cm or even more than 10 cm as the seedlings grow. In principle, the roots cannot be damaged, so the seedlings cannot be bruised or hoed off.

7.4.2 Chemical Weeding

Chemical weeding refers to using chemical agents, such as herbicides or rust removers, to control weeds in a nursery. Its advantages are speed, high efficiency, and low cost (as compared to manual weeding). Unlike the increasingly expensive labor cost and inefficiency, as well as the high recurrence rate of weeds, chemical weeding can save 60% to 80% of labor, reduce 40% to 80% of cost, and provide advantages such as high weeding efficiency and thorough weeding effect.

However, chemical weeding also has prominent disadvantages. While killing weeds, chemical agents may also harm other organisms and have adverse effects on humans, animals, and biodiversity. Therefore, we should be careful when using chemical weeding. Use herbicides scientifically and minimize the damage to biodiversity while ensuring human and animal safety.

7.4.2.1 Characteristics of Chemical Weeding

(1) High Technical Requirements

There are many types of herbicides, and the use requirements are strict. Only by choosing herbicides correctly and mastering their usage technology, can we achieve the goal of weeding and seedling raising. Otherwise, the seedlings may be killed together with the weeds.

(2) Long Duration of Weeding Effect

Manual weeding has only a short-duration effect, for example, only 12 days in the south before the same area again needs to be weeded. Chemical weeding can last for several months, however, perhaps even the entire growing season. According to experimental reports, glyphosate's effect can last for at least 2 months in the nursery of Weihe River Basin (Figure 7-27).

Figure 7-27 Seedling cultivation land after herbicide application (Photo by Liu Yong).

(3) Easy to Use and Good Effect

Generally, sprayer, sprinkler, fine-mesh watering, and pesticide-clay mixtures can be used in herbicide application in nursery. The speed of application is fast, and weeding is timely and thorough. According to experimentation, more than 90% weeds on cultivated land of *Pinus tabuliformis* and *Populus* seedlings can be killed by glyphosate and gramoxone.

(4) Reduce Pests and Diseases

Some herbicides and pesticides are mixed prior to application. For example, the mixture of herbicide propanil and insecticide carbaryl in specific proportions can control weeds and kill insects, while also increasing the activity of herbicides. In addition, because of the elimination of weeds, some diseases and insect pests that have weeds as their intermediate host lose the conditions of transmission and spread, therefore reducing their occurrence. According to the investigation, when nitrofen is applied to the seedbed, mole crickets will be greatly reduced.

7.4.2.2 Types of Herbicides

Herbicides are available in many types and with different properties. Its product properties, chemical structure, mode of action, suitable objects, and usage methods are various. Classifying them from different aspects is the basis of correctly understanding and mastering their working principles and usage methods.

(1) Classification According to Chemical Structure

Organic Herbicide. This type is synthesized primarily by benzene, alcohol, fatty acids, organic amines, organic phosphorus, and other organic compounds. Commonly used varieties include oxyfluorfen, trifluralin, paraquat, atrazine, glyphosate, napropamide, 2,4-D, haloxyfop, and sulfometuron methyl. Because of its small dosage, wide applicability, and good effect, it is the main herbicide used all over the world. Some of them are banned because of safety problems such as poisoning and pollution.

Inorganic Herbicide. This type of herbicide is made of natural minerals, which does not contain organic carbon compounds, but instead contains inorganic elements such as copper, iron, sodium, potassium chlorate, sodium chlorate, sodium chloride, arsenious acid, copper sulfate, and sodium arsenite. Because inorganic herbicides have stable chemical properties, do not readily decompose, and are generally soluble in water, they are usually systemic herbicides and sterilant herbicides. They are not only harmful to plants, people, and animals but also easily lost in the soil, causing pollution to the environment. Therefore, they are currently rarely used in production.

(2) Classification According to Action Characteristic

Contact herbicide can kill only the parts of plants that are in contact with the herbicide. It has poor effect on the underground parts or weeds with a subterraneous stem. Such herbicides include oxyfluorfen, nitrofen, paraquat, sodium pentachlorophenol, propachlor, and caproanilide.

Systemic herbicide can be absorbed by roots, stems, leaves, and coleoptiles of the plant and can be transmitted to the whole plant through the transfusion tissue, damaging the physiological balance of the internal structure of the plant and resulting in plant death. Systemic herbicides include 2,4-D, glyphosate, atrazine, pucaolong, and diuron.

(3) Classification According to Selective Characteristics

Selective herbicide can kill weeds and is harmless to seedlings. Products include nitrofen, weed killing ether, chlornitrofen, haloxyfop, prometryne, simazine, gol, and quizalofop-ethyl.

Sterilant herbicide is toxic to all plants. The seedlings and weeds will be killed as long as a green portion of the plant is touched. Such herbicides include glyphosate, paraquat, and sodium pentachlorophenol.

(4) Classification According to Usage

Stem and Leaf-treated Herbicide. Weeds are evenly sprayed with small droplets of a solution of water and an herbicide, such as haloxyfop, oxyfluorfen, glyphosate, quizalofop-ethyl, or paraquat.

Soil Treatment Herbicide. By atomizing, spraying, spraying powder, pesticide-clay mixture, pouring, and scattering granules, the medicament is applied to the soil to form a layer of "medicine", which can contact the seeds, seedlings, young roots of the weeds, or be absorbed by them, and kill the weeds. Herbicide includes nitrofen, propachlor, ethaprochlor, trifluralin, simazine, prometryne, or chlornitrofen.

These two methods are relatively used, because some herbicides can be used for both stem and leaf treatment and soil treatment, such as 2, 4-D, atrazine, and chlorotoluron.

7.4.2.3 Principles of Chemical Weeding

Herbicides can disturb and destroy one or several physiological and biochemical links of weeds, making the whole biochemical process disordered and thus inhibiting growth and development or leading to death of weeds. The principle of killing weeds can be summarized as follows.

(1) Inhibiting Photosynthesis

The inhibition of herbicides on photosynthesis take effect through the inhibition on photoreaction and dark reaction of photosynthesis. Herbicides such as paraquat, atrazine, lasso, propachlor, and diuron can prevent the Hill reaction at very low concentration, and strongly hinder the photosynthetic phosphorylation of chloroplasts. The stronger the light, the faster the action, making the plant leaves show fading, chlorosis, and wilting. The above herbicides can block Hill reaction, which is closely related to their chemical structure. NH groups occur in these compounds, which easily form hydrogen bonds with zymoprotein. In addition, herbicides with CO or CN groups may also form hydrogen bonds with zymoprotein. These hydrogen bonds hinder the normal process of biochemical activities in photosynthesis, affect the light reaction, and make the second step reaction of photosynthesis —the process of fixing carbon dioxide in the dark reaction—unable to proceed normally. Plants die of hunger because of the failure of normal synthesis of carbohydrates.

(2) Interfering with Respiration

The respiration of plants is not completed at one time, but rather through multiple steps of degradation and conversion. When one of the important links is destroyed, the survival of plants is threatened. Different types of herbicides have different effects on plant respiration. Herbicides such as pentachlorophenol sodium destroy the oxidative phosphorylation process in the respiration process, so that the energy released by respiration cannot be used, and the life of plants cannot be maintained; while nitrofen inhibits the dehydrogenase, so the formation of high-energy bond compounds ATP cannot be realized in the respiration of plants, which makes weeds die.

(3) Interfering with Normal Hormone Action

Some chemical herbicides are hormones themselves, such as 2, 4-D, 2, 4, 5-T, dicamba, and MCPa. Hormones have two opposite effects on plants. On the one hand, they can promote growth at a proper low concentration; on the other hand, when the concentration is too high, the growth of plants is out of balance and inhibited. For example, 2, 4-D has a strong contact killing effect at high concentration, which mainly causes abnormal cell proliferation and abnormal changes in nucleic acid and protein metabolism and synthesis, thereby forming a deformed tumor that blocks the conducting tissue, hinders the transport of organic matter, and causes the roots of plants to die from lack of nutrients.

(4) Interference in Nucleic Acid Metabolism and Protein Synthesis

Some herbicides can interfere or inhibit nucleic acid metabolism, protein synthesis, and enzyme activity. For example, glyphosate can inhibit the synthesis of aromatic amino acids in plants, thus hindering the synthesis of proteins; 2,4-D, MCPa, and dicamba can reactivate the sensitive cells in plants, cause excessive synthesis of nucleic acids and proteins, make plant tissues proliferate, and form abnormal growth.

7.4.2.4 Herbicide Selectivity

When herbicides are applied in a nursery, we expect they will eliminate weeds and do no harm to seedlings. Some herbicides have a certain specific selectivity, while others have no selectivity or poor selectivity, but we can use some of their characteristics or the differences between seedlings and weeds to achieve the purpose of weeding. The principle of herbicide selectivity can be divided into biological and non-biological.

(1) Biological Selection

This type of selection utilizes the differences of germination morphology, structure morphology, and physiological differences between seedlings (seeds) and weeds. For example, when the *Larix gmelinii* seedbed is treated with nitrofen, it is generally carried out after sowing and before the seedlings emerge from the ground. After application, a layer of nitrofen is applied on the surface of the soil, which will be touched by the buds when weed seeds germinate and buds will be killed under the light. The top shell of *Larix gmelinii* unearths during germination while the young buds are wrapped in the seed shell, and the seed shell opens and pushes the poisonous soil, so the young buds are not harmed. Another example, when the seedlings of Scots pine (*Pinus massoniana*) are unearthed, their growth points are not exposed during a certain period, but they are surrounded by cotyledons, on which is a surface with a very thick waxiness layer the contains oil substances. Hence, herbicides do not easily penetrate. On the contrary, for some weeds, such as pigweed, the top bud is exposed after germination, and the cuticle on the leaf surface is thin, so once it is dipped in the liquid herbicide, the plants are killed through the quick penetration of herbicide.

(2) Non-biological Selection

This type of selection can be divided into place difference selection and time difference selection. Place difference refers to the difference in the distribution depth of the roots of seedlings and weeds in the soil. For example, the distribution of the roots of the reserved bed seedlings and the changed bed seedlings is deeper than that of some 1-year-old weeds. The surface weeds are killed by chemicals, such as simazine or prometryne, with little movement into the soil, while the seedlings are protected from damage due to lack of contact with the chemicals. Time difference refers to the use of different germination stages of seedlings and weeds. In many cases, the germination of weed seeds is earlier than that of forest tree seeds. When the weeds have germinated and the seedlings have not yet sprouted, the germinated weeds can be eliminated by using a contact herbicide, such as pentachlorophenol sodium, to treat the stems and leaves.

For many broad-leaved trees, because they are sensitive to herbicides, seedling contact with a herbicide should be avoided. For example, in the seedling stage, inter-row application of herbicide or herbicide-clay mixtures can be used (particles can also be used) because they do not easily adhere to the aboveground part of the seedlings. The herbicide or herbicide-clay mixtures do drop down into the surface of the soil, so it will prevent and eliminate the weeds between rows and plants.

7.4.2.5 Usage of Herbicides

(1) How To Use Herbicides

Because of the different properties, dosage forms, and purpose of herbicides, the methods used in a nursery are different. At present, stem and leaf treatment and soil treatment are generally used.

Stem and Leaf Treatment. This method refers to spraying or applying herbicide directly on the stems and leaves of weeds. According to the application period, it can be divided into stem and leaf treatment before sowing and stem and leaf treatment after seedling. The treatment of stem and leaf before sowing is to spray the herbicide on the weeds that have already grown but sowing has not yet been done, or before seedling transplantation, which requires a wide-spectrum herbicide with poor selectivity and no residue. The commonly used herbicides include paraquat, glyphosate, and oxyfluorfen. This method can eliminate the weeds that have already grown, but it is difficult to control the weeds that may occur later. The treatment of stem and leaf after seedling refers to the method of applying herbicide after germination and excavation of tree seeds. If no protective measures are taken during application, weeds and seedlings will be exposed to the liquid herbicide. Therefore, it is necessary to take protective measures for seedlings when using stem and leaf treatment after emergence, for example, covering the seedlings and/or washing the seedlings with water immediately after spraying, so as to prevent damage to the seedlings. For wide-row seeding seedlings, sterilant herbicides can be used to eliminate weeds between rows. However, there must be protective plates in the sprayers so that the liquid herbicide cannot reach the seedbed.

Spraying method is generally used for stem and leaf treatment, and herbicide dosage forms suitable for spraying include wettable powder, emulsifiable concentrate, and water agent (in which herbicides are dissolved in the water for use). The diameter of the spray mist point must be less than 100 to 200 μm. A mist point that is too large in diameter has poor adhesion, making it easy for the herbicide solution to run off; a mist point that is too thin is easy to be blown away by the wind, which reduces the amount of adhesion. The water quantity should be 450 to 750 kg/hm^2.

Water tanks, buckets, filter gauze, and mixing tools are used for dispensing. To make the dosage accurate and error free, it is necessary to determine the container, dosage, and quantity of water. According to the size of the container, the needed dose should be weighed by a balance or calibrated scale and packed separately in advance; prepare one package each time. A quantitative water line should be marked on the fixed bucket for consistency each time. The weighed herbicide should be packed with gauze and dissolved with a small amount of water. Remove the residue in the gauze because the dilution is accurate only if all that residue is included. Add the required amount of water to dilute the herbicide liquid. The quantitative herbicide can also be dissolved in a small amount of water, fully stirred into a paste, and then added into the measured quantity of water to stir evenly. Liquid herbicide should be used right after it is prepared and should not be stored for a long time to avoid losing its needed effects.

Soil Treatment. This step refers to spraying, pouring, sprinkling herbicide-clay mixtures, and other methods to apply herbicides to the soil to form a certain thickness of pesticide layer, which can kill weeds by contacting weed seeds, buds, or other parts of the weeds (such as coleoptiles). Soil treatment is generally used to control and remove the annual weeds germinated by seeds or some perennial weeds. It is most effective to treat the soil after the tree seeds are sown but before emergence of seedlings, when the soil moisture is suitable and many weeds germinate. Herbicide can also be sprayed on the soil surface before sowing. The mist point can be thicker and the dosage can be larger when using the spiketooth harrow and evenly dispersing the herbicide into the soil layer 3 to 5 cm deep.

Toxic soil is a mixture of herbicide and fine soil. The herbicide formulations suitable for making toxic soil include powder, wettable powder, and emulsifiable concentrate. The fine soil is generally sieved by a #10-20 sieve. It cannot be too dry or too wet; it is suitable when it can be kneaded it into a ball by hand and loosened by a gentle touch to disperse automatically. The amount of soil is subject to the uniform distribution, generally 225 to 375 kg/ha. The powder can be directly mixed with soil; in terms of emulsifiable concentrate, it is diluted with water first, then sprayed on fine soil and mixed evenly. If the dosage of the herbicide is less, it can be thoroughly mixed with a small amount of soil first, and then with the full amount of soil. Each batch of toxic soil should be prepared and used at one time, and it should not be stored for a long time. The application of toxic soil should be uniform and appropriate.

(2) Herbicide Mixing

The correct mix among herbicides or herbicides with pesticides and chemical fertilizers can save labor, reduce costs, expand the weed-killing spectrum, and serve multiple functions of weeding, insecticide, disease prevention, and top-dressing.

Mixed Use among Herbicides. Reasonable mixed use of herbicides can improve the control effect of weeds; reduce the times of application of herbicides; play the complementary role among herbicides in the function, property, speed, and effect of weeding; and increase the adhesion and distribution performance of herbicides or enhance the permeability to plants. In the case of synergism between two herbicides, the dosage can be reduced to save the cost of weeding.

Principle in Mixed Use of Herbicides.

i. Do not Affect the Chemical Properties of the Agentia. No chemical reaction occurs between the components, and the chemical properties of various effective components cannot be changed. For example, organophosphorus herbicides are prone to alkaline hydrolysis under alkaline conditions. Therefore, they are not suitable to mix with alkaline substances, or they can be prepared and used only at the same time. For the same reason, acid herbicides cannot be mixed with alkaline substances. Many herbicides with metal ions such as copper, manganese, and zinc form soluble metal salts under alkaline conditions, which can cause herbicide damage or failure.

ii. Do not Damage the Physical Properties of the Herbicides. The original physical properties of the involved herbicides in a mix, such as emulsification, dispersion, wetting, or suspension, will not disappear or decline, but rather will strengthen.

iii. Reduce the Toxicity. The best case is reducing toxicity for humans, livestock, fish, bees, natural enemy insects, and other beneficial organisms.

iv. Do not Decrease Efficacy. It is better to improve the effect. The mix of an extremely strong contact herbicide and a systematic herbicide sometimes reduces the herbicidal effect. For example, when a large proportion of paraquat is mixed with glyphosate, paraquat can kill the leaves of plants quickly and make them lose their absorptive capacity, which means that glyphosate cannot be fully absorbed by the leaves of weeds. If this occurs, it reduces the herbicidal effect.

v. Do not Cause Herbicide Damage. Mixed herbicides sometimes produce a substance that causes harm to seedlings due to chemical reactions. Before mixing herbicides, we should first understand the physical and chemical properties of herbicides. Only when no antagonistic effect can occur between them, can we obtain the ideal effect.

The mixed herbicides that have been successfully applied in production include pentachlorophenol sodium + 2,4-D, pentachlorophenol sodium + nitrofen, ethaprochlor + nitrofen, haloxyfop + oxyfluorfen, Lasso + linuron, Lasso + atrazine, and trifluralin + 2,4-D, Lasso + paraquat.

Mixture of Herbicide, Insecticide, and Fungicide. Such a mixture can play a role in weeding, killing insects, and preventing diseases. For example, in a nursery of *Pinus tabuliformis* and *Pinus massoniana* seedlings, the use of nitrofen 750 g + thiram 375 g per hectare not only controls and removes 1-year-old weeds but also inhibits damping-off in the seedlings.

Mixed Use of Herbicide and Chemical Fertilizer. The mixture of herbicide and chemical fertilizer is an effective method. For example, when 2,4-D sodium salt and ammonium sulfate are mixed, its effect of killing weeds is significant. When 2,4-D sodium salt is added into ammonium sulfate, the surface tension and pH value of the solution can be reduced, thus the adhesion of the agent can be increased. Because of the acidification of the herbicide solution, 0.5 to 0.8 kg ammonium sulfate or 10 kg superphosphate or 6 to 8 kg ammonium nitrate are added into the free 2,4-D sodium salt solution per hectare, which can improve the efficacy and promote the growth of seedlings.

7.4.2.6 Remedies after Herbicide Damage

If seedlings are damaged by herbicide, corresponding remedial measures should be taken according to the situation (Table 7-5). For a seedling plot with very serious herbicide damage and an estimated final yield loss above 60% or even more, the plot should be destroyed immediately and re-cultivated (replanted) or replaced (replanted) with other tree species, so as to avoid greater loss due to the delay of the farming season. The following remedial measures can be taken for the plot with less pesticide damage.

(1) Wash with a Large Quantity of Water

If herbicide damage is found quickly, a large amount of clear water can be sprayed repeatedly on the damaged leaves of the seedlings, with the goal of washing off the herbicide on the surface of the plants as much as possible. An application of phosphorus and potassium fertilizer should be increased along with intertillage and soil loosening to promote the development of the root system, which will enhance the recovery ability of the seedlings. When using a large amount of water to wash, the seedlings absorb and increase the water content in their cells, which dilutes the herbicide concentration in the seedlings, therefore to a certain extent reducing the herbicide damage.

(2) Increase the Application of Quick-acting Fertilizer

To enhance the growth vigor and recovery ability of the seedlings, quick-acting fertilizer such as urea should be rapidly applied to the seedlings damaged by herbicide. This measure works well on the seedlings with a minimum of damage.

(3) Spray Medicine to Relieve the Damage

Depending on the nature of the herbicide causing the damage, spraying certain chemicals can alleviate the potential damage to seedlings, for example, the damage caused by oxyfluorfen, 2,4-D, MCPa, glyphosate, and pyrazosulfuron can be alleviated by spraying gibberellin, foliar fertilizer, or cytokinin.

Table 7-5 Usage methods and symptoms of herbicide damage of several herbicides commonly used in nurseries

Name	Type	Scope of application	Usage method	Symptoms of herbicide damage
Oxyfluorfen	Contact type	Conifer, populus, and willow cutting nursery	After sowing or before sprouting of cuttings; 40 days after seedlings sprout or spraying between rows	Spots appear on the surface of leaves; local tissues or the whole plant is dry
2,4-D	Sterilant contact type	Road and leisure area	Stem and leaf treatment in spring and summer	Leaves and stem tips are curly; roots are coarser and shorter, and root hairs are short and damaged in the shape of "brush hairs"
Trifluralin	Systemic type	Populus and willow cutting nursery	Soil treatment before cutting	Root tip swells and thickens; fibrous roots are few, and the root system is easy to pull out
Glyphosate	Inactivation	Sowing land, road, and leisure land	Stem and leaf treatment in germination period of weeds before sowing	Entire plant is chlorotic, necrotic, and dry, and it does not harm non-green parts

(4) Remove the Parts of Plants with Severe Damage

After the occurrence of the herbicide damage, the branches and leaves of the seedlings with severe damage shall be removed quickly to avoid further transportation, transmission, and penetration of the herbicide in the seedlings; the damaged plots shall be irrigated quickly to prevent the scope of the herbicide damage from expanding.

7.4.2.7 Precautions for Herbicide Application

(1) Correct Selection of Herbicides

Herbicides are highly selective. Different plants have different sensitivity to herbicides, so select effective herbicides according to the types of seedlings and weeds. For example, coniferous trees have strong resistance, so nitrofen, haloxyfop, and oxyfluorfen can be used; broad-leaved trees have poor resistance, so nursery weed sealing, nursery weed cleaning, butralin, and prometryne can be selected. Other factors that affect herbicide selection include soil types, soil temperature, soil moisture, soil pH, organic matter, weeds or seedlings under stress (or not), herbicide use method, herbicide retention on leaves or soil surfaces, and spray volume. In the same nursery, if a single type of herbicide is used for the same type of weeds year after year, it will not only induce the weeds to produce resistance to the herbicide but also promote the original minor weeds to gradually rise to be the dominant weeds. As the weed population changes, nurseries may experience increasing difficulty in controlling the weeds. Therefore, herbicides can be used in combinations or interchangeably.

(2) Reasonable Control of Dosage

Compared to other pesticides, herbicides have no strict requirements on the concentration of solution, but they have strict requirements on the usage amount and uniformity of the unit area. The reasonable use amount is affected by the type and age of seedlings, the application time, and the environmental conditions. Generally speaking, the dosage of coniferous species can be larger and that of broad-leaved species should be less. With the increase of seedling age, the amount of herbicide can be increased accordingly. For example, to control the Poaceae weeds of pine and fir seedlings, 50 mL of 23.5% oxyfluorfen missible oil can be applied singly per mu, or 30 mL of 23.5% oxyfluorfen missible oil is used in a mix with 100 mL of 50% acetochlor missible oil. If it is used for broad-leaved seedlings, the dosage should be reduced appropriately. For 1-year-old weeds, the recommended dosage can be used; for perennial malignant weeds and perennial weeds, the dosage should be increased appropriately.

Soil with a high content of organic matter has fine particles, large adsorption capacity for herbicides, and more soil microorganisms with vigorous activities, which means that if the herbicide dosage is diluted, the amount applied should be appropriately increased. Sandy loam soil has coarse particles with limited adsorption capacity for chemicals, and the chemical molecules are mostly in the free state between the soil particles with strong activity, which means the chemical is more likely to cause herbicide damage, so the dosage can be appropriately reduced. In addition, under the condition of high temperature and rain, the dosage should be reduced; the lower limit of dosage should be used when weeds are small, and the upper limit should be used when weeds are large. Increasing the dosage at will, spreading the mixed herbicides unevenly, spraying with too little water, or spraying with too much local concentration after re-spraying will cause herbicide damage to plants. In the herbicide preparation, it should be handled according to the recommended dose and concentration, according to the proportion, and weighed with measuring tools to ensure the accurate dosage.

(3) Correct Choice of Application Time and Method

The sensitivity of weeds to herbicides differs across growth periods. When the herbicides are used in the most sensitive period, the best weed control effect can be achieved. In general, weeds are most sensitive to soil treatment herbicide during germination, and the two to three leaf stage is most sensitive to stem and leaf treatment herbicide. For soil treatment herbicides, they must be used before the germination of weeds. After the weeds grow, resistance increases, and the weeding effect is poor. For stem and leaf treatment herbicides, we should grasp and implement the principle of "removing the weed early when it is small". The older (and larger) the weeds, the stronger the resistance. In normal years, when approximately 90% of weeds emerge, the tissue is young, weakly resistant, and easy to be killed.

One of the basic conditions for achieving ideal weeding effects is to choose the correct method for herbicide application and improve the herbicide application technology. According to the nature of herbicides, correct use methods should be determined. For example, for sterilant herbicides such as glyphosate, directed spray must be used; otherwise, it will cause damage to seedlings. Trifluralin needs to be mixed with soil, otherwise it is easy to cause photolysis and failure. Uniform application of herbicides is the basic requirement of herbicide application. In addition, spraying should be avoided when it is windy, so as not to endanger adjacent plants. Spray appliances, such as sprayers, are best using exclusively for one chemical. If that is not possible, after using the sprayer, rinse with bleaching powder.

(4) Pay Attention to the Eluviation and Persistence of Herbicides

Generally, the solubility of herbicides in a soil treatment agent is small, but in sandy soil, when precipitation is greater, a small amount of herbicide will be leached into the deep layer of the soil, which may then cause damage to seedlings. Therefore, in the above circumstances, the dosage should be reduced appropriately.

Under normal conditions, the persistence of herbicides vary, for example, 5 to 7 days for pentachlorophenol sodium, 20 to 30 days for nitrofen, and half a year for simazine. For herbicides with short effects, they should be applied in the germination period of weeds; for herbicides with persistent effect, the safety of subsequent seedlings should be considered (Table 7-6).

Table 7-6 Effective duration of herbicides in soil

Herbicide	Effective dose /(kg/hm^2)	Treatment time	Duration /d
2,4-D	0.750-1.500	Before or after seedling (growth period)	7-30
MCPa	0.555-1.125	Before or after seedling (growth period)	7-30
Sodium pentachlorophenate	5.595-18.000	Before or after seedling (growth period)	7-35
Amerol	1.950-10.005	Before or after seedling (growth period)	21-35
Simazine	1.125-1.500	Before emergence of seedlings	90-180
Phenacolone	0.750-1.500	Before budding of weeds	90-360
Monuron	0.750-1.500	Before emergence of seedlings	90-180
Diuron	0.750-1.500	Before emergence of seedlings or budding of weeds	90-180

(Continues)

Herbicide	Effective dose /(kg/hm^2)	Treatment time	Duration /d
Neburon	0.750-2.400	Before emergence of seedlings or budding of weeds	90-180
Propham	4.485-8.955	Before emergence of seedlings or budding of weeds	15-40
Chlorpropham	4.485-8.955	Before emergence of seedlings or budding of weeds	21-30
Nitrofen	1.500-2.250	Before and after emergence of seedlings	25-40
Prometryne	0.495-1.500	Before and after emergence of seedlings	100-180
Trifluralin	0.600-1.245	Before weeds budding	20-30
Lasso	1.950-3.450	Before weeds budding	45-60
Chlornitrofen	1.500-2.250	Before and after emergence of seedlings	30-75
Oxyfluorfen	0.150-0.300	Before and after emergence of seedlings	60-90

(5) Safety Issue of Herbicides

Although herbicides bring convenience to weeding in a nursery, it inevitably causes a series of safety problems. For example, trifluralin and Lasso have carcinogenic effects on human beings and animals; sodium pentachlorophenol and paraquat are highly toxic; the smell of MCPa is an irritant to human beings and causes herbicide damage to seedlings; 2,4-D can cause floating pollution, which has caused severe herbicide damage to large areas of vegetables, cotton, and golf course trees. Therefore, herbicides such as nitrofen, diphenyl ether, chlornitrofen, dalapon, trifluralin, Lasso, pentachlorophenol sodium, and paraquat have been banned in China since 2000. Many countries in Europe prohibit MCPa, 2,4-D, and others.

Some herbicides commonly used in nurseries and their application technology are shown in Table 7-7.

Table 7-7 Common herbicides in nurseries and their application technology

Commodity name	Dosage form	Reference dosage	Applicable object	Usage method	Suitable tree species	Remarks
Oxyfluorfen	24% missible oil	675-900 mL/hm^2	Broad-spectrum	Stem, leaf treatment, soil treatment before budding	Conifer	Contact kill
Haloxyfop	10.8% missible oil	450-750 mL/hm^2	Gramineous weeds	Stem and leaf treatment	Broadleaf, conifer	Contact kill

(Continues)

Commodity name	Dosage form	Reference dosage	Applicable object	Usage method	Suitable tree species	Remarks
Sencaojing	70% wettable powder	5-50 g/hm^2 250-900 g/hm^2	Broad-spectrum	Stem, leaf treatment, soil treatment before budding footpath, large seedling, etc.	Broadleaf, conifer (except for *Cunninghamia lanceolata* and *Larix gmelinii*)	Systemic
Trifluralin	48% missible oil	2100 mL/hm^2	Gramineous weeds, small seeds, and broadleaved weeds	Soil treatment before budding	Broadleaf, conifer	Contact kill
Napropamide	20% missible oil	1500-3750 g/hm^2	Broad-spectrum, but no effect on perennial weeds	Soil treatment before budding	Broadleaf, conifer	Systemic
Acetochlor	50% missible oil	900-1125 mL/hm^2	Broad-spectrum	Soil treatment before budding	Broadleaf	Contact kill
Glyphosate	10% water aqua	100-450 g/hm^2	Broad-spectrum	Stem, leaf treatment, soil treatment before budding	Conifer	Systemic
Prometryne	50% wettable powder	500-1500 g/hm^2	Broad-spectrum	Soil treatment	Broadleaf	Systemic
Butachlor	60% missible oil	1350-1700 mL/hm^2	Gramineous weeds	Soil treatment before budding		Systemic
Paraquat	20% water aqua	100-300 mL/hm^2	Broad-spectrum	Stem and leaf treatment	Broadleaf, conifer	Contact kill
Atrazine	40% suspension concentrate	450-750 mL/hm^2	Broad-leaved weeds	Stem and leaf treatment	Conifer	Contact kill
Quizalofop-p-ethyl	5% missible oil	600-3000 mL/hm^2	Gramineous weeds	Stem and leaf treatment	Broadleaf	Systemic

(Continues)

Commodity name	Dosage form	Reference dosage	Applicable object	Usage method	Suitable tree species	Remarks
Diuron	25% wettable powder	2750-4500 g/hm^2	Broad-spectrum	Soil treatment before budding	Broadleaf, conifer	Systemic
Paracetamol	12.5% engine oil, missible oil	200-400 g/hm^2	Gramineous weeds	Stem and leaf treatment	Broadleaf, conifer	Systemic
Simazine, 2,4-D	25% wettable powder	1500-3750 g/hm^2	Gramineous weeds	Soil treatment before budding	Conifer	Systemic
Butylparaben	72% missible oil	600-900 mL/hm^2	No effect on gramineous weeds	Stem and leaf treatment	Conifer	Systemic

7.5 Pest and Disease Control

7.5.1 Seedling Diseases

7.5.1.1 Disease and Its Symptoms

When seedlings are infected by pathogens and other biological parasitism or environmental factors, they will present as diseased in physiology, tissue structure, and morphology, which results in the decline of yield, variety deterioration, yield reduction, and even death. Collectively, we refer to such symptoms and outcomes as seedling diseases. Seedling diseases are caused by pathogens. Generally, according to the pathology, diseases can be divided into two categories: Infective diseases and non-infective diseases. Diseases caused by fungi, bacteria, viruses, mycoplasma, nematodes, and parasitic seed plants are called infective diseases or parasitic diseases; seedling damage caused by poor environmental conditions and/or improper nursery operations are called non-infective diseases or physiological diseases.

In a nursery operation, people tend to think of seedling diseases as only the infective diseases, which can propagate and transmit rapidly under suitable conditions. The occurrence of seedling disease is based on certain pathological processes in the seedlings. For all diseases, a physiological activity of some sort that cannot be observed by the naked eye causes internal changes, and then the cells and tissues also begin to change. Finally, a range of abnormal characteristics, namely symptoms, are displayed in the external morphology, which is composed of disease symptoms and plant symptoms. The characteristics of the pathogens in the infected parts are attributed to disease symptoms, and the abnormal state of the plant after being infected with the disease displays as plant symptoms. Common types of symptoms are as follows.

(1) Types of Disease Symptoms

Powder (white powder, black powder, rust spot). The pathogen covers the surface of seedling organs and forms white, black, or rust-colored powder, such as rose (*Rosa*) leaf and clove (*Syzygium aromaticum*) leaf powdery mildew, Chinese lime (*Tilia tuan*) leaf and bamboo leaf smut disease, and *Populus* and *Larix gmelinii* leaf rust.

Mildew. The mildew covers the infected parts, such as the mold layer on Chinese chestnut (*Castanea mollissima*), Manchurian ash (*Fraxinus mandshurica*), and Manchurian walnut (*Juglans mandshurica*) leaves.

Bituminous Coal. The pathogen forms a layer of bituminous coal on the surface of seedling organs, such as Japanese camellia (*Camellia japonica*) bituminous coal disease, crape myrtle (*Lagerstroemia indica*) sooty mould disease.

Bacterial Ooze. The infected part exudes a thick, gelatinous substance containing a large number of pathogens that weaken the growth of seedlings or can cause death; for example, peach gummosis and Korean pine root rot.

(2) Types of Plant Symptoms

Cells and tissues of the diseased parts seedlings die off and cause rotting. The secretase of the pathogen dissolves the substances in the cells of the seedling tissues, causing the tissues to soften and disintegrate, and fluids flow out. The roots, stems, leaves, flowers, and fruits of the seedlings could all be infected.

Deformity. With the invasion of pathogens, some or all of the seedling tissue displays abnormal morphology. The exact nature of the deformity differs according to the disease and the seedling species; examples include flowering peach leaf curl and oriental cherry witches broom. A deformity does not necessarily kill the seedling, but it may affect the seedlings' form and function.

Wilting. Root rot or necrosis of the stems of the seedlings cause damage to the internal vascular tissues and impede the conduction, causing localized areas or entire seedlings to wither, such as Persian silk tree (*Albizia julibrissin*) blight. Wilting caused by disease is irrecoverable and poses a great threat to seedlings. Wilting may also be caused by poor cultivation conditions such as drought, high temperature, or flooding; when these adversities or stress factors are relieved, the seedlings may recover.

Discoloration. The pigment of the cells changes in the diseased portion of the seedling, but the cells usually do not die. The root rot of Korean pine (*Pinus koraiensis*) can cause the needles to turn yellow. The main symptom of larch needle cast fungi is that the needles turn yellow and then fall off. Leaf discoloration is not unique to infectious diseases. Many non-infectious diseases, such as lack of oxygen, lack of water, the presence of soil toxic substances, and low temperatures cause discoloration; moreover, many seedling leaves have discoloration before natural shedding. When discoloration appears, however, we should pay attention to determine the cause.

7.5.1.2 Primary Diseases and Their Control

Based on the damaged portion of the plant, seedling diseases can be divided into three types: Root damaged (root disease), leaf damaged (leaf disease), and stem damaged (stem disease).

After the root is diseased, the cortex of the root and the root collar will rot, forming tumors; sometimes, white filaments, purple mats, and black dots form in the diseased position, such as when infected by seedling damping-off, neck rot, purple streak, and southern blight. If the leaves or the young shoots are diseased, spots in different shapes, sizes, and colors form; or yellow-brown, white, and black powdery, filiform, and punctate substances form on them, suggesting leaf spot, anthracnose, rust, powdery mildew, and sooty mold. If the stems are diseased in the seedling stage, the diseases often continue to occur in the young trees and large trees, for example, Paulownia witches broom, jujube witches broom, and poplar canker.

Control of Powdery Mildew. The powdery mildew is commonly seen on broad-leaved tree leaves. Seedlings should be kept away from lengthy exposure to high temperatures and humidity and are healthier if kept in a ventilated and cool environment. Diseased branches and leaves should be detached and destroyed in a centralized way to avoid the spread of powdery mildew. They can be also sprayed with 50% of ambam solution, made with 1 part ambam to 1000 parts water, or 75% chlorothalonil solution made with 1 part chlorothalonil to 800 parts water.

Rust Control. The rust is heteroecious (different stages of the life cycle occur on alternate or even unrelated hosts), and the seedlings cannot be planted adjacent to or mixed with the host; when the disease occurs, the 50% solution of ambam adding 1000 times of water can be used to spray for control. Zineb, chlorothalonil, and lime sulphur are also good fungicides.

Control of Anthrax. The plants should be spaced in a reasonable density to allow ventilation around the seedlings and have access to light. The source of disease should be removed promptly to prevent its spread.

Prevention and Control of Damping-off Disease. Prevention includes conducting timely soil disinfection, avoiding continuous cropping, and using fully decomposed organic fertilizer. In the southern nursery, 300 kg quicklime per hectare can be applied to disinfect the soil; in the northern nursery, zineb can be used for soil treatment or can sprayed after the disease occurs.

7.5.2 Seedling Pests

Many types of insects can damage seedlings in the nursery. They can be divided into root (underground) pests, trunk-boring pests, branches and leaves pests according to the location on the seedling and their mode of harm.

(1) Root (Underground) Pests

Under the soil surface or near the ground, this type of insect bites the buds and the roots of seedlings, which harms the seedlings sown in the same year, the seedlings with slow growth, and the seedlings maintained for several years of some varieties, therefore reducing the outplanting quality. Common pests include grubs, mole crickets, cutworms, wireworms, and crane flies, among which the grubs and mole crickets are the most common.

Grubs. Commonly known as the strong ground worm, white soil silkworm, the grub is the beetle larva. Grubs do harm to young buds and roots, and adults do harm to leaves, flowers, and fruits, and the adults are more difficult to control. Many birds and poultry like to eat the larvae, which can be found during the incubation period of adults. Combined with a plow attracts birds and poultry to eat the grubs; adults can be killed by light at night; for chemical control, refer to the control methods for the mole cricket; and pyrethroids, omethoate, and other insecticides can also be used.

Mole Crickets. Living in the soil for its whole life, this pest endangers mainly the roots of young trees and seedlings. Adults or nymphs bite the roots and young stems close to the ground to make them become irregular filaments. They often eat the seeds just sown or newly sprouted; they also dig crisscross tunnels on the surface of the soil, leading to the separation of the fibrous roots of seedlings from the soil and causing seedling death due to wilting. In the nursery, *Gryllotalpa unispina* and *Gryllotalpa africana* are common. At the adult emergence stage, they can be trapped and killed by light at night. They can also be killed by poison bait, for example, mix together phoxim 0.5 kg, water 0.5 kg, and 15 kg of half-cooked millet, and evenly broadcast the mixture on the seedbed or bury it in the soil at night. Horse manure put in a nearby pit will attract them, at which point they can be captured. Birds such as hoopoe and magpie can also be attracted to eat or kill the pests.

(2) Trunk Borers

Trunk borers burrow into the branches and shoots of the seedlings to eat the cambium and xylem tissue, which causes the branches of the nursery seedlings to wither and/or to become wind fall. This pest will reduce the qualified rate of the seedlings being sold from the nursery. The common and seriously damaging pests include the clearwing, longicorn, scale insect, and the moth *Dioryctria splendidella*.

Prevention and Control of *Parathene tabaniformis*. Clearwing harms all species of poplar and willow. Adults are very much like wasps and are active during the day. Larvae eat stems and terminal buds, forming tumors, affecting the transport of nutrients, and thus affecting the development of seedlings. During the adult pest activity period, the male adult pests are trapped by the synthetic sex hormone of *Populus tomentosa*; or the larvae are killed by injecting omethoate, fenitrothion and other pesticide dilutions into the wormhole; the injured branches are cut and burned in time to avoid transmission.

Prevention and Control of Longicorn. The poplar borer *Saperda populnea* endangers all species of poplar and willow. When the larvae eat the trunk, especially the branch tips, the harmed parts form spindle-shaped tumors that hinder the normal transportation of nutrients. The branches become dried up or broken by wind, resulting in tree trunk deformity and baldness. If they harm the medulla of young trees, the whole plant may die. Adult pests can be caught, and the pest eggs on the trunk can be eliminated artificially; spraying the trunk with a 100 times diluent of fenitrothion can effectively control the larvae and adults. Natural enemies such as woodpeckers are attracted to eat pests. Nursery workers should collect the damaged trunks and destroy them.

(3) Branch and Leaf Pests

This type includes many pest species, which can be divided into two categories: Piercing-sucking mouthparts and chewing mouthparts.

Piercing-sucking Mouthpart Pest Control. The pests use piercing-sucking mouthparts to suck the sap from plant tissues, which results in withering of branches and leaves, dehydration of branches, and even death of the whole plant. At the same time, a virus is also transmitted. Juice-sucking pests are small in size, and the early symptoms of harm are not obvious, so they are easy to miss. However, this kind of pest has strong fecundity and rapid spread, so it is difficult to achieve satisfactory control effects if the best timing for control is missed. This group is made up of primarily of aphids and mites.

Aphids are widely distributed and comprise many kinds. Almost all plants in the nursery are its hosts. It often damages a broad range of garden seedlings, such as flowering plum (*Prunus triloba*), apples (*Malus*), pear (*Pyrus*), Chinese hawthorn (*Crataegus pinnatifida*), peach (*Prunus*), cherry blossom, and cherry plum (*Prunus cerasifera*). Aphids cluster on young leaves, twigs, and buds, and the damaged leaves curl to the back. Control efforts should focus on prevention, and attention should be paid to the dynamics of aphids, for example, cut off the diseased branches at their first occurrence to prevent the spread of aphids. Spray irrigation and heavy rain can also eliminate aphids. When the damage is serious, 1 part of 40% of omethoate, or 1 part of 50% of fenitrothion, can be mixed with 1000 parts water for a spray. In spring, a ditch can be dug at the root of seedlings, and 3% carbofuran granules can be added into the ditch, covered with soil, and irrigated to effectively control the occurrence of aphids.

Mites, also known as red spiders, harm conifers and broad-leaved trees. The regular occurrence of mites should be understood and noticed, so as to control it in advance. The dead branches and leaves should be cleaned up to eliminate the places mites can live in winter. When the harm from mites is serious, 40% kelthane of 1000 times or omethoate of 1000 times liquid can be used to spray on the back of leaves for control.

Chewing Mouthpart Pest Control. Many kinds of these pests occur, and they feed on the leaves and buds. The common ones are inchworms, slug caterpillar moths, tent caterpillars, and lappet moths.

Two kinds of inchworms may be problematic: *Semiothisa cinerearia* and *Sucra jujuba Chu*. *Semiothisa cinerearia* damages primarily locust trees and Chinese pagoda trees. The pupa overwinters in the loose soil under the tree and matures into its adult form in the middle of April. They lay eggs at night. The larvae damage the leaves; when they crawl, they arch up in the middle of their bodies, like an arch bridge, with the habit of spinning and sagging. Control is most effective with a light trap to kill the adult pest. Or, spray with 1000-fold solution of 50% fenitrothion emulsion or 1500-fold solution of 75% phoxim before the third instar in mid-to-late July.

Questions for Review

1. What procedures are included in soil tillage?
2. What is the principle of fertilization in nursery?
3. Expound on the concept of symbiosis, symbiotic bacteria, mycorrhiza, root nodule, and mycorrhizal seedling cultivation.

4. What is the mechanism of mycorrhizal and root nodule action on host?

5. What is the difference between ectomycorrhiza and endomycorrhiza in the concept and morphology?

6. What is the key problem in mycorrhizal application? What are the inoculation methods of ectomycorrhizal fungi?

7. What should be paid attention to in mycorrhizal application?

8. What are the characteristics of chemical weeding?

9. What are the types of chemical herbicides?

10. Why do herbicides kill weeds rather than seedlings?

11. What is the right way to use herbicides?

12. What are the main diseases in a nursery? How would you prevent and control them?

13. What are the main pests in a nursery? How would you prevent and control them?

References and Additional Readings

BAI S L, 2011. Mycorrhizal research and ectomycorrhizal resources in Daqingshan Mountain, Inner Mongolia[M]. Hohhot: Inner Mongolia People's Publishing House.

BRUNDRETT M, BOUGHER N, DELL B, et al., 1996.Working with mycorrhizas in forestry and agriculture[M]. Canberra: ACIAR Monograph.

CHEN C, LIU Y, LI G, et al., 2015. Effects of sub-irrigation gradients on growth and nutrient status of containerized seedlings of *Quercus variabilis*[J]. Scientia Silvae Sinicae, 51(7): 21-27.

CHEN W, WANG E, 2011. Rhizobia in China[M]. Beijing: Science Press.

CHENG F, 2012. Nursery of landscape plants[M]. Beijing: China Forestry Publishing House.

DUMROESE R K, LUNA T, LANDIS T D, 2009. Nursery manual for native plants: A guide for tribal nurseries [M]. Washington , DC: US Department of Agriculture, Forest Service.

GONG M, CHEN Y, ZONG C, 1997. Mycorrhizal research and application[M]. Beijing: China Forestry Publishing House.

INGESTAD T, ÅGREN G I, 1995. Plant nutrition and growth: Basic principles[J]. Plant and Soil, 168/169: 15-20.

JIA H, ZHENG H, 1991. Theory and techniques for steady-state mineral nutrition of plants[J]. Plant Physiology Communications, 27(4): 307-310.

JIN T, 1985. Seedling cultivation technology[M]. Heilongjiang: Heilongjiang People's Publishing House.

LI G, LIU Y, ZHU Y, et al., 2011. Study of techniques for steady-state nutrition supply of seedlings[J]. Scientia Silvae Sinicae, 35(2): 94-103.

LIU R, CHEN Y, 2007. Mycorrhizology[M]. Beijing: Science Press.

SALIFU K F, JACOBS D F, 2006. Characterizing fertility target sand multi-element interactions in nursery culture of *Quercus rubra* seedlings[J]. Annals of Forest Science, 63(3): 231-237.

SALIFU K F, TIMMER V R, 2003. Optimizing nitrogen loading of *Picea mariana* seedlings during nursery culture [J]. Canadian Journal of Forest Research, 33(7): 1287-1294.

SHEN H, 2009. Seedling cultivation[M]. Beijing: China Forestry Publishing House.

SUN S, LIU Y, et al., 2013. Forest seedling technology[M]. Beijing: Jindun Press.

WANG C, 2015. Effect of exogenous abscisic acid and its inhibitors in tea plant chilling resistance under chilling stress[D]. Nanjing: Nanjing Agricultural University.

WU F, WANG H, et al., 2015. Research progress on the physiological and molecular mechanisms of woody plants under low temperature stress[J]. Scientia Silvae Sinicae, 51(7): 121-122.

WU J, CHEN Y, et al., 2016. Effects of calcium on Ca^{2+}-ATPase activity and lipid peroxidation level of loquat seedling under low temperature stress[J]. Journal of Northwest A&F University, 44(2): 124-126.

XU H, CHEN J, XIE M, et al., 2009. The role of dehydrin in plant response to cold stress[J]. Acta Botanica Boreali-Occidentalia Sinica, 29(1): 199-206.

ZHAI M, SHEN G, 2016. Silviculture[M]. 3rd edition. Beijing: China Forestry Publishing House.

ZHANG J, WANG H, et al., 2017. The effect of exogenous abscisic acid on physiological and biochemical indexes of *Camellia* leaves under chilling stress[J]. Journal of Anhui Agricultural University, 44(1): 144-145.

ZHANG S, LIU C, XU S, 2014. Garden plant disease occurrence and control[M]. Beijing: China Agricultural University Press.

ZHENG H, JIA H, 1999. Study and prospects of theory and techniques for steady-state mineral nutrition of plants [J]. Scientia Silvae Sinicae, 35(1): 94-103.

Chapter 8   Bare-rooted Seedling Cultivation

Ying Yeqing, Cao Banghua

Chapter Summary: This chapter introduces the technical links of bare-root seedling sowing and cultivation. It includes the causes of seed dormancy and germination promotion methods, seedling raising methods and seedling bed preparation, sowing season and sowing amount, sowing methods, and nursery and seedling management. This chapter is the basic technology of seedling cultivation. Many large-sized seedlings are cultivated from sowing seedlings.

Bare-root seedlings have bare roots with no soil or other attachments. Compared with container seedling, it has the advantages of small weight, easy seedling raising, labor-saving planting, and more convenient packaging, transportation, and storage. It is the most widely used type of seedling for afforestation at present. Bare-root seedlings can be cultivated in the open air or in a manually controlled environment, on natural soil, on artificial substrate. In production, cultivation in open-air natural soil is the most widely used, which is traditional, effective, low-cost, and has relatively low technical requirements. The seeding and seedling raising of bare-root seedlings includes the selection of seedling raising mode, soil disinfection, and seed treatment; the determination of sowing date, sowing amount, and sowing mode; and the management of sowing land before and after seedlings emergence.

8.1   Seed and Seedbed Preparation

8.1.1   Seed Dormancy

8.1.1.1   Overview

In the process of long-term adaptation to nature ecology as well as artificial cultivation and selection, forest tree seeds show different responses to the environmental conditions suitable for germination after maturity. Some seeds can germinate as long as the external conditions are suitable, while other forest tree seeds cannot germinate. Seed dormancy refers to the phenomenon that the seeds with vital force cannot germinate under the suitable environment conditions because of internal reasons in the seeds.

According to the strict definition of dormancy, not all seeds have the characteristics of dormancy. Seeds that germinate quickly under suitable conditions are called quiescent seeds. However, some tree species, such as mangrove, exhibit neither dormancy nor quiescence, and directly germinate after the seeds mature on the parent body, which is the so-called vivipary.

Seed dormancy can be divided into primary dormancy and secondary dormancy according to the condition. Most forest tree seeds can remain dormant for a period after harvest, which is called primary dormancy. When dormant seeds are released from dormancy, but if environmental conditions suddenly change before they germinate, such as an unsuitable temperature, low water potential, lack of oxygen, or high carbon dioxide content, they will enter dormancy again, which is called secondary dormancy. The relationship between primary dormancy, secondary dormancy, and germination is illustrated in Figure 8-1.

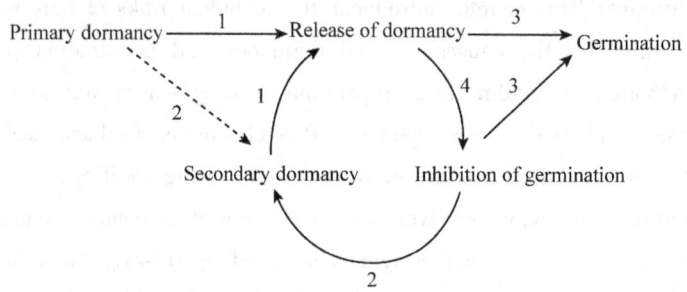

Figure 8-1 Relationship between the release of primary dormancy and secondary dormancy and germination (From Liang, 1995).
1. factors releasing dormancy; 2. factors inducing dormancy; 3. factors required for germination; 4. non-germination conditions or factors inhibiting germination.

Seed dormancy is a temporary suspension phenomenon in the process of plant development, and it is a beneficial biological characteristic for plants themselves. It is also the result of long-term natural selection of plants, exhibiting a biological adaptability to environmental and seasonal changes obtained through long-term evolution. Dormant seeds have high resistance to adverse environments, can endure long-term storage and still maintain their viability, can survive in severe seasons, have strong adaptability, and are beneficial to the seed storage and allocation. Nonetheless, dormant seeds have adverse effects on seedling cultivation and afforestation. If the dormant seeds are seeded, they usually germinate in succession within 1 to 2 years, which leads to uneven growth of the seedlings, prolongs the seedling cultivation, and direct seeding afforestation is prone to failure. Therefore, the study of seed dormancy has important theoretical and practical significance.

8.1.1.2 Types and Causes of Seed Dormancy

(1) Types of Seed Dormancy

There are many patterns of seed dormancy, which can be divided into three types: Compulsive dormancy, physiological dormancy and comprehensive (mixed) dormancy.

Compulsive Dormancy. In many types of seeds, compulsive dormancy is caused by the existence of coverings. If the growth inhibition factors are removed, seeds can germinate and grow rapidly just like seeds without dormancy. Compulsive dormancy is often caused by physical factors other than seed embryos.

Physiological Dormancy. The primary causes for physiological dormancy are embryo growth standstill and the physiological conditions, and they are all attributable to the internal physiological issues of seeds. Breaking dormancy allows embryo to grow/develop and is a consequence of a physiological process.

Comprehensive Dormancy. Comprehensive dormancy has the characteristics of both of the above two types of dormancy, some of which may have three to four types of dormancy combined. These seeds are the most difficult to germinate and are often known as "two-year seeds".

(2) Causes of Seed Dormancy

Mechanical Obstruction of the Seed (Pericarp). Mechanical obstruction affects seed dormancy from three aspects: Imperviousness, imporosity, and mechanical obstruction to an embryo. In addition, the seed coat contains inhibitory substances also related to dormancy.

Seed coat prevents water from entering the seed. Imperviousness of a seed coat is usually caused by the sclerenchyma cells arranged closely together on the surface of the testa, which also contains oils and waxes. This type of obstruction is often known as hard seed in agriculture and forestry. Such seeds can be found in legumes, as well as in Malvaceae, Chenopodiaceae, Liliaceae, and Solanaceae. The substances in the seed coat that hinder the passage of water varies with to the plant species. Removing the seed coat can significantly improve the germination rate.

Imporosity or the low oxygen permeability of the seed coat can cause a mechanical obstruction. For example, increasing the oxygen supply to the seeds of European ash (*Fraxinus excelsior*), linden tree (*Tilia americana*), and the common apple (*Malus domestica*) can promote germination, and the dormancy of such seeds is caused by the limitation of oxygen supply by the seed coat or endosperm.

Mechanical obstruction pertains to the seed coat blocking access to the embryo. Generally, the hard wooden husk, or the coriaceous membranous seed coat, has a mechanical binding effect on embryo growth to various degrees. When the seed has insufficient thrust generated by water absorption, the radicle cannot penetrate the seed coat, and germination cannot proceed. This blockage is common in the seeds of many woody plants, such as Rosaceae and Proteaceae.

Immature Embryo (Embryo Needs After-ripening Period). Seeds of many plants do not germinate even after removing the seed coat. Most of these seeds need to finish the after-ripening process at a certain temperature and higher humidity, during which organ differentiation and substance metabolism are completed. According to the condition, it can be divided into two types.

Incomplete organ differentiation of embryo (morphological after-ripening): The embryo of a seed is formed by differentiation of a zygote. A complete embryo consists of cotyledon, radicle, plumular axis, and bud. It is similar to the embryonic form of an adult plant. When collecting seeds (fruits) of some plants, however, such as ginkgo (*Ginkgo biloba*), Manchurian ash (*Fraxinus mandshurica*), Chinese ash (*Fraxinus chinensis*), multiflora rose (*Rosa multiflora*), Oriental holly (*Ilex chinensis*), and winged euonymus (*Euonymus alatus*), it looks mature from the outside of the seeds, but their embryos have not yet differentiated well, so they need to continue to complete organ differentiation under suitable conditions.

Embryo does not have growth ability while its differentiation is completed (physiological afterripening): This type of embryo dormancy is that the embryo shape seems to be mature and the morphological differentiation is completed, but it is not fully mature in physiology. With suitable conditions, even if the seed coat (pericarp) is stripped, it cannot germinate, such as the seeds of Rosaceae (*Malus domestica*, *Pyrus* spp., *Cerasus pseudocerasus*, *Prunus persica*, *Prunus armeniaca*, plums) and Coniferae. These seeds usually need low-temperature and humid conditions for several weeks, even several months before they complete their physiological after-ripening, germination, and growth. This treatment method is called stratification.

Substances Inhibiting Germination. Some tree seeds cannot germinate because certain substances inhibit germination in the fruits or seeds. These substances are primarily organic acids (salicylic acid, ferulic acid, abscisic acid) and alkaloids (caffeine). Inhibiting substances vary with tree species, and their distribution and action modes also differ. The inhibitor in the seeds of hawthorn is hydrocyanic acid; seeds of peach and apricot contain amygdalin, which is decomposed continuously under humid conditions, releasing hydrocyanic acid to inhibit germination. When amygdalin is completely decomposed and no longer releases hydrocyanic acid, the seeds can release the dormancy. The juice of European mountain-ash (*Sorbus aucuparia*) fruits contains sorbic acid (unsaturated lactone), which also has an inhibiting effect on seed germination. The testa of some tree seeds contains testa phenols, which can react with oxygen in the air and consume oxygen, thus affecting the oxygen supply to embryos and causing dormancy.

Recently, studies have confirmed that inhibitors in embryos are mainly abscisic acid. The level of abscisic acid in an embryo is directly proportional to dormancy depth.

Germination Needs Strict Environmental Conditions. Many tree species cannot germinate under suitable conditions of water, oxygen, and temperature; only after a certain temperature is reached can they germinate. For example, some wild and semi-wild plant seeds have evolved and survived under different historical conditions, such as 25 ℃, which is far from meeting the special needs of germination; 30 ℃ to 35 ℃ is required for the germination of citrus seeds. Seeds with strict temperature requirements are dormant by the force of genetic heredity. Photo germination and dark germination are essentially regulated by phytochrome.

Note that more than one reason can account for dormancy of many seeds, and comprehensively dormant tree species may have the characteristics of both compulsive dormancy and physiological dormancy. For example, the seed dormancy causes of *Fraxinus excelsior* include the embryo is immature; after the embryo grows to its proper size, it is limited by the seed coat; and low temperature treatment is required.

8.1.1.3 Seed Germination Process

Seed germination refers to the phenomenon that the embryo begins to grow and the radicle and plumule tissue break through the seed coat and grow outward. In the process of germination, the seeds have a variety of changes in the external morphology and structure along with a series of complex physiological and biochemical changes internally. At this stage, they also show a high sensitivity to the external environmental conditions. Understanding the physiological metabolism in the process of seed germination is very important in facilitating seed germination and utilization.

Seed germination goes through three overlapping stages: i. Absorbing water and swelling, eventually breaking the seed coat; ii. Enzyme activation, respiration, and assimilation rate intensify, indicating that the nutrients are transported to the growing regions; and iii. The cells enlarge and divide, the radicle sprouts, and the seed germinates. These three processes and their physiological and biochemical changes are summarized as follows.

(1) Water Swelling

The volume of seeds expands after absorbing water. This is the first step of seed germination. Seeds that cannot absorb water will never germinate. Therefore, water swelling is the basis of seed germination.

The cells in the dry seeds do not have vacuoles, and the water absorption does not rely on the permeation of living cells but rather on the swelling of the organic matter hydrocolloids in the seeds. Therefore, the process of water absorption and swelling of seeds is a physical process, also known as the physical stage. The drier the seeds are, the stronger the imbibition capacity.

After initial water absorption, the seed coat softens, which not only ensures that the water continues to infiltrate into the whole seed and cause swelling but also increases the permeability to oxygen and carbon dioxide, promotes gas exchange, improves the breathing of seeds, and contributes to the substance transformation and energy supply. The softened seed coat is also beneficial to the extension of the embryo.

Seeds with viability and being able to germinate have the characteristics of three-stage water absorption of "fast-slow-fast". The third stage does not appear in dead seeds and dormant seeds, through which the dead and alive seeds can be distinguished. Compared with dormant seeds, "dropsy" is often observed in water absorption in dead seeds, which is related to protein denaturation, cytomembrane destruction, with a large amount of free water filling all the space inside and outside the cell.

(2) Budding

Once the seeds are provided with the suitable external conditions, the activities within the seeds will intensify, and material metabolism and energy metabolism change from a relatively static state to an intense active state. The external manifestation of this movement is budding, which refers to the phenomenon that the radicle penetrates the seed coat as embryo cells divide and elongate. Subsequently the embryo volume increases to a certain size as the seeds absorb water and swell. In agriculture and forestry production, this phase is generally called "exposed white", which indicates that the white embryo tissue begins to appear from the seed coat cracks and marks the end of the budding stage. During seed germination for most tree species, the radicle breaks through the seed coat first. Because the radicle tip is facing the germinating hole, and the seeds absorb water mainly through the germinating hole, the radicle firstly obtains water and its cell metabolism and division are also the first; therefore, the radicle emerges first. However, when too much water is present, the plumule will emerge first. This alternative occurs because the water quickly meets the needs of the plumule; in addition, under the condition of oxygen deficiency, the inhibition of the plumule is less than that of the radicle. Therefore, when conducting the germination experiment, the water regime of the germinating bed can be judged according to the status of radicles and plumules.

Budding is the second stage of seed germination, also known as the biochemical stage. This phase is the stagnant stage of water absorption, and the water content in seeds does not increase much, but the biochemical changes of swollen seeds begin to intensify, entering a new physiological process. The changes of physiological metabolism during this period are shown in three aspects. First, substance transformation. The rate of substance transformation is directly related to the enzymatic activity. Studies have shown that many enzymes in the seeds can be readily reactivate upon water absorption, leading to a significant increase in their activity. In addition, ribosome activation in an embryo plays an important role in protein synthesis and embryo growth. Second, respiration intensity greatly increases. The energy needed for seed germination is supplied primarily by respiration, and respiration during the budding period is mainly anaerobic respiration. The change of respiration is related to the change of mitochondrial structure and the increase of relevant enzyme activity. A part of the energy generated by respiration is used for new cell construction, some is used for growth and movement, and the remaining energy is released in the form of heat. In addition, many intermediate products produced by organic-matter degradation through respiration can be used as raw materials for building up new cells. The life activities of seeds all depend on the nutrients stored in the endosperm (or cotyledon). Therefore, large and full seeds have stronger germination ability (germination potential). Third, the change of plant hormones. Non-germinated seeds usually do not contain auxin. When the seeds germinate, the endogenous hormones begin to form and change constantly, regulating the metabolism process and growth of the embryo. For example, after stratification treatment, the growth inhibitor content of larch seeds gradually decreased, while the gibberellin content gradually increased, thus promoting germination. In addition, studies have found that cytokinins and auxin also play an important regulating role in seed germination, and different hormones are needed for different plant species.

In brief, under suitable external conditions, for seeds in the budding stage, the physiological activity of cells is reactivated and enzyme activity increases; the rapid increase of respiration provides the material and energy source (ATP) for protein synthesis; protein synthesis provides the material basis (zymoprotein, structural protein, and soluble protein) for germination; and by using the transformation substance in seeds, the radicle breaks through the seed coat and enters the visible germination stage.

(3) Germinating

As a result of physiological and biochemical activities inside the seed, the embryo cells divide and grow rapidly, the embryo volume increases, and the radicle breaks through the seed coat at a certain time, starting another process of individual life. Germination refers to the status when the seed grows rapidly based on budding, and the radicle (or plumule) reaches a certain length. The sign to judge the end of germination varies with the requirements. When sowing and raising seedlings, the so-called end of germination, for one seed, means that the budlet sprouts out of the soil surface; for another type of seed it means that the cotyledons or euphyllas unfold, forming a seedling that can produce and absorb nutrients, and no longer rely on the storage materials in the seeds. For a whole nursery of one-time sowing, the so-called end of germination refers to the time point that most seeds germinate and form seedlings. In the determination of seed germination, when the extending length of the radicle reaches half of the seed length, it is regarded as the standard sign of seed germination.

During the germination stage, the metabolism internal the seeds is particularly vigorous; the respiratory intensity reaches the maximum; seeds are highly sensitive to the external environment; and resistance to external adverse environment conditions reduces. Therefore, special attention should be paid to providing good environmental conditions at this stage, especially ensuring the oxygen supply and preventing the occurrence of hypoxia breathing.

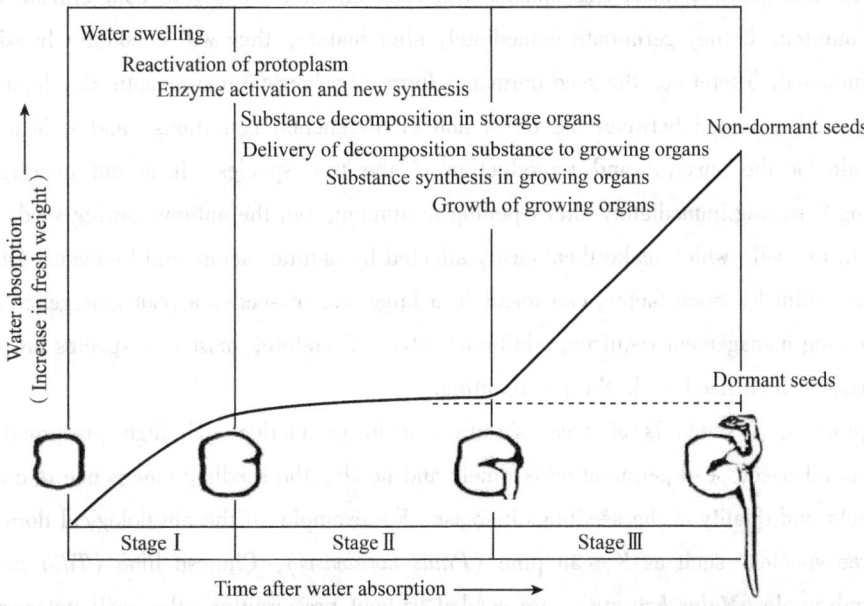

Figure 8-2 Three stages of seed germination (From Liang, 1995).

8.1.2 Presprouting of Seeds

8.1.2.1 Concept and the Significance of Presprouting of Seeds

Seed dormancy is a temporary suspension phenomenon in the tree development process. For trees, it is a way to adapt to the environment and to continue to survive. From the perspective of production, seed dormancy is also beneficial to seed storage, but it brings some difficulties to sowing and germinating.

Seed dormancy is reversible. Dormancy can be released by a series of comprehensive treatments at a certain time. However, when the dormancy is released, if seeds encounter adverse environmental conditions before germinating, they often enter the dormancy status again, that is, a secondary dormancy. The reversibility of seed dormancy makes it possible to control seed dormancy artificially. Controlling seed dormancy includes two aspects: Prolonging dormancy period and shortening dormancy period. Prolonging dormancy period is an important research topic upon seed storage, while the measures of shortening dormancy period and breaking dormancy are called presprouting in seed processing technology.

Presprouting is a tool to release seed dormancy and promote seed germination. For the seeds without dormancy, the presprouting technology can promote germination and improve the germination rate in the nursery. Generally, the seeds of evergreen broad-leaved tree species in South China have no dormancy period, or the dormancy period is very short. As long as the seeds are slightly dried by air, they can be sowed immediately, and can germinate at any time under the conditions of proper temperature, water, and ventilation. However, dormancy is common in the seeds of the northern tree species, and seeds of most tree species will soon encounter a severe cold climate after they mature in autumn. If they germinate immediately after mature, they will encounter freezing injury and be eliminated. Therefore, the seed dormancy formed by long-term systematic development is the result of natural selection between the forest and environmental conditions, and it is a favorable characteristic for the survival and reproduction of the tree species. It is not necessary to do presprouting if sowing immediately after ripening in autumn, but the autumn-sowing seeds stay for a long time in the soil, which make them easily affected by natural factors and biological factors such as birds and animals. Such factors can result in a large loss of seeds, a poor emergence rate, and time-consuming management requiring additional labor. Therefore, most tree species are chosen to sow in spring, which need to do the presprouting.

Presprouting of seeds is of great significance in production. Through presprouting, seed dormancy is released; seed germination is timely and neatly; the seedling emergence rate improves; and the yield and quality of the seedlings increase. For example, if the physiological dormant seeds of some tree species, such as Korean pine (*Pinus koraiensis*), Chinese lime (*Tilia tuan*), and Chinese crab apple (*Malus baccata*), are seeded without presprouting, they will not germinate or only a few seeds will germinate in the same year. However, a high germination rate of nursery seeds can be achieved through presprouting. Taking Japanese zelkova (*Zelkova serrata*) as an example, the germination rate of the seeds without presprouting in the nursery was 3%, and it increased to 94% after stratification for 60 days at low temperature (1-3 ℃). For forced dormant seeds, such as larch seeds, low-temperature stratification also significantly improved the field germination rate and seedling yield. After 30 days of low-temperature stratification, the field germination rate increased by four times, and the average seedling height and dry weight of the whole plant greatly improved compared with the control (Table 8-1).

Table 8-1 Germination experiment on *Larix olgensis* seeds

Treatments	Laboratory germination rate/%	Field germination rate/%	Average emergence number in one meter of seed furrow	Average seedling height /cm	Dry weight of the whole plant/g
30 days of low-temperature stratification	86.5	55.5	233.2	8.43	17.58
Untreated	64.7	11.1	98.1	4.83	12.36

Source: Liang (1995).

8.1.2.2 Conditions for Seed Germination

The necessary conditions for seed germination include sufficient moisture, suitable temperature, and oxygen. Some tree species also need light to germinate.

(1) Moisture

Water is the first condition needed for seed germination. Seeds have to absorb a certain amount of water to germinate because it can soften the seed coat and therefore facilitate the gas exchange between inside and outside of the seed. Softened seed is beneficial to respiration and convenient for the radicle and germ growth to break through the seed coat. Water entering into seeds changes the protoplasmic colloid from the gel status to the sol status; further improves the physiological activity; increases the enzyme activity; and accelerates the biochemical reaction rate. Organic substances can be transformed and transported in seeds only when dissolved in water; and water is also needed for embryo cell division and elongation.

(2) Temperature

Life activities in the seeds require a certain temperature to proceed. Seeds can only germinate under suitable temperature conditions. Germinating rate also varies with temperature.

Generally, the temperature range for seed germination is relatively wide. Some tree species can germinate at a range of temperatures without significant differences. Nevertheless, temperature requirements for most tree species can be assigned into three points: The lowest, the optimum, and the highest. These so-called three critical points of temperature for seed germination vary with species and geographical origin. The temperature range of seed germination is narrow for tree species originating from high latitude areas in the north, but the range is wider for tree species originating from low latitude areas in the south. The optimum temperature mainly varies with tree species. For example, for tree species of Eucalyptus, the optimum temperature of *Eucalyptus regnans* is 15 ℃, that of *E. robusta* is 20 ℃, and that of *E. citriodora* and *E. globulus* is 25 ℃; that of *E. camaldulensisc* is 30 ℃. The optimum germination temperature of most tree species is approximately 25 ℃.

Under nature conditions, the temperature changes between the day and the night. Studies have shown that the changing temperature conditions are beneficial to seed germination.

(3) Oxygen

When seeds transform from relatively static status to germination, the respiratory intensity greatly increases, which requires a continuous oxygen supply. Most tree seeds can germinate normally when the oxygen concentration is above 10%. When the oxygen concentration is below 5%, many seeds cannot germinate.

(4) Light

Generally seed germination for most tree species is not affected by light. However, some species need light for seed germination, and the seeds are correspondlingly are called light seeds (or light-loving seeds), such as European birch. Germination of other types of seeds is generally inhibited by light, and they are called light-inhibited seeds (or dark-loving seeds). Seeds that are not affected by light during germination are called medium-light seeds.

For the light seeds, the most favorable light exposure time is 8 to 12 hours. Experiments indicated that changing-temperature or low-temperature sand storage can sometimes alter the light requirement for some seeds.

8.1.2.3 Essence of Seed Pregermination

Certain technical methods can be adopted for different dormancy reasons. For example, for poor penetration of a seed coat or pericarp, mechanical abrasion and acid erosion can be adopted to break that limitation; for embryo dormancy and seeds that contain germination-inhibiting substances, stratification treatments and plant hormone treatments can be adopted to release the inhibition and to break embryo dormancy. Dormancy can be released by seed pregermination to create a suitable condition for seed germination. The general metabolism direction in the process of seed pregermination is consistent with germination.

Seed germination needs to go through three stages: Water swelling, budding, and germination. Under natural conditions, seed germinations are affected by low temperature in winter and spring to pass through the first two stages before germination, however, some tree species can take more than 2 years. The essence of seed pregermination is to create artificial conditions, break seed dormancy, and facilitate moving the seed quickly through the first two stages to germination with rapid and healthy emergence to obtain high seedling quality and yield.

8.1.2.4 Methods and Technical Points of Seed Pregermination

Seed dormancy may be caused by one or several complex reasons. For example, the dormancy of hawthorn and linden seeds is mainly caused by physiological reasons, but the impermeability of the seed coat or pericarp also contributes to it. Therefore, the reason for dormancy, especially the leading factors, should be clarified first, and then the proper method of pregermination should be determined. At the present, several commonly used methods can facilitate pregermination.

(1) Mechanical Abrasion

Mechanical abrasion is a method to treat the imperviousness of seed coat or pericarp. It is widely used in the seeds of agricultural fodder crops such as alfalfa and clover. In forestry, it is used in "hard seeds" with impermeable seed coat or pericarp, and other tree species with a tough testa, such as juniper bush and hawthorn.

Mechanical abrasion treatment changed the physical properties of seed coat and increased the permeability. Many physical methods can abrade the testa, such as using a rasp across the seed, a hammer to smash, sandpaper to rub, stone rolling to roll, or artificial peeling, but the most convenient method is to mix the seeds with a volum of sand three to four times greater than the volume of seeds. After mixing, gently pound or roll the mixture, and the seed coat can be cut to assist with the absorption of water to aid germination. A seed abrader can also be used.

The seed coat of Chinese honey locust (*Gleditsia sinensis*) is hard so the seeds can more readily germinate by crushing the seed coat with stone. The seed coat of Japanese oak (*Lithocarpus glaber*) is very hard and the seeds do not easily germinate. After removing the seed coat, seeds can germinate early and the germination rate is more than 75%. However, for some plant seeds without after-ripening, dwarf seedlings are often produced after seed-peeling germination. This phenomenon needs further study to address the mechanisms.

Abrasion treatment is simple and easy to accomplish, with no need to control the environment, but the seeds can be injured with excessive treatment, resulting in the loss of germination ability because of damage from pathogenic bacteria. Therefore, the required degree of abrasion must be determined. Generally, seeds can be soaked first and the swelling stutus and the seed coat can be monitored with a magnifying glass. The abrasion degree is appropriate when the seed coat becomes dark, but the pockmark should not be deep nor should seed cracks expose the inside.

(2) Acid Etching

Acid etching is a common chemical method to increase the permeability of the seed coat or pericarp, but the damage to the embryo should be strictly prevented when this method is used. The germination rate of *Gleditsia sinensis* seeds can be increased by soaking seeds with 98% concentrated sulfuric acid, then washing them with water, and then soaking them in water at 40 ℃ for 86 hours. Good results can also be obtained when the seeds of black locust (*Robinia pseudoacacia*) are soaked with sulfuric acid solution(specific gravity of $1.84 \ g/cm^3$) for 25 to 60 min. Seeds of the Persian silk tree (*Albizia julibrissin*), elaeagnus (*Elaeagnus angustifolia*), Chinese lacquer tree (*Toxicodendron vernicifluum*), *Robinia pseudoacacia*, bush clover (*Lespedeza bicolor*), ironwood tree (*Erythrophloeum fordii*), and royal poinciana (*Delonix regia*) can also be soaked with sulfuric acid. With the appropriate extension of sulfuric acid treatment time, the proportion of hard seeds decreased, and the germination rate increased. However, too long time treatment, such as seeds 30 min of treatment for sulfuric acid, it will cause damage to the seeds and reduce the germination rate.

The specific method of acid etching is to use a net made of corrosion-resistant materials to hold the seeds while soaking them in the acid at 18 to 25℃. Soaking time varies with tree species, but generally can be for 10 to 60 min. Then the net is lifted and drained for a while; finally, wash the seeds with plenty of water for 5 to 10 min. To avoid seed drainage, acid etching time should be determined carefully, and the most suitable soaking time of each batch should be determined by a small number of sample tests. If the seeds are not sown immediately, they can be air-dried for use at a later time. Sulfuric acid for acid etching can be reused. Seeds treated by acid etching can be stored for 1 month or even longer, and the seeds are not in the status of imbibition the way soaking and scalding seeds would be; therefore, it is helpful to arrange the sowing time.

In the process of acid treatment, seeds can be taken out at set intervals for inspection. If lots of pits and scars have appeared on the surface of the seeds, or the endosperm has been exposed, it indicates that soaking is excessive; if the seed coat is still glossy, it means the treatment time is not enough; and when the seeds have been treated appropriately, the seed coat is dim and dark without deep potholes.

(3) Water-soaking Pregermination

Water-soaking pregermination is used primarily in the seeds with compulsive dormancy. Generally, the process is divided into two steps: Soaking and sprouting. First, soak seeds with water to soften the seed coat, absorb enough water for germination, and to kill bacteria and remove inhibitors. Then place seeds in an appropriate place to sprout.

Soaking. The water absorption rate varies according to the structure of seed coat, composition of inclusions, water content, and the temperature. For example, seeds with thin seed coats absorb water faster than those with thick and hard seed coats; seeds with low water content can absorb water faster than those with high water content; and seeds at high temperature absorb water faster than those at low temperature. The water absorption ability of seeds depends on the seed characteristics (e.g., seed size, inclusions) and soaking time. Seeds with more protein have higher water absorption than those with more starch. Generally, seeds with water absorption of 25% to 75% of their own dry weight can start to germinate. Excessive water absorption is unfavorable because it leads to poorly ventilated air, and the seed respiration is hindered. The key in handling soaking is to master the following points.

Temperature of Seed-soaking Water. Seed soaking in cold water (0-10 °C) has successfully accelerated the germination of some seeds with compulsive dormancy (e.g., Dahurian larch or fir), but the treatment time is long and the effect is not as effective as that of low temperature stratification. Therefore, generally, seed soaking is in warm water or hot water. The water temperature varies greatly for different tree species.

i. For tree seeds with thin seed coat and higher water content, such as poplars and willows, water temperature for seed soaking should not be higher than 20 °C.

ii. For tree seeds with a thin seed coat and lower water content, such as paulownia, oriental plane, and mulberry, water temperature is approximately 30 °C for seed soaking.

iii. The common tree seeds with thicker seed coat or pericarp, such as Chinese pine (*Pinus tabuliformis*), Korean pine (*Pinus densiflora*), Japanese black pine (*Pinus thunbergii*), Oriental arborvitae (*Platycladus orientalis*), Chinese fir (*Cunninghamia lanceolata*), Chinese red pine (*Pinus elliottii*), casuarina (*Casuarina equisetifolia*), tree-of-heaven (*Ailanthus altissima*), yellowhorn (*Xanthoceras sorbifolia*), can be soaked in warm water at 40 to 50 °C, while seeds of Shantung maple (*Acer truncatum*), Chinese wingnut (*Pterocarya stenoptera*), chinaberry (*Melia azedarach*), Sichuan chinaberry (*Melia toosendan*), date-plum (*Diospyros lotus*), and false indigo bush (*Amorpha fruticosa*), can be soaked in hot water at 60 °C.

iv. For any tree species with hard seed coat and hard seeds, such as black locust (*Robinia pseudoacacia*), Chinese honey locust (*Gleditsia sinensis*), Persian silk tree (*Albizia julibrissin*), and acacia petit feuille (*Acacia confusa*), hot water greater than 70 ℃ can be used.

For the hard seeds of some tree species, the effect of soaking seeds with successively increasing temperature is better. For example, *Robinia pseudoacacia* seeds were soaked in hot water at 70 ℃, while the water cooled down, the seeds continued to be soaked for 1 day; after the seeds swelled, they were selected for pregermination; the unexpanded seeds were soaked further in 90 ℃ water and cooled for 1 day, then the swollen seeds were selected for pregermination; the same method was repeated one or two times. It is safe and effective to accelerate germination by soaking seeds in batches and at successively increased temperature.

Ratio of seeds to water(v : v). The volume ratio of seed to water should be 1 : 3 when soaking seeds. Water is poured into the container that already contains the seeds, paying attention to stirring while pouring the water. Let the water cool naturally.

Seed-soaking time. The soaking time depends on the characteristics of the seeds. Common seeds are 1 to 3 days; thin seeds are shortened to several hours. Seeds with a hard seed coat or pericarp, such as walnuts, should be prolonged to 5 to 7 days. To check the water absorption for large-sized seeds, cut the seeds to observe from the cross section. Generally, water absorption is considered to be complete when 3/5 of the seed volume is saturated with water.

Water temperature affects the soaking time, and tree seeds with high water temperature requirement can be soaked for a longer time. Generally, the thickness of the seed coat is considered in pregermination for broad-leaved trees.

If the soaking time is longer than 12 hours, the water should be changed to remove impurities in seeds, to reduce carbon dioxide, and to replenish oxygen. Water can be changed one to two times a day. The practice in production has proved that in the process of seed soaking for small-sized seeds with many impurities (such as *Paulownia*) and easy-to-release sticky material, attention should be paid to washing and rubbing the seeds until the wash water is clear; otherwise, the effect of pregermination is not as effective.

Pregermination. There are two methods of seed pregermination after soaking.

"Bean Sprouting" Method. Wet seeds are put in a mud plate without glaze, covered with wet gauze, and placed at a warm place for pregermination. Wash seeds two to three times a day until the pregermination reaches the maximum. This method is used when the seed quantity is small.

Mixed with Sand Stratification. Mix the soaked seeds with wet sand at a ratio of three times the volume of the seeds(humidity is 60% of its saturated water content), and then cover the mixture to maintain moisture. Place at a warm location to accelerate germination.

For any method of pregermination, attention should be paid to the temperature (25 ℃ in general), moisture, and ventilation condition. The pregermination of seeds with a fast germination rate takes 2 to 3 days (such as *Robinia pseudoacacia*), and those with a slow germination rate take 7 to 10 days (such as dragon tree [*Paulownia* spp.] and *Melia azedarach*). When 1/3 of the seeds are "cracked and exposed in white", they can be sown.

(4) Stratification

Stratification is the method that in a certain period, seeds and moist substances are mixed or placed in layers to promote seed germination. It is simple and easy to operate, with very good effect. For some tree species, it not only accelerates germination but also is a way to store seeds.

Stratification is suitable for many seeds, such as Korean pine, Chinese juniper, larch, fir, *Ginkgo biloba*, linden, Chinese ash, *Fraxinus mandshurica*, *Acer truncatum*, Amur cork tree, walnut, filbert, chestnut, oak, maple, chinaberry, hardy rubber tree (*Eucommia ulmoides*), *Pterocarya stenoptera*, glossy privet (*Ligustrum lucidum*), golden rain tree (*Koelreuteria paniculata*), Walter's dogwood (*Swida walteri*), Chinese wild peach (*Prunus davidiana*), Siberian apricot (*Prunus sibirica*), cherry-apple tree, smoketree (*Cotinus coggygria*), *Malus baccata*, hawthorn, prickly ash (*Zanthoxylum bungeanum*), torch tree, and *Euonymus alatus*.

Effect of Stratification.

Releasing Dormancy. The stratification softens the seed coat and increases the permeability; under low temperature conditions, oxygen solubility increases, which ensures the oxygen supply to the embryo as a necessary step for the seed respiratory activity, leading to release of the dormancy.

When the seeds were treated with low temperature, the content of the inhibitive substance (such as abscisic acid) in the seeds decreased significantly, and the ability of inhibiting germination decreased, thus the dormancy was broken. At the same time, the growth stimulants for germination increased, which promoted germination. Khan put forward the three-factor theory of hormone, which holds that GA is the main growth promoter, ABA is the inhibitor, and cytokinin is the substance to overcome the inhibition from inhibitor and promote gibberellin to stimulate. The effect of three hormones in germination and dormancy is shown in Figure 8-3.

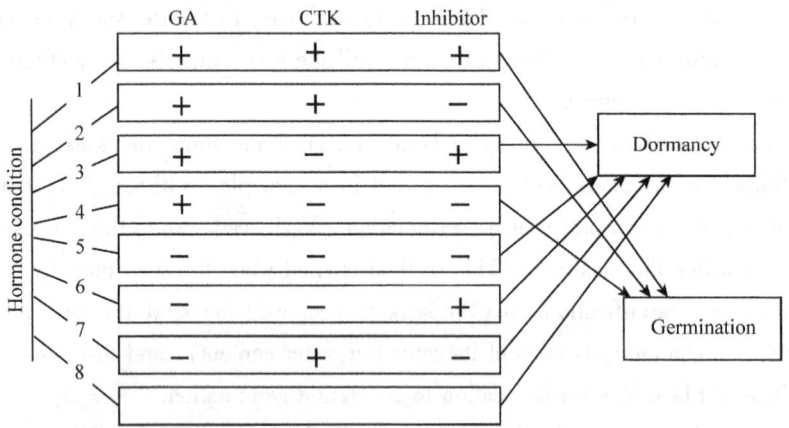

Figure 8-3 The relationship between growth regulators and dormancy as well as germination (From Khan, 1971, 1975)

For the tree seeds that need to undergo morphological after-ripening, such as *Ginkgo biloba* seeds, during the stratification process, the embryo grows obviously. After a certain period, the embryo grows to its proper length, and the after-ripening process is completed, then the seeds can germinate. This type of tree species also includes *Euonymus alatus,* wild roses, and Chinese torreya.

In the process of stratification, the general direction and processes of seed metabolism are consistent with those of germination.

The biological mechanism of stratification to overcome seed embryo dormancy has not been fully researched and explained. A large number of studies have shown that activities appear in the process of low-temperature stratification, including the improvement of seed water absorption, increase of enzyme activity, acidity increase and transformation of complex storage substances. These may not mean the releasing of factors controlling dormancy mechanism, but are predictions the improvement of seed germination ability. The general direction and process of metabolism are consistent with those of germination.

In the process of stratification, the seed embryo experienced a stage similar to vernalization, which was ready for germination.

Conditions in Stratification.

i. Temperature. Temperature plays a dominate role in the effect of pregermination. Most forest seeds (especially the seeds in the northern area) need certain low-temperature conditions. Generally, the low limit is slightly higher than 0 ℃ (above the freezing point), the upper limit is 10 ℃, and the suitable temperature is 2 ℃ to 5 ℃. For some tree species with a frost-crack seed coat, such as peach, apricot, and *Swida walteri*, a temperature below 0 ℃ can also be used. Temperature conditions for some individual tree species can extend to 15 ℃. Low temperature is conducive to morphological development of embryo and oxygen absorption, and it can reduce microbial activity, be beneficial to some biochemical processes, and prevent the germination of after-ripening seeds from completing at the same time.

ii. Moisture. The moisture in dry-storage seeds is low, so that the seeds should be soaked before pregermination. To ensure the water supply in the process of pregermination, the seeds and interlayer substances are often mixed or placed in layers to create a humid environment for the seeds. The common interlayers are peat, vermiculite, and sand. A 60% saturated water content is suitable for wet sand. (To test by hand: When holding tightly a handful of wet sand, it should not drop water; and when opening your fingers, sand will not disperse.) If wet peat is used, the water content can reach saturation degree.

iii. Ventilation. To meet the oxygen demand from seeds during pregermination, provide the interlayers with sufficient ventilation and accommodate water retention. Ventilation devices are necessary for stratification to circulate air. With a small number of seeds, straw should be used as the vent, and with a large number of seeds, a special vent should be built.

Certain Pregermination Days. The time required for stratification differs across tree species. When time is too short, pregermination requirements are not satisfied. See Table 8-2 for the number of days of stratification of some tree species.

Table 8-2 Time required for low-temperature stratification of seeds

Tree species	Time/d	Tree species	Time/d
Acer saccharum, Larix gmelinii, Malus hupehensis, Pinus tabuliformis	30	*Prunus avium, Prunus salicina, Swida walteri, Xanthoceras sorbifolia*	90-120
Dragon spruce, *Eucommia ulmoides*, fir, *Pinus sylvestris, Platycladus orientalis*	30-60	*Cotinus coggygria*, hazelnut, *Pinus bungeana*	120
Elaeagnus angustifolia, Ligustrum lucidum, Phellodendron amurense, Pyrus betulaefolia, Zelkova schneideriana	60	*Castanea mollissima, Juglans mandshurica, Tilia tuan*	150
Acer negundo, Amygdalus davidiana, Armeniaca sibirica, Betula spp., *Fraxinus chinensis*	60-90	*Cerasus vulgaris, Fraxinus mandshurica, Pinus koraiensis*	100-180
Celtis sinensis, Juglans regia, Malus baccata, Zanthoxylum bungeanum		*Cerasus tomentosa, Crataegus pinnatifida, Sabina chinensis*	180-210

Methods of Stratification.

i. Low-temperature Stratification. When treating a large number of seeds, stratification can generally be carried out by digging a pit outdoors. Select a high and dry terrain with good drainage, providing shady and leeward surroundings. The depth of the pit should be determined according to the local climate conditions. In principle, the bottom of the pit is below the frozen layer and above the groundwater level, so that the optimum temperature of the seed-sand mixture can be maintained for pregermination. For example, in Beijing, a pit depth of 60 to 80 cm is usually optimal. The pit bottom width should be 0.5 to 0.7 m, and the maximum width should not be more than 1 m, so as to keep the mixture temperature consistent. The length of a pit depends on the number of seeds. Pave the bottom of the pit with approximately 10 cm of wet sand (or other bedding beneficial to drainage).

Dry seeds can be soaked and disinfected before pregermination, mixed or placed in layers with wet sand according to a certain proportion. The large-sized seeds and the humid substance are generally placed in layers: Place a thick layer of seed for 3 to 4 cm, then place 3 to 5 cm of wet sand, one upon another. Medium- and small-sized seeds are generally placed in layers with the humid substance, and for every 3 to 4 cm of seed, place 3 to 5 cm of wet sand. The medium- and small-sized seeds are fully mixed with the humid substance evenly (volume ratio of 1:3), and then put into the pit. Ventilation facilities should be set up, one for every 0.7 to 1 m, from the bottom of the pit to the 20 cm above the pit top (the part close to the seed-sand mixture is preferably made of wire entanglement). The thickness of the mixture should be 50 to 70 cm to prevent uneven temperature. Wet sand of approximately 10 cm is added on the seed-sand mixture, then cover with soil to form a "ridge" to facilitate drainage. Small ditches should be dug around the pit for drainage (Figure 8-4).

The temperature, humidity, and ventilation condition should be checked regularly and adjustments may be needed during stratification. If the seed pregermination intensity does not meet the requirements, the seeds should be taken out 1 to 2 weeks before sowing (depending on the seed

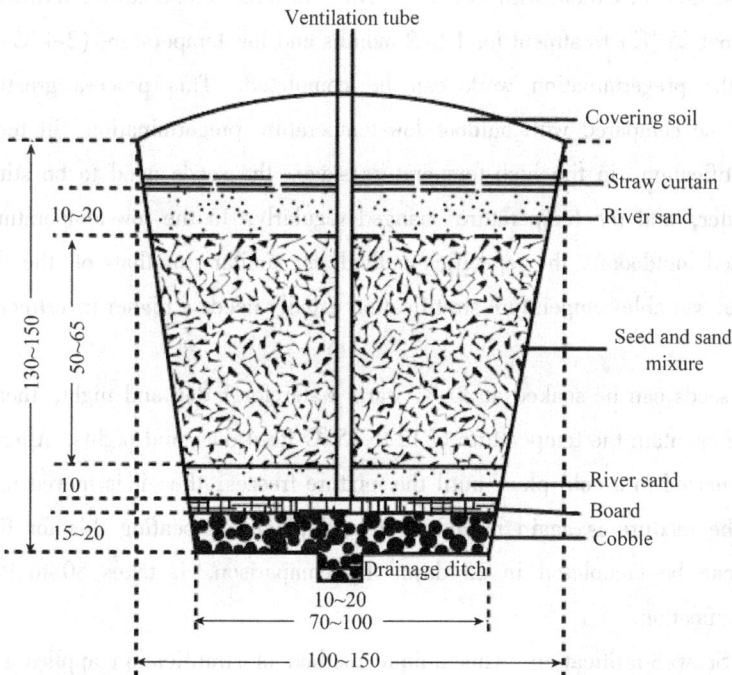

Figure 8-4 Schematic diagram of low-temperature stratification pit.

status) and transferred to a warm place (generally 15-25 ℃) to accelerate germination. The other way to make the time to finish pregermination is to dig an outdoor pit at a leeward and sunny place. Transfer those seeds into the pit until the seeds reach the pregermination intensity (1/3 of seeds are exposed in white). This method is also called a "transfer cellar".

  ii. Variable-temperature Stratification. This method alternates high and low temperatures to promote germination. Specifically, start the process with a high temperature (15-25 ℃) and follow with a low temperature (0-5 ℃), and then use the high temperature again for a short time, if necessary. For example, the seeds of *Pinus koraiensis*, *Fraxinus mandshurica*, hawthorn, and *Sabina chinensis* need a long time at low temperature for pregermination. Using the variable-temperature stratification can dramatically improve the result and shorten the pregermination time. For example, the germination of the seeds of hornbeam (*Carpinus*), mulberry (*Morus*), and hazelnut can be promoted by low-temperature stratification, but the results are not as good as that of variable-temperature stratification. The temperature and duration of the variable-temperature stratification vary with the tree species. Generally, apply a shorter period of high temperature and follow with a longer time of low temperature. For some specific tree species, such as *Fraxinus mandshurica*, the time length of high temperature and low temperature are almost the same; they are all approximately 3 months.

  Seeds of *Pinus koraiensis* can be soaked in warm water at 45 ℃ for 3 to 5 days and nights. Change the water once a day with 45 ℃ water in the beginning and let it cool naturally, then disinfect the seeds and mix them with wet sand. After variable-temperature stratification at the high temperature (about 25 ℃) treatment for 1 to 2 months and low temperature (2-4 ℃) treatment for 2 to 3 months, the pregermination work can be completed. This process greatly shortens the pregermination time compared with outdoor low-temperature pregermination. In terms of variable-temperature stratification, in the high-temperature stage, the seeds need to be stirred frequently, sprayed with water, and the temperature changed regularly. In the low-temperature stage, if the seeds are treated outdoors, the specific method is similar to that of the low-temperature stratification. So, variable-temperature stratification usually needs a higher investment of labor, with higher cost.

Smoketree seeds can be soaked at 30 ℃ warm water for 1 day and night, then mix the seeds with wet sand to maintain the temperature of 12 to 25 ℃ for 4 days and nights. Afterward, the seed-sand mixture is moved to a cold place until the mixture freezes, then it is moved to a warm place. After 4 days, the mixture is again moved to a cold place. Repeating this for five times, seed pregermination can be completed in 25 days. By comparison, it takes 80 to 90 days in low-temperature stratification.

iii. Mixing Snow Stratification. This unique method of stratification applies snow as the wet interlayer matter to create a low temperature along with moist and adequate ventilation conditions. The process can be applied in a place that has snow cover in the winter. Mixing snow stratification is suitable for most temperate coniferous species, such as larch, *Pinus sylvestris*, spruce, and fir.

A place with good drainage, a low underground water level, and shade should be selected to dig the pit. The depth is 40 cm, with length and width being determined by the number of seeds. The pit should not be too deep, because a too-deep pit might lead to freezing of seeds and would make it difficult to take out seeds. It should not be too shallow either, because a too-shallow pit is close to the ground level, which makes the seeds vulnerable to freezing damage in severe cold climates, and may also makes the upper seeds to easily germinate when the surface temperature is high in spring. When the soil is not frozen, the pit should be dug first. When the snow has not yet melted in winter, the pit bottom can be covered with several centimeters of snow. Mix the seeds and snow evenly according to the volume ratio of 1:3 (or the seeds, sand and snow should be mixed according to the ratio of 1:2:3). Afterward, place the mixture in the pit, build a snow cover on top of the mixture, and cover it with straw mulch. Gradually remove the snow cover in spring, and the mixture is still covered with the straw mulch, 1 to 2 weeks before sowing, the seeds should be taken out and placed in a warm place until the snow melts, and then the seeds can be soaked in snow water for 1 to 2 days. At this stage the seeds should begin high-temperature pregermination. Seeds can be sown when they reach the pregermination intensity.

As a special wet interlayer, the snow creates a good condition for low temperature, humidity, and ventilation. The advantages to using snow water include low water content with few inhibiting effects, high physiological activity because it is easily absorbed by seeds, high nitrides content which have certain nutritional values. Therefore, mixing snow stratification has beneficial effects: It promotes early seed germination, increases germination rate, and supports healty and strong ceedling growth.

(5) Other Physical and Chemical Treatment Methods

Phytohormone Treatments. Gibberellin can replace the low-temperature stratification condition required by dormant seeds. It is also a necessary substance for seed germination. Addition of an appropriate amount of gibberellin can promote seed germination and enhance germination ability. The dosage of gibberellin differs among tree species, but generally the effective range is wide, from 5 to 500 mol/L. For example, after gibberellin treatment of 10 to 30 mol/L, the germination rate of larch seeds reached 50% to 58%, while that of the control was only 43%. Cytokinin can eliminate the inhibition of abscisic acid (ABA), so it is very necessary for the germination of dormant seeds. The dosage of cytokinin should be controlled at less than 10 mg/L, otherwise the adverse effect can be induced with higher concentrations. Ethylene also has an effect on dormancy release and seed germination and is often released through ethephon at a working concentration of 1 to 1000 mg/L with a wide range of action. In addition, indoleacetic acid (IAA), indolebutyric acid (IBA), and naphthoxyacetic acid (NAA) promote seed germination.

The conventional way to treat seeds with hormones is to soak seeds in an aqueous solution. In recent years, this process has been improved upon for use in other countries, using an organic solvent permeation method. Common organic solvents are acetone and dichloromethane. Research has shown that the solvent itself is harmless to the seeds, and the treatment time was greatly shortened, sometimes with beneficial synergism.

Treatment of Microelements and Other Chemical Agents.

Microelement treatments can strengthen the physiological and biochemical processes in the seeds, promote seed germination, improve the growth and development of seedlings after germination, and improve stress-resistance ability. A large number of studies have documented the effective microelements for seed treatment inducing copper, manganese, zinc, boron, and molybdenum. The concentration of the agent for seed soaking varies with seed species and element species, but it is generally 5 to 500 mg/L.

Commonly used and effective chemical agents for breaking seed dormancy and promoting seed germination include potassium nitrate (KNO_3) (0.1%), potassium cyanide (KCN) (0.01 mg/L), sodium nitrate ($NaNO_2$) (0.5 mg/L), mercaptoethanol (0.05 mg/L), hydrogen peroxide (H_2O_2) (1%), hydroxyamine (NH_2OH), methyl blue, triphenyltetrazolium chloride (TTC), thiourea, and various sulfhydryl compounds. These chemical agents are used as respiratory inhibitors or as electron receptors, which is beneficial to the pentose phosphate pathway, which releases dormancy and promotes germination. For seeds that contain inhibitory substances, ethyl alcohol plus mercury bichloride($HgCl_2$) can break seed dormancy.

In addition, when oil and wax are outside the seed coat, such as seeds of *Swida walteri*, Chinese pistachio (*Pistacia chinensis*), Chinese tallow tree (*Excoecaria sebifera*), and mountain ash, soaking the seeds in soda water for 12 hours can remove the oil and wax, soften the seed coat, and promote the metabolism of the seed embryo. Treating larch seeds with 1% zinc sulfate solution can improve the germination potential of the seeds, and treating *Pinus tabuliformis* with 0.2% potassium nitrate has a significant promoting effect on seed germination. Such chemical agents also include calcium chloride, carbolic acid, sodium sulfate, potassium bromide, and hydroquinone.

Treatment of Other Physical Factors. The application of X-ray, radioisotope, laser, ultrasonic, magnetized water, ultraviolet light, and infrared rays in seed treatment can break the limitation of the seed coat and promote germination in some cases. For example, when the seeds of *Excoecaria sebifera* were soaked in magnetized water for 40 h, the emergence rate was more than 90% at 55 days after sowing, whereas emergence rate was only 35% at 60 days after sowing when using warm water to soak the seeds.

8.2 Seedling-raising Pattern and Seedbed Preparation

Seedling-raising pattern, also known as operation pattern, is divided into seedbed seedling and field seedling.

Seedbed seedling is the primary approach for nursery production. Some tree species grow slowly, especially the small seeds, which need careful management, such as Scots pine (*Pinus massoniana*), poplar, and dragontree or Empress tree (*Paulownia*), and are generally sown in the seedbed. The time to make the seedbed should be closely matched with the sowing time, and it should be completed within 5 to 6 days before sowing.

According to the form of the seedbed, it can be divided into two types: High seedbed and low seedbed (Figure 8-5).

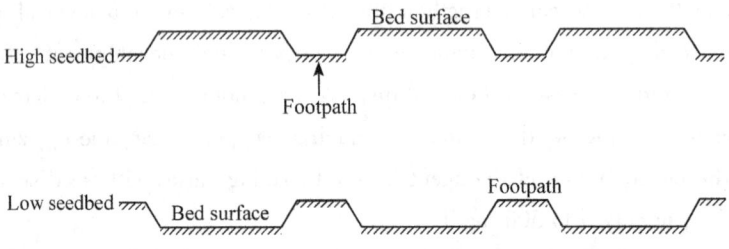

Figure 8-5 Profile of seedbed.

8.2.1 Seedbed Seedling

(1) High Seedbed

The seedbed surface is generally raised 15 to 25 cm above the level of the ground. The width of the seedbed should be designed in a way that is convenient for operations, generally 1.0 to 1.2 m, with a footpath width of 30 to 50 cm. The length of the seedbed should be determined according to the practical situations in the nursery land. Considering maximum land utilization and the type of sprinkling irrigation, the seedbed length can be 10 to 20 m or more. If canal irrigation is adopted, 10 to 15 m is generally suitable, because if the seedbed is too long, it cannot easily be leveled, and the high land area cannot be watered. In places with high terrain and good drainage, the seedbed surface can be slightly lower, while in the nursery with poor drainage, the seedbed surface should be higher (Figure 8-5).

A high seedbed has good drainage and permeability, along with high ground temperature, which is convenient for side irrigation, and the seedbed surface will not be hardened. High seedbeds have good aeration and a thick, fertile soil layer, so seedlings can grow very well. High seedbeds are suitable for nursery land made up of clay soil that has a large amount of precipitation and poor drainage. These seedbeds are used primarily for tree species that are sensitive to soil moisture or that have difficulty in sprouting and need to be carefully managed.

When making a seedbed, the earth should be moved up from the footpath line and added onto the seedbed. Pat the soil to form a smooth bed. The edges of the seedbed should be filled with earth and also patted down. Make four straight sides for the seedbed, and use the shovel to cut the edges of the seedbed. Finally loosen the soil within the seedbed. A nursery bed machine can be used to improve the efficiency of the operation and the quality of seedlings.

(2) Low Seedbed

In this scenario, the seedbed surface is lower than the footpath. A footpath is usually 15 to 18 cm higher than the seedbed surface. The width of the seedbed surface is 1 to 1.5 m, and that of the footpath is 30 to 40 cm. The seedbed length is the same as that of the high seedbed (Figure 8-5).

A low seedbed retains moisture well, saves labor, and uses less irrigation water, but it is not good for drainage. The seedbed surface may easily become hardened after irrigation.

Low seedbeds are suitable for areas with less precipitation, or no water retention in rainy seasons. They are good for the hygrophilous tree species that are not affected by a little ponding water, such as most broad-leaved tree species and some coniferous tree species.

When making the seedbed, first form a line according to the width of the seedbed surface and footpath. Then pad the footpath with the soil from the seedbed surface line, and pat down the soil to prevent the footpath from collapsing into the seedbed. After the footpath is complete, the seedbed surface is plowed loosely and leveled. Many machines are suitable for making a low seedbed.

8.2.2 Field Seedling

Field seedlings can be produced using ridge culture and flat culture, both of which can be mechanized.

Generally, both forms can be used for tree species with fast growth and low management technology requirements. Ridge culture plants seedlings on a ridge above the ground. Building a ridge can thicken the fertile soil layer, improve the soil temperature, facilitate the transformation of soil nutrients, and provide the seedlings with sufficient light and ventilation, which results in strong growth. Ridge culture is also convenient for mechanized operations, improving labor productivity and reducing the seedling cultivation cost. However, the management of ridge seedlings is not as meticulous as seedbed seedlings, and the seedling yield is also lower than that of seedbed seedlings.

(1) High Ridge

The specification of a high ridge is generally 60 to 80 cm in ridge spacing, 15 to 20 cm in ridge height, 20 to 40 cm in ridgetop width, and the length depends on the terrain or tillage method. When making the high ridge, first draw a line according to the specified ridge distance, then the soil is turned along the line to both sides, forming the back of the ridge, and then the ridgetop is scraped with a plank to create the desired ridge height and a consistent ridgetop width, making it convenient for sowing seeds. The ridge surface is furrowed, and seeds are sown by hand or with a machine seeder.

High ridges have the characteristics of high seedbeds, making them suitable for tree species with middle and large seeds, providing strong growth potential for seedlings, and needing only minimal management after sowing. Wide ridges are effective for arid areas because they are beneficial to soil moisture conservation, whereas narrow ridges are good for humid areas to facilitate drainage.

(2) Low Ridge

Low ridge, also known as flat ridge or flat work, is the seedling cultivation method in which seeds are sown directly after the nursery land is leveled. Its advantages include simple operation, minimal land requirements, and a high seedling yield per unit area. The drawback of this method is that it is not conducive to irrigation and drainage. The adoption of multi-row strip configuration can improve the land utilization rate, which is suitable for tree species with large or medium seeds that have strong germination.

8.3 Sowing Season and Sowing Quantity

8.3.1 Sowing Season

Timely sowing is one of the important measures in the cultivation of strong seedlings. Selecting and adjusting seeding time can indirectly improve the environmental conditions and growth time of seedlings to increase the seedling yield and quality. The basis of determining seeding time includes local climate conditions and tree species characteristics. In China, seeding can be carried out in spring, summer, autumn, and winter in some areas in the South, while seeding is primarily during the spring in the North.

(1) Spring Seeding

Spring seeding is the primary season of sowing used in production. Sowing seeds in springtime takes shorter time for seed staying in soil and reduces the management of the sowing land. However, in spring, when the agriculture-related schedule is busy, seed storage and pregermination of some tree species are still needed, so this might create a higher workload at the time.

The earlier the seeding, the better the effect, as long as the seedlings will not suffer from low temperature damage (late frost) after being unearthed. When the ground temperature at the depth of

5 cm of soil is stable at approximately 10 ℃, or the average temperature over a period of 10 days is 5 ℃ (reaching the lowest temperature of most seeds' germination), seeds can be sown. In Inner Mongolia and Northwest China, spring seeding is generally in late April, while in Shandong and Henan it is from late March to early April. Early sowing prolongs the growth period for seedlings, and when the hot summer comes, the seedlings have a certain degree of resistance to the heat, so the quality of seedlings is better. Too early or too late spring seeding can significantly affect the growth of seedlings (Table 8-3). In areas with severe spring drought in the north of China, protective measures such as plastic film coverage and greenhouse seedlings can significantly improve the seedling quality.

Table 8-3 Effect of sowing time on seedling growth of *Pinus koraiensis*

Sowing day and month	Stem		Root system			Ground temperature when sowing/℃			
	Seedling height /cm	Root collar diameter /cm	Length of axial root/cm	Length of lateral root/cm	Lateral root number	0	5 cm	10 cm	15 cm
04-03	5.5	0.22	11.8	4.06	9	13.5	7.0	4.0	2.7
04-11	5.3	0.23	9.9	4.03	9	15.1	9.3	6.0	5.2
04-19	5.8	0.23	13.6	6.24	9	11.0	10.4	6.6	6.1
04-27	5.1	0.23	0.23	5.15	8	25.5	19.7	14.1	9.0
05-05	4.8	0.22	9.1	4.33	7	24.5	18.7	14.0	6.7

Notes: From Xie Wei et al. (2011).

(2) Summer Seeding

Some seeds of tree species, such as poplar, willow, elm, mulberry, birch, and eucalyptus, mature in summer and can easily lose their vitality. Seed storage can be moitted when seeds are collected and directly sown, thus prolonging the growth period of seedlings.

The specific time of sowing in summer depends on the soil and climate conditions. When sowing, the soil should be wet to ensure the successful germination and that seed unearths easily. Attention should be paid to sunburn to prevent high temperature damage.

(3) Autumn Seeding

Autumn seeding is mainly suitable for the seeds with a long dormancy period, such as Korean pine and *Tilia tuan*, and the large seeds with a hard seed coat, such as Manchurian walnut (*Juglans mandshurica*), *Prunus davidiana*, and *Prunus sibirica*. A long sowing time in autumn makes it easy to arrange for nursery labor needs. The seeds with a long dormancy period complete the process of pregermination during winter in the field, sprout early in the next spring, and germinate quickly, leaving out the work of seed storage.

When autumn sowing is carried out in areas that experience freezing injury in winter, ensure that no germination occurs in the autumn of that year, although early sowing can be carried out if the dormancy period is long. For areas where soil moisture conservation is poor in the seeding field and without irrigation conditions, seeds can be sown in early autumn, and the seedlings can survive through winter by soil burying after germination and excavation.

(4) Winter Seeding

In the areas of South China where the soil is not frozen in winter, Chinese fir, Masson pine, and eucalyptus can be seeded in winter. The advantage of seeding in winter includes early sprouting, a long growth period, and strong resistance of seedlings.

8.3.2 Sowing Quantity

The sowing quantity is the weight of the seeds sown per unit of area (or per unit length). Sowing quantity of large seeds is sometimes expressed by the number of seeds. If sowing too much, the density of seedlings is high, which wastes seeds and increases labor given that thinning may be required; if sowing too little, the density of seedlings is too low. Therefore, based on reasonable density, we can calculate the sowing quantity scientifically. To determine the sowing quantity, one approach is based on the empirical data of production practice; the other is to calculate the sowing quantity according to the planned seedling density, seed quality index, and loss coefficient of seedlings. The following formula can be used:

$$X = \frac{A \times W}{P \times G \times 1000^2}(1+C)$$

In which:

X is the actual required sowing quantity per unit of area (or per unit length), kg;

A is the planned number of seedlings per unit of area (or per unit length);

P is the seed purity;

G is the germination potential;

W is the thousand seed weight, g;

C is the loss coefficient; and

$\frac{1}{1000^2}$ converts thousand seed weight to the weight of each seed, kg.

The loss coefficient C varies with tree species, nursery environment, and seedling technology level. The smaller the seed size, the higher the C value. The C value of very small seeds is more than 4, for example, that of poplar seed is 10 to 20; the C value of medium and small seed is between 0 and 4, for example, that of *Pinus tabuliformis*, *Pinus sylvestris*, dragon spruce, and Dahurian larch is 0.4 to 0.8; the C value of large-sized seeds is close to 0. The C value is high in areas with poor environmental conditions, whereas it is low under intensive management.

When calculating the sowing quantity of a production area with the planned number of seedlings per unit area, it should be according to the "net area" of seedling cultivation (excluding footpath, bed frame, field ditch, and furrow).

8.4 Sowing Method

8.4.1 Three Types of Sowing

Sowing is an important part of nursery work, directly affecting the field germination rate of seeds, emergence speed, and uniformity. Sowing methods can be divided into three types: Drilling, broadcast sowing, and dibbling. Effective sowing methods can be selected according to seed size, seed characteristics, and nursery conditions.

(1) Drilling

Drilling is a sowing method of sowing seeds evenly in sowing ditches according to a certain row spacing. It is the most widely used sowing method, suitable for all tree species, especially for small and medium seeds.

Drilling has fixed row spacing, which is convenient for machine operation and seedling management, such as loosening soil, weeding, and topdressing. Seedlings are evenly exposed to light, well ventilated, and grow robustly.

The row spacing of drilling depends on the characteristics of the tree species and the soil conditions. The width of the seeding row is called the sowing width, which is generally 2 to 5 cm. It is beneficial to overcome the weakness of drilling and improve the seedling quality by widening the sowing width properly. For example, the sowing width of most tree species can be widened to 10–15 cm. The ideal direction of a sowing row is in a north-south direction in which the seedlings accept light evenly, but it also varies according to the direction of the seedbed. To facilitate irrigation and other tending measures, the longitudinal-row drilling parallel to the long side of the seedbed is often used. While, when raising seedlings in a high seedbed, adopting horizontal-row drilling is beneficial to lateral irrigation.

To facilitate mechanized operations, strip drilling can be used in field seedling cultivation. Several sowing rows form a belt, increasing the belt spacing and reducing the row spacing.

(2) Broadcast Sowing

Broadcast sowing is a method of sowing seeds evenly on the seedbeds or ridges. This sowing method is suitable for very small seeds, such as poplar, willow, and eucalyptus.

The advantages of broadcast sowing include high seedling yield and maximum utilization of land nutrition area. While, the seed quantity in broadcast sowing is increased. Because fixed row spacing is nonexistent, it is very inconvenient for soil loosening, weeding, and other tending management. At the same time, the density of seedlings is high and they are not in line, resulting in poor light and ventilation and affects the growth of seedlings.

(3) Dibbling

Dibbling is a method of digging holes and sowing seeds on the seedbed or field according to a certain planting spacing. Generally, it is used to raise seedlings of large seeds, such as walnut, Chinese chestnut, peach, apricot, *Quercus*, and ginkgo.

The advantage of dibbling is to save seeds, similar to the use of drilling sowing, but the yield is less than that of the other two sowing methods.

8.4.2 Key Points of Sowing Technology

Sowing technology is related to whether the seedlings can be excavated at the appropriate time, and the cultivation of strong seedlings can be realized only through careful attention throughout all aspects of seeding. Taking artificial drilling as an example, the seeding work can be divided into four steps.

(1) Artificial Seeding

Lining. The position of the seeding ditch is determined according to the planned row spacing and the direction of the seeding row.

Ditching. Ditching should be carried out according to the sowing width, which needs to be straight, uniform in depth, and the depth of the ditch should be determined according to the seed size.

Seeding. The sowing quantity should be strictly controlled, and the seeds should be uniformly sown. Mix small seeds with fine sand, which ensures greater uniformity of sowing.

Soil Covering. Soil covering creates good conditions for seed germination. The thickness of soil cover has a great influence on the soil moisture around the seeds, so as to on the germination rate of the field, the emergence time, and the evenness of the seedlings. When the soil cover is too thick, the temperature is low and oxygen is insufficient, which is harmful to seed germination and the excavation after germination. If the soil cover is too thin, seeds are easily exposed and cannot obtain enough water, which is also harmful to seed germination and allows the seeds to be vulnerable to bird and animal scavenging (Table 8-4).

Table 8-4 Soil cover thickness and seed germination status

Soil cover thickness /cm	Sprout quantity in each meter of sowing ditch (individual plant)		
	Dahurian larch	Dragon spruce	European pine
0.5	—	—	130
1.0	22	198	173
1.5	24	197	148
2.0	23	194	145
3.0	20	—	13
4.0	0	13	0
5.0	0	—	0

Source: From Lu et al. (1989).

Figure 8-6 Artificial seeding steps of Korean pine seeds (Photos by Liu Yong).
(a) Carefully level the seedbed surface; (b) Sow with a special sowing frame to ensure the regular sowing width and row spacing; (c) Note the regular seedbed surface after sowing; (d) Press seeds into the soil; (e) Cover the soil. In this method, the conventional furrow seeding is changed to direct seeding on the flat seedbed surface, and then the seeds are pressed into the soil, making the seeding more regular, soil covering more even, and the emergence uniformity better.

The basis for determining the soil cover thickness is:

i. Characteristics of Tree Species. The soil cover thickness of large seeds should be thicker than that for small seeds. In practice, the soil cover is thick for hypogeal (below soil surface) germination, and thin for epigaeous (aboveground) germination.

ii. Climate Conditions. Soil cover thickness should be thick in dry climates and thin in wet climates.

iii. Soil Conditions. Soil cover thickness should be thick in loose soil and thin in heavy soil.

iv. Sowing Season. Seed sowing should be thick for autumn seeding and thin for spring and summer seeding.

v. Soil Covering Materials. For small and very small seeds, use loose materials such as sand and peat soil. If covered with original soil, the layer tends to be over covering.

After seeds undergo pregermination, when sowing in soil with optimum moisture, the soil cover thickness should be two to three times the seed diameter from the short axis.

For very small seeds, soil cover thickness is enough as long as seeds are invisible. To ensure that seeds and sowing ditches are wet, be sure to dig the ditch, sow seeds, and cover the soil at the same time (Figure 8-6).

(2) Mechanical Seeding

Using a seeder or sower can greatly improve the working efficiency of seedling raising. Ditching, sowing, soil covering, and compacting are completed at one time, which is the trend of nursery seedling development in the future. It will not only use less manpower but also ensure the seedlings are excavated regularly. Before application, check whether the performance of the machine is easy to control, such as seed quantity, sowing depth, and soil covering thickness. Also check whether it can damage the seed. The requirements of each technical point are the same as those of artificial sowing (Figure 8-7).

(a)   (b)

Figure 8-7 Mechanical sowing (a) and status after emergence (b) (Photo on the left provided by RK Dumroese; Photo on the right by Liu Yong).

8.5 Nursery Soil and Seedling Management

8.5.1 Management of Sowing Land

The management of sowing land refers to the technical measures undertaken on sowing land before seedling emergence. The goal is to create conditions that will ensure seeds have timely excavation and that germination is uniform and even. The ultimate purpose is to improve the field germination rate and seedling emergence rate.

(1) Mulching to Preserve Soil Moisture

Effects of Mulching. Mulching can preserve soil moisture, prevent soil hardening caused by irrigation, and improve soil temperature. Especially when seeding is in early spring, the benefits of mulching on seed germination and seedling excavation are very obvious. Mulching should be emphasized especially for the sowing land with thin soil covering.

Mulching Materials and Methods. Common mulching materials are thin plastic film, rice or wheat straw, crop curtain, tree branches, and crop straw. Attention should be paid to weed seeds, diseases, and insect pests in the mulching materials. Disinfect the mulching material, if necessary.

Plastic film mulching can preserve soil moisture and improve soil temperature. During seedling emergence, the temperature under the film should be monitored around noontime to avoid heat damage. The film can be broken to expose the seedlings when the film hampers the growth of seedlings. Mulching with other materials provides a certain degree of improvement on water conservation, but the effect is minimal. If mulching is too thick, it will affect the emergence of seedlings. The appropriate mulching thickness is to cover the ground evenly and to not see bare land.

Mulch Removal. When a large number of seedlings are excavated, the mulch should be removed in a timely manner because at this stage mulching can affect the light and the normal growth of seedlings. Removal of mulch should be conducted gradually, so that the seedlings can adapt to the external environment. If necessary, the timing should cooperate with the cooling measures.

After sowing, a soil surface heating agent can be sprayed to form a uniform gelatinous film, which has the same effect as the plastic film mulching, making the seeds germinate 3 to 5 days earlier than without the gelatinous film.

(2) Irrigation

Seed germination requires certain water conditions. In production, water should be adequately provided before sowing to keep good soil moisture content. Ideally, no further irrigation should be applied before emergence. If the soil moisture is insufficient at the seedling stage, however, irrigate the soil properly to keep the soil moist. Sprinkling irrigation is the best for the soil surface. The water volume of ridge irrigation or high seedbed irrigation should not be too much. The water surface should not be over the back of the ridge (seedbed), preventing the soil from hardening.

For small-sized seeds and thin soil covering, spray or drip irrigation can be adopted; for sowing land with fast emergence and mulching, the irrigation can be reduced or excluded. During irrigation, the amount of water should be controlled to prevent the seeds from being exposed due to water scouring the seedbed.

(3) Loosening Soil and Weeding

Weeding measures should be taken if weeds appear in the sowing land. To prevent the soil from hardening, loosen the soil promptly. Soil loosening and weeding are usually carried out together. Soil loosening in the sowing land should be shallow to avoid touching the seeds. It is necessary to loosen the soil and pull the weeds out before they emerge, especially with autumn seeding.

8.5.2 Tending Management of Seedlings

Seedling tending includes from the beginning of seedling excavation to the beginning of lifting the seedlings. The purpose of seedling tending is to provide better nutrient and growth conditions for seedlings. These measures ensure the rapid growth, high yield, and high quality of seedlings. Tending management includes soil management (such as irrigation, fertilization, soil loosening, and weeding), seedling management (such as thinning and root cutting), and seedling protection (such as shading, pest control, and wintering and frost protection measures). Among these, soil management and seedling protection improve the environmental factors of seedling growth. As for seedling management, it builds on itself. Each measure will directly affect the ultimate quality and yield of seedlings.

8.5.2.1 Shading

The newly unearthed seedlings are tender and not resistant to high temperature and drought. To prevent the seedlings from scorching, shading can be provided at the time that the mulch is removed. Shading can reduce the surface temperature in seedling land, decrease the evaporation from soil water, and decrease the transpiration of seedlings, which will benefit seedling growth. Nonetheless, shading can also reduce the photosynthetic intensity of seedlings because of the lack of light, resulting in a decline in the quality of seedlings. Therefore, appropriate light transmittance should be controlled when taking shading measures.

Figure 8-8 Shade shed of seedlings (Photo provided by R. K. Dumroese).

Shade net, reed curtain, or a bamboo curtain can be used to set up a movable shade shed horizontally above the seedlings, with a light transmittance of 50% to 80%, generally shading from 9:00 or 10:00 to 17:00 every day (Figure 8-8). The shading duration varies according to tree species and local climate conditions. Based on the principle of protecting seedlings from high temperature scorching, shading is usually set up for 1 to 2 months from the time of mulch removal at the beginning of the early growth stage to the beginning of seedlings' fast-growing time. For summer seeding and seedling cultivation, shading should be stopped after the lignification at the seedling base or after relieving the high temperature and drought conditions.

8.5.2.2 Irrigation

(1) Significance of Reasonable Irrigation

Water is not only the basic element in seedling life but also an important component of soil fertility factors (water, fertilizer, gas, heat). The physiological activities of seedlings cannot take place without water, and soil nutrients need to be dissolved in water to be absorbed and utilized. Insufficient or too much water in the soil/substrate will inhibit the normal activities of seedlings and even cause disasters such as drought and waterlogging. When water content is between 60% and the field capacity, the soil can maintain a good ratio of solid, liquid, and gas states. The soil water supply capacity and root absorption capacity can be maintained at a higher level. When the water content of soil is 60% to 70% of the field capacity, it reaches the most favorable conditions for root water absorption and microbial activities in the soil. When the soil water content exceeds the field capacity, because of poor ventilation conditions in the soil, cell respiration weakens, root pressure reduces, and the water absorption and fertilizer absorption of the root system are blocked. The worst scenario is that the root system would die, and the growth of seedlings would stop or generate toxic elements.

Generally, rainfall and underground water alone cannot meet the requirements of the whole growth process of seedlings. Although excessive irrigation can inhibit the growth of seedlings and cause secondary salinization of soil, especially in arid and semi-arid areas, irrigation is an indispensable measure for seedling cultivation, so it is necessary to carefully conduct irrigation.

(2) Principles of Proper Irrigation

Irrigation should be conducted at the proper time, that is, the optimal irrigation period and a rational irrigation amount should be selected. It should be specifically determined according to local climate conditions, biological characteristics of the tree species being grown, and the developmental stage of seedlings. Taking the field capacity as the upper limit of irrigation, nursery workers should adhere to the following principles.

Irrigate According to the Climate and Soil Conditions. When the climate is arid or the weather is dry, water can be consumed quickly, so the irrigation amount should be increased accordingly. The interval of irrigation for soil with strong water-holding capacity can be extended, whereas for sandy soil with poor water-holding capacity, the interval of irrigation needs to be shorter.

Irrigate According to the Biological Characteristics of Cultivated Trees Species. Conifer species require less water than broad-leaved species, and slower growing species require less water than faster growing species. For example, seedlings of poplar, willow, birch, and larch need more water and irrigation times need to be increased; followed by *Fraxinus chinensis*, *Acer truncatum*, and the elms (*Ulmus pumila*); and *Pinus tabuliformis* and *Platycladus orientalis* need even less water so irrigation times can be appropriately reduced during their seedling cultivation.

Irrigate According to Growth Stages of Seedlings. Irrigation intensity differs according to the water requirement of specific growth stages of seedlings. Sufficient water irrigation to the bottom soil before sowing is conducive to the germination and excavation of seedlings. At the seedling stage and early growth stage, the tissues of the seedlings are tender, and the root system is small and shallow. Although the water demand is small, the seellings are very sensitive, so they should be irrigated accordingly. However, at the early growth stage, the growth of seedlings should be restrained to allow for root development, which means controlling the water supply to enhance the drought resistance ability of seedlings, promote the growth of root system, and lay a good foundation for the vigorous growth at the fast-growing stage. At the fast-growing stage, seedlings need the most water within the growing season, and the water utilization rate is also that the highest, so that they should be irrigated regularly. At the later stage, to prevent seedlings from excessive growth, the irrigation should be stopped. The cold resistance of seedlings can be enhanced by a one-time irrigation before overwintering.

(3) Technology of Rational Irrigation

Irrigation Time. The specific irrigation time is in the morning or at dusk for surface irrigation. At those specific periods, evaporation is minimal and the difference between water temperature and ground temperature is small.

Irrigation Intensity. For each application, irrigation should reach to the farthest root distribution area.

Irrigation Continuity. Ensure that the soil water in the seedling areas is always in a suitable status to prevent severe drought, which can reduce seedling quality. Soil topdressing should be combined with irrigation. After soil top-dressing, irrigation should be conducted immediately.

Irrigation Water Quality Standards. The content of soluble salt in irrigation water is generally required to be less than 0.2% to 0.3%.

Water Temperature Requirements for Irrigation. Water temperature for spring and autumn irrigation should be above 10 to 15 ℃, and temperature for summer irrigation should be above 15 to 20 ℃. Water temperature that is too low or too high should be adjusted by appropriate measures. A reservoir of water is usually used to raise the water temperature.

(4) Irrigation Methods

Irrigation methods include side irrigation, flood irrigation, sprinkler irrigation, and drip irrigation. See Chapter 6 and section 7.3.2 in Chapter 7.

8.5.2.3 Intertillage, Weeding, and Drainage

Intertillage is a tillage method used during the growth of seedlings. During this time, the soil appears to harden after irrigation and rainfall, which is harmful to the development of seedling root systems. Through the combination of middle plowing and weeding, the surface soil layer can be broken and the capillaries can be cut off, which can reduce the evaporation of soil water, increase the aeration performance, improve the living conditions of soil microorganisms, accelerate the decomposition of fertilizer, and improve the utilization of effective nutrients in the soil. By weeding, unwanted plants can be eliminated, competition with seedlings is reduced, and the growth environment of seedlings is improved.

Intertillage and weeding are usually combined and carried out after irrigation or rainfall. In addition to eliminating weeds, intertillage needs to reach a certain depth to produce the desired effects. In the early stage of growth, loosen just the topsoil, generally 2 to 4 cm, and then gradually deepen to 8 to 12 cm. To promote the lignification of seedlings, the work of loosening soil should be stopped when irrigation is stopped in the later growing period.

Intertillage and weeding should be carried out when the soil is not too dry or too wet and the shallow roots growing on the ground can be cut properly. Pay strict attention to avoid damaging the seedlings. The times and duration of intertillage and weeding can be determined according to the local conditions and the growth characteristics of seedlings. The depth of intertillage depends on the depth of the seedling root system.

Artificial intertillage and weeding is a target-orientied with good effect, but the method is outdated and the work efficiency is low. Mechanical intertillage and weeding is more advanced and efficient than the artificial intertillage and weeding, but suitable machinery should be selected according to the method of seedling raising. If heavy weed growth is present on the nursery land, the of herbicides should be considered to improve the weeding efficiency. The types of appropriate herbicide can be selected according to the specific weed types; see 7.4.1 in Chapter 7. If ponding has occurred in the nursery land after rainfall, the water should be drained as soon as possible. See section 7.2.4 in Chapter 7.

8.5.2.4 Fertilization

The primary fertilization method is combining base fertilizer with top-dressing. The base fertilizer can be combined with soil preparation before sowing. Fertilizer can be applied in two ways at the seedling stage: Soil top-dressing and foliage top-dressing.

Quick-acting fertilizer, such as plant ash, ammonium sulfate, urea, muriate of ammonia, superphosphate, or muriate of potash, is generally used for top-dressing of seedlings, and the principle of "a small amount for many times, right time and right amount" should be followed. The mobility of nitrogen fertilizer in the soil is fast, so a shallow application can penetrate the root distribution layer and be absorbed by seedlings. The mobility of potassium fertilizer is slower, and that of phosphorus fertilizer is the slowest, so it is better to make a deep application into the dense distribution area of the root system.

Different soils contain different types and quantities of nutrient elements, so different fertilizers are applied. In calcareous soil, phosphorus deficiency often occurs, therefore, it is necessary to increase the number of applications and increase the amount of phosphorus fertilizer. Nitrogen fertilizer is commonly applied to most soils. If the soil has high levels of nitrogen, the proportion of phosphorus and potassium should be increased. For soil that has the ability to maintain its nutrients and fertility, the amount of top-dressing can be increased for each time and the application times can be less. For sandy soil with poor fertility-keeping ability, top-dressing times should occur more frequently, but the top-dressing amount for each time should be less.

The first top-dressing of annual seedlings should be in the first half of the early growth stage. In the period of emergence, the primary nutrition source for seedlings comes from the seed, and only minimal amounts of fertilizer are needed. At the seedling stage, seedlings are sensitive to nitrogen and phosphorus, so more applications should be implemented. The fast-growing stage is a period of vigorous seedling growth, and so the demand for nutrients increases. At this time, the amount and frequency of nitrogen fertilizer should be increased. Phosphorus and potassium fertilizer should be applied also and in proper proportion. To promote the lignification of seedlings and improve the resistance of seedlings in the later growth stage, the application of nitrogen fertilizer should be stopped at the right time, and the application of overall fertilizer should be stopped at the lignification stage to improve the resistance of seedlings.

At the early growth stage in spring, top-dressing should be started for reserved bed seedlings, and the phosphorus fertilizer should be put in place at one time. Nitrogen fertilizer should be applied several times with the most applications during the fast-growing period. To promote diameter growth and root system growth, apply nitrogen fertilizer one more time at a later growth stage for those seedlings with early-stage growth pattern.

Foliar topdressing, also known as foliage spray, is a method of spraying quick-acting fertilizer solution directly on the leaves of seedlings during the growth period. Generally, young leaves and leaf backs absorb water faster than old leaves and leaf surfaces. So, take care to spray on the leaf backs and young leaves to facilitate absorption. Select a sprayer with greater pressure and strictly control the concentration. The concentration of urea solution should be 0.2% to 0.5%, and that of superphosphate can be approximately 0.3%. The best spraying should be conducted on cloudy days or on sunny days after 10:00 a.m. to prevent evaporation of the liquid at higher temperatures, leading to fertilizer damage.

For more information on fertilization, see section 7.1.3 in Chapter 7.

8.5.2.5 Thinning and Seedling Transplantation

(1) Thinning

Thinning is a measure to adjust the density of seedlings to a reasonable range. In production practice, the sowing quantity is often higher than that of ultimately needed, and uneven sowing is hard to avoid. An appropriate nutritional area is critrical for each seedling to ensure the yield and quality. When seedlings are allowed to grow too densely grouped together, there can be insufficient light, poor ventilation, and inadequate nutrition for each seedling. Therefore, it is necessary to thin out seedlings and fill the gaps with seedlings.

The thinning timing depends on seedlings' density and growth rate, following the principle of "early thinning, late final singling". Early thinning should be conducted for seedlings with higher density and faster growth rate. In general, for broad-leaved tree species, thinning should be started at the beginning of the early growth stage, and the second thinning should be done after 10 to 20 days. The final singling should be near the end of the early growth stage. Coniferous trees grow slowly and prefer a dense growth environment, so the time of thinning is later than that of broad-leaved trees. Generally, thinning is started later at the early growth stage, and final singling is at the beginning of the fast-growing stage. In areas where high temperature may comes to be potential stress, final singling should be conducted after the high temperature period.

The principle of thinning is "keeping the superior and removing the weak, keeping the ones in a sparse area and removing ones in the dense area". Thinning includes the seedlings affected by diseases and insect pests, mechanical damage, abnormal growth, and poor growth. Those dense seedlings should also be removed to avoid growth competition.

The intensity of thinning depends on the tree species. For most broad-leaved tree species, such as *Robinia pseudoacacia*, *Ulmus pumila*, and *Ailanthus altissima*, the seedlings grow quickly with strong resistance, so thinning can be completed once after the seedlings have sent up two true leaves. For most conifer species, the seedlings grow more slowly (as compared to the broad-leaved trees), and thinning can be completed over two to three sessions in combination with weeding, with an interval of 10 to 20 days each time. At final singling, seedlings should be evenly distributed to reach a reasonable density of approximately 5% more than the planned seedling yield.

It is preferable to carry out thinning after rainfall or irrigation, or in combination with weeding. Attention should be paid to protect the reserved seedlings. If root pores have occurred between seedlings, irrigation and silting should be done immediately after thinning.

(2) Seedling Transplantation

For seedlings that need intensive management, precious tree species with fewer seeds, and broad-leaved trees with rapid growth (such as eucalyptus and paulownia), reasonable seedling density can be achieved by transplanting seedlings to avoid seedling waste. Transplanting also can promote the growth of multiple lateral roots and fibrous roots. The ideally transplantation atian of seedlings is during their early growth stage. For most broad-leaved tree species, when the seedlings have sent up 2 to 5 true leaves, seedling transplantation can be conducted, and the survival rate is relatively high. Generally, transplanting is carried out during rainy days, and timely irrigation and appropriate shading are needed after transplanting.

8.5.2.6 Pest Control of Seedlings

The principle of "prevention is more important than treatment" should be followed in the pest control work at a seedlings nursery, achieving "early control, complete control". The basic principle of pest control is "prevention first, then comprehensive control." To control pests, properly use cultivation, biological, physical, chemical methods and other ecological methods at the lowest level so as not to cause damage. In every link of seedling technology, we should pay attention to controlling diseases and insect pests. This approach includes the selection of nursery land, treatment to the soil and seed, and all steps that can effectively prevent the occurrence of diseases and insect pests. Once diseases and insect pests occur in a certain area, they should be controlled without delay.

(1) Common Diseases in a Nursery

Common diseases include damping-off, leaf spot, powdery mildew, rust disease, anthracnose, root rot, leaf blight, trunk rot, and gummosis. Commonly used medicaments include carbendazim, chlorothalonil, dichloroisocyanuric acid, Green-Shield, Yeku, Yebanqing, and others. Damping-off is the most common and severe disease in a nursery, and seedlings, such as *Pinus* seedlings, are usually severely damaged within one month after they emerged from the ground. This disease can be controlled by spraying Bordeaux mixture and Cercobin. Powdery mildew often damages the leaves of broad-leaved tree seedlings. The application of Bordeaux mixture can effectively control the disease.

(2) Common Pests in a Nursery

Leaf-eating pests, trunk pests, scale insects, aphids, mites, and soil insects are common. The aboveground pests should be prevented and comprehensively controlled. For example, high-efficiency cypermethrin can be used for leaf pests; acetamiprid can be used for scale insect; and pymetrozine can be used for aphid. The common soil insects in a nursery include scarab, cutworms, and mole cricket. Disinfection of seeds and soil before sowing can prevent damage from soil insects.

For more information on pest control, please refer to section 7.5 in Chapter 7.

8.5.2.7 Root Cutting

The method for cutting the roots of seedlings or nursery stock is called root cutting. It can promote the development of more lateral roots and fibrous roots, so as to expand the absorption area in the root system. Furthermore, root cutting can inhibit the growth of stems and leaves, so as to increase photosynthate supply to roots and to increase the root to treetop ratio, which is beneficial to the later growth of seedlings. Root cutting can also reduce the damage to roots when lifting seedlings and improve the survival rate of seedling transplantation.

Root cutting has the same effect as transplanting cultivation, but it is faster in terms of seedling recovery from the cut and reviving seedlings as compared to bed-changing transplanting. It is easy to accomplish and saves labor. Reports have shown that, after cutting roots in early spring, the ground diameter, number of lateral roots, fresh underground weight, and the rate of qualified seedlings of *Larix principis-rupprechtii* in reserved beds are all significantly higher than those of the control after a growing season. The survival rate of afforestation and the height increment within the same year are also significantly higher than those of the control.

(1) Timing of Root Cutting

For tree species such as camphor trees, walnuts, *Quercus*, and Chinese parasol, with developed main roots and stunted lateral roots, root cutting should be carried out when two true leaves are spread out in seedling stage. For the 1-year-old or more than 1-year-old seedlings of common tree species, if they do not come out of the nursery in the same year, root cutting can be conducted at the beginning of the later growth stage in autumn, when height growth slows down and tends to stop but the growth of the root system is still vigorous. Root cutting at this stage is conducive to incision healing and the occurrence of new roots.

(2) Depth of Root Cutting

The depth of root cutting depends on the root distribution depth of the tree species. Ensure that the reserved part of main roots can produce enough lateral fibrous roots. Root cutting depth at the seedling stage is generally 8 to 12 cm. The root cutting depth of 1-year-old seedlings is 10 to 15 cm.

Special tools, such as a root cutting shovel and a root cutting cutter, are used for manual root cutting (Figure 8-9); for mechanical root cutting, a special root cutting plow can be used, and after root cutting, irrigation should be carried out immediately to ensure the contact between roots and soil

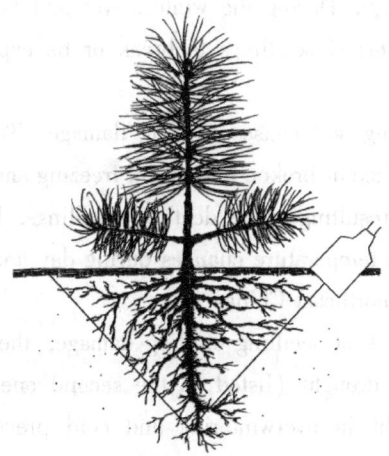

Figure 8-9 Manual root cutting shovel.

Figure 8-10 Mechanical root cutting plow (Photo and illustration provided by RK Dumroese).

(Figure 8-10).

8.5.2.8 Overwintering and Cold Prevention of Seedlings

(1) Expressions of Seedling Chilling Damage

Freezing to Death. Also known as freezing damage, it generally occurs when the seedlings are not fully lignified in late autumn. Once seedlings encounter continuous low temperatures, the water in the seedlings freezes and the cell protoplasm will dehydrate and freeze, leading to cell damage, so that seedlings would not be able to carry out normal physiological activities. This will result in the death of seedlings. Another scenario is in early spring when seedlings sprout prematurely, suddenly encounter extremely low temperatures from early and late frost, and their cells or intercellular space freezes, which can also cause seedling death.

Withering. This type of chilling damage is caused by physiological drought. It happens mainly in the northwest and North China during winter and spring. The dry and windy weather is potential to cause a large amount of transpiration and water loss in the aboveground part of the seedlings. At the same time the physiological drought occurs in the underground part of seedlings because the root system cannot absorb water from the freezing soil. At the end, the seedlings lose their water balance, and the shoots wither and die.

Ground Fissuring Damage. During the winter, wet and heavy soil can freeze and crack, which will cause the root systems of seedlings to break or be exposed to the wind and wither to death.

Lifting by Frost. Freezing soil causes chilling damage. The root system of seedlings was pulled out of the excavated surface or broken due to soil freezing and swelling. Then it was blown by wind and exposed to sunlight, resulting in the death of seedlings. Lifting by frost usually occurs in the seedling land with dramatic temperature changes during day and night with higher soil moisture content, such as the region in northeast China.

Among the above four types of seedling chilling damage, the greatest percentage of seedling deaths is due to physiological drought (listed as the second one), so it is critical to focus on preventing physiological drought in overwintering and cold prevention of seedlings in northern China.

(2) Overwintering and Cold-proof Measures

Low temperature is a major environmental factor limiting plant growth, development, and distribution. According to the degree of low temperature and plant damage, it can be divided into two categories: Cold injury and freeze injury. Cold injury refers to plant damage caused by cold temperatures that are still above 0 ℃; freeze injury refers to the plant damage caused by temperatures below 0 ℃. Overwintering protection is to improve the ability of plants to resist cold stress through their own regulation and adaptation to the adversity, as well as the external protective measures implemented by human beings.

Damage to Seedlings by Low Temperatures in Winter. Low temperature in winter is one of the abiotic stress factors that affect plant growth and development as well as yield. It directly affects the process of biothermodynamics, the stability and function of biomolecules, and the normal cells. After overwintering, the seedlings of some tree species will die because they are more susceptible to freeze injury or "physiological drought" during winter into early spring. The drought wind in early spring causes a large amount of water loss in the aboveground part of the seedlings. At this time, because of the freezing soil and weaker root activity, the aboveground portion of the seedlings cannot be supplied with water, resulting in local branches withering and some damage to branches. In severe cases, the entire plant will die. Similarly, under the condition of low temperature stress, the accumulation of reactive oxygen species (ROS) in plants damages the plant membrane system, which leads to the peroxidation of membrane lipids and the destruction of cell membrane structure and function.

Effects of Low Temperature on Seedling Physiology in Winter. When plants are subjected to low temperature stress, a series of physiological and biochemical reactions and cell responses occur internal the plants, aiming to reduce the damage of harmful substances produced in the metabolic process to cells, and to improve the plants' cold resistance, so that to reduce the damage on seedlings. Under low temperature stress, a large amount of reactive oxygen species (ROS) was accumulated in plants. To cope with this change, antioxidant protection systems are initiated, including activity increase of superoxide dismutase (SOD), catalase (CAT), ascorbate peroxidase

(APX), glutathione reductase (GR) and dehydroascorbate reductase (DHAR) in enzymatic antioxidant systems, and content increase of ascorbic acid (AsA) and reduced glutathione (GSH) in non-enzymatic antioxidant systems, which eliminates the accumulated oxygen free radicals in plants and reduces the damage of harmful substances produced by cell metabolism.

Low temperature can enhance plant hydrolysis, increase the content of soluble sugar, reduce water potential, and increase water-holding capacity. Dehydrin in plants began to play a role in binding with membrane lipids to prevent excessive water loss in cells, maintain the hydration protection system of membrane structure, prevent the decrease of membrane lipid bimolecular layer spacing, prevent membrane fusion, and prevent biological membrane structure damage. Soluble protein in plant cells decreased at first and then increased because cold shock induces protein production to protect cells from ice crystal damage. Protein chaperone and dehydrating protein prevent biological macromolecules from denaturation and inactivation in the process of cold adaptation. Uncoupling protein enables plants to maintain growth below the freezing point for a period of time, in an effort to adapt to the cold and to reduce the chance of plant death due to cell dehydration and protoplast freezing under low temperature conditions. During long-term evolution process, plants have formed the ability to adapt to low temperature stress through the physiological and molecular changes of themselves.

Effects of Low Temperature on Seedling Morphology in Winter. During the early winter and early spring, the temperature differentiation between day and night can be large. The cortex tissue of seedlings would start to move with the increase of solarization temperature. When the temperature drops sharply at night, seedlings will be frozen. For example, when the temperature drops suddenly in early winter, the cortical tissue in the seedling trunk rapidly shrinks, and the stress in the xylem will open the bark, causing the trunk to freezing crack damage. When the branches of the seedling are relatively thin and the duct is not yet developed, the branch accumulates snow, and the snow water infiltrates the bark. At that point, the tissue softens, and the sudden drop of temperature at night may cause damage, such as cortex discoloration, necrosis, and depression; or, a vertical crack may occur along the trunk, forming a branch freeze injury.

The root system is part of the seedling that stops growing the latest in autumn and starts growing the earliest in spring, so the cold resistance can be poor. The root neck and the root near the surface are vulnerable to the damage caused by low temperatures. The great temperature change can freeze the cortex, especially in winter, sandy land with less snow and potential drought makes the plants more vulnerable. Root damage is not easy to detect in time to prevent it from occurring. For example, some trees have sprouted in spring, but after a period of time, they suddenly die. This situation is primarily caused by the roots having been frozen.

Overwintering Protection of Seedlings. Safe overwintering of seedlings depends on the time of freezing resistance and dormancy of genotypes, the size of frost resistance in midwinter, the speed of de-hardening, the ability of re-hardening, and the timing of bud sprouts. The production loss can be reduced by determining the suitable planting site, selecting the tree species most likely to resist freezing, and applying the appropriate cultivation techniques. Many temperate species have gradually evolved the ability to increase their frost resistance under the conditions of long-term cold in autumn without freezing despite temperature and photoperiod changes. Existing studies have shown that chemical induction, such as exogenous abscisic acid, can also enhance the cold resistance of seedlings. Through the selection of cold-resistant tree species, cold-resistance hardening, and artificial technical measures, the cold resistance of seedlings can be improved, ultimately ensuring the plant safely overwinters.

Technical Measures for Cold-resistance Induction of Seedlings. Abscisic acid (ABA), as a signal transduction substance that triggers plant response to low temperature, participates in the regulation of plants to low temperature stress. ABA is a major participant in the regulation genes and participates in the regulation of plant stress response genes through the transcriptional activation of ABA-dependent transcription factors. ABA treatment can improve the level of the antioxidant stress system, improve the ability of scavenging reactive oxygen species under stress, and reduce the level of membrane lipid peroxidation so as to maintain the stability of membrane structure and function.

Calcium prevents membrane damage and cytoplasmic leakage, so that plays an important role in maintaining the structure and function of cell walls and membranes. When plants sense the signal of low temperature, Ca^{2+} in the cytoplasm increases rapidly. The temperature that induces the increase of Ca^{2+} concentration is the same temperature as that of cold acclimation. Studies have shown that exogenous calcium had a regulating effect on low temperature stress. For example, calcium treatment can increase loquat leaf cell CaM content and activate the activity of Ca^{2+}-ATPase, promote the production of Ca^{2+}-CaM, activate the CAT and SOD under low temperature stress, thas reduce the accumulation of intracellular reactive oxygen species, thus reduce the damage of membrane lipid peroxidation and improve the frost resistance of loquat.

Technical Measures for External Protection of Seedlings

i. Seedling Management in Spring and Autumn. Timely irrigation and sufficient fertilizer and watering during the spring season can enhance the conversion and utilization of photosynthesis, which helps to store nutrients in the seedlings and promotes the development of branches and new shoots on the seedlings. Applying appropriate amounts of phosphorus and potassium fertilizer in autumn, along with deep plowing and weeding, can promote the finish of vegetative growth of branches as soon as possible, enrich and lignify tissues, and prolong the nutrient accumulation time so as to increase the cold resistance ability.

ii. Timely Winter Irrigation. Winter irrigation conducted from the last 10 days of October into the middle of November can help the seedlings absorb sufficient nutrients and water, and it can increase the ground temperature by approximately 2 ℃, thus reducing the occurrence of freeze injury of seedlings. During the period of winter irrigation, soil piles with a diameter of 50 to 80 cm and a height of 30 to 50 cm should be built around the roots of seedlings to reduce the evaporation of soil water and to increase the soil temperature around the roots, so as to enhance the cold resistance of trees.

iii. Anti-freezing and Heat Preservation Measures.

Burying Method. Burying seedlings with soil have potentials to prevent cold damage, especially to prevent physiological drought of seedlings. It is suitable for small-sized seedlings of most tree species in North China. Generally, after the seedlings enter dormancy and before the soil freezing, soil is taken from the footpaths and ridges of the seedbed to bury the seedlings 3 to 10 cm. For the seedlings with a little larger size, they can be buried after being overwhelmed by the cold. In the following spring, remove the soil cover at the time of seedling lifting or before seedling growth. The specific time of soil burying and soil removing is the key to this method. If it is not well mastered, it will affect the cold-proof effect or make the covered seedlings worse.

Covering with Straw. Cover seedlings with straw or fallen leaves in the fall and the remove the mulch the following spring. The mulch can reduce the wind speed on the surface of seedlings, prevent physiological drought, and reduce the damage of strong solar radiation to seedlings.

Wind Barrier Method. Use covering cloth or a straw curtain and other windproof materials to build wind barriers around the seedlings or in the direction perpendicular to the main wind. The barrier can reduce wind speed, reduce seedling transpiration, prevent physiological drought, improve the ground temperature and air temperature on the leeward or sunny side of the wind barrier, therefore prevent or reduce the freeze injury of seedlings. When wind barriers are set up, the first row with high and narrow wind barriers is generally set at the windward side, 1 to 1.5 m away from the first seedbed, and the distance between the wind barriers is generally about 15 times the height of the wind barriers.

White Coating Method. Apply the specially prepared white pigment to the stems and branches of seedlings to prevent insect oviposition, rot, and canker; to reduce the absorption of solar radiation by the aboveground part of the seedlings; to delay the germination period of buds; to effectively prevent the drastic change of temperature in early spring; to avoid the occurrence of freeze injury of branches and buds; to reduce the local temperature sudden increase; and to prevent the damage of sunburn. Generally, from late October to mid-November, the white coating agent should be evenly applied on the seedling trunk with a brush, and the height should be 1.0 to 1.5 m above the root collar diameter. The common white coating agent is the lime sulfur made of copper sulfate, quicklime, and water in the ratio of 1:20:60 to 80. In preparation, copper sulfate is fully dissolved in a small amount of boiling water, then diluted with 2/3 of the total amount of water. The quicklime is slowly dissolved into a concentrated lime milk by adding 1/3 of the total amount of water. After the two solutions are fully dissolved and the temperature of the two solutions is the same, the copper sulfate is poured into the concentrated lime milk. The combined solution must be continuously stirred to be used (Figure 8-11).

Wrapping Method. Straw rope or straw can be used to wrap the trunk, which is suitable for larger-size seedlings (Figure 8-12).

Figure 8-11　Trunks treated with white coating method (Photo by Liu Yong).

Figure 8-12　Seedling trunk wrapped with straw rope (Photo by Liu Xiaojuan).

Questions for Review

1. What is seed dormancy? What is seed pregermination? How to determine pregermination methods according to seed dormancy types in production?

2. How many ways to cultivate seedlings? What are the applicable areas and conditions?

3. What are the main causes of seedling death in overwintering?

4. What are the main technical measures of sowing and cultivating of bare-root seedlings?

5. What factors should be considered in calculating the sowing quantity?

6. What are the sowing methods?

7. What are the main measures of seedling management?

8. What is the annual growth rule, and what are the key points of seedling raising technology of 1-year-old seedlings?

References and Additional Readings

KHAN A A, 1971. Cytokinins: Permissive role in seed germination[J]. Science, 171: 853-859.

KHAN A A, 1975. Primary, preventive and permissive roles of hormones in plant systems[J]. Botanical Review, 41: 391-420.

LIANG Y T, 1995. Seedling science[M]. Beijing: China Forestry Publishing House.

LU X Y, 1989. Technical manual of forest tree seedling in north China[M]. Shenyang: Liaoning Science and Technology Press.

SHEN H L, 2009. Seedling cultivation[M]. Beijing: China Forestry Publishing House.

SU J L, 2003. Landscape seedling[M]. Beijing: China Agriculture Press.

XIE W, ZHANG L M, et al., 2011. Study on the sowing and seedling cultivation technology of *Pinus koraiensis*[J]. Journal of Liaoning Forestry Science & Technology, 1: 39-41.

ZHAI M P, 2011. Theory and practice of modern forest cultivation[M]. Beijing: China Environmental Science Press.

Chapter 9 Container Seedling Cultivation

Li Guolei

Chapter Summary: This chapter focuses on container seedling cultivation and introduces container, substrate, fertilizer, water, and other production factors, as well as temperature, humidity, light, and other environmental factors and their control methods. Concepts and characteristics of the developmental stages of container seedlings are discussed. And the key control techniques for seedling quality, such as water, nutrient, temperature, and light, are introduced in combination with the developmental stages of seedlings.

9.1 Characteristics of Container Seedling Production

Container seedling production is a method of cultivating seedlings by filling solid matrix in the container, directly sowing seeding seeds, and cutting or transplanting the seedlings; these cultivated seedlings are called container seedlings. The term solid matrix refers to some form of soil-like mixture that is firm and allows for root growth while also supporting the aboveground portion of a plant. According to the conditions of radicle emergence and caulicle growth, container seedling cultivation can be divided into three categories: Directly sowing, planting germinant, and transplanting germinants. Container seedling production can be accomplished outdoors, especially for cultivating large container seedlings or seedlings with multiple growth seasons. Container seedling production can also be implemented in greenhouses. A comparison of container seedlings and bare-rooted seedlings in terms of stock quality assessments can be found in Grossnickle and El-Kassaby (2016).

Container seedling production is a widely used technology in industrialized seedling production, and compared with other seedling production methods, industrialized container seedling production has the advantages of shorter seedling production cycle, high seedling yield per unit area, fewer seeds required relative to other production methods, higher seedling production efficiency, and remarkable afforestation effects. Although many advantages are associated with container seedlings and industrialized container seedling production, some countries and regions have some issues with the process, such as the high cost of facilities and the relatively complex seedling production technology.

Container seedling production industrialization is a new development of intensive seedling production in forestry-developed countries, and it is also the development direction of container seedling production in various countries. Industrialized seedling production is a large-scale seedling production mode that adopts modern biotechnology, soilless culture technology, environment control technology, and information management technology to achieve standardized production, such as specialization, mechanization, and automation, to produce high-quality seedlings efficiently and stably under the excellent environmental conditions created artificially. Compared with the traditional seedling production methods, the industrialized seedling production technology has some advantages, such as less seed consumption, less land occupation, shortened seedling age out of the nursery, shortened seedling production time, reduced occurrence of diseases and insect pests, improved seedling production efficiency, reduced cost, and is conducive to enterprise-style management and rapid promotion of new technology. At present, only a few countries in the world, such as Finland, Canada, and Japan, have realized or partially realized the industrialized production of container seedlings. Since the late 1980s, China has successively introduced foreign industrialized forest seedling production technology; successfully implemented research projects of industrialized production of excellent clones or excellent individual plants of tissue culture seedlings of eucalyptus, Scots pine (*Pinus massoniana*), poplar, and paulownia; and established several forest tissue culture seedling factories in Beijing, Guangxi, Guangdong, and Hainan, from which they have made a batch of woody plants, such as eucalyptus and poplar that have entered large-scale industrialized production. At present, however, container seedling production in China is primarily taking place in the field and in plastic greenhouses.

9.2 Seedling Containers

9.2.1 Seedling Container Types

According to material types, seedling containers can be divided into non-woven fabric containers, plastic containers, straw-clay containers, and more; according to the combination modes of containers, they can be divided into single containers, plug seedlings, and so forth; and according to the utilization times, seedling containers can be divided into disposable containers and recyclable multiple utilization containers. In China, the disposable containers are primarily light matrix mesh bags; in other countries, recyclable multiple utilization containers are the most common. The same container, however, can be classified into more than one type according to the classification methods. As the container is a comprehensive embodiment of the elements, such as the materials and combination methods, it is summarized in Table 9-1.

Table 9-1 Common seedling containers

Types	Volume /mL	Height /cm	Top diameter /cm	Remarks
Single standing type				
Polybags	1474-15 240	10-20	15-20	Non-woven fabrics
RediRoot™ singles	9020-46 900	23.1-34.1	24.6-47.0	Rigid plastic
RootMaker® singles	3110-18 440	19.0-25.4	16.5-33.7	Rigid plastic

(Continues)

Types	Volume /mL	Height /cm	Top diameter /cm	Remarks
Round pots	1474-73 740	15-45	15-35	Rigid plastic
Treepots™	2310-30 280	24-60	10-28	Rigid plastic
Gallon pots	3785-56 781	17.0-41.5	16.5-45.0	Rigid plastic common in China
Root control container		20-100	20-60	Plastic common in China
Root control bag		18-70	12-100	Non-woven fabrics common in China
Plastic film container		10-16	5-10	Soft plastic common in China
Single needing auxiliary pallet type				
Ray Leach Cone-tainers™	49-164	12-21	2.5-3.8	Rigid plastic
Deepots™	210-983	7.6-36.0	5.0-6.4	Rigid plastic
Jiffy® Forestry Pellets	10-405	2-7	2.0-5.6	Non-woven fabrics disposable
Zipset™ Plant Bands	2065-2365	25-36	7.5-10.0	Rigid plastic
Light matrix mesh bags		3.5-5.0	8-10	Non-woven fabrics common disposable in China
Plug seedling containers				
Forestry Trees	98-131	12.2-15.2	3.8-3.9	Rigid plastic
"Groove Tube" Growing System™	28-192	6-13	3.3-5.8	Rigid plastic
Hiko™ Tray System	15-530	4.9-20.0	2.1-6.7	Rigid plastic
IPL® Rigi-Pots	5-349	4-14	1.5-5.8	Rigid plastic
RediRoot™	173-956	9.7-16.5	5.6-9.7	Rigid plastic
Root Maker®	26-410	5.0-10.2	2.54-8.6	Rigid plastic
Styroblock™ and Copperblock™	17-3200	7-18	1.8-15.7	Polystyrene
Plug seedlings	32-288 holes			Rigid plastic visible in China

Source: Dumroese et al. (2009).

9.2.2　Selection of Seedling Containers

　　Selection of seedling containers needs to comprehensively consider container characteristics, seedling production density, seedling quality, tree species characteristics, seedling cultivation time, afforestation site characteristics, seedling production cost, nursery management level, and more. Therefore, careful consideration is needed to select appropriate containers for specific tree species. Because the following chapters include discussions on the cultivation of large-scale container seedlings, this chapter involves only the theory and technology of small-scale container seedling cultivation, and the specifications of containers used for seedlings that are not more than 50 000 mL and are mostly within the 1000 mL range (Figure 9-1 to Figure 9-3).

(a) Polybags

(b) RootMakersingles® Photos from www.stuew.com

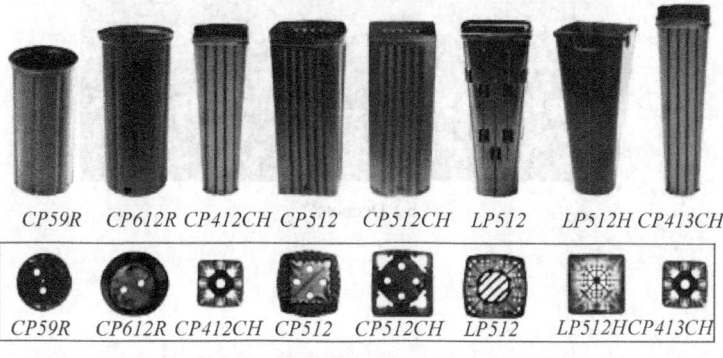

(c) Treepots™ Photos from www.stuew.com

Figure 9-1 Single-Standing Type Containers.

(a) Ray Leach Cone-tainers™

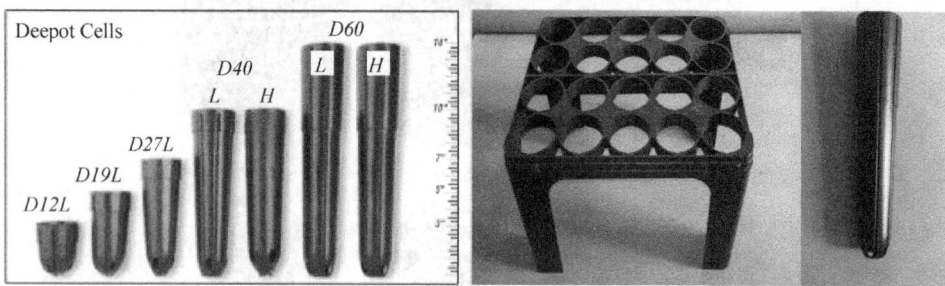

(b) Deepots™

(c) Jiffy® Forestry Pellets

(d) Zipset™ Plant Bands Photos from www.stuew.com

(e) Light matrix net bag

Figure 9-2 Single planting pots need auxiliary pallet-type containers.

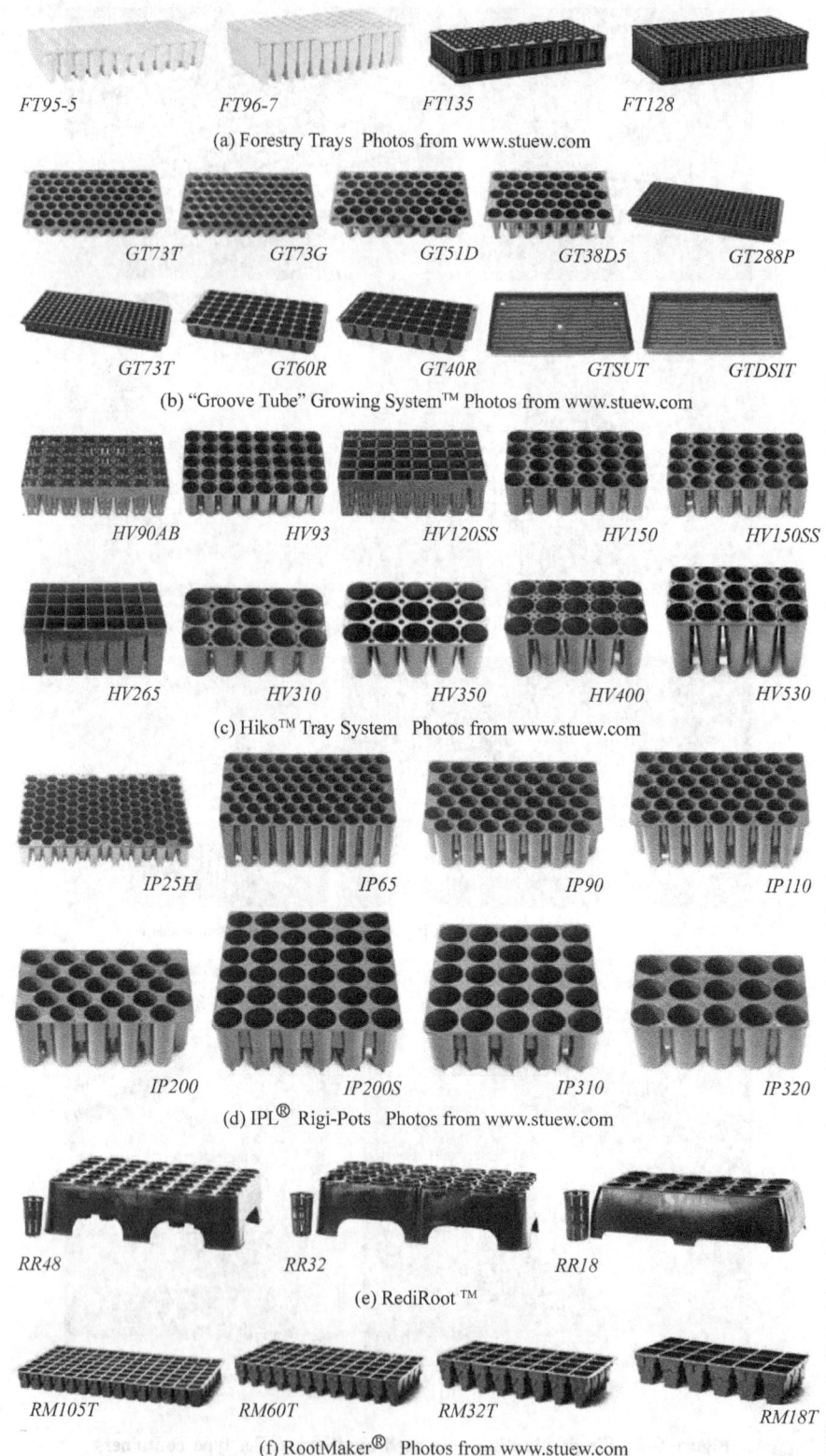

FT95-5 FT96-7 FT135 FT128
(a) Forestry Trays Photos from www.stuew.com

GT73T GT73G GT51D GT38D5 GT288P
GT73T GT60R GT40R GTSUT GTDSIT
(b) "Groove Tube" Growing System™ Photos from www.stuew.com

HV90AB HV93 HV120SS HV150 HV150SS
HV265 HV310 HV350 HV400 HV530
(c) Hiko™ Tray System Photos from www.stuew.com

IP25H IP65 IP90 IP110
IP200 IP200S IP310 IP320
(d) IPL® Rigi-Pots Photos from www.stuew.com

RR48 RR32 RR18
(e) RediRoot™

RM105T RM60T RM32T RM18T
(f) RootMaker® Photos from www.stuew.com

(g) Styroblock™ and Copperblock™

(h) Hole tray

Figure 9-3　Plug seedling containers.

Within an appropriately sized container, plant biomass increases by 43% with each doubling of container volume (Pooter et al., 2012), so choosing a suitable container volume is very important. If the container volume is too small, there will be root system malformation or even some root system death near the end of a growing season, which affects the quality of the seedlings. If the container volume is too large, the root system and the matrix are not closely connected during the early growth period and the root cluster cannot be formed; the price of the container itself and the matrix required to fill the container is higher than is necessary; seedling production density will be reduced; transportation costs will be higher; all of which results in an overall increase in the cost of seedling production. Container volume first depends on the depth of the containers (Figure 9-4). Container diameter (usually expressed as upper diameter) primarily affects seedling-raising density, but also affects seedling quality. Using containers with a smaller diameter can reduce the cost of containers and matrix, and improve the yield of seedlings per unit area, but the smaller diameter container can also cause the decline of seedling specifications (such as seedling size) (Table 9-2). Therefore, when selecting a container, consider the characteristics of the tree species, the seedling quality, and the seedling-raising effect should be fully considered.

　　Root system characteristics are related to the selection of container type. Many nursery growers recommend the use of rigid plastic containers or non-woven fabric containers of a suitable length instead of plastic film containers for species with developed main roots. Because the texture of plastic film containers is soft, and the root guide rib is not prominent, it cannot effectively guide the root to grow downward. Plant growth can result in abnormal roots, such as a twisted root or a spring

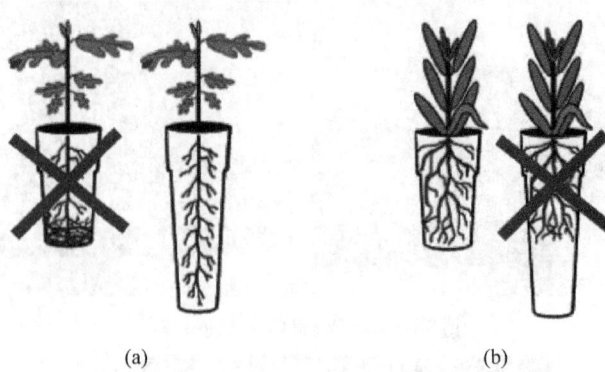

Figure 9-4 Container depth and seedling development (From Dumroese et al., 2009).
(a) frizzy root caused by insufficient container depth; and
(b) root loss below caused by container depth being too deep

Table 9-2 Relationship between container seedling density and 1-year-old seedling specification

Tree species	Scientific name	Seedling-raising density (plant/m^2)	Container volume (cm^3)	Seedling height (cm)	Root collar diameter (mm)
Loblolly pine	Pinus taeda L.	535	94	25	4.5
		530	122	30	5.0
		364	165	36	6.0
		284	221	51	8.0
Douglas fir	Pinus menziesii var. menziesii Mirb. (Franco)	756	65	19	3.0
		530	80	24	3.2
		364	125	28	3.7
		284	220	35	4.4
		213	336	42	4.8
White spruce× Engelmann spruce	Picea glauca (Moench) Voss × P. engelmannii Parry ex Engelm.	756	54	17	2.9
		681	60	18	3.0
		530	95	19	3.2
		364	80	20	3.3
		284	220	27	4.2
		213	336	35	4.8

Source: Grossnickle and El-Kassaby (2016).

root, affecting the afforestation effect (Figure 9-5a). A protruding root guide rib on the side wall of the rigid plastic container can guide the root system to grow downward. Container length is also critical (Figure 9-5b) because a short container will cause a certain degree of frizzy root (Figure 9-5c). The stronger air-cutting ability of non-woven fabric containers can effectively avoid excessive root system growth (Figure 9-5d), and the container itself has a low price and a smaller volume, which requires less matrix. The disadvantage of non-woven fabric containers, however, is that the stronger air permeability makes water loss occur more quickly, and the seedlings need to be irrigated frequently. Non-woven fabric containers should be placed on the ground or on a seedbed covered with plastic ground cloth, and the containers should not make direct contact with the ground to prevent the root system from growing into the soil. Non-woven fabric containers should not have contact with each other; a plastic tray with spacer holes can be used to separate the non-woven fabric containers and allow airflow to prune the roots effectively (Figure 9-5e). The higher cost of the trays can be offset by improved seedling quality and reduced input of labor.

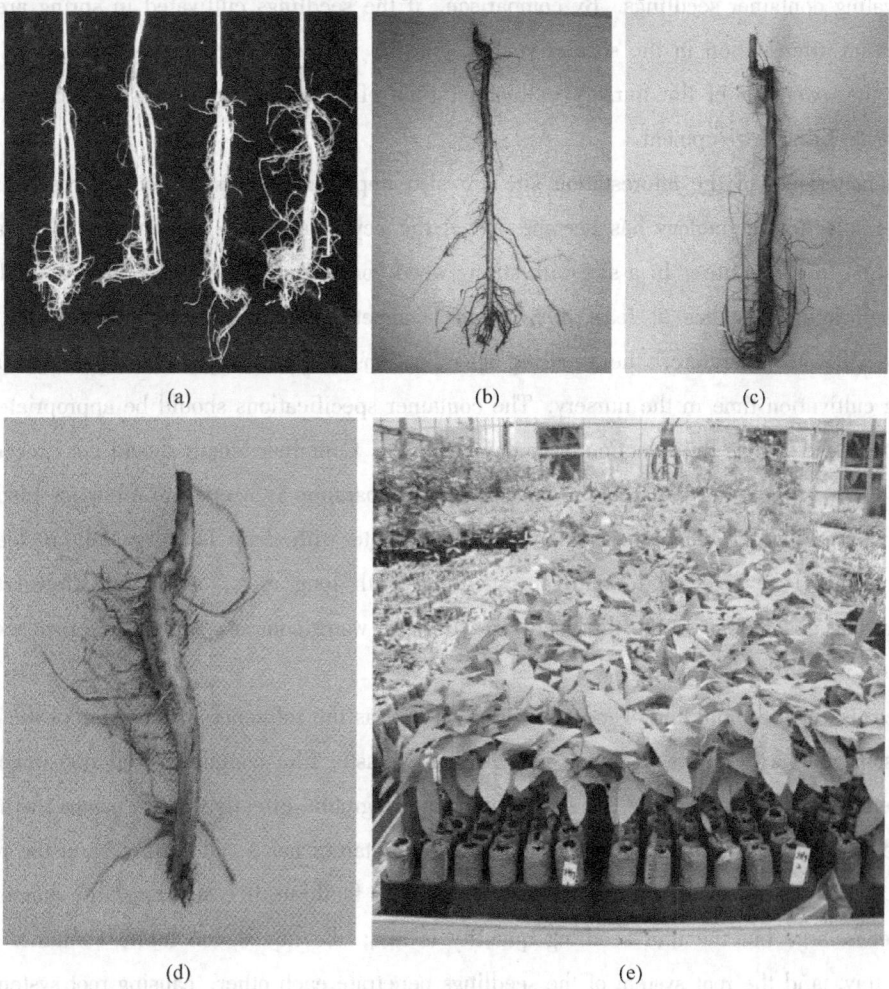

Figure 9-5 Root system of *Quercus variabilis* cultivated in different seedling containers (Photo by Li Guolei).

(a) root system cultivated in plastic film; (b) root system cultivated in rigid plastic D40 with the length of 25 cm; (c) root system cultivated in rigid plastic D60 with the length of 36 cm; (d) root system cultivated in Jiffy Forestry Pellets; and (e) non-wovens with pallets to space non-woven fabric containers.

Spring is one of the most common seasons for afforestation in China. Seedlings can be cultivated in the nursery for one season or even for several seasons. Seedlings can be transplanted to larger containers as the root system develops by checking the surrounding matrix of roots from time to time throughout the growing season. After the seedlings stop height growth in autumn, root growth enters the peak period. Because summer afforestation seedlings lack this development stage, the root development of the seedlings is relatively weaker. If the container is too large, the root system cannot form the root mass. Even if the seedlings are carefully taken out of the containers, the matrix could be scattered, and "container seedlings" become bare-rooted seedlings, thus losing the purpose of cultivating container seedlings. By comparison, if the seedlings cultivated in spring are used in rainy-season afforestation in the second year of growth, the container can be appropriately larger, because the extension of the nursery cultivation time of the seedlings needs to accommodate the space for seedling development.

Characteristics of the afforestation site are also important factors in container selection, and container selection technology has become one of the key measures to improve the survival rate of afforestation in harsh sites. In a site with strong weed competition, seedlings can receive light only when their height reaches at least 80% of the competitive shrub height (Grossnickle and El-Kassaby, 2016). Therefore, the seedling specifications require large seedling height and long seedling cultivation time in the nursery. The container specifications should be appropriately large. Soil depth also restricts the selection of container length. Container length should not exceed the soil depth, otherwise, cutting part roots or special land preparation is needed to allow for placement of the entire seedling into the afforestation hole. In a site with deep but dry soil, a deep (tall) container should be selected to cultivate seedlings with long roots; and after afforestation, the seedlings can absorb water from the deep soil with high water content, so as to improve the drought resistance of seedlings.

Because of the limitation of mechanization as well as the influence of the price of the container itself, some nursery growers in China primarily use plastic film containers and non-woven fabrics containers. These containers are typically placed on the ground directly, which means the air rooting effect of the container is not ideal. The plastic film container has a soft texture, and the root guide rib has no obvious effect on guiding the root system. To facilitate free standing, the diameter of the container is large but the unit seedling quantity is less; the non-woven fabric container generally lacks a tray, and the root system of the seedlings penetrate each other, causing root system damage and loss when moving the containers, for instance, when moving them out of the nursery. With industry upgrading and pressure-driven labor costs, the seedling industry will gradually upgrade in China; thus, the rigid plastic containers characterized by high seedling yield per unit area, high seedling quality, and large one-time investment but able to be recycled will be more widely used.

9.3 Seedling Matrix

9.3.1 Matrix Components

The seedling matrix needs to provide support, ventilation, and water-holding functions. The matrix can be composed of one or several kinds of peat, vermiculite, perlite, soil, and organic compost. The size, bulk density, permeability, water-holding capacity, pH value, and cation exchange capacity (CEC) of the matrix formed should meet the requirements of plant development. In China, soil that is abundant and low cost is generally used to prepare the matrix; however, the seedling-raising cycle is longer, and the heterogeneity of soil nutrients is more great. In recent years, in some places, a machine process has been used to realize the light-matrix seedling cultivation with non-woven mesh bags (Figure 9-6). In forestry-developed countries, a soilless matrix, such as peat, vermiculite, and perlite, is mostly used for seedling raising (Figure 9-7). See Table 9-3 for the properties of peat, vermiculite, and perlite.

Figure 9-6 Production line of light-matrix mesh bag containers(Photo by Li Guolei).
(a) matrix blender and (b) light matrix filling and non-woven fabric container cutting.

Figure 9-7 Common light-matrix components(Photo by Li Guolei).
(a) peat; (b) vermiculite; and (c) perlite.

Table 9-3 Physicochemical property of common light-matrix components

Matrix components	Dry bulk density /(kg/m³)	Porosity /%		pH	Mineral nutrients	Cation exchange capacity	
		Ventilation	Water-holding			Weight /(meq/100 g)	Volume /(meq/100 m³)
Sphagnum peat	96.1–128.2	25.4	58.8	3.5–4.0	The least amount of all minerals	180.0	16.6
Vermiculite	64.1–120.2	27.5	53.0	6.0–7.6	K–Mg–Ca	82.0	11.4
Perlite	72.1–112.1	29.8	47.3	6.0–8.0	None	3.5	0.6

Source: From Landis (1990).

Seedling farmers and enterprises can also use bark, sawdust, pruned branches and leaves of garden trees, and abandoned fungus bags (that were used to produce mushrooms) to produce an organic matrix. Doing so replaces certain peat components, improves recycling of waste, and alleviates environmental pollution. Decomposed organic waste is the decomposition process in which insects, fungi, and bacteria participate. A suitable length of time to create waste organic matter, a reasonable carbon-to-nitrogen ratio, good permeability, and a humid decomposition environment are very important for accelerating the decomposition rate and obtaining high-quality organic products. If the bark and pruning branches of garden trees are too long, they need to be crushed into smaller particles by machinery with a closed cabin (to reduce dust) before composting. If the organic waste contains high carbon content, waste leaves and weeds (25%-50%) with high nitrogen content can be added to reduce the carbon to nitrogen ratio of the mixture; or, a nitrogen fertilizer solution can be sprayed to accelerate decomposition (Figure 9-8). Of particular importance is to keep the air permeability and humidity during the decomposition process. Thus, to avoid flying dust, the pile of decomposition materials can be covered with a sunshade net with a large pore diameter. Air circulation can also ensure the respiratory needs of fungi, bacteria, and insects involved in the decomposition. During the decomposition process, water can also be properly sprayed to improve the decomposition speed, and it is ideal to keep the humidity of organic matter at 50% to 60%.

When the color of the organic matter in the stack retting turns black, it indicates that it has been thoroughly decomposed (Figure 9-9). Decomposition can also be verified through seed germination testing.

9.3.2 Matrix Preparation

The propagation method is the first factor to consider in the preparation of matrix. Because seed germination needs energy to grow out of the matrix, the matrix used for seedling raising needs to be thin with fewer nutrients, and matrix particles need to be small. Because spray irrigation is frequent during the process of cuttings taking root, the permeability of the matrix used in cutting propagation

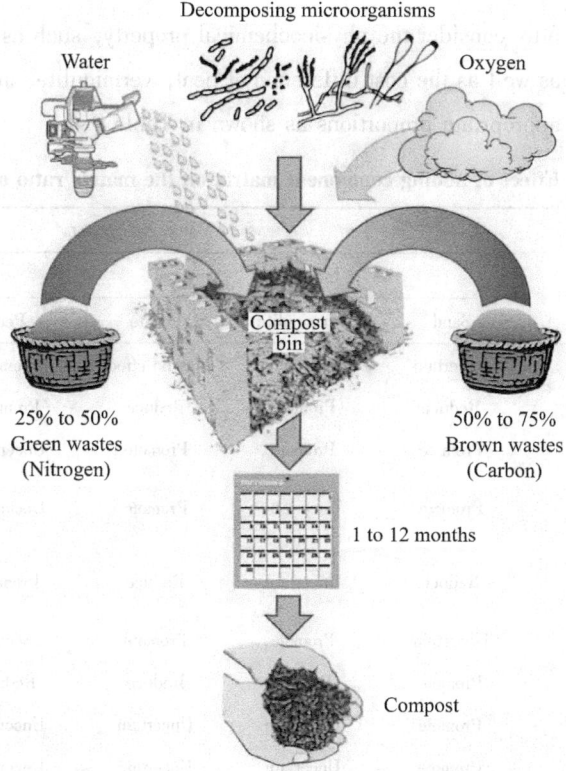

Figure 9-8 Decomposition process of organic matter (From Dumroese et al., 2009).

Figure 9-9 Color of decomposed organic matter (Photo by Li Guolei).

needs to be high so as to avoid excessive humidity having a negative effect on rooting. The selected particles of matrix used for transplanting need to be larger, and 10% to 20% soil can be added to utilize the formation of mycorrhizal fungi in cultivating local tree species.

Matrix components are another consideration in matrix preparation. When preparing the matrix, nursery growers need to fully consider the physicochemical property, such as air permeability, water holding, and nutrients, as well as the cost difference of peat, vermiculite, and perlite, to select the matrix components with appropriate proportions as shown in Table 9-4.

Table 9-4 Effect of adding component matrix on the matrix ratio of seedlings

Matrix properties	Matrix components				
	Inorganic			Organic	
	Sand	Vermiculite	Perlite	Peat	Saw powder or bark
Subacidity pH	Uncertain	Has no effect	Has no effect	Promote	Uncertain
High cation exchange capacity	Reduce	Promote	Reduce	Promote	Promote
Low fertility	Promote	Promote	Promote	Uncertain	Promote
High porosity facilitating air permeability and drainage	Promote	Uncertain	Promote	Uncertain	Uncertain
Low porosity not conducive to water retention	Reduce	Uncertain	Reduce	Promote	Promote
No pests	Uncertain	Promote	Promote	Uncertain	Uncertain
Bulk density	Promote	Reduce	Reduce	Reduce	Reduce
Material source convenience	Promote	Uncertain	Uncertain	Uncertain	Promote
Saving cost	Promote	Uncertain	Uncertain	Uncertain	Promote
Component homogeneity	Uncertain	Promote	Promote	Uncertain	Uncertain
Long-time storage	Promote	Promote	Promote	Promote	Uncertain
Volume change	Promote	Promote	Promote	Uncertain	Uncertain
Facilitating mixing	Promote	Promote	Promote	Uncertain	Uncertain
Facilitating formation of root mass	Reduce	Promote	Reduce	Promote	Uncertain

9.3.3 Matrix Filling

Matrix filling refers to the process of loading the prepared matrix into a specific container, using either mechanical operation and/or manual filling. Mechanical filling has the benefits of fast speed and high efficiency; in addition, it allows for uniform compactness of the matrix across containers, which is convenient for later water management. Mechanical filling equipment includes the specific filling equipment (shown in Figure 9-6) needed for the light matrix of the non-woven fabric mesh bags and the general filling equipment (Figure 9-10). Manual filling of the matrix is common in China, which avoids the purchase of an expensive piece of equipment, but the cumulative input of labor cost is high. The difference in matrix compactness among containers has an impact on the water infiltration, drainage, and evaporation, which increases the heterogeneity of seedlings. Therefore, to fill the matrix quickly and well by hand is highly technical work.

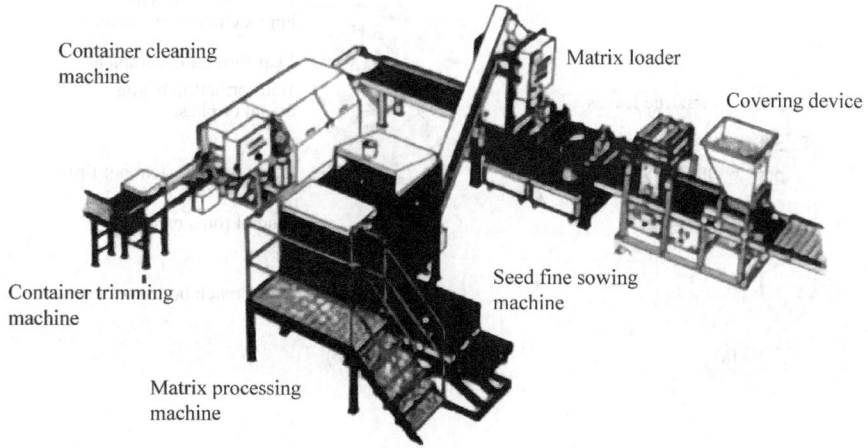

Figure 9-10 Fully automatic loading and sowing production line for container seedling raising (manufactured in Sweden by the BCC Company) (From Zhai and Shen, 2016).

9.4 Sowing and Seedling Management

9.4.1 Development Stage of Container-seeded Seedlings

The growing season of container seedlings in China can be divided into four stages: Emergence, establishment, rapid growth, and hardening; internationally, it can be divided into three stages: Establishment, rapid growth, and hardening. Internationally, it is collectively known as the establishment period before the formation of the first true leaves, and the rapid growth stage starts after the true leaves have formed. However, in China, the seedling stage is defined as the period from the formation of true leaves to when the seedling height starts substantially growing; fast-growing stage refers to the period from the acceleration of seedling height growth to the decline in the rate of height growth. In comparison with China, the international division of stages is more scientific and convenient for applying in production.

The establishment period is defined as the period from sowing the seeds to the formation of true leaves (Figure 9-11a), going through germination (the radicle protrudes from the seed coat), emergence (the hypocotyls extend out of the soil), and formation of true leaves or primary needles. The rapid growth stage is defined as the period from the formation of true leaves to the terminal bud formation and height growth basically stopping (Figure 9-11b). Hardening stage is from the height growth stopping to seedling storage (Figure 9-11c), and the growth center turns from shoot to root again. The development process, growth characteristics, and cultivation techniques of each period differ from each other, as shown in Table 9-5.

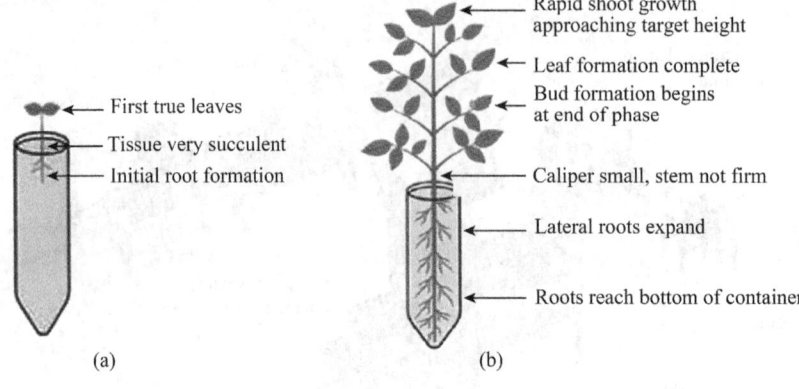
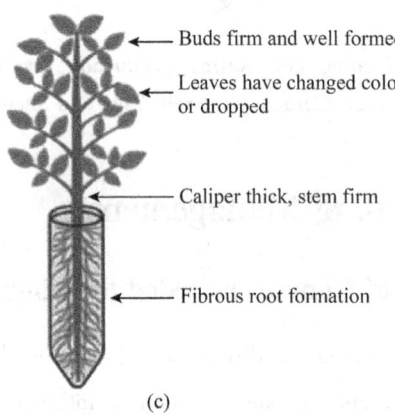

Figure 9-11 Development stage of container-seeded seedlings (From Dumroese et al., 2009).

(a) establishment stage, (b) fast-growing stage, and (c) lignification stage

Table 9-5 Characteristics and management technology of container-seeded seedlings according to development stages

Items	Stages		
	Establishment	Fast-growing	Lignification
Time of duration	Excavation needs 14-21 days; early growing for 4-8 weeks	Varies with tree species, generally 10-20 weeks	Varies with tree species, 1-4 months
Objectives	Regularity of germination is the highest; survival rate is the highest; damping-off is the least	Promotes growing of overground part, maintains the optimum environmental conditions, seedlings reach target height, root system is completely full in the container	Seedling height stops growing, promotes the growth of root and ground diameter, promotes seedling dormancy, adapts to outdoor environment, endures stress
Special requirements	Avoid sunburn, maintain suitable environment, irrigate in low intensity and high frequency, little or no fertilization	Avoid stress, implement the most suitable temperature, regular irrigation, appropriate fertilization, proper light supplement	Reduce irrigation, lower temperature, shorten duration of day, grow in outdoor temperature and humidity, apply low N, high P, and K fertilizer

Source: From Dumroese et al. (2009).

9.4.2 Container Seedling Sowing

Transplanting small container seedlings into larger containers is called miniplug transplants. In addition to that, container seedling sowing can be divided into three types according to the condition of the seed radicle exposed and the growth of caulicle: Seeds directly sowing, seeds with the radicle after sowing, and seedling transplanting. Direct seeding is the most common, and the number of seeds to be sown in each container needs to be determined according to the seed germination rate (Table 9-6). Too many seeds in each container will increase the workload of thinning. Seeds with an exposed radicle can often be found in the seeds with physiological dormancy, such as white-barked pine (*Pinus bungeana*), yellowhorn (*Xanthoceras sorbifolia*), and linden, or in the seeds with high germination heterogeneity, such as Chinese cork oak (*Quercus variabilis*) and Mongolian oak (*Q. mongolica*).

Table 9-6 Relationship between seed germination rate and number of seeds seeded

Seed germination rate/%	Number of seeds seeded	Proportion of at least seedling per container/%
90+	1-2	90-100
80-89	2	96-99
70-79	2	91-96
60-69	2	94-97
50-59	2	94-97
40-49	2	92-97

Source: From Dumroese et al. (2009).

Carry out stratification before sowing because of the physiological dormancy characteristics of the seeds (refer to Chapter 8 for the stratification time of various tree species). Pay close attention to placing different provenances or other different treatments at the same depth when accelerating germination by stratification, because different stratification depths will lead to temperature differences, causing different germination promotion effects and seeding emergence time differences, and finally covering up the treatment differences.

The sowing depth is twice the short axis of the seed (Figure 9-12); Too deep or too shallow is not suitable (Figure 9-13). If the mulch is too thin, seeds dry out easily and lose water, or seeds can be eaten by birds and rodent animals. If the mulch is too thick, the seed will not be able to germinate out of the soil. The mulch can be granite particles, pumice stone, coarse sand stone, or vermiculite, for example. Perlite is too light, making it easy to drift away with the water after watering, so it cannot be used as the mulch (Figure 9-14, Figure 9-15).

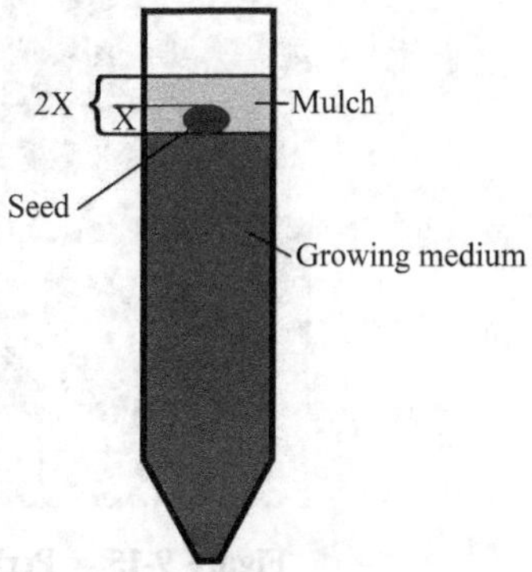

Figure 9-12 Sowing depth (From Dumroese et al., 2009).

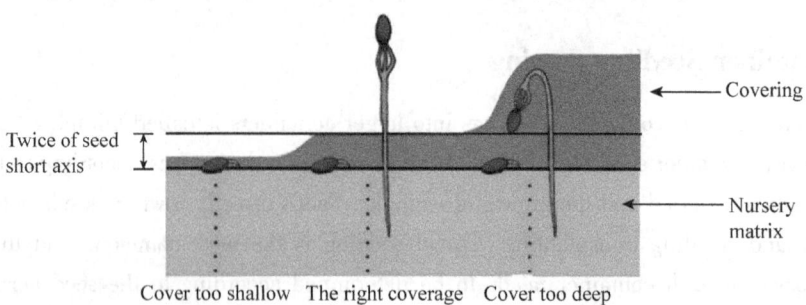

Figure 9-13 Effects of mulch thickness on emergence (From Dumroese et al., 2012),

Figure 9-14 Granite sand as the mulch of *Quercus variabilis* seedlings (Photo by Li Guolei).

Figure 9-15 Perlite is difficult to use as the mulch of *Pinus tabuliformis* seedlings (Photo by Li Guolei).

9.4.3 Container Seedling Irrigation

Water quality of irrigation water is the first consideration of container seedling water management. Table 9-7 lists the water quality requirements of seedling irrigation water. Monitor water quality irregularly in production and scientific research.

Table 9-7 Quality requirements of nursery irrigation water

Indexes	Optimum	Acceptable	Unacceptable
pH value	5.5–6.5		
Salinity/(μS/cm)	0–500	500–1500	>1500
Na/(mg/L)			>50
Cl/(mg/L)			>70
B/(mg/L)			>0.75

Source: From Dumroese et al. (2009).

Container seedling irrigation is based on the percentage of saturated water in matrix, namely irrigation parameters. The saturated water weight of the matrix is usually calculated according to the saturated water weight of the air-dried matrix for the convenience of production application. By comparison, the saturated water of drying matrix is used to calculate the weight in scientific research, and the specific operation methods refer to Dumroese and others (2015). It is necessary to determine the saturated water weight of the matrix of the specific container when filling the matrix, and to then determine the irrigation parameters according to the development stage (Table 9-8). Irrigation parameters vary among tree species in the same development stage, so it is particularly necessary to determine the irrigation parameters of common tree species through scientific research. The principle of irrigation is that the matrix should be fully saturated in each irrigation, and the time of irrigation should be determined by the method of weighing a certain number of containers from time to time (Figure 9-16). The weighing time can be predicted by referring to the irrigation parameters of each development stage, for example, more irrigation times and more frequent weighing occurs during the germination stage. When the irrigation is too much, moss will grow on the surface of the matrix (Figure 9-17). Since moss will slow down the infiltration of irrigation water, lack of water in the matrix will cause seedling quality to decline.

Table 9-8 Irrigation parameters in developmental stages of container seedlings

Developmental phase	Establishment stage		Fast-growing stage	Lignification stage
	Germination stage	Early-growth stage		
Irrigation parameters/%	90+	55–80	55–80	50–65

Figure 9-16 Weighing method determines irrigation time (Photo by Li Guolei).

Figure 9-17 Excessive irrigation causes moss to grow on the matrix surface (Photo by Li Guolei).

Two main irrigation methods can be used: Subirrigation (Figure 9-18) and overhead irrigation. The latter is the most common, and it can be divided into three types: Manual watering, stationary-type spray system, and self-propelled watering machine. Among them, the self-propelled watering machine (Figure 9-19) is the most efficient. At present, container seedling cultivation mainly adopts overhead irrigation. Given the closure of seedling leaves (Figure 9-20), 49% to 72% of irrigation water is not used by the plants. And given that fertigation technology is adopted primarily in the nursery, the loss of irrigation water leads to a great waste of nutrients. The nitrogen and phosphorus in the leaching solution of the seedling-raising matrix can reach 11% to 19% and 16% to 64%, respectively. Water rich in mineral nutrients continuously flows to the surface or into an underground water system for a long time, which causes water body eutrophication and drinking water pollution, posing a threat to the ecological environment and human health. Many states in the United States have legislated to limit the discharge rate of nursery irrigation water. With the enhancement of public awareness of resource and environmental protection, problems such as the waste of water and fertilizer resources and environmental pollution in the process of container seedling cultivation in nurseries have become increasingly prominent. Therefore, whichever methods can save energy and reduce emissions as well as protect the environment have become key to the sustainable development of container seedlings.

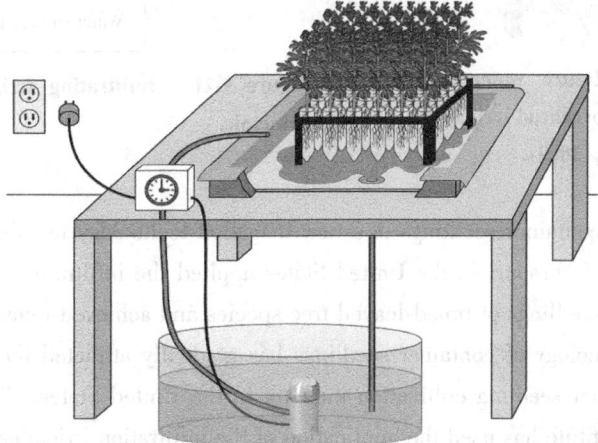

Figure 9-18 Subirrigation system (From Dumroese et al., 2007).

Figure 9-19 Irrigating with self-propelled watering machine (Photo by Li Guolei).

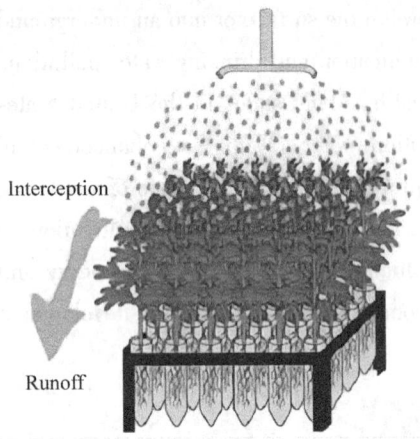
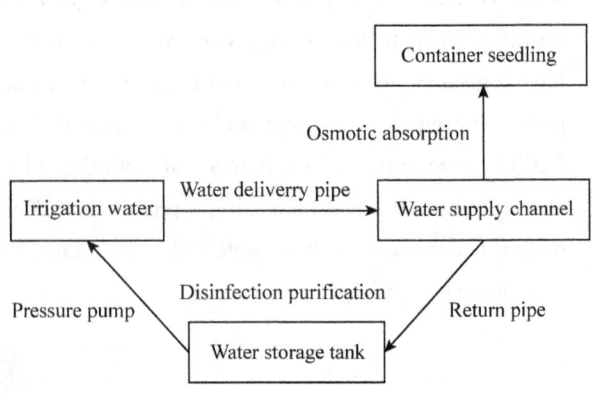

Figure 9-20 Leaf closure water and irrigation water loss of overhead irrigation (From Dumroese et al., 2009).

Figure 9-21 Infiltrating irrigation system operating principle.

Subirrigation of container seedlings is a new irrigation technology to solve the above problems. Since the University of Missouri in the United States applied the infiltration irrigation technology to the cutting container seedlings of broad-leaved tree species and achieved remarkable results in 2002, the subirrigation technology of container seedlings has gradually attracted the attention of the forest protection and container seedling cultivation industry in the United States. The US Forest Service's Forestry Research Institute has used the application of the infiltration irrigation system on a variety of tree species and demonstrated that the system can reduce the water used for seedling and nutrient leaching and produce seedlings of equal or better quality than those produced by overhead irrigation. The subirrigation system comprises a water storage tank, pressure pump, water delivery pipe, water supply channel, and return pipe (Figure 9-21). During infiltration irrigation time, the irrigation water enters into the water supply channel through the water delivery pipe. When the bottom pore of the container gets to the water, the seedling matrix in the container will supply water to the seedlings from the bottom to the top through capillary action. When the seedling matrix reaches the field water capacity, the unused water will move from the return pipe to the water storage tank for recycling. Among these elements, the porosity and type of the seedling matrix are the decisive factors of the saturated water height and water absorption speed of the matrix in the container. Water evaporation and plant transpiration will cause the water consumption in the system, and the water in the water storage tank will need to be supplemented regularly. The system can control the irrigation amount by using the electromagnetic valve to control the working time of the pressure pump. Beijing Forestry University introduced and improved the infiltration irrigation system, which has been applied to *Quercus variabilis*, Chinese pine (*Pinus tabuliformis*), and a variety of Gmelin larch (*Larix principis-rupprechtii*) with better effects.

9.4.4 Container Seedling Fertilization

After outplanting the seedlings from the nursery to the afforestation site, the physiological activities of the root system have not been fully recovered, and the survival and growth of the seedlings depend on the remobilization of nutrients (sources) stored in the seedlings (sink). In the process of nursery cultivation, sufficient fertilizer must be applied to the seedlings so they have enough nutrient storage, which plays an important role in improving out-planting performance. In addition, fertilization after afforestation can promote seedling development, but it can also promote the growth of competing species. The importance of fertilization theory at this stage on both the performance of container seedings in the nursery and at the afforestation site cannot be overemphasized. Therefore, the key to fertilization theory at the present stage is to simultaneously pay attention to the performance of container seedlings in the nursery and afforestation site.

As discussed in Chapter 7, 16 types of elements occur in plants, among them, the content of carbon (C), hydrogen (H), and oxygen (O) account for approximately 96% of the dry weight of plants; 13 types of elements, such as nitrogen (N), potassium (K), and calcium (Ca), account for approximately 4% of the dry weight of plants. The proportion of elements in the natural plants could provide some reference for plant nutrient diagnosis and fertilization technology (Table 9-9).

The nutrient content of the container seedling matrix is lower than the soil, so the element content, element proportion, and plant growth stage should be fully considered in fertilizer selection. Two types of fertilizer are used: Water-soluble fertilizer (fast-acting fertilizer) and controlled-release fertilizer (slow-release fertilizer), both of which contain macronutrients and macronutrients needed by the plants. Fertilizer contains major elements such as calcium, magnesium, and sulfur and micronutrients such as iron and zinc, and are usually marked with three numbers, such as XX-YY-ZZ, respectively, representing the proportion of N-P-K, that is, the proportion of nitrogen, phosphorus pentoxide (P_2O_5), and potassium oxide (K_2O). For example, water-soluble fertilizer 20-10-20 means the content of nitrogen (N), phosphorus (P_2O_5), and potassium (K_2O) in the fertilizer account for 20%, 10%, and 20%, respectively. Controlled-release fertilizer slowly releases nutrients for plant development after one-time application. Although the resin material and processing technology used to make controlled-release fertilizer also make its price slightly higher, one-time application can save labor. The selection of slow-release fertilizer should fully consider three factors: Nutrient content, nutrient release time, and nutrient release mode. Nutrient content is expressed in the same way as water-soluble fertilizer, for example, 16-9-12, 15-9-12. The key factors for determining the rate of controlled-release fertilizers are the coating material and thickness, temperature, and humidity of the seedling-raising site, and the time frame needed; for example, 3 to 4 months, 5 to 6 months, 8 to 9 months, 12 to 16 months, and 16 to 18 months are common in production. The fertilizer release mode is another factor to consider in choosing slow-release fertilizer. The standard slow-release fertilizer, which releases nutrients evenly over time within the valid period; or, the low start-up fertilizer, which begins at a low level and increases after a period of time. The low start-up fertilizer is suitable for autumn and winter fertilization, releasing no nutrients at first, but releasing nutrients intensively later. Nutrient content, nutrient release time, and nutrient release mode of the controlled-release fertilizer all need to be considered when choosing the characteristics of tree species and cultivation cycle.

Table 9-9 Proportion of nutrients in natural plants

Nutrients	Element	Logogram	Proportion/%
Macronutrients	Carbon	C	45
	Oxygen	O	45
	Hydrogen	H	6
	Nitrogen	N	1.5
	Potassium	K	1.0
	Calcium	Ca	0.5
	Magnesium	Mg	0.3
	Phosphorus	P	0.2
	Sulfur	S	0.1
Micronutrients	Ion	Fe	0.01
	Chlorine	Cl	0.01
	Manganese	Mn	0.005
	Boron	B	0.002
	Zink	Zn	0.002
	Copper	Cu	0.0006
	Molybdenum	Mo	0.00001

Source: From Dumroese et al. (2009).

Water-soluble fertilizer and controlled-release fertilizer are produced by professional companies. These commercial fertilizer manufacturers pay attention to not only the reasonable and diverse proportion of elements, such as nitrogen, phosphorus, potassium, and the incidental adding of micronutrients, but also the nitrogen form for the selection and utilization of plant tissue. Nitrate nitrogen is conducive to the development of the root system, and ammonium nitrogen is conducive to the development of aboveground plant parts (Figure 9-22). Nitrogen composition includes nitrate nitrogen, ammonium nitrogen, and urea. For example, Osmocote controlled-release fertilizer with a release cycle of 5 to 6 months and standard of 15-9-12, of which the nitrogen content, the nitrate nitrogen content, and the ammonium nitrogen content account for 15%, 6.6%, and 8.4%, respectively. In the 20-20-20 water-soluble fertilizer produced by Peters Professional, nitrogen content is 20%, ammonium nitrogen content is 4.8%, nitrate nitrogen content is 5.4%, and urea content is 9.8%. The plant development stage is also a considering factor for the selection of fertilizers composed of N, P, and K. The low-N, high-P, and low-K fertilizers should be selected at the seedling establishment stage; the high-N, moderate-P, and moderate-K fertilizers are selected at the fast-growing stage, and the low-N, low-P, and high-K fertilizers need to be selected at the hardening stage (Table 9-10). Selection of fertilizer type is the first factor to consider in fertilization technology, and the fertilization amount and method are also considered—elements that are often overlooked during production and some research studies.

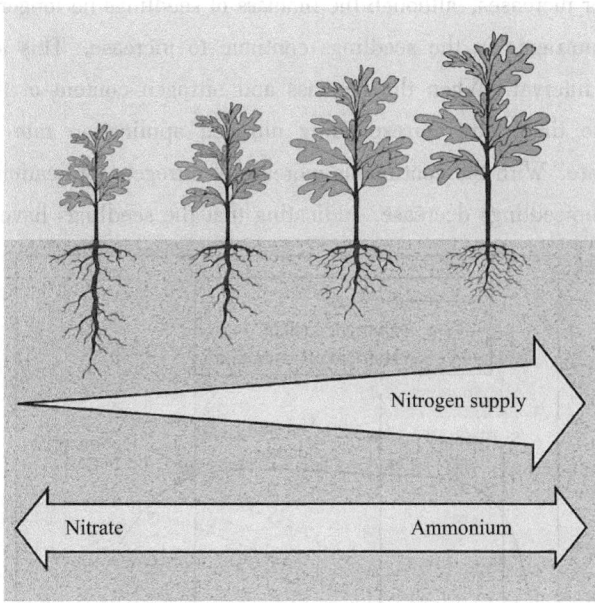

Figure 9-22 Effects of nitrogen supply and nitrogen form on seedling development (From Dumroese et al., 2009).

Table 9-10 Relationship between seedling development stage and fertilizer type selection and nitrogen application

Development stage	Nitrogen application	Proportion		
		N	P	K
Establishment stage	Moderate intensity	Moderate	High	Low
Fast-growing stage	High intensity	High	Moderate	Moderate
Hardening stage	1/4 intensity	Low	Low	High

Source: From Dumroese et al. (2009). *Notes*: Nitrogen solution with the concentration of 100 mg/L can be used for the high intensity.

Fertilization application of container seedlings is usually determined by seedling growth (biomass), nutrient content, and concentration *in vivo*; for example, the relationship between nitrogen application rate and growth and the nitrogen content of the whole plant (Figure 9-23). With nitrogen application rate increasing, the biomass, nitrogen content, and nitrogen concentration of seedlings increase rapidly; but, when the nitrogen application rate continues to increase, the biomass of seedlings increases slowly, while the nitrogen concentration and content of seedlings continue to increase, and when the biomass begins to reach the maximum, the corresponding nitrogen application rate becomes the sufficient nitrogen application rate. When the nitrogen application rate further increases, although the biomass of seedlings no longer increases, the content and concentration of nutrients in the seedlings continue to increase. This stage is called nutrient loading, which is an interval. When the biomass and nitrogen content of the seedlings reach the maximum at the same time, the corresponding nitrogen application rate becomes the optimum nitrogen application rate. With the continuous increase of nitrogen application rate, the biomass and nitrogen content of the seedlings decrease, indicating that the seedlings have been poisoned.

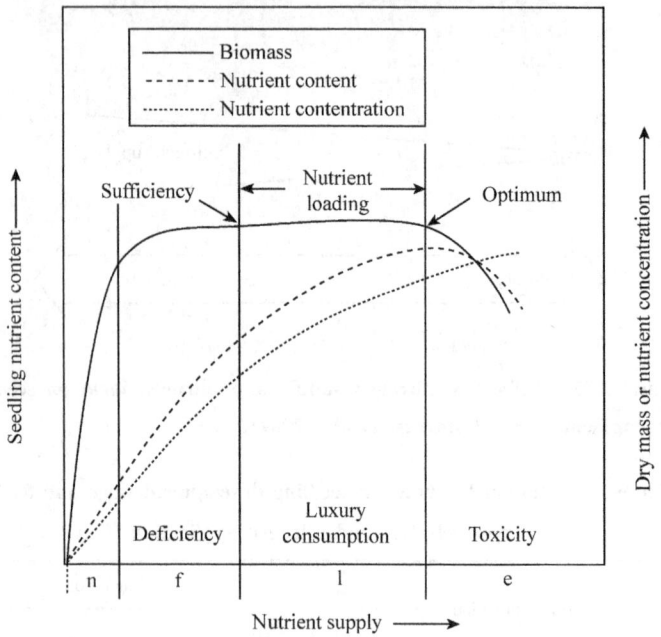

Figure 9-23   Relationship between nitrogen application rate and seedling growth, nutrient content, and concentration (Modified from Timmer and Aidelbaum, 1996).

The fertilization amount should be determined by the tree species and development stage. The fertilization amount needed for seedlings is less in the establishment stage, high in the fast-growing stage, followed by the hardening stage. The fast-growing stage has the most concentrated fertilization research and the most extensive application in production; the seedlings need more fertilizer and have a longer growth cycle. If water-soluble fertilizer is used, the concentration of water-soluble fertilizer can be adjusted according to the growth rate of seedlings to meet the requirements of seedling fertilizer. Exponential fertilization is widely proved to be an efficient way of nutrient loading, because the nutrient-adding rate coincides with the growth rate. However, it is rarely used in production, because this technology needs fertigation equipment to overcome the labor pressure caused by 12 to 24 or more times of fertilization. This method also needs professional technicians to calculate the frequently changing fertilization concentration.

If controlled-release fertilizer is used, the formula, release rate, and release mode of controlled-release fertilizer should be fully considered by the development law of tree species. In autumn, when the seedlings enter the hardening stage, the terminal buds gradually form and the height growth slows down, while the biomass of seedlings, especially the biomass of roots, continues to grow. If the fertilization is stopped at this time, the available nutrients from the soil and the nutrient concentration in the seedlings will be reduced, which will affect the quality of seedlings and the afforestation effect in the next year. To avoid the nutrient dilution effect caused by the increase of biomass in the hardening stage of seedlings, human-applied, appropriate amounts of fertilizer to the seedlings, that is, fall fertilization, is used. In Europe and America, fall fertilization is regarded as an important means of nutrient loading. The research on fall fertilization has been carried out in the seedlings of commonly grown tree species, such as pines (*Pinus*), oaks (*Quercus*), and spruces (*Picea*), and this technology has been widely used in production practice. China is currently cautious about fall fertilization, taking up the research on tree species such as *Pinus tabuliformis*, *Quercus variabilis*, *Larix olgensis*, *Larix principis-rupprechtii*, and *Populus tomentosa*, and researchers have obtained certain positive effects thus far. Fall fertilization should be conducted after the formation of a terminal bud. Fall fertilization should be conducted 1 to 2 times a week for 5 to 8 weeks. Generally, water-soluble fertilizer with low nitrogen, low phosphorus, and high potassium should be selected, and the fertilization concentration should be approximately a quarter of that used in the fast-growing period.

Water-soluble fertilizers are dissolved in water in correct proportions and applied by manually fertilizing with a stationary-type spray system or self-propelled fertigation. Controlled-release fertilizer can be mixed into the matrix (incorporation) before filling the container; it can also cover the surface of the matrix (top-dressing).

9.5 Greenhouse and Environment Control

The growth of seedlings can be accelerated by extending the light application time during the rapid growth period. The light application time of seedlings is usually 16 to 18 hours, including times of natural light and artificial supplemental lighting, but it is not necessary to supplement the light when the natural light is sufficient in the daytime. The best light sources for greenhouse lighting are metal-halide lamps with a wide spectrum and/or high-pressure sodium lamps with a lower price and wide usage. Fluorescent lamp quality is also very effective, but more lamps need to be installed to achieve the same light intensity as other light sources. Incandescent lamps work on seedlings in low light. When the plants enter the hardening stage, light supplement needs to stop. To speed up the formation of terminal buds, shading nets can be used to shorten the light time to 10 to 12 hours, and the short-day time of individual tree seedlings can be 6 to 8 hours, with the short-day period of 2 to 3 weeks. Excessively shortening the short-day time will inhibit seedling growth and affect the quality of seedlings.

For most tree species, the optimal temperature in the greenhouse is 20 to 25 ℃ in the daytime and 16 to 20 ℃ in the nighttime. Two methods can control the ambient temperature: One is to regulate the temperature with facilities and equipment, the other is to regulate the rhizosphere temperature with cultivation methods. It is very important to choose a greenhouse with good heat preservation and lighting performance that is equipped with heating equipment for the cold zone. At present, greenhouse heating is based primarily on high-efficiency, hot-blast stoves that use coal, oil, or gas as fuels, followed by geothermal hot spring heating in areas where conditions permit. In summer, greenhouse cooling is more important, especially in the South. When the outdoor temperature is too high, natural ventilation to reduce the temperature should be the first consideration. If it fails to meet the requirements, external shading, internal shading, wet curtain, micro spray, and other measures need to be used in combination. The rhizosphere temperature of seedlings is very important for the growth and development of seedlings. First, the rhizosphere temperature can be provided by bed-frame seedlings; second, the excellent seedling matrix can be used, and the water temperature can be raised through a special water reservoir and then be irrigated by spray irrigation.

Air humidity directly affects the growth and development of plants. When the air humidity is too low, the plant will close the stoma to reduce transpiration, indirectly affecting photosynthesis and nutrient transport; when the humidity is too high, the plant will grow weakly, causing excessive growth and mildew. Ventilation is adopted to reduce humidity, whereas water and spray are used to increase humidity.

Questions for Review

1. What are the main stages of sowing seedling development? What is the effect on seedling quality control?
2. What are the main types of containers? How to choose suitable containers for seedling raising?
3. What is the significance of irrigation parameters for container seedling irrigation? How to determine irrigation parameters?
4. What is the theoretical basis of seedling fertilization? What is steady-state nutrient loading?
5. From the perspective of fertilizer type, fertilizer nutrient element ratio, fertilizing amount, fertilizer application method, and fertilizer application time, combined with seedling development stage, describe the key technology of container seedling fertilization.
6. What is the key technology of container seedling light control?
7. What are the advantages and disadvantages of container seedlings?
8. What is the composition and proportion of container seedling culture medium?
9. What are the water and fertilizer control technologies for raising container seedlings?

References and Additional Readings

CHENG F Y, 2012. Nursery of landscape plants[M]. Beijing: China Forestry Publishing House.

DUMROESE R K, JACOBS D F, DAVIS A S, et al., 2007. An introduction to subirrigation in forest and conservation nurseries and some preliminary results of demonstrations. National Proceedings: Forest and Conservation Nursery Associations. Proceedings RMRS-P-50 [C]. Fort Collins (CO): US Department of Agriculture, Forest Service, Rocky Mountain Research Station.

DUMROESE R K, LUNA T, LANDIS T D, 2009. Nursery manual for native plants. Agriculture Handbook 730[C]. Washington, DC: US Department of Agriculture, Forest Service.

DUMROESE R K, LUNA T, LANDIS T D, 2012. Raising native plants in nurseries: Basic concepts. General Technical Report RMRS-GTR-274[C]. Fort Collins (CO): US Department of Agriculture, Forest Service, Rocky

Mountain Research Station. 84 p.

DUMROESE R K, MONTVILLE M E, PINTO J R, 2015. Using container weights to determine irrigation needs: A simple method[J]. Native Plants Journal, 16(1): 67-71.

GE H Y, JIANG S D, 2003. Seed and seedling production in the tray[M]. Beijing: China Forestry Publishing House.

GROSSNICKLE S C, EL-KASSABY Y A, 2016. Bareroot versus container stocktypes: A performance comparison [J]. New Forests, 47: 1-51.

LANDIS T D, TINUS R W, MCDONALD S E, et al., 1989. The container tree nursery manual volume 4, seedling nutrition and irrigation. Agriculture handbook 674[C]. Washington, DC: US Department of Agriculture, Forest Service.

LANDIS T D, 1990. Containers and growing media. The container tree nursery manual volume 2. Agriculture handbook 674[C]. Washington, DC: US Department of Agriculture, Forest Service. p 41-85.

LANDIS T D, TINUS R W, BARNETT J P, 1998. The container tree nursery manualvolume 6, seedling propagation. Agriculture Handbook 674[M]. Washington, DC: US Department of A griculture, Forest Service.

LI G L, LIU Y, ZHU Y, 2011. Review on advance in study of fall fertilization regulating seedling quality[J]. Scientia Silvae Sinicae, 48(11): 166-171.

LI G L, LIU Y, ZHU Y, et al., 2012. A review on the abroad studies of techniques in regulating quality of container seedling[J]. Scientia Silvae Sinicae 48(8): 135-142.

POOTER H, BOHLER J, VAN DUSSCHOTEN D, et al., 2012. Pot size matters: A meta-analysis of the effects of rooting volume on plant growth[J]. Functional Plant Biology, 39: 839-850.

SHEN H L, 2009. Seedling cultivation[M]. Beijing: China Forestry Publishing House.

TIMMER V R, AIDELBAUM A S, 1996. Manual for exponential nutrient loading of seedlings to improve outplanting performance on competitive forest sites[C]. Natural Resources Canada, Canadian Forest Service, Great Lakes Forestry Centre, Sault Ste Marie, Ontario.

ZHAI M P, SHEN G F, 2016. Silviculture[M]. 3rd edition. Beijing: China Forestry Publishing House.

ZHU Y, LIU Y, LI G, L, et al., 2013. Current development and prospect on sub-irrigation in container cultivation of forest trees[J]. World Forestry Research, 26(5): 47-52.

Chapter 10　Vegetative Propagation

Hou Zhixia

Chapter Summary: In this chapter, three kinds of vegetative propagated seedlings are introduced: cutting seedlings, grafting seedlings, and tissue culture seedlings. The cutting seedling cultivation section introduces the characteristics, species, and methods of cutting breeding, survival mechanism, factors affecting the survival of cutting, and methods to promote root growth of cuttings. For grafted seedling cultivation, we introduce the utilization value of grafted seedlings, the principle and influencing factors of grafting survival, the relationship between rootstock and scion, the selection and cultivation of rootstock and scion, common grafting methods, the key to improve the survival rate of grafted seedlings, and the management of grafted seedling breeding. For tissue culture seedling breeding, this chapter introduces the principles, types, application fields, and construction of tissue culture rooms; and the culture conditions, types and preparation of medium, preparation of explants, and the primary process of tissue culture seedlings.

Vegetative reproduction, also known as asexual reproduction, refers to the use of plant vegetative organs (roots, stems, branches, leaves, and buds, etc.), or plant tissues, cells, and protoplasts as reproductive materials for seedling breeding. The seedlings propagated by vegetative method are called vegetative seedlings. Many methods are possible for vegetative propagation, including cutting, grafting, layering, and tissue culture. The outstanding characteristic of vegetative seedlings is that the reproduction material adopts plant vegetative organs and does not go through the process of pollination and fertilization to form seeds. It has less genetic variation and can inherit excellent characters of the mother, which is conducive to the reproduction of excellent varieties and types. Vegetative seedlings often result in early flowering and fruit, which makes it possible to obtain the target traits and products to cultivate improved varieties. Vegetative seedlings also allow for selection of new varieties of forest trees that have important significance, such as those widely used in cultivation of improved varieties of garden plants, fruit trees, and multipurpose tree species.

10.1　Cutting

10.1.1　Cutting Propagation and Its Characteristics

Cutting is an important method of plant propagation. A part of the vegetative organs, such as roots, stems (branches), and leaves, are made into cuttings and struck into the matrix to produce adventitious roots or adventitious buds under certain conditions and finally to be cultivated as independent new plants. Stocks obtained by this cutting method are called cutting stock. Cutting propagation is easy to accomplish, with the cuttings being quick to grow, early to bloom, made up of maternal qualified characteristics, easy to propagate with low cost, can be propagated in many seasons, and can be widely used in stock cultivation. Cutting propagation also has some disadvantages, however. For example, compared with trees from seedlings, ramet seedlings, and grafted seedlings, generally the lifespan of trees from cuttings with the same germplasm is shorter, and the root system is weaker with shallow distribution, and the adaptability is relatively poor.

10.1.2 Types and Methods of Cutting Propagation

According to the sampling location and types of cuttings, cutting propagation of forest trees primarily includes stem cutting and root cutting.

10.1.2.1 Stem Cutting

According to the properties of cuttings, stem cutting can be divided into hardwood cutting (during the dormancy period) and softwood cutting (also known as greenwood cutting, during the growing period).

(1) Hardwood Cutting

This widely used seedling raising method in production uses fully lignified branches as cuttings to raise seedlings by cutting. The method is simple, with low cost, and is suitable for many tree species.

Cutting Selection. Choose 1-year-old robust branches in the cutting orchard as cuttings, which will best keep the qualified characteristics of the seed tree with strong vitality; the cuttings are of high quality and readily survive. Without access to a cutting orchard, select 1- to 2-year-old branches with full development from robust, young, seed trees with fast growth, straight shape, no disease, and no insect damage.

Cutting Collection Time. For deciduous tree species, the collection time of cuttings should be in late autumn and early winter (after defoliation) or before sprouting in the spring, at these times, the branches are full of nutrition and easy to root. For evergreen tree species, cuttings should be collected before budding in spring, when the rooting rate is high and cuttings are not likely to rot.

Making the Cuttings. The principle of making cuttings is to be sure they contain a certain amount of root primordia, nutrients, and water; and are an appropriate cutting length for saving cuttings and easy striking. For branches with different positions, the cutting rooting rate differs. Generally, the rooting rate of the middle and lower branches with full development is high. The making of cuttings can be carried out before striking, first cutting off the immature shoots, and then making the cuttings according to the specifications. The length and cut shape of the cuttings, and the reserved amount of buds and leaves, should be determined according to the rooting characteristics, growth types, the moisture of the cutting soil, and the environment. The cutting length of broad-leaved tree species is 10 to 20 cm and that of coniferous tree species is 5 to 15 cm. Making cuttings should be carried out in shade and leeward to the wind (that is, out of the wind). The cut should be smooth and not split; the top cut is mostly flat, usually 1 to 2 cm from the top of bud or bud scale scar; the bottom cut is mostly oblique, of which the location varies according to plant species. Many parenchyma cells are near the node, with fast cell division, developmental nutrition, and easy-to-form callus and root; therefore, the bottom cut should be close to the node, 0.3 to 1 cm below the bud or bud scale scar. Cuttings should be made according to the length requirements, paying attention to protecting the buds at the top of the cuttings during cutting. To prevent water loss in the cuttings of evergreen trees, leaves and terminal buds should be properly preserved. After completing the cut, the cuttings should be banded according to the thickness and should be promptly struck or stored.

Cutting Storage. For the collected cut branches and cuttings, prevent them from drying and losing water. If they are not struck immediately, they need to be stored. Cutting storage can be divided into short-term storage and overwintering storage. Short-term storage is used for cuttings that cannot be struck quickly enough, and they can be buried in cool, wet sand. Overwintering storage is suitable for cuttings or cut branches collected in autumn and winter and used for cutting in the next spring; these are often stored in a cellar or an outdoor ditch. Select a place with high, dry terrain and good drainage to dig a ditch or build a cellar; the cuttings or cut branches are buried and stored in wet sand with a temperature of 0 to 5 ℃.

For overwintering storage, the cuttings can be tied in bundles of 50 or 100 cuttings and placed vertically in the ditch in layers; if the cut branches are to be stored, they can be cut 60 to 70 cm in length, tied in bundles of 50 or 100 cuttings, and buried horizontally in layers. Information such as variety, collection date, and collection location should be indicated for cuttings and branches. During burial, a 10-cm layer of wet sand is added to each layer of cuttings (or branches). The ditch is filled with wet sand and then sealed with soil to form a roof ridge. To prevent the scions from overheating, a vent hole should be set every 1 m or so. In early spring, when the temperature rises, attention should be paid to prevent the branches sprouting too early, that is, in advance of when they will be planted.

Soil Preparation and Cutting. Conduct fine soil preparation for the cutting seedlings, and in general, the depth of cultivated land should reach 25 to 30 cm. Ridge culture and bed culture are used for the hardwood cutting. Ridge cultivation is convenient for mechanical operation during seedling management. For easy-to-root tree species, the broad ridge seedling in the field can be used, and the low bed cutting can be used in arid areas and for flower shrubs.

Cutting Technology. Two kinds of hardwood cutting can be used: Straight cutting and oblique cutting. Straight cutting is used in the majority of cases. When the soil is heavy, oblique cutting could also be used. When cutting, pay attention to the polarity of the cuttings. The apical side should upturn and not be inverted, and the lower cut should be closely connected with the soil to avoid scratching the cortex of the lower cut. Generally, cutting depth is better to expose a bud on the overground part. In windy, dry areas and sandy nurseries, the whole cutting can be inserted into the soil, with the upper end flush with the ground, and the cuttings should be compacted after cutting.

Nursery Land Management. After cutting, the nursery land should be irrigated immediately so as to bring the soil into close contact with the cuttings, which will meet the water demand of cuttings and help with their survival. Follow-up water should be timely replenished according to soil moisture. In production, a cutting bed with plastic film mulch is often used to assist cutting, which can play a role in increasing the ground temperature and moisture conservation and promote the survival of cuttings and the growth of seedlings.

(2) Softwood Cutting

Softwood cutting uses young shoots with strong growth capacity or semi-lignified leafy branches as cuttings to propagate seedlings in the growing period, also known as greenwood cutting. The shoots are rich in growth hormone, soluble sugar, and amino acids, of which the tissues are tender and the meristem cells have strong ability to divide, which is beneficial to the formation of callus and rooting. Generally, softwood is easier to root than is hardwood, but softwood has strict requirements for air humidity and soil humidity, which requires certain equipment and careful management. For example, for indoor misting cutting propagation, leaves are covered with a layer of water film to reduce transpiration, enhance photosynthesis, and reduce respiration so as to maintain the vitality of the difficult-to-root cuttings for a longer time and to facilitate rooting and growing. With the improvement of production facilities, softwood cutting has become one of the main techniques of large-scale propagation of seedlings, which has been applied in more and more tree species.

Cutting Time. Softwood cutting in open fields is mostly in the growing season, and the time of cutting varies according to the region, environmental conditions, and tree characteristics. Generally, it can be carried out in spring, summer, and autumn in the south, and mainly in summer in the north. For example, cutting of sweet osmanthus (*Osmanthus fragrans*) and Chinese ilex is better in mid-May, while that of ginkgo, oriental arborvitae (*Platycladus orientalis*), camellia, port wine magnolia (*Michelia figo*), and Chinese hawthorn (*Photinia serrulata*) is better in mid-June.

Cut Branch Collection. In terms of softwood cutting, cuttings were collected, followed by striking. The cut branch should be selected from a strong, young seed tree, and the semi-lignified young shoots are the best as they contain sufficient nutrients with strong vitality and can easily regenerate adventitious roots. After branch picking, keep the branches moist by spraying water on them promptly or by placing them directly into a bucket or water tank for temporary storage. Then quickly make cuttings in the shade.

Making Cuttings. The cutting length should be determined according to the tree species, branches, and planting bed matrix, in combination with factors such as saving cuttings and convenient cuttings. Generally, cuts should include 2 to 4 nodes and be 5 to 20 cm long. To reduce transpiration, the lower leaves of the cuttings should be removed properly, while part of the upper leaves should be reserved for photosynthesis to ensure nutrition supply, and then, when the leaves are larger, part of them can be cut off. The lower cut can be flat or oblique, and the upper cut should be a smooth plane.

Striking Technology. Striking is generally carried out on the cutting bed, following the principle of "collecting, followed by making, and striking". The matrix can be made up of peat, vermiculite, coarse sand, sandy loam, or a special composite matrix, all of which are loose and ventilative with better moisturizing effect. The striking depth is 2 to 5 cm. If the environmental conditions can be controlled manually and high air humidity can be maintained, the striking depth can be shallower and striking vertically is more commonly used. The cutting density is the best when the leaves of two cuttings just overlap, which allows for sufficient photosynthesis and does not interfere with the ventilation and light transmission of each portion. The density of cuttings in the cutting bed is high, so it is better to transplant them into the nursery in time for cultivation after rooting.

Management After Cutting. Softwood cutting requires strict conditions of temperature and humidity, which means the environment must have temperature and humidity controls. In terms of tree species that are difficult to root, softwood cutting is carried out primarily in a greenhouse, plastic greenhouse, or low tunnel equipped with a spraying device; sunshade measures are needed to prevent the temperature from being too high. In the early stage of cutting, air humidity should be kept above 95%, and it can be reduced to between 80% and 90% after the callus of the lower cut grows. Generally, humidity can be controlled by spraying water two to three times a day, or even three to four times a day when the temperature is high, and the amount of water spray each time should not be too high so as to achieve the purpose of cooling. The goal is to increase the air humidity without making the cutting soil too wet. No water should accumulate in the cutting soil to prevent the cuttings from rotting. The temperature should be controlled at 18 to 28 ℃, and measures such as ventilation, water spraying, and sunshade should be taken immediately when the temperature exceeds 30 ℃. The rooting of cuttings needs the supply of synthetic substances from the leaves of the cuttings and suitable light conditions. Therefore, when shading and cooling, the shading degree should not be too high, so as not to affect the photosynthesis of leaves. At present, the all-optical, automatic, intermittent spray is used to improve the survival rate of softwood cuttings (Figure 10-1).

(a)　　　　　　　　　　　　(b)　　　　　　　　　　　(c)　　　　　(d)

Figure 10-1 Softwood cutting seedlings of *Prunus humilis* in sunlight greenhouse.
(a) softwood cutting in nutrition cup on seedbed; (b) 40 days after cutting; (c) rooting in matrix; and (d) root system status.

Acclimatization and Transplant. After rooting by cutting, if the seedlings are raised in a plastic shed, the ventilation and light transmittance should be increased gradually to make the cuttings adapt to the natural conditions. After the cuttings survive, they should be transplanted into the nursery or into a larger container for further cultivation.

(3) Burying Propagation

Burying propagation method involves burying whole 1-year-old branches in the soil to root, germinate, and finally form many plants. This special method of cutting is similar to layering propagation in operation technology. It is most suitable for the tree species with easy cortex rooting, and sometimes for the tree species difficult to root by cutting. The burying time is generally in the spring, and because the branches are longer, once one part is rooted, the entire branch can survive so as to improve the survival rate of cuttings. For example, poplar can be propagated by this method in production, and good effects can be achieved in the propagation of London planetree (*Platanus acerifolia*) and paulownia.

The cuttings should be collected after defoliation and before sprouting, and a soaking treatment needs to be carried out within 3 to 5 hours before the cuttings are buried. Generally, there are two burying methods: Plane and point. The plane burying method is to make the trench on the seedbed (usually on a low seedbed in the north-south direction) according to a certain row spacing, with a trench depth of 3 to 5 cm and width of 4 to 6 cm. Branches are placed in the trench horizontally and covered with fine soil at a thickness of 1 to 2 cm. When placing the branch, it should be properly matched according to the thickness, length, and bud condition of the branch, and most buds should be upward or on both sides of the branch. The point burying method purposely does not cover some sprouting parts with soil, so that the buds are exposed to the outside environment, so as to facilitate growth. A trench with a depth of approximately 3 cm should be dug according to a certain row spacing, and the cuttings should be laid in a flat manner. Then, a long circular mound with the length of 20 cm, width of 8 cm, and height of about 10 cm should be built across the row every 40 cm. Two to three buds should be on the exposed branches between the mounds; the buds can germinate and grow at the higher outside temperature and take root at the mounds.

For the plane burying method, do not let the soil cover be too thick or it will affect sprout excavation; for the point burying method, after burying the mounds, compact them to prevent the mounds from collapsing during irrigation. The emergence of the buds is faster and more orderly by the point burying method than by the plane burying method, and the plant spacing will be more regular than that of plane burying, which is conducive to the final singling out of seedlings. Water retention performance is better in the point burying method, but the operation efficiency is lower, requiring a lot of labor.

After the cuttings are buried, irrigation should be applied immediately, and the soil needs to be kept moist. Generally, irrigation should be carried out every 5 to 6 days during establishment and before rooting. Before rooting and sprouting, it is also necessary to check the soil cover frequently to remove soil when it appears to be too thick and to bury the exposed cuttings.

10.1.2.2 Root Cutting

Root cutting is a propagation method that uses a portion of the roots of an existing plant as a cutting to propagate a new plant. The key to its survival is to grow adventitious buds on the root cuttings and support further development into a complete plant. Root cutting propagation is a type of selection for trees that are difficult to root by stem cutting. Similar to stem cutting rooting, root

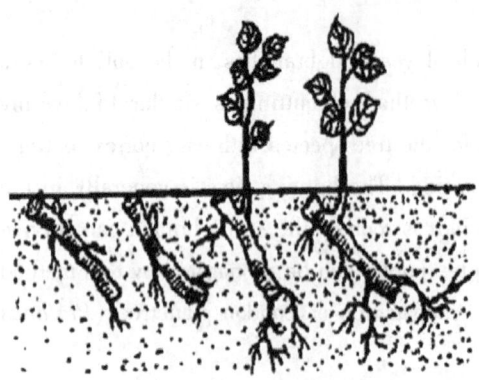

Figure 10-2 Root cuttings.

cutting sprouting involves the process of organ differentiation, of which the physiological basis is complex and varies across plants. Generally speaking, tree species prone to producing root suckers are more likely to adopt root cutting propagation, for example, Chinese mahogany (*Toona sinensis*), paulownia, Chinese white poplar (*Populus tomentosa*), black locust (*Robinia pseudoacacia* cv. Idaho). Tree species that have difficulty surviving by stem cutting or are slow to root, such as jujube, persimmon, walnut, hickory, and yellowhorn (*Xanthoceras sorbifolia*), are likely to survive by root cutting. Tree stock, such as birchleaf pear (*Pyrus betulaefolia*), quince (*Cydonia oblonga*), Chinese crab apple (*Malus baccata*), and plum-leaf crab, can be propagated by root cutting using root segments cut from seedlings during outplanting or using residual roots left in the ground. The root segment should be 0.3 to 1.5 cm thick and be cut approximately 10 cm long. The top cut should be flat and the bottom cut should be cut obliquely. The root segment can be struck vertically or horizontally; it germinates readily by vertical striking, but the cutting's polarity should not be upside down (Figure 10-2).

10.1.3 Mechanism of Cutting Surviving

10.1.3.1 Types of Cutting Rooting

According to the type of cuttings, the survival of the cuttings involves the formation of adventitious roots and adventitious buds. The survival principle of stem cutting and root cutting differs. Stem cutting rootings need to form root primordium in the cambium and vascular bundle sheath of branches, so as to develop adventitious roots and form root systems. Root cuttings need to generate adventitious buds in the cortex parenchyma of the roots and to then develop into stems and leaves. The adventitious bud develops primarily from the pericycle near the vascular bundle cambium on the young root and from the cork cambium or ray proliferation callus on the old root.

The survival of cuttings by stem cutting depends primarily on whether cuttings can root. According to the position and mechanism of cutting rooting, it can be divided into two basic rooting types: Cortex rooting and callus rooting.

(1) Cortex Rooting Type

For the tree species that take root mainly from the cortex or first from the cortex (Figure 10-3a), the root primordium has been formed in the cuttings of the tree before cutting, which grows and develops to form adventitious roots under the condition of *in vitro* cutting induction. The cortex of cuttings is composed of special parenchyma cell groups, with dense cytoplasm and close arrangement, and it typically locates at the junction of the widest pith ray and cambium in branches, of which the outer end leads to the lenticel where oxygen can be obtained and nutrients are obtained from myelocyte. The origin and developmental degree of a root's original body vary across tree species. Usually, the root's original body is in the dormant state before the branches are cut off from the seed tree, and it further develops to form the primordium under certain environmental conditions. After the cuttings are inserted into the matrix, the root primordial body obtains nutrition and oxygen. Under suitable temperature and humidity conditions, the root primordial body tip grows and develops continuously, and it generates adventitious roots through phloem and cortex, absorbing water and mineral nutrition from the soil. The earliest rooting positions of cuttings are at a depth with suitable soil temperature, humidity, and aeration conditions (3 to 8 cm from the surface), from which one to two adventitious roots can develop to ensure the survival of cuttings. Subsequently, callus appears in the lower portion of the cutting, and a large number of adventitious roots are produced nearby.

Generally, the better the cambium develops, the more nutrients the root original body obtains, the better the growth and development will be, and the faster the rooting will occur. Because of a polarity effect, the distribution of roots on the cuttings will be in the lower part of the cuttings and near the cut. The more root included from the original body of a tree species or branches, the higher the survival rate will be. This rooting type can root quickly, and it has a wide rooting area, which is the characteristic of an easy-to-root tree species.

(2) Callus Rooting Type

For the tree species that take root primarily from callus, the adventitious roots are formed by callus induction after the cuttings are separated from the seed tree and cut (Figure 10-3b). First of all, the parenchyma cells of the cambium divide at the lower end of the cuttings or near the cambium, and the callus gradually forms on the surface of the lower cut, which further forms the xylem, phloem, cambium, and other tissues related to the corresponding tissues of the cuttings; eventually the cutting fully heals. Callus and its neighboring cells are very active in the process of rooting, forming growing points. Under suitable temperature and humidity conditions, the root original body can be differentiated from growing points or cambium, which can further grow and develop to produce adventitious roots. Survival can be difficult for cutting seedlings of the tree species that take root mainly by callus rooting or first from the callus. Callus rooting needs a longer time.

Some tree species have the ability to grow from both rooting types, which are known as comprehensive rooting types.

(a)   (b)

Figure 10-3 Cortex rooting and callus rooting types.
(a) mainly cortex rooting for mulberry; and (b) mainly callus rooting for loquat.

10.1.3.2 Physiological Basis of Cutting Rooting

In the aspect of cutting rooting theory, many scholars have worked diligently and put forward many hypotheses and techniques from different perspectives to guide the cutting practice, which has achieved certain effects.

(1) Auxin and Cutting Rooting

Cutting rooting and callus formation of plants is controlled and regulated by auxin, which is also related to cytokinin and abscisic acid. Endogenous auxin of branches can promote the formation of the root systems, and it is primarily synthesized in the young buds and leaves of branches, and then moves to the base to participate in the formation of the root system. At present, artificially synthesized auxin applied in production includes IBA, NAA, NAD, ABT, and HL-43, which can improve the rooting rate and shorten the rooting time after application to the cutting base. Many experiments and production practices have also proved that auxin is not the only substance needed to promote rooting of cuttings to grow adventitious roots; a special kind of substance produced in buds and leaves is also needed.

(2) Rooting Inhibitors and Cutting Rooting

A rooting inhibitory substance is found in plants and has a hindering effect on rooting. It often exists in tree species or cuttings that are difficult to root. Studies have shown that the content of inhibitory substances in old trees is higher, while it is the highest in the dormancy period in the annual growth cycle of trees. In production practice, the corresponding measures, such as water elution, low temperature treatment, dark treatment, and other steps, can be taken to eliminate or reduce the inhibitory substances, so as to facilitate rooting.

(3) Nutrients and Cutting Rooting

A certain relationship occurs between the survival of cuttings and their nutrients, especially the content and relative proportion of carbon and nitrogen. Generally speaking, higher C/N ratio is favorable for the induction of adventitious roots. Researchers have proved that supplementing carbohydrate and nitrogen to cuttings can promote rooting. Generally, the rooting rate can be increased by soaking the bottom of cuttings with sugar solution or by spraying nitrogen, such as urea, on the cuttings. However, special attention should be given to use of exogenous carbohydrate supplementation because it readily causes cut decay.

(4) Plant Development Situation and Cutting Rooting

Practice has shown that the rooting ability of cuttings is closely related to their developmental state, especially weakening with the increase of seed tree age. According to this feature, for rare and precious tree species or for tree species that are difficult to propagate, various measures can be taken to rejuvenate or improve the survival rate of cuttings: Hedgerow Scion Collecting. The seed tree selected for gathering scions needs to be strongly cut to make it sprout many new branches as cuttings. Continuous Cutting Propagation. Rooting ability of new shoots increases significantly after two to three consecutive cuttings. Continuous Grafting Propagation Using Young Stock. The scions collected from the old seed tree can be grafted to the young stock. After grafting repeatedly and continuously, the branches or needle leaf bundles of the grafts are used for cutting. Using Budding at the Base of a Tree as Cuttings. A new sprouting branch produced in the young part can be used by cutting off the old trunk.

(5) Anatomical Structure of Cuttings and Cutting Rooting

In some cases, the occurrence and growth of adventitious roots in cuttings are related to the anatomical structure of the cortex. It is difficult to take root if one or more layers of annular sclerenchyma, which is composed of fibrous cells, occurs between the phloem and cortex and outside the initial site of adventitious roots. If no annular sclerenchyma occurs, or discontinuous sclerenchyma is present, rooting will be easier. In production practice, the method of cutting the cortex can be used to destroy the annular sclerenchyma to promote rooting. To a certain extent, the survival rate of cuttings can be improved by longitudinal cutting for most cuttings. However, for most tree species that are difficult to root by cutting, rooting is related to the genetic potential and physiological status of root original body formation, and the structure of cuttings does not affect the potential rooting.

The cutting rooting mechanism is very complex, and no one theory can explain all the problems in different plants. The rooting ability of cuttings in cutting propagation is affected by the genetic characteristics of tree species and the cultivation environment. Therefore, plant propagators need to make full use of the characteristics of the plant itself to overcome rooting difficulties, and at the same time, with necessary technical measures, to maximize the survival rate of cuttings.

10.1.4 Factors Affecting Cutting Survival

Whether the cuttings can root and survive after striking is closely related to the internal factors of cuttings and external environmental conditions.

10.1.4.1 Internal Factors

(1) Genetic Characteristics of Tree Species

Genetic characteristics vary with tree species, and the survival rate of cuttings varies greatly. According to the difficulty of rooting, the tree species can be divided into four types:

Tree Species Easy to Root. These species typically include the 1- to 2-year-old cuttings collected from young seed trees. When striking cuttings in the field, they will root within 5 weeks, and the rooting rate is more than 80%. For example, this response occurs in most poplars (except *Populus davidiana* and *Populus tomentosa*), most willows, Chinese fir (*Cunninghamia lanceolata*), false indigo bush (*Amorpha fruticosa*), honeysuckle, and grape.

Tree Species Relatively Easy to Root. The 1- to 2-year-old cuttings collected from young seed trees need to be shaded in a special cutting soil, such as sand or vermiculite, and the rooting time is within 6 to 7 weeks, with a rooting rate of more than 80%. These trees include species such as boxwood, Japanese kerria (*Kerria japonica*), kiwi fruit, and Chinese ash (*Fraxinus chinensis*).

Tree Species Relatively Difficult to Root. The 1- to 2-year-old cuttings collected from the young seed trees were cut in sand, vermiculite, or other matrix with a certain amount of shading and spray to moisturize after treatment with hormones. Rooting time is often still more than 5 to 7 weeks, and the rooting rate might be approximately 50%. This circumstance includes species such as Atlas cedar (*Cedrus atlanticus*), hoop pine (*Araucaria cunninghamii*), gingko, camellia, European olive (*Olea europaea*), hazelnut, and jujube.

Tree Species Extremely Difficult to Root. The 1- to 2-year-old cuttings collected from young seed trees are treated with a hormone, and in the special ideal cutting matrix and the more suitable culture environment, the rooting time was more than 7 to 10 weeks, with relatively low rooting rate and seedling rate. These outcomes are seen in species such as larch, Korean pine (*Pinus densiflora*), Chinese pine (*Pinus tabuliformis*), Mongolian oak (*Quercus mongolica*), persimmon, Manchurian walnut (*Juglans mandshurica*), and white birch.

With the deepening of scientific research and the improvement of cutting technology, some tree species that are difficult to root may become easier to root by cutting. Some tree species do not readily produce adventitious roots by stem cutting, but their roots easily form adventitious buds. Root cutting propagation can be adopted for this kind of tree species.

(2) Growth Characteristics, Age, Branch Age, and Position of Seed Trees

Generally, shrubs are easier to root by cutting than arbors, and the rooting of creeping types are easier than those of upright types. Tree species distributed in high-temperature and wet areas are easier to root than those in low-temperature and dry areas. Cuttings obtained from young trees are easier to root than those obtained from old trees; the sprouting branches at the root collar are easier to root than the 1-year branches at the top of crown; and the cuttings obtained from the main branches are easier to root than those from the lateral branches, especially the lateral branches that branch many times. The rooting ability of cuttings collected from various parts of the same branch differ, and deciding which part is better needs consideration of the rooting type and maturity of the branches. Generally speaking, the middle and upper branches of evergreen trees are better given vigorous metabolism and adequate nutrition, which is beneficial to rooting. The middle and lower branches are better in the hardwood cutting of deciduous tree species, given they have developed fully and have stored more nutrients; whereas the middle and upper branches are better in softwood cuttings, in which the auxin content is higher with vigorous cell division, which is favorable for rooting.

(3) Nutrient Content in Cuttings

The main source of nutrients needed for the formation of new organs and the early growth stage after cutting depends on the nutrients stored in cuttings, especially carbohydrates. Generally, cuttings with a high content of carbohydrates, nitrogen compounds, and amino acids are easy to root, and when the C/N ratio is high, cuttings are easier to root. Therefore, the branches with strong growth, full tissue, and plump leaf buds are easier to root than the fine branches with insufficient nutrients.

(4) Growth and Inhibitor Substance Content

When the content of inhibitor is higher, cuttings are more difficult to root; when the content of endogenous growth hormone is higher, such as heteroauxin and gibberellin, the cuttings are easier to root.

(5) Leaves and Buds of Cuttings

The buds on the cuttings are the basis for the formation of stems and trunks, and buds and leaves can provide the necessary nutrients, growth regulators, and vitamins for rooting of cuttings, especially for conifers, evergreen broad-leaved trees, and all kinds of softwood cuttings. Some leaves should be on the cuttings. The number of leaves remaining on the cuttings depends on the specific circumstances, but generally two to four leaves. If a spray device is available that can regularly and quantitatively moisturize, more leaves can be left on the cutting to speed up rooting.

(6) Diameter of Cuttings

The suitable diameter of cuttings varies with tree species. Most coniferous species have a diameter of 0.3 to 1 cm, and that of broad-leaved species is 0.5 to 2 cm. If the cuttings are too thin, the nutrients stored are too few, making it difficult for the cutting to root and survive. If the cuttings are too thick, the bottom cut will not heal completely for quite some time after cutting, which will affect the growth of seedlings.

(7) Length of Cuttings

Generally, the survival rate of long cuttings is high, and the growth condition of seedlings is also better. The adventitious root sprouting and growing leaves need to consume the nutrients. If the cuttings are too short, the survival rate of cuttings will be low because the lack of nutrients in the cuttings lead to their death. Alternatively, if the cuttings are too long, it requires more labor to collect and make the scion, and it is inconvenient for striking cuttings. The length of cuttings depends on two aspects: One is the speed of rooting of cuttings (tree species characteristics), that is, the cuttings can be appropriately shorter for the tree species with faster rooting speed, otherwise the cuttings should be appropriately longer; the other aspect considers the environmental conditions (wet conditions of the soil), that is, cuttings can be appropriately shorter under wet conditions and should be appropriately longer under dry conditions. On the basis of ensuring survival, the shorter cuttings are better.

(8) Water Content of Cuttings

Too much water loss in cuttings is not conducive to rooting and will reduce the survival rate. After making the cuttings, the water loss of the cuttings should be reduced as much as possible. If necessary, the cuttings can be soaked with water to increase the water content and maintain the water balance of the cuttings.

10.1.4.2 Environmental Factors

External environmental conditions that affect rooting of cuttings include primarily temperature, humidity, ventilation conditions, and light. Organic coordination among various factors can meet the requirements of rooting, improve rooting rate, and cultivate high-quality seedlings.

(1) Temperature

The cutting matrix temperature and air temperature play an important role in the rooting speed of cuttings, and the suitable temperature varies with tree species. Usually, 15 to 25 ℃ is appropriate for most tree species, and about 20 ℃ is the optimum temperature. In spring, 15 to 20 ℃ is suitable for hardwood cutting for common tree species, and above 20 ℃ is required for some tree species such as Rose of Sharon (*Hibiscus syriacus*) and pomegranate (*Punica granatum*). In summer, a temperature of about 25 ℃ is usually suitable for softwood cutting. Different tree species have different ecological habits and suitable temperatures, and in general, the suitable temperature for rooting of tropical plants is higher than that of temperate trees. A matrix temperature that is 3 to 5 ℃ higher than the air temperature is more favorable for rooting.

(2) Humidity

In addition to keeping the soil or cutting matrix with appropriate water content to facilitate the rooting of cuttings, it is necessary to control the relative humidity of the air, especially for the softwood cuttings, for which the relative humidity of the air should be above 90% so as to reduce the transpiration intensity and keep the cuttings from wilting. Water spray and intermittent spray can be used to improve relative humidity of air during production. With the rooting of cuttings, gradually reducing the air humidity and matrix humidity is beneficial to promote root growth and cultivate strong seedlings.

(3) Ventilation

Cuttings keep breathing during the rooting process, and oxygen supply is needed, so the permeability of the cutting matrix has a great influence on the rooting of cuttings. The matrix should not be waterlogged, so as to avoid insufficient oxygen supply and affect the occurrence and growth of adventitious roots in cuttings. Therefore, matrixes such as loose sandy soil, peat soil, vermiculite, and perlite should be selected, which have good permeability and are favorable for rooting and development.

(4) Light

Cutting rootings should be under certain light conditions, especially for softwood cuttings. Sufficient light can promote the production of photosynthetic products in leaves and promote rooting; especially in the late stage of cutting, light conditions are needed after rooting. In the early stage of cutting, however, it is necessary to avoid direct and hard light to prevent excessive evaporation of water, which can result in wilting or burning of leaves and affect rooting and root growth. In production practice, measures such as water spray cooling or appropriate shade can be taken when the sunlight is too bright. The all-optical automatic intermittent spraying can be used with summer cuttings, which ensures not only soil and air humidity but also that the cuttings receive sufficient light, beneficial to growth.

(5) Rooting Medium

Only when the selection of rooting medium can meet the requirements of water and aeration conditions, can it be beneficial for rooting. At present, the commonly used rooting medium can be divided into three types: solid, liquid, and gas. Solid medium is the most commonly used in production, generally including river sand, vermiculite, perlite, quartz sand, furnace ash, moss, peat soil, peanut shells, and so forth, with good ventilation and drainage. The most suitable medium should be selected according to the requirements of tree species. In the open field, sandy loam with good drainage is usually selected. In liquid medium, cuttings are placed into water or a nutrient solution to root and survive, which is called liquid striking or water striking, and this method is commonly used for tree species that are easy to root. When the nutrient solution is used as the medium of water striking, the cuttings easily rot, so diseases should be prevented. The gaseous medium uses air misting and hangs the cuttings in the mist to make them root and survive; this process is also known as mist striking or gas striking. For mist striking, as long as the temperature and relative humidity of the air are well controlled, the space can be fully utilized, the cuttings can root quickly, and the culture period can be shortened. Because the cuttings take root under high-temperature and high-humidity conditions, hardening off will become an important part of cutting survival when using mist striking.

10.1.5 Methods to Promote Rooting of Cuttings

10.1.5.1 Treatment before Scion Collecting

(1) Mechanical Treatment

In the growing season, the base of the branch can be girdled and cut, or tied with iron wire and hemp rope, so as to cut off the path of nutrient transport from the upper part of the branch to the lower part, with the goal of accumulating nutrients in the treatment part. After the injured part of the branch has expanded, the branch will be cut from the base during the dormancy period for cutting. Concentrated storage of nutrients is beneficial to rooting, which not only improves the survival rate but also facilitates seedling growth.

(2) Etiolation Treatment

In general, during the growing season, before making the cuttings, the branches to be used as cuttings are wrapped with black cloth or soil to block the sunlight. Doing this allows them to grow in dark conditions, form tender tissues, and make some changes in nutrients contained in the branches. When the branches and leaves grow to a certain extent, the cuttings are made, and the rooting rate is significantly increased. For some tree species difficult to root and containing pigment, vegetable fat, camphor, rosin, and other inhibiting substances, after the etiolation treatment, the rooting effect of cuttings is significantly improved.

(3) Rejuvenizing and Germination Promotion Treatment

It is an effective measure to promote rooting of cuttings by rejuvenizing the cuttings so the seed tree produces a large number of sprouting branches for collection, but this is a longer pretreatment process that needs to be carried out according to the plan in advance.

10.1.5.2 Treatment before Cutting

(1) Treatment of Growth Regulators

The cuttings can be treated with auxin. The commonly used auxins include NAA, IAA, IBA, and 2,4-D. The bottom end of cuttings can be soaked in a solution made with a low concentration (e.g., 50-200 mg/L) for 6 to 24 hours, and rapid processing (several seconds to 1 min) can be used for a solution made with a high concentration (e.g., 500-10 000 mg/L). The dissolved auxin can also be mixed with talcum powder or charcoal powder evenly, which is dried in the shade and then made into powder. The bottom end of wet cuttings can be dipped into the powder. Alternatively, the powder can be diluted with water to become paste, in which the bottom end of the cuttings can be dipped; or, the powder can be made into mud, and the bottom end of the cuttings can be buried. The treatment time and the concentration of solution vary with the tree species and the type of cuttings. Generally, the concentration needs to be higher for difficult-to-root species, lower for those species that are easy to root; higher for hardwood, lower for softwood.

(2) Rooting Promoter Treatment

The treatment of rooting promoters is one of the measures often used in cutting propagation at present. Rooting promoters include all synthetic substances that can promote rooting, such as endogenous auxin, or compounds that can promote the formation of endogenous auxin and inhibit the decomposition of auxin, as well as substances that can promote adventitious root formation or root system growth by inhibiting the evaporation of leaf water and reducing water loss in leaves. At present, rooting promoters relatively widely used in China include the ABT Rooting Powder series developed by the Chinese Academy of Forestry, broad-spectrum Plant Rooting Agent HL-43 developed by Huazhong Agricultural University, Genbao developed by Shanxi Agricultural University, and 3 A-series Rooting Promoting Powder developed by Kunming Institute of Landscape Architecture, all of which can effectively improve rooting rate of various trees.

(3) Treatment of Reducing Rooting Inhibitors

Generally, elution treatment, such as warm water, running water, and alcohol, is used, which can reduce the inhibitory substances and increase the water content in the branches, and it is the easiest way to promote rooting of cuttings.

Warm Water Eluting Treatment. Soak the bottom portion of cuttings in warm water at a proper temperature. For some gymnosperms, such as pine and spruce, the formation of cut callus often hinders rooting, and rooting is also inhibited due to the pine resin contained in the branches. To eliminate the pine resin, cuttings can be treated with warm water for 2 hours.

Running Water Eluting Treatment. Cuttings are put into running water for several hours, and the specific time differs according to the tree species. Most treatments are within 24 hours; others can reach 72 hours or even longer. It has a positive effect on removing the inhibitory substances in the cuttings of some tree species.

Alcohol Eluting Treatment. This method can effectively reduce the inhibitory substances in the cuttings, generally with a concentration of 1% to 3%, or with a mixture of 1% alcohol and 1% diethyl ether, soaking for approximately 6 hours.

(4) Nutrient Treatment

Treating cuttings with vitamins, sugars and other nitrogen is one of the measures beneficial to rooting, which is often more effective when combined with auxin. For example, pine and cypress can be treated with sugars, and the lower end of the cuttings can be soaked with sucrose solutions of 4% to 5% for 24 hours, with a well cutting effect. Available nutrients also include glucose, fructose, urea, etc.

(5) Chemical Agent Treatment

Some chemical agents can also promote rooting of cuttings, such as acetic acid, phosphoric acid, potassium permanganate, manganese sulfate, magnesium sulfate, etc. The cuttings were soaked in potassium permanganate solution of 0.05% to 0.1% for 12 hours in production, which not only promoted rooting, but also inhibited the development of bacteria and played a role in disinfection.

(6) Low-temperature Storage Treatment

Putting the hardwood into low-temperature condition of 0 to 5 ℃ for a certain period of time can make the inhibiting substances in the branches decompose and transform, which is beneficial to the rooting of cuttings.

10.1.5.3 Treatment after Cutting

Increasing the temperature of the cutting bed is a universal measure to promote rooting of cuttings. Hardwood cutting is often carried out in early spring when the temperature rises rapidly. Cuttings sprout and leaves expand more readily at that time, and they consume the nutrients stored in the cuttings. At the same time, however, transpiration increases and cuttings may experience water loss. In addition, ground temperature is still low at this time, which cannot support rooting demands and may cause the cuttings to die and/or reduce the survival rate. Generally, a ground temperature of 3 to 5 ℃ higher than the air temperature is favorable for the rooting of cuttings. In production, measures such as using heating wire (electric heating hotbed), hot water pipes, or horse manure (brewing hotbed) can be used to improve the ground temperature, combined with the use of a greenhouse, tunnel, or plastic greenhouse, which can not only create the temperature environment suitable for rooting but also prolong the time suitable for cutting propagation and seedling growth, so as to achieve the purpose of promoting rooting and cultivating strong seedlings.

Be aware that reasonable control of temperature, humidity, and light may be necessary. Softwood cutting is primarily carried out in summer and autumn. The branches are tender under high-temperature conditions in summer and can easily lose water and die. Therefore, at this time, the key to cutting seedling survival is to increase the relative humidity of the air, reduce the transpiration intensity of leaves in the cuttings, improve the survival rate of the leaves *in vitro*, and then improve the survival rate of rooting. Cutting management is based on how to maintain a supportive humidity in the cutting environment and ensure the survival of cuttings. Common methods include using a plastic low tunnel, installing spray facilities such as water pipes and spray head on the vault, gas cutting in greenhouse and plastic greenhouse, and all-optical automatic intermittent spray cutting.

10.1.6 Cutting Season

Generally speaking, cutting propagation can be carried out in all seasons of the year, but the practice has proved that cutting at a suitable time has a high survival rate of rooting. Because of different cutting environments, climate conditions, and tree species, cutting methods and cutting time varies greatly. For example, the hardwood cutting of deciduous trees can be carried out in spring and autumn, but mostly in spring, and the survival rate is higher when cutting is carried out as early as possible before the sprouting of buds in spring. In northern, it can be carried out when the soil begins to thaw, generally from late March to late April. Autumn cutting should be carried out before soil freezing; the cuttings are collected and are closely followed by cutting. Autumn cutting is widely used in warm areas of South China. In the dry and cold areas, such as the areas with little snow in winter in North China, the autumn cuttings dry out and freeze, so the cuttings should be covered with soil after cutting, and the soil cover then needs to be removed when sprouting in spring. To solve the difficulties of autumn striking and reduce the overwintering work, such as soil covering, cuttings can be stored until the next spring for cutting, which is relatively safe in terms of cutting survival. At the same time, combined with storage, various treatments, for example, applying auxin, can be carried out to promote rooting. The softwood cutting of deciduous trees is typically carried out in the stable period after the end of the first vigorous growth period in summer. Cutting of the evergreen trees in the south is mostly conducted in the plum rain season.

The development of science and technology—including the use of various traditional and modern facilities and technologies, such as cold frames, tunnels, greenhouses, and the application of all-optical sprays, along with seedbed warming—has made continuous improvement in the success of plant cutting propagation. It has also expanded the cutting season and application scope, greatly improving the efficiency of cutting propagation and supporting various important technical measures.

10.2 Grafting

10.2.1 Grafting Propagation and Its Characteristics

Grafting is a type of propagation method connecting the branches, buds, and other organs or tissues of one plant to the appropriate parts of the branches, stems, and roots of another plant so that they can grow together and form a new plant. Seedlings propagated by grafting are called grafting seedlings. The branch or bud used for grafting is called the scion, and the plant bearing the scion is called the stock. The combination of stock and scion is called stock-scion combination. Grafting propagation is widely used in seedling production and provides many advantages.

(1) Keeping Qualified Characteristics of Female Parent

The part above the interface of the grafted seedling provides the continuous growth and development of the scion and shows the biological characteristics of the parent plant of the scion. Grafting technology can stabilize the high quality characteristics of the scion germplasm.

(2) Early Blossoming and Bearing Fruit

Using the root system of the stock plant, grafting can supply sufficient nutrients for the scion, support vigorous development, accumulate more carbohydrates at the stock-scion interface, and promote the blossoming and bearing of fruit. Because the grafting scions are usually collected from mature plants, compared with the seedlings, the grafting plants blossom and bear fruits earlier. In addition, grafting young scions onto mature stock can promote early blossoming. Therefore, the grafting method is often used in forest tree breeding, which can promote early fruit bearing in the hybrid seedlings, shorten the time of selecting target traits, identify the value of breeding materials in the early stage, and accelerate cultivating new varieties.

(3) Enhancing Stress Resistance and Adaptability

Utilizing the influence of the stock on the scion, and selecting the stock with stronger stress resistance and adaptability, can improve the adaptability of the grafting plants to the environment. Therefore, the characteristics of tree dwarfing, cold resistance, drought resistance, waterlogging resistance, saline-alkaline resistance, and disease and insect resistance of the stock are often used to enhance the adaptability and stress resistance of the scion varieties and to regulate the growth potential, which is beneficial in expanding the cultivation scope and improving the planting mode.

(4) Overcoming the Difficulty of Reproduction

Some tree species have good quality vegetative characteristics, but they have poor fructification ability and significant variation of seed propagation, so they are not suitable for seed propagation purposes. In addition, some tree species or varieties are difficult to propagate by cutting, layering, and division and are seedless or a low-seed variety. For these tree species or varieties, a large number of seedlings can be propagated by grafting.

(5) Cultivating New Varieties

Bud mutation is one of the ways of breeding new varieties of forest trees. Grafting can fix and propagate bud mutation characters in time to cultivate new varieties. Grafting is usually the main technical means for the clonal-expanding propagation and the cultivation of new varieties when a single plant with qualified characteristics is selected from the progeny of seed propagation.

(6) Adjustment and Conservation of Trees

Grafting can be used to select scions of excellent varieties for high branch grafting in trees of inferior varieties, to quickly obtain trees of excellent varieties, and to realize tree improvement, that is, "high grafting for excellence". "Bridge joint" can be carried out to repair branches of trees with diseases, weakness, and damage, so as to rejuvenate the tree body and restore the tree potential. Grafting can also be used to enrich the inner space of hollow trees to make the tree form complete and make full use of space; and for a planting garden made up of a single variety, pollination variety grafting can be adopted to meet the requirements of effective pollination. In addition, grafting technology is often used in the cultivation of special garden seedling types, for example, selecting dwarfing stock and vigorous stock to cultivate modeling trees; and utilizing multiple variety grafting on the homophytic seedlings to achieve a tree with multiple flower colors and multiple fruit types is also widely used.

In recent years, grafting technology has also been used in microbody grafting and virus-free seedling detection. Using grafting technology to study the polarity of plant tissue; the interaction and affinity between stock and scion; the absorption, synthesis, transfer, and distribution of nutrients in the tree body; and the application of endogenous hormones to the tree growth, flowering, and physiological activities of roots are also increasingly in-depth and extensive.

10.2.2 Principle and Influencing Factors of Grafting Survival

10.2.2.1 Healing and Survival Process of Grafting Union

In addition to the stock-scion affinity, whether the two sides of a graft can survive depends on whether the cambium of the stock and the scion can be closely connected, so that both sides produce callus that can heal and be integrated into one body and differentiate to produce new conducting tissues. The faster the callus proliferation, the earlier the stock-scion union heals and the greater the possibility of grafting survival. The grafting union healing process can include the following stages:

　i. The cell inclusion of stock-scion interface oxidizes and protoplasm condenses to form isolation layer.

　ii. The callus formation and proliferation of stock and scion break through the isolation layer.

　iii. Conducting tissue differentiates, and the vascular bundle differentiates in the callus of the stock and scion, which connects the stock and scion.

　iv. The grafting symbiont forms.

10.2.2.2. Factors Affecting Grafting Survival

The many factors influencing grafting survival can be divided into internal and external factors. Internal factors primarily include the affinity and quality of stock and scion, and external factors primarily include grafting technology and external conditions when grafting.

(1) Internal Factors That Affect Graft Survival

Grafting Affinity. Grafting affinity, that is, the affinity between stock and scion, refers to the ability of stock and scion to heal, grow, and develop normally after grafting. Specifically, it refers to the ability of both sides of the graft to connect and grow normally because of the same or similar internal tissue structure, physiological metabolism, and genetic characteristics of both. Correspondingly, the phenomenon of partial or complete healing failure between stock and scion can be caused by various reasons and is called graft incompatibility. Grafting affinity is the key factor and basic condition for grafting survival, the strength of which is a characteristic of plants formed in the process of phylogeny, meaning the genetic relationship, genetic characteristics, tissue structure, physiological and biochemical characteristics, and the influence of virus on stock and scion. It is generally believed that the closer the genetic relationship, the stronger the affinity, and the greater likelihood the grafting will survive.

Grafting affinity between the same species or the same variety is the strongest, which is called "self-specie stock grafting", for example, Chinese chestnut (*Castanea mollissima*) grafting using *Castanea mollissima*, Xinjiang walnut (*Juglans regia*) grafting using *Juglans regia*, yellow peach (*Amygdalus persica*) grafting using *Amygdalus persica*, and so forth. Grafting affinity between different species in the same genus varies with tree species, for example, common apple (*Malus domestica*) is grafted on Asiatic apple (*Malus spectabilis*) or Chinese crab apple (*Malus baccata*); pear is grafted on birchleaf pear (*Pyrus betulaefolia*); persimmon is grafted on date-plum (*Diospyros lotus*); red-leaf peach (*Prunus persica*) is grafted on Chinese wild peach (*Prunus davidiana*), and in these cases the grafting affinity is strong. The affinity between the different genera in the same family is mostly weak, but some species have an affinity for good production performance, for example, *Juglans regia* grafted on Chinese wingnut (*Pterocarya stenoptera*). Distant grafting among plants in different families has not been widely used in production, although the literature records some successful examples. The expression of grafting affinity is complex and diverse. Generally, grafting affinity can be divided into:

i. Good Affinity. The growth of stock and scion is consistent, and the graft union heals well, with normal growth and development.

ii. Poor Affinity. Although grafting can survive, it has various adverse manifestations, for example, the tree body is weak after grafting, or the graft union heals poorly, becoming expanded or nodular; the binding site is uneven in the diameter of upper and lower parts, namely the so-called big foot or small foot.

iii. Short-term Affinity. The graft dies back after it has survived for several years.

iv. Incompatibility. After grafting, no callus is produced and the graft soon dies back. Although affinity is closely related to the genetic relationship, some special cases occur. It is of great importance to study the affinity between plants and to look for the stock with high affinity with production significance.

Quality of Stock and Scion. The process of stock-scion healing requires that both sides store enough nutrients to facilitate the cambium of both sides dividing and healing normally and surviving successfully. Survival after grafting is highly likely if stock and scion are rich in tissue and nutrition. Especially, the quality of the scion (nutrients and water content) plays the more important role. Therefore, the branches with full tissue and buds should be selected as scions. Across the tree species, water content requirements for grafting survival differ, but most of them show that the more water loss of the scion, the less callus formation and the lower grafting survival rate.

Physiological Characteristics of Stock and Scion. Physiological characteristics of stock and scion are the main factors affecting the affinity and grafting success or failure. In grafting propagation, affinity is strong and the graft typically survives when the absorption and consumption of water and nutrients, the active period of cambium, the root pressure and other physiological characteristics of stock and scion are similar. If the root pressure of stock differs from that of scion, for example, or if the root pressure of stock is higher than that of scion, and the physiological activity is normal, the graft can survive; otherwise, it cannot survive. Metabolites of some plants, such as phenols (tannins), resins, and gums, can affect healing of the graft. The root pressure is higher in tree species such as *Juglans regia* and in grapes, and bleeding is more likely to appear in the wound site in the aboveground part after the root system starts to activate. Therefore, grafting trees such as *Juglans regia* or grapes during the spring is prone to having a large amount of bleeding from the grafting union, which in turn suffocates cell respiration at the cut and affects survival. In addition, the gummosis of the wound site in peach and apricot, the resin outflow from the cut in conifer and cypress, and the water-fast tannin compound formed by oxidation of tannin in the cut cells of *Juglans regia* and persimmon can affect the formation of callus and reduce the survival rate.

Polarity of grafting. Callus has obvious polarity, and callus polarity of stock and scion can affect normal growth in the grafting union. Any stock and scion have upper and lower ends in morphology, and callus initially develops in the lower part, which is called vertical polarity. In conventional grafting, the morphological lower end of the scion should be inserted into the morphological upper end of the stock (heteropolar grafting), which is necessary for grafting union healing and survival. For example, in a bridge joint, the polarity of the scion will be inverted. Although it can survive for a period of time, the scion cannot conduct thickening growth. But if the polarity is correct, the grafted scions will thicken normally (Figure 10-4).

(2) External Factors Affecting Grafting survival

Environmental Factors.

i. Temperature. Activities of cambium and callus can be carried out normally only within a certain temperature. The activity level of the meristem of stock and scion is closely related to air temperature and soil temperature. An important condition of ensuring survival is to choose the

Figure 10-4 Thickening situation of the scion polarity in the bridge joint indicating scion reverse polarity grafting in the middle (Hartman and Kestor, 1975).

season with suitable temperatures for grafting. In early spring, the temperature is lower, and the cambium can begin activity, with slow healing; if too late, the temperature rises, and the scion buds germinate, which is adverse to healing and survival. In production practice, the optimum temperature for callus activity of various tree species differs, but most of them are in the range of 20 to 25 ℃, which is related to the natural germination and growth habits of the tree species.

ii. Humidity. Soil moisture and grafting union humidity are important environmental conditions that affect grafting survival. Formation and growth of callus require a certain humidity within the environment. Generally, it takes 3 to 4 weeks after branch grafting and 1 to 2 weeks after bud grafting for the stock and scion to heal. In this period, the key to ensure the survival of grafting is to ensure the humidity of the grafting site. Soil water affects the growth potential of stock and the active state of cambial meristematic cells. When the bark of stock readily detaches and the water content is sufficient, the cambium differentiation ability of both the stock and scion is stronger, with faster healing and combining, and the transfusion tissue of the stock and the scion is most apt to connect. When the soil is dry and short of water, the activity of the cambium of the stock is slow, which affects the survival rate of grafting. The grafting union humidity creates suitable conditions for the union and connection of the stock and scion and promotes the formation of callus. Therefore, it is necessary to use moisturizing materials to bind, bag, or earth to maintain the humidity of the union and promote the survival of grafting.

iii. Light. The presence of light has an obvious inhibitory effect on callus. In dark conditions, more callus is produced, showing tender milky white, and the stock and scion heal well; in light conditions, the less callus is produced and the outer layer is hardened, showing light green or brown, and the stock and scion do not heal well. In production, the use of opaque materials for binding, or ridging, is beneficial to the formation of callus and an increased survival rate of grafting.

iv. Air. Oxygen is one of the necessary conditions for callus growth. In grafting propagation, when using all kinds of binding or materials to maintain moisture, the air demand for graft healing must be considered so as not to cause insufficient oxygen supply or suffocation of callus. Therefore, it is important to coordinate the relationship between humidity and air in grafting propagation.

Grafting Technology. Normal and skilled grafting technology is an important condition for the survival of grafting. When stock and scion have a smooth cut surface and a tight junction of cambium, with quick and accurate operation as well as tight binding of the union, grafting survival rate is high; otherwise, a rough-cut surface, dislocation of cambium, a larger union gap, and loose binding will reduce the survival rate. Attention also should be paid to the polarity of the stock and scion when grafting. The morphologic lower end of the scion is connected with the morphologic upper end of the stock, and the reverse grafting generally cannot survive. Improper binding methods and materials often lead to grafting failure.

In the grafting propagation system, many factors affect the survival of grafting. Many internal and external factors coordinate and restrict one another, which comprehensively affects the grafting results.

10.2.3 Relationship between Stock and Scion

After a graft survives, both the stock and the scion will heal and become a new plant, but they will continue to influence each other in the growth process.

10.2.3.1 Effects of Stock on Scion

After grafting, stock and scion heal and become a new plant, in the growth process, they will continue to affect each other.

(1) Impact on Growing and Bearing

Some stock can promote the growth of the tree body, which is called vigorous stock. For example, *Malus spectabilis* is the vigorous stock of *Malus domestica*, *Pyrus betulaefolia* is the vigorous stock of pear, and *Prunus davidiana* and *Prunus sibirica* are the vigorous stocks of peach. Some other stocks can make the tree body dwarf, which is called dwarfing stock, such as apple dwarfing stock M_9, M_{26}, M_{27} and semi-dwarfing stock M_6, M_7, MM_{106} introduced from abroad. The different growth potential of stock obviously affects the total increment of scion branch. The stock also has an impact on the life span of the tree body, and in general, the lifespan of vigorous stock is long and that of dwarfing stock is short. In addition, the phenological phase of the grafted trees, such as germination stage and abscission period, is significantly affected by the stock.

The stock has some influence on the time of the grafted tree species reaching the fruiting stage, fructescence, fruit quality, yield, and storability. For example, the apple grafted on the dwarfing or semi-dwarfing stock enters the fruiting period early, whereas the apple grafted on the Siebold's crabapple (*Malus sieboldii*) stock enters the fruiting period late, relative to other grafted stock.

(2) Impact on Stress Resistance and Adaptability

The wild or semi-wild tree species are mostly used as the stock, which has a wide range of adaptability, showing different degrees of cold resistance, drought resistance, waterlogging resistance, salt and alkali resistance, and disease and insect resistance. Using these stocks can improve the stress resistance and adaptability of the grafted trees and expand the cultivation area of the trees. For example, peach trees grafted on *Prunus davidiana* have strong cold resistance and drought resistance ability, and peach trees grafted on *Amygdalus persica* have strong moisture resistance ability.

Although many effects of the stock are on the grafted tree species, these effects are generated on the basis that the heritability of the scion has been stable, and the effects pertain to physiological effects and do not involve changes of a genetic basis. So, they will not change the inherent characteristics of the scion germplasm.

10.2.3.2 Effects of Scion on Stock

The growth of root systems associated with grafted trees depends on the organic nutrition produced by the overground part, so the scion will also have a certain impact on the growth of stock. For example, suppose that the Kaido crab apple (*Malus micromalus*) is the stock, grafted with the scions of three apple varieties, 'Summer Pearmain', 'White Winter Pearmain', and 'Ralls', and the fibrous roots of the root system of apple trees formed are significantly different, among which the apple 'Summer Pearmain' has the most fibrous roots. If *Pyrus betulaefolia* is used as the stock of pear and *Pterocarya stenoptera* is used as the stock of *Juglans regia*, the root distribution of the plant is often shallower, with more tillering. In addition, the content of starch, carbohydrate, total nitrogen, albuminous nitrogen, and the activity of catalase in the root system of the stock changed under the influence of scions.

The relationship between stock and scion is complex. From the existing data, many studies have been conducted on the mechanism of effects of stock on scion, which primarily focus on nutrition and transportation, endogenous hormones, anatomical structure, and metabolic relationships.

10.2.4 Selection and Cultivation of Stock and Scion

10.2.4.1 Selection and Cultivation of Stock

The quality of stock is one of the bases of cultivating excellent grafted seedlings. It is very important for the cultivation of grafting seedlings to select and cultivate the stock. Obvious differences are seen in the adaptability of the stock to climate, soil type, and other environmental conditions, as well as the effects on the scion. In general, the following conditions should be met when selecting the stock:

i. Good affinity with scion, good healing abilities, and high survival rates.

ii. Strong adaptability to the environmental conditions in the cultivation area, such as drought resistance, waterlogging resistance, cold resistance, salt and alkali resistance, disease and insect resistance, with a developed root system and strong growth.

iii. Good effect on the growth, fruiting, and ornamental value of scion, such as strong growth, high yield, bearing early, improving quality, and/or long lifespan.

iv. Having abundant material sources or easy to propagate.

v. Having some special required characters, such as dwarfing.

Seedlings have the characteristics of strong stress resistance and adaptability, long life span and easy to propagate, so seed propagation is usually adopted to cultivate the stock. For tree species with insufficient seed sources or that are difficult to propagate by sowing, the method of cutting and sometimes layering and suckering can also be adopted to cultivate the stock. The suitable stock seedling for grafting is 1 to 3 cm in root collar diameter, 1 to 2 years old, and 3 years old for slowly growing trees. If used for top grafting in large size seedling, stock seedlings should be cultivated with a certain height of the trunk. In the process of seedling raising, besides the normal management of fertilizer and water, a method such as pinching can be adopted to control the growth of seedling height, promote the growth of seedling diameter, and make the seedling reach the required diameter as early as possible. In the commercial grafting propagation system, selection and cultivation of stock is related to the grafting quality and effect; therefore, it is necessary to establish a special cultivating nursery for stock and to carry out production management according to standardized requirements.

10.2.4.2. Selection, Cultivation, Collection, and Storage of Scion

Scion is the main object of grafting propagation. To ensure the consistency of grafted seedling quality, scions should be selected from the clonal cutting orchard or cutting orchard. If a cutting orchard is not available, scions should be taken from an identified superior tree. The scion germplasm should have stable, high-quality characteristics with marketing potential, and the seed tree providing the scion should be a mature plant with characteristics of strong growth, high yield, stable yield, high quality, and with no quarantine objects. The scion should be 1-year-old branches in the middle and upper part of the crown periphery, and the scion branches should be strong and full with buds, fully lignified, well-proportioned and smooth, with no diseases or pests.

The collection of scions varies with the grafting methods. Generally, the shoots of the developmental branches in the same year are selected and collected for bud grafting, followed by grafting. To reduce the water loss of branches, young shoots and leaves of the scion collected from the seed tree should be cut off immediately. Petioles should be reserved to protect the axillary buds from damage. Collected scions should be promptly wrapped with wet cloth or other moisturizing materials to prevent water loss. If the collected scions are not used on the same day, they should be soaked in water or wrapped with moisturizing materials and stored in a low-temperature environment for a short time. Storage time is related to the tree species of the scions and the quality of branches. Ideally, scions should be grafted as soon as possible to avoid storage. Scions used for grafting are usually collected from the branches during a period of dormancy in winter and spring, and then grafted before sprouting in spring. Collected branches should be bundled and marked clearly first, then stored by sand-storage method or by wax-sealing method and placed in a 0 to 5 ℃ environment on standby. The main purpose is to protect the moisture and vigor of branches and to avoid water loss.

10.2.5 Grafting Method

Many grafting methods and modes can be used in production. The commonly used grafting methods include bud grafting, branch grafting, and root grafting.

10.2.5.1 Bud Grafting

Bud grafting is a method of cutting buds (called grafted buds) from scions, with little or no xylem; inserting the buds into the cut on the stock; binding them, to bring them close to each other for healing; and nurturing them so they survive and germinate as new plants. The advantage is that the scion can be used economically, and the stock seedlings sown in that year can be used for bud grafting. Moreover, bud grafting is easy to operate and master, with high work efficiency, long grafting periods, firm union, and fast seedling rates; and for the grafted seedlings that did not survive, it is easy to fill the loss because bud grafting can propagate seedlings in large quantities, which is the most commonly used grafting seedling raising method in modern seedling production. The new buds of the annual shoot are primarily used as the scion for bud grafting, which are picked along with the grafting, and the leaves should be cut off immediately for preservation. However, bud grafting with xylem can use the buds of 1-year-old shoots collected in a dormancy period. Bud grafting period is in spring, summer, and autumn, and as long as the cortex is easy to peel off, the stock reaches the diameter required by bud grafting, and the grafted buds develop fully, bud grafting can be carried out. In the North, because of the cold winter, the budding period is mainly from the beginning of July to the beginning of September, and the bud germinates and grows in the second year after survival. If the bud grafting is carried out too early, the grafted buds are apt to germinate in the same year, which makes them susceptible to freezing injury in winter. If bud grafting is carried out too late, the cortex is not easy to peel, and grafting survival rate is low. In East China and Central China, bud grafting is usually carried out from mid-July to mid-September. In South China and Southwest China, the grafting survival rate is higher from August to September for deciduous trees and from June to October for evergreen trees. Specific budding times should be determined according to the characteristics of different tree species and local climate conditions. In recent years, to speed up seedling raising, greenhouse and other facilities have been used to sow early, graft early, and grow seedlings in the same year. The common methods of bud grafting include T-shaped budding, square budding, and plate budding.

(1) T-shaped Budding

T-shaped budding is a commonly used budding method in production (also known as "ding-shaped" budding or shield budding) (Figure 10-5) that is easy to operate and has a high survival rate. Shoots with strong and full buds are selected as scions, of which the leaves should be cut off and the petioles left in place. They are wrapped with wet straw curtains or soaked in water until needed later in the process. The method of cutting bud is to cut transversely at approximately 0.5 cm above the bud, deep into the xylem, and to then cut the bud obliquely upward from 1 to 1.5 cm below the bud, inserting the knife into the xylem and cutting it up to the transverse cut. Cut it into wide upper and narrow lower shield-shaped bud slices, holding the bud by hand and gently moving it to the side, so as to separate the bud slices and shoots and take down the bud slices for later use.

A T-shaped cut should be made (depending on the size of bud slices) at the smooth part of the stock, 2 to 5 cm from the surface, deep into the xylem. During grafting, a knife is used to pry off the intersection of the T-shaped cut, inserting the bud slice into the cut, aligning the upper edge of the bud slice with the T-shaped cut, and then binding it tightly with plastic tape (Figure 10-5).

In production, the T-shaped budding can be changed into a transversely touching budding method, that is, in the center of the transverse cut of the stock, the point of a knife is used to gently touch without cutting. After touching, it will be pulled left and right to pry off the cortex union, then insert the bud slice into the cortex and push slowly. The cortex of the stock will fracture with the pushing of the bud until the upper end of the grafted bud is aligned with

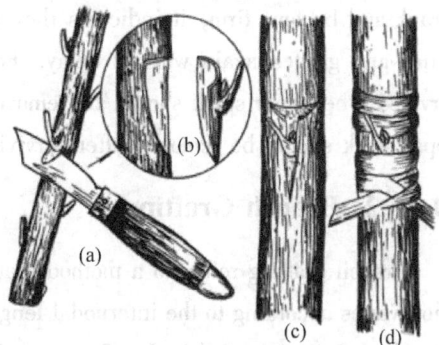

Figure 10-5 T-shaped budding.
(a) cutting the bud slice; (b) removed bud slice; (c) inserting the bud slice; and (d) binding.

the transverse cut. In this way, the bud slice and the xylem of the stock are closely combined, and the cortex of the stock also tightly wraps the grafted bud slice. This method has the advantages of fast grafting with high survival rate. In addition, according to the characteristics of tree species, bud slices can be cut into different shapes, such as the commonly used square-shaped bud slices, and the stock should be cut into "□-shaped" or "⊥-shaped" cuts with the same size bud slices, which is called square budding. It has good application effect in tree species such as *Juglans regia*.

(2) Plate Budding

Plate budding is a method of budding with xylem in the bud slice, which is used for tree species with angular or grooves on the branch tip of scions, such as chestnut, jujube, and citrus, or when the scion and stock are not easy to peel the cortex. The scion branch should be taken reversely when cutting the budder. First, use the knife to cut obliquely downward approximately 1 cm above the bud, then cut obliquely at an angle of 30° below the bud. Push forward to cut to the bottom of the first cut and take off the bud slice, which is the scion bud, at a length of 2 to 3 cm. In a similar way, make a cut on the stock that is slightly longer than that of the bud slice. After inserting the bud slice, one line-width of the stock cortex in the upper end of the bud slice needs to be exposed, and then bind tightly, as shown in Figure 10-6.

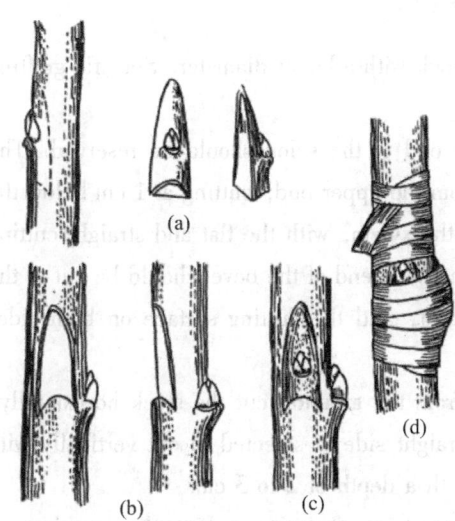

Figure 10-6 Plate budding.
(a) cutting the grafted bud; (b) cutting the stock; (c) inserting the bud slice; and (d) binding.

Approximately 15 days after grafting, if the grafted buds are fresh and the petioles fall when touching, it means that the buds have survived. But if the bud slice shrank and withered, and the petioles shrank and became firm, it indicates they did not survive. Those that failed to survive should be untied and grafted again without delay. For the plants grafted in the previous summer and have survived, the upper stock should be removed the following spring; for those grafted in spring, the upper stock should be removed after surviving.

10.2.5.2. Branch Grafting

Branch grafting refers to a method that uses a branch segment as the scion. The length of the scion varies according to the internodal length of different tree species and varieties; generally, each scion has two to four full buds. To save scions, using a single bud for branch grafting can also be adopted. Compared with budding, the operation technology of branch grafting is more complex and the working efficiency is relatively low. However, the branch grafting method is more advantageous when the stock is relatively thick, the cortex is not easy to be peeled off during the dormancy period of the stock and scion, the top grafting for excellence is carried out for young trees, or the seedlings are used to build an orchard. According to the lignification degree of scions, scion grafting can be divided into hard branch grafting and soft branch grafting. The fully lignified developing branch in the dormancy period is generally used as the scion for hard branch grafting, and the grafting is carried out from the circulation period of sap to before the vigorous growth period. Soft branch grafting is in the growing period, in which the lignified or semi-lignified branches in the middle and end of the growing period are used as scions. The suitable period of scion grafting differs across tree species.

Commonly used methods of branch grafting include cut grafting, cleft grafting, notch grafting, side grafting, whip grafting, inarching, pith-cambium grafting, and close grafting.

(1) Cut Grafting

The cut grafting method is generally applicable to stock with a larger diameter. Specific grafting steps are as follows.

Cutting the Scion. The 2 to 3 buds at the upper end of the scion should be reserved. The upper end of the scion to be cut should be 1 cm away from the upper bud, cutting at 1 cm below the back of the lower bud of the scion to a depth of 1/3 of the xylem, with the flat and straight cutting surface and a length of approximately 2 to 3 cm. Then the back end of the bevel should be cut to the bevel of approximately 0.5 cm (the cut surface is shorter), and the cutting surface on both sides should be smooth.

Cutting the Stock. At approximately 5 cm away from the ground, cut the stock horizontally, and the cutting surface is flattened. The smooth and straight side is selected to cut vertically with the xylem at the diameter of 1/5 to 1/4 of the stock, with a depth of 2 to 3 cm.

Inserting the Scion. The cut scion should be inserted into the cut, making the cambium on both sides of the long cutting surface of the scion and the cambium on both sides of the cut of the stock align and close to each other, until the 2 to 3 mm of the upper end of the large cutting surface is slightly exposed. If the scion is thinner, the cambium on one side must be aligned. The cambium of the stock and scion should not be staggered, otherwise the grafting will not survive.

Binding. A plastic strip is used to tightly bind the grafting place from bottom to top. When binding, special attention should be paid not to move the cut, to keep away from the buds on the scion, and to seal the upper end cut of the scion while tying. If the top of the scion has been sealed with wax, there is no need to bind and cap the upper surface of the scion (Figure 10-7). The moist soil can also be used to bury the whole scion to promote graft survival.

(2) Cleft Grafting

The suitable scion should be selected and approximately 3 cm should be cut down the bottom of the scion to make a wedge-shaped cutting surface on two sides. Generally, the cutting surface should be below the two sides of the upper bud of the scion, so that the wound

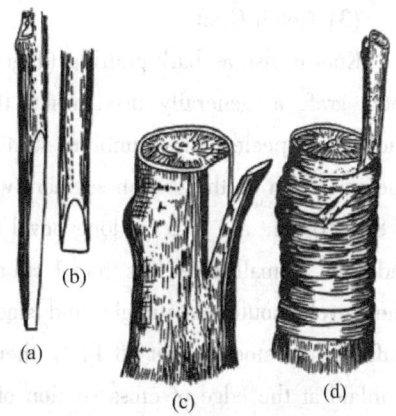

Figure 10-7 Cut grafting.
(a, b) long and short cutting surface of the scion; (c) the cut stock; and (d) binding.

is far away from the bud and so as to reduce the impact on the germination of the bud. If the stock is thicker, the scion should be cut into a flat wedge shape. The side of the upper bud of the scion is thicker and the other side is thinner, so as to facilitate the clamping of the stock. If the diameter of the stock and scion is similar, the scion can be cut into a regular wedge shape, which is beneficial not only to clamping but also to the healing of the stock and scion because of the large contact surface. According to the size of the stock, it should be cut off at an appropriate position with the flat cutting surface to facilitate healing; then cut down vertically in the middle of cutting surface of the stock, with the cleft cut length of 3 to 5 cm. After cleaving the stock, gently pry up the cleft cut with a cleaver and quickly insert the cut scion. If the scion is thinner than the stock, the scion should be close to one side to ensure the cambium on at least one side between the scion and stock is aligned. A thick stock can also be inserted with two scions on both sides, and a robust scion is reserved after sprouting. After jointing, the plastic strip should be used to bind it tightly; or, cover the graft with earth to keep in moisture, so as to avoid cambium staggering of scion and stock (Figure 10-8).

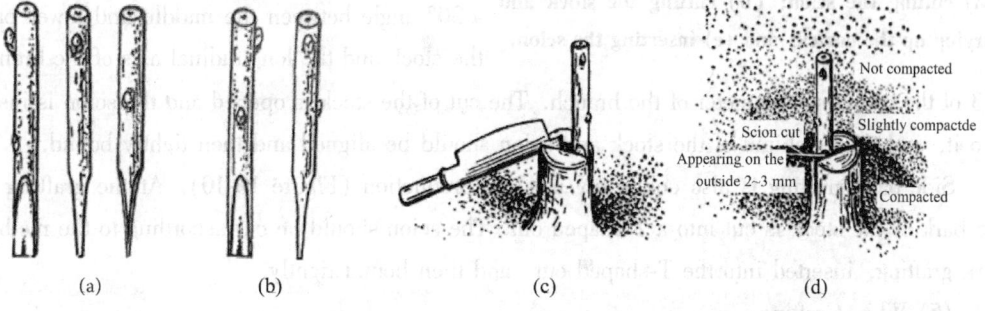

Figure 10-8 Cleft grafting.
(a) side, front, and back of flat-wedge-shaped scion; (b) front and side of regular-wedge-shaped scion; (c) inserting the scion by cleft grafting; and (d) covering with earth.

(3) Notch Graft

Known also as bark grafting (Figure 10-9), a notch graft is generally used when the stock is thicker and peeled (the cambium starts to move). The scion can be the branch segment with a length of 8 to 10 cm, a 3 to 6 cm long bevel at the lower end, and a small bevel less than 1 cm at the back. The bevel should be straight and smooth. When grafting, the stock is cut off first, then the cortex is inlaid at the edge of cross section of the stock, making the long cutting surface of the scion directed at the xylem of the stock and then insert between the cortex and xylem of the stock, with the upper part exposed approximately 2 mm. The scion should be inserted quickly and tightly, and then bound firmly from top to bottom with binding materials. This method is easy to accomplish and survival is high, because the scion is inserted into cortex inside with large contact surface of the cambium of stock and scion. However, the scion is easy to loosen and the callus will only develop on one side with poor firmness, so it needs to be bound tightly and a pillar needs to be set up to support binding during the growth period to avoid wind-breakage. On this basis, the methods of bark whip graft and side bark grafting were developed in production.

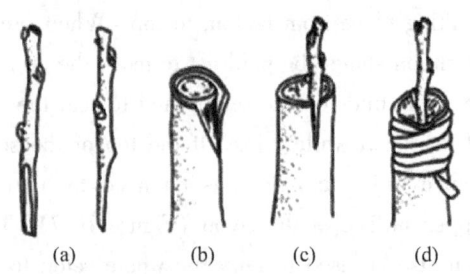

Figure 10-9 Notch graft.
(a) cutting the scion; (b) prying up the cortex; (c) inserting the scion; and (d) binding.

(4) Side Grafting

This type of grafting can be grafted on the branches of the crown without cutting off the crown of the stock, thus called side grafting. The upper branches should be cut off after the survival of the graft. The length of scion is 5 to 8 cm, with two full buds, and two cutting surfaces at the base with one side thick (the same side as the terminal bud) and the other side thin. Cut obliquely along a 30° angle between the middle and lower part of the stock and the longitudinal axis of the branch to 1/3 of the transverse diameter of the branch. The cut of the stock is opened and the scion is inserted into it, and the cambium of the stock and scion should be aligned and then tightly bound.

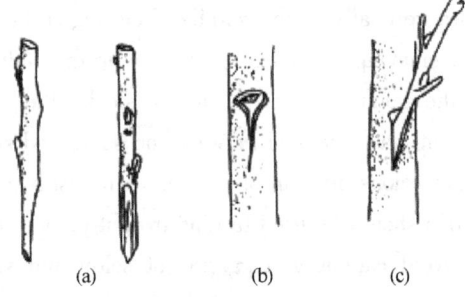

Figure 10-10 Side bark grafting.
(a) cutting the scion; (b) cutting the stock and prying up the cortex; and (c) inserting the scion.

Side bark grafting is also commonly used in production (Figure 10-10). At the grafting site, the bark of the stock is cut into a T-shaped cut. The scion should be cut according to the method of bark grafting, inserted into the T-shaped cut, and then bound tightly.

(5) Whip Grafting

A whip graft is usually used for the tree species that have difficulty surviving scion grafting, and it requires that the diameter of the stock and scion be about the same. The back of the lower bud of the scion is cut into an oblique cutting surface of approximately 3 cm, and then a cut of about 1 cm is made up along the scion at 1/3 of the cutting surface from the lower to the upper to form a ligule. The stock is also cut into at about a 3 cm cutting surface, and a 1 cm long cut is made downward into 1/3 of the cutting surface from the upper to the lower, corresponding to the oblique cutting surface of the scion, so as to cross and clamp with the scion. Then, the cutting surfaces of the stock and scion are inserted into each other to make the ligule sites of both crossing and the cambium aligning, and then tightly bind them together (Figure 10-11).

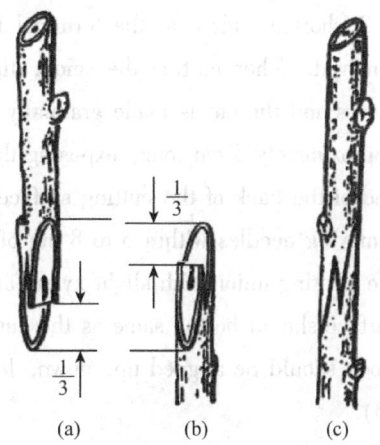

Figure 10-11 Whip graft.
(a) cutting the scion; (b) cutting the stock; and (c) joining the scion and stock.

(6) Inarching

Some trees with poor affinity do not survive well by general grafting methods, in which case an inarching method can be used. Inarching is to make the stock and scion close. Cut a 3 cm long cutting surface on the stock to expose the cambium, and then cut a corresponding surface on the scion to expose the cambium, or cut to the pith. Then bind the two together during the growing season (Figure 10-12). Inarching can also be carried out using whip graft method (see Figure 10-11), which is to cut the corresponding cutting surface on one side of the stock and the scion, cut them into the ligule, and then insert and join each other. More than one month after grafting, the graft is completed by cutting off the original branches of the stock at the upper end of the healing part and the original branches of scion at the lower end. In terms of inarching, the survival rate is high, but it requires that the stock and scion have a root system, which is cut off after healing, which can be a cumbersome operation.

(7) Pith-cambium Grafting

Pith-cambium grafting is used primarily for coniferous trees. Because the contact surface between the pith of scion and the cambium of stock is larger and easier to fit closely, and the pith ray and pith parenchyma cells also play an active role in the healing process, the healing can be accelerated, which improves the survival rate. The optimum grafting period is when the stock buds began to expand in spring, and grafting can also be carried out in summer and autumn when the new shoots of the stock and scion are lignified. The approximately 10 cm long, 1-year-old shoots with terminal buds are taken down from the scion. More than 10 bundles of needles and 2

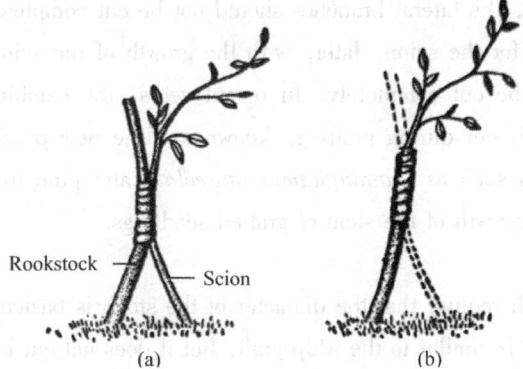

Figure 10-12 Inarching.
(a) binding the stock and scion; and (b) cutting off the upper end of the stock and scion.

to 3 whorled buds near the terminal buds are reserved, and the rest of the needles and buds are removed. When cutting the scion, the knife is inserted at about 1 cm below the reserved needle leaves and the cut is made gradually straight down through the pith center, with a cutting surface approximately 5 cm long, exposing the pith center. Then cut a small, oblique section at the lower end of the back of the cutting surface. The middle trunk tip of the stock is utilized, which means removing needles within 6 to 8 cm of the slightly thicker position than the scion, and then cutting the grafting union with slight xylem and exposing the cambium. The length and width of the cutting surface should be the same as the cutting surface of the scion. The cutting surface of the scion and stock should be aligned up, down, left, and right, and then tightly bind them together (Figure 10-13).

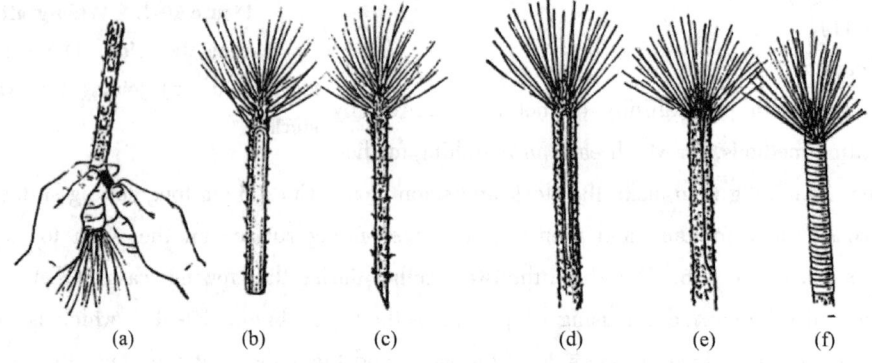

Figure 10-13 Pith-cambium grafting.
(a) Cutting scion; (b) Front side of scion; (c) Side of scion; (d) Cutting stock;
(e) Fit the scion to the rootstock; and (f) Binding.

After grafting, for more than 1 month of healing, the branches of the stock near the top of the grafting union were cut off and replaced by the scion. Meanwhile the vigorous lateral branches of the stock are cut off or cut short, especially the large lateral branches near the scion, to ensure the main position of the scion branches. In the same year, the lateral branches should not be cut completely so as to keep the function of providing nutrition for the scion; later, with the growth of the scion, the lateral branches of the stock will gradually be cut completely. In recent years, the cambium pairing grafting has been improved. The stock is cut during grafting, known as "the new pairing grafting method", which is used for tree species such as *Cunninghamia lanceolata* and pine trees with good effects. It is beneficial to the upright growth of the stem of grafted seedlings.

(8) Close Grafting

Close grafting is suitable for saplings, which require that the diameter of the stock is basically the same as the scion. The close grafting method is similar to the whip graft, but it does not cut into ligule on the cutting surface, and only the oblique cutting surfaces of both are joined together and bound tightly. The specific method is to cut the stock off at approximately 10 cm near the ground and to then cut it into a bevel. The scion with the same diameter as that of the stock is chosen, of which the lower end is cut into a bevel with the same angle and length as that of the stock. The bevels of the scion and the stock are closely joined together, and then the grafting union is bound tightly with a plastic band.

10.2.5.3 Root Grafting

Root grafting is the grafting propagation method with root segments as the stock. It uses a branch of an excellent variety that is difficult to take root as the scion, and the root of a tree species with a close genetic relationship as the stock for grafting. The process is carried out primarily by cleft grafting, cut grafting, or reverse side grafting. Root grafting should be carried out during dormancy, and polarity should not be reversed. If the root segment is thinner than the scion, 1 to 2 root segments can be inserted into the lower part of the scion by reverse side grafting. The grafting union should be bound tightly after the root grafting is completed (Figure 10-14).

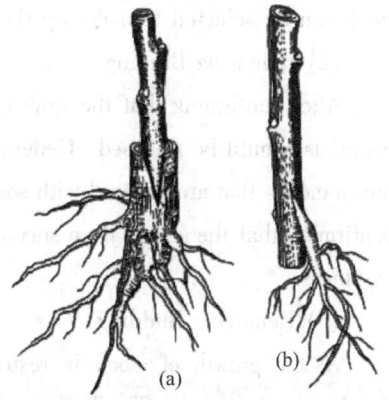

Figure 10-14 Root grafting.
(a) cleft grafting; and (b) reverse side grafting.

10.2.6 Key Links to Improve the Survival Rate of Grafting

For either grafting method, the following key points should be strictly grasped and followed.

i. When grafting, the cambium should be aligned and the cutting surface should be enlarged properly; that is to say, the junction surface of the cambium between scion and stock should be enlarged and the healing area should be enlarged.

ii. The grafting speed should be fast. Whether branch or bud grafting, the longer the cutting surface is exposed to air, the easier it will oxidize and discolor, which will affect the formation of meristem.

iii. The joint of the two should be bound tightly to promote survival.

iv. For the grafted union after grafting, certain temperatures and humidity should be kept to create ideal conditions for union healing. At present, the commonly used methods in production, such as binding with plastic strips, covering with plastic strips, or sealing with wet soil, are all effective measures for heat and moisture preservation.

10.2.7 Management after Grafting

(1) Check Survival

Generally, at 20 to 30 days after branch grafting and 7 to 15 days after budding, the survival rate can be checked. The buds on the grafted scions that have survived are fresh, full, or have germinated. When the graft has survived, the grafted bud is in a fresh state, and the original reserved petiole can be abscised by touch. If the bud turns black and the petiole does not readily fall off, it does not survive. For those that did not survive the branch grafting, one robust branch can be selected from the sprouted branches of the stock for cultivating, to be used for grafting, and the rest can be cut off. For those that did not survive by bud grafting, a suitable position on the stock can be selected to make up the grafting immediately.

(2) Releasing Binding

After confirming that the grafting has survived and the union has healed firmly, the binding materials should be released. Generally, it can be unbound 3 to 4 months after the grafting. As for the branches that are covered with soil to maintain moisture after branch grafting, after checking and confirming that the scions have survived and sprouted, the covered soil should be removed gradually by stages.

(3) Removing Budding

As the growth of stock is restrained after grafting, a large number of sprouts are likely to develop on the stock. The buds and sprouts of stock should be cut off in time according to the circumstances.

(4) Cutting the Stock

After grafting and surviving, for the plants grafted by bud grafting and other grafting methods, the stock branches above the grafted buds should be cut off to ensure the growth of scions, which is called stock cutting. In spring and summer, the stock can be cut immediately after bud grafting and surviving. In summer and autumn, after surviving the bud grafting, the stock will not be cut in the same year so as to prevent the grafted buds from germinating in the same year and having overwintering difficulty. In that case, it will be necessary to cut the stock before germination during the next spring. After the grafted plants by scion grafting with reserved stock crown, such as side bark grafting, survive, the stock should soon be cut, according to the specific situation.

(5) Binding the Shoot

The new shoots grow quickly with tender branches that are easily broken by the wind. Therefore, when the new shoots are approximately 10 cm long, a pillar should be set up to bind the new shoots onto with a rope or plastic belt.

(6) Shearing in the Nursery

The basic cultivation tasks of the tree form of seedlings should be completed in the nursery.

(7) Other Management

Tasks related to disease and pest control, fertilization, irrigation, and drainage of grafted seedlings are similar to those of other seedling raising methods.

10.3 Tissue Culture

Tissue culture is a type of propagation method that inoculates the plant organs, tissues, cells, protoplasts, and other materials *in vitro* on artificial culture media by aseptic technique, which makes them grow and develop into complete plants under the controlled conditions of artificial prediction. The various operations of tissue culture under the aseptic conditions are collectively referred to as following aseptic techniques; the plant materials cut from plants for *in vitro* culture are called explants; the seedlings produced under the aseptic conditions by using tissue culture technology are called tissue culture seedlings, which has become a form of large-scale seedling production.

10.3.1 Principle of Tissue Culture

10.3.1.1 Cell Totipotency

The most important basis for the establishment of tissue culture technology is the theory of plant cell totipotency, which means that every cell of a normal organism contains all the genetic information of the species and has the potential to develop into a complete individual under certain conditions. Theoretically, every living cell of an organism should have totipotency. The highest expression of cell totipotency is the oosperm (zygote), and the formation of buds or roots (which can produce complete plants) in tissue culture is also a typical expression of plant cell totipotency. In a complete plant, some somatic cells of some parts show only certain shape and function, which is attributable to their own genetic material and bound by the specific organ environment. Their genetic potential, however, has not been lost. Once they are separated from the original organ or tissue, they become an *in vitro* state. Under the effect of certain nutrients, hormones, and other external conditions, they may show totipotency and develop into complete plants. The expression of cell totipotency should meet two conditions: one is to free these cells from the inhibitory effects of the rest of the plant, that is, making these cells behave as *in vitro*; the other is to give them necessary stimulation, such as providing certain nutrients or a growth-regulating substance. The realization of this process under artificial conditions is plant tissue culture.

10.3.1.2 Plant Cell Regeneration

(1) Differentiation

Differentiation refers to the process in which cells with the same genetic composition produced during fertilization differ in morphology, structure, chemical composition, and physiological function, and in turn form different types of specialized cells. Put another way, it is a change of cell structure and function or a change of development mode caused by the division of cells. From the point of view of molecular biology, the essence of cell differentiation is differential gene expression, synthesis of specific proteins, and cells producing stable genetic phenotype. Differentiation is relative, and the degree of differentiation varies among multicellular plants. According to the degree of cell specialization and division ability, the cells of a plant body can be divided into three types: the first type, such as stem tip, root tip, and cambium cells, always maintain a strong mitogenetic ability from one cell cycle to another; the second type, highly specialized cells, such as the guard cells of sieve tube, duct, and stomatal apparatus, always lose the mitogenetic ability; the third type, such as epidermal cells and various parenchyma cells, do not usually undergo DNA synthesis and cell division, but they can restart DNA synthesis and cell division after being properly stimulated.

(2) Dedifferentiation

Dedifferentiation is the reversal of differentiation, which means that cells lose their existing differentiation characteristics and return to the relatively undifferentiated meristem. Dedifferentiation is also the process that a series of physiological and biochemical changes, based on the selective expression of genes, makes the differentiated cells become embryogenic cells.

(3) Redifferentiation

Redifferentiation means that the dedifferentiated cells and tissues get the characteristics of various differentiation degrees again. It is generally believed that callus is the product of dedifferentiation. In the practice of tissue culture, it can be seen that some callus has the full potential to produce various types of cells, tissues, organs and even plant bodies of this kind of plant, which is a more thorough dedifferentiation, almost restored to the state of oosperm. However, some of them can differentiate reproductive organs and regenerate vegetative buds under the control of different kinds and concentrations of exogenous hormones, which indicate that this kind of callus or explant cells only partially dedifferentiated and only partially recovered the totipotency of cells. Therefore, the products of redifferentiation can reflect that dedifferentiation is also relative and staged, rather than completed at one time.

Plant regeneration refers to the ability of parts (organs or tissues) separated from the plant body to restore the rest of the plant. When organs or tissues, such as roots, stems, leaves, and so on, separated from plants are cultured under appropriate conditions, it is possible to develop into new complete plants. Dedifferentiated cells have two primary ways to undergo redifferentiation. One is organogenesis mode, that is, the stem, bud, and root are formed independently in different parts of the callus, and they are unipolar structures in which each vascular bundle is connected with the callus, but no common vascular occurs between the adventitious bud and the adventitious root to connect the two. The other is embryogenesis mode, that is to say, a lot of embryoids, or known as somatic embryos, are formed on the surface or inside of the callus, and they are bipolar structures, with common vascular bundles running through the two poles. They can be separated from the callus and can germinate independently on the hormone-free medium to form a complete plant.

10.3.2 Types of Tissue Culture

10.3.2.1 According to the Source of Explants

(1) Plant Culture

Plant culture refers to the method of *in vitro* culture of seedlings or larger plants with complete plant morphology.

(2) Embryo Culture

Embryo culture refers to the method of *in vitro* culture of mature or immature embryos of plants. The commonly used embryo culture materials include immature embryo, mature embryo, endosperm, ovule, and ovary.

(3) Organ Culture

Organ culture refers to the method of culturing various organs and organ primordium of plants *in vitro*. The commonly used organ culture materials include root (root tip, root segment), stem (stem tip, stem segment), leaf (leaf primordium, leaf, cotyledon), flower (petal, stamen, pistil), fruit, and seed etc.

(4) Tissue Culture

Tissue culture refers to the method of *in vitro* culture of tissues in various parts of plants or induced callus. The commonly used tissue culture materials include meristem, cambium, epidermis, cortex, parenchyma cells, medulla, xylem, etc.

(5) Cell Culture

Cell culture refers to the method of *in vitro* culture of an *in vitro* single cell or smaller cell mass and single pollen cell obtained by liquid shake culture of callus that can keep better dispersity. The commonly used cell culture materials include sexual cells, mesophyll cells, root tip cells, phloem cells, and so forth.

(6) Protoplast Culture

Protoplast culture refers to the method of culturing protoplasts without the cell wall *in vitro*.

10.3.2.2 According to Cultivation Process

(1) Primary Culture

Primary culture is the process of culturing an explant separated from a plant for the first few generations. Its purpose is to establish aseptic culture material, induce axillary bud or apical bud germination, or produce adventitious bud, callus, or protocorm, for example. It is usually a difficult stage in plant tissue culture, also known as start-up culture and induction culture.

(2) Subculture

Subculture is the process of redividing the culturematerial induced by primary culture and transferring it to fresh culture medium for further culture. Its purpose is to make the culture material mass propagation, also known as enrichment culture.

(3) Rooting Culture

Rooting culture is the process of inducing rootless plantlets to produce roots and form complete plants. The purpose is to improve the survival rate of tissue culture seedlings after transplanting in the field.

In addition, according to the type of culture medium, it can be divided into solid culture (mostly with agar or Carrageenan, as the support to solidify the culture medium), semi-liquid and semi-solid culture, and liquid culture.

10.3.3 Application of Tissue Culture

(1) Rapid Propagation of Qualified Seedlings

Rapid propagation by tissue culture has the most potential application in production. Rapid propagation technology is not limited by seasons and other conditions, and the growth cycle is short. In rapid propagation, the stem tip and stem segment can be used to propagate axillary buds in quantity; adventitious buds can be produced by direct induction of roots, leaves, and other organs or through callus culture induction; and the number of plants can be rapidly expanded in a short period of time. The obvious characteristics of tissue culture are that it can propagate plants significantly faster and in greater volume than is possible with traditional seedling production. Therefore, use of this technology is of great significance for the propagation of some famous, excellent, and special plant varieties that have a low propagation coefficient and cannot be propagated by seeds. At present, some or most of ornamental plants, horticultural crops, economic trees, vegetative propagation crops, and so forth can adopt *in vitro* rapid propagation to provide seedlings, and the test-tube plantlet has appeared in the international market and formed industrialization.

(2) Virus Removal

Plant detoxification and rapid propagation *in vitro* are the most widely used and effective aspects of plant tissue culture. Almost all plants suffer from varying degrees of virus diseases, and some species even suffer from several kinds of viruses at the same time. Especially many horticultural plants rely on vegetative propagation methods to breed, and if parental generations are infected with the virus, it will be passed down from generation to generation, and the more severe the infection will become. As early as 1934, White found that the virus concentration near the plant growth point was very low, and at some points was even without the virus. Virus-free seedlings can be obtained by micro-stem tip culture, which has become one of the important ways to solve the harm of virus diseases and is often combined with heat treatment to improve the effect of virus-free culture.

(3) New Variety Cultivation

Plant tissue culture technology has opened up a new way for the cultivation of excellent crop varieties. The breeding work can be carried out more effectively under these new conditions, and the technology has allowed for significant progress in many aspects. Consider these examples: Using anther or pollen culture to carry out haploid breeding has the advantages of high speed, high efficiency, and one-time genotype homozygosis; using embryo, ovary, and ovule to culture plant embryo *in vitro* can effectively overcome the obstacle of distant hybridization and obtain hybrid plants; and using protoplasts for somatic hybridization and gene transfer can obtain somatic hybrids and can select and culture cell mutants.

(4) Conservation of Germplasm Resources

Working with a cryoprotectant, such as liquid nitrogen, can effectively reduce the metabolic level of tissue culture material, such as test-tube plantlets and callus, which is beneficial to preserving them for a long time. Using tissue culture to preserve plant germplasm resources has many advantages, such as protecting small volumes, preserving in large quantity, controlling conditions, avoiding reinfection of diseases and pests, and saving manpower and land. It is an economical and effective method of germplasm conservation. Some progress has been made in the research of "artificial seeds". The so-called artificial seed refers to the embryoid as the material; through artificial film packaging of the embryo, in the appropriate conditions, it germinates and grows into seedlings. China has successfully developed artificial seeds of rice. At present, more than 100 types of plants can obtain a large number of embryoids through tissue culture, which provides a basis for the production of artificial seeds.

(5) Plant Bioreactor

Tissue culture was used as a bioreactor to produce secondary metabolic products. Secondary metabolites of plants are important raw materials for many medicines, foods, pigments, pesticides, and chemical products, for which the demand increases year by year. Tissue culture technology provides an effective way for the production of plant secondary metabolites. Because *in vitro* cells of plants still have the ability to synthesize drug components under artificial culture, the content and quality of drug components can be effectively improved by adjusting the culture conditions. Therefore, drug production by tissue culture has become one of the main applications of tissue culture. Using tissue culture technology, a large number of secondary metabolites can be produced to meet the current demand, and wild medicinal plant resources can be protected to minimize growing shortages.

(6) Genetic Transformation

Using tissue culture technology to establish the genetic transformation system of plant materials is the basis of molecular biology research, especially gene function.

Tissue culture has become a conventional experimental technology that is widely used in plant detoxification, rapid propagation, gene engineering, cell engineering, genetic research, production of secondary metabolites, and industrialized seedling production, for example.

10.3.4 Construction of Tissue Culture Room and Culture Conditions

The ideal tissue culture room or large-scale facility that is capable of many scientific research tasks should be in an area that is quiet, clean, far away from congested traffic conditions but with convenient traffic access, and in the upwind direction of the primary annual wind direction in the city to avoid pollution contamination. The overall layout of a tissue culture laboratory should be convenient for isolation, operation, sterilization, and observation. A well-designed workspace includes a preparation room, sterilization room, aseptic processing room, culture room, observation room, auxiliary laboratory, as well as a domestication room and greenhouse. For a small-scale, economical laboratory, the whole process can be completed in an indoor, partitioned off space. A commercial tissue culture room or factory generally requires two to three test rooms with a total area of not less than 60 m^2, which is divided into a preparation room, buffer room, aseptic processing room, and culture room, and if necessary, the test-tube plantlet domestication room and greenhouse need to be added. The tissue culture process is carried out under strictly aseptic conditions, and to achieve the aseptic condition, certain facilities, equipment, and appliances are needed. A standard tissue culture area should include preparation room, weighing room, aseptic processing room, culture room, and domestication room. In practice, some of these areas can be combined within existing conditions.

(1) Preparation Room

This area is used to carry out all the preparations related to an experiment and to complete the storage, weighing, dissolution, preparation, and medium packing of various drugs used. The primary equipment includes a medicine cabinet, dustproof cabinet (in which the culture container is placed), refrigerator, balance, distilling apparatus, acidimeter, and common glass instruments for medium preparation. The area must be spacious and bright, with good ventilation conditions, and the floor should be easy to clean and have some type of anti-skid surface treatment.

(2) Washing and Preparation Room

This area is used for washing, drying, preservation, sterilization of culture medium, and so forth. Necessary equipment includes water tank, operating floor, high-pressure steam sterilization pot, and drying sterilizer (such as drying oven).

(3) Aseptic Processing Room (Inoculation Room)

This area is used for plant material disinfection, inoculation, transfer of culture material, subculture of test-tube plantlets, preparation of protoplasts, and all the technical procedures of aseptic technique that need to be carried out. Equipment includes ultraviolet (UV) light source, clean bench, sterilizer, alcohol lamp, and inoculation equipment such as inoculation tweezers, scissors, scalpel, and needles. The inoculation room should not be too large, depending on the production capacity, generally 7 to 8 m^2 is recommended. The floor, ceiling, and four walls need to be airtight and as smooth as possible so they are easy to clean and disinfect. A sliding door should be configured to reduce air disturbance when opening and closing the door. The aseptic processing room needs to be dry, quiet, clean, and bright, and one to two ultraviolet sterilization lamps should be hoisted at appropriate positions for irradiation and sterilization. It is good idea to install a small air conditioner to control the room temperature, so that the doors and windows can be closed tightly to reduce convection with the outside air.

The inoculation room should have a buffer room with an area of 3 to 5 m^2. Before entering the aseptic processing room, clothes and shoes should be changed here to reduce the level of miscellaneous bacteria being brought into the inoculation room when entering. A UV sterilization lamp needs to be installed in the buffer room for irradiation and sterilization. For maximum effectiveness, the UV light needs to be turned on for at least 20 min before entering the room, and formaldehyde and potassium permanganate should be used to fumigate the indoor space regularly. These general principles are to ensure aseptic and convenient operation and pollution prevention.

(4) Culture Room

This area is where the inoculated materials are cultured and grown. Control of the light, temperature, and humidity is required for optimum outcomes, and to prevent microbial infection, the culture room should be kept dry and clean. The room will also need air interchangers, suitable culture racks, and lighting sources. The size of the culture room can be determined according to the size and number of culture racks needed and other auxiliary equipment. The principle is to make full use of space and to save energy, and the surrounding walls should provide heat insulation and fire protection. The main equipment in the culture room includes the culture racks, shaking table, incubator, and UV light source. The culture materials are placed on the culture racks, which typically have an overall height of 1.7 m. Within the racks are five layers, generally, with the lowest layer approximately 10 cm above the ground, and the other layers with an interval of about 30 cm each. Its length can be designed according to the specification length of the lighting source. For example, if 40 W fluorescent lamps are used, the length is 1.3 m; if 30 W fluorescent lamps are used, the length is 1 m, both with a typical width of 60 cm.

The most important factor of the culture room is the temperature, which is generally kept at about 20 to 27 ℃ with a heat production device and an air conditioner. Because the temperature requirements differ across plant types, it is better to have different culture rooms to accommodate the various temperatures. Indoor relative humidity should be 70% to 80%, which can be maintained by installing a humidifier. A timing switch can be installed to control the lighting. Generally, the room needs 10 to 16 hours of lighting every day, and some need continuous lighting. In the design of a modern tissue culture laboratory, sunlight is usually the preferred energy, and on this basis, necessary light sources are added to save energy and improve the extent of growth and domestication of tissue culture plantlets. Natural sunlight can be supplemented by lamplight on rainy days.

Figure 10-15 Tissue culture seedling facilities.
(a) preparation room; (b) aseptic inoculation room; (c) culture room; and (d) domestication and transplanting greenhouse.

(5) Auxiliary Laboratory

Cytology Laboratory. This room is used for observation and analysis as well as for the counting of culture materials. The main equipment includes binocular stereomicroscope, microscope, and an inverted microscope. Small instruments and equipment include the magnetic stirrer, low-speed table centrifuge, dispenser, blood cell counter, pipette, filter sterilizer, and electric furnace and other heating appliances.

Biochemical Analysis Room. In a laboratory with the main purpose of culturing cell products, the corresponding analysis laboratory should facilitate the sampling and inspection of the culture material components at any time.

Domestication and Transplanting Room. This area comprises the transplanting greenhouse and/or greenhouse of test-tube plantlets.

10.3.5 Type and Preparation of Culture Medium

10.3.5.1 Composition of Culture Medium

A culture medium is required that will provide all of the nutrients needed for the growth and differentiation of culture materials. A perfect culture medium should at least include inorganicnutrients (including major elements and microelements), organic nutrients (such as vitamins, amino acids, and sugars), growth regulating substances (various plant hormones), and a curing agent (generally agar) within a solid culture.

(1) Inorganic Nutrients

Inorganic nutrients refer to the mineral elements that are very important in plant growth and development. *In vitro*, the nutrient elements needed for plant growth and development include major elements and microelements, just like the plants growing naturally in the field. The way to provide them is included in the culture medium. According to the recommendations of the International Society of Plant Physiology, the major elements are those with a concentration required by plants that is greater than 0.5 mmol/L, and the microelements are those provided at concentrations less than that. Iron is often provided in the form of a mixture of $FeSO_4$ and Na_2-EDTA to avoid precipitation, that is, having the insoluble substances in the solution sink to the bottom layer *in vitro*. A medium that contains only major elements and microelements is often called the minimal medium, and the differences in its formulation lay the foundation for other types of medium.

(2) Organic Nutrients

To ensure good growth and development of the culture materials, it is often necessary to add some organics to the minimal medium. Many kinds of organic nutrients can be in the culture medium, such as carbohydrates, amino acids, vitamins, and organic additives.

Carbohydrates. Carbohydrates are indispensable in plant tissue culture. They act as a carbon source for growth of tissue *in vitro* and supplement the culture medium so it can maintain a certain osmotic pressure. Sucrose is the most commonly used (1%-5%), sometimes glucose and fructose are used.

Vitamins. Vitamins can obviously promote the growth of tissue *in vitro*. The vitamins in the culture medium are primarily B vitamins. Generally speaking, thiamine (VB_1) is an essential component, and other vitamins such as pyridoxine (VB_6), nicotinic acid (VB_3), calcium pantothenate (VB_5), and inositol can also significantly improve the growth of plant tissue. Ascorbic acid (VC) has strong reducing ability and is often used to prevent tissue from oxidation and browning.

Amino Acids. Glycine is the most commonly used amino acid in culture medium, followed by arginine, glutamic acid, glutamine, and alanine, which are high quality organic nitrogen sources.

In addition, inositol, adenine, and other complex nutrients with unknown chemical components, such as protolysate, coconut milk, corn endosperm, tomato juice, and yeast extract, are also used as organic components of culture medium.

(3) Growth Regulators

Growth regulators play an irreplaceable role in the differentiation, growth, and development of tissues or organs in tissue culture. The type and quantity of plant growth regulators added to the culture medium are often the key to the success of plant tissue culture. Common plant growth regulators include the following.

Auxins. Auxin affects the elongation of plant stems and internodes, orientation, apical dominance, leaf abscission, and rooting. It is used for inducing cell division and root differentiation *in vitro*. IAA, IBA, NAA, NOA, and 2,4-D are commonly used auxins. Among them, IBA and NAA are widely used for rooting and can cooperate with cytokinin to promote stem proliferation. 2,4-D is very effective for callus induction and growth.

Cytokinins. The main function of cytokinins is to promote cell division and induce bud differentiation. It is helpful for eliminating the inhibition of apical dominance on axillary buds and can be used for stem proliferation. The commonly used cytokinins include benzyladenine (BA), isopentennyladenine (2-iP), kinetin (KT), and zeatin (ZT).

Gibberellins. There are many kinds of gibberellins, of which GA_3 is commonly used in tissue culture, and its main physiological function is to promote the growth, especially the elongation of the stem. It can also replace the low temperature requirement that releases the dormancy of relevant tissues or organs.

Abscisic Acids. Abscisic acids have an indirect inhibitory effect on the culture materials in tissue culture. They can inhibit the formation of cell embryoids in explants and are often used in plant embryoid culture.

In addition, other growth regulators such as folic acid, salicylic acid, polyamine, and fulvic acid, for example, are also used in tissue culture, and good effects are achieved in certain plants.

(4) Supporter

Agar is the most commonly used support (curing agent) in tissue culture, at a concentration of 0.7% to 1.0% generally, depending on the medium hardness and climate. Often 0.8% is used in summer and 0.7% in winter. Agar is not an essential component of culture medium. When agar or other solid agents are not added, the culture medium is liquid, and is called liquid culture. It would be used primarily for callus culture and embryoid culture.

(5) Activated Carbon

Activated carbon is added to culture medium, mainly for its adsorption capacity to reduce the impact of some harmful substances, such as preventing phenolic substance pollution from causing tissue browning and death. In addition, activated carbon can make the medium black, which is beneficial to rooting of some plants. However, activated carbon is non-selective in the adsorption of substances, so it should be used with caution.

10.3.5.2 Culture Medium and Preparation Method

(1) Selection of Medium Components

Culture medium can be divided into solid medium and fluid medium according to its phase state: induction medium, proliferation medium, and rooting medium according to functional components; minimal medium and complete medium according to nutritional level.

The key link of tissue culture is to carefully compare the characteristics of various formulations and combine the species of culture plants and explants with the culture goals to determine the appropriate formulation of culture medium. Medium formulations have their own characteristics. Before selecting, carefully consult relevant information or conduct testing. For example, some mediums have low content of mineral elements, such as the White culture medium commonly used for woody plant culture; some are rich in mineral elements, such as Murashige and Skoog (MS) culture medium with rich ammonium salt; while B_5 and N_6 culture medium contain rich nitrate. Nitsch medium is most widely used in angiosperms; B_5 medium was originally designed for legume culture and is now widely increasingly used; woody plant medium (WPM) is suitable for woody plants such as *Tilia* and *Rhododendron*; N_6 medium is more suitable for Gramineae (the grass family). Some mediums are rich in vitamins, such as B_5 medium. In contrast to B_5 medium, Nitsch medium does not contain any organic additives or vitamins.

(2) Preparation of Medium

For culture medium preparation, the appropriate formula should be chosen and a series of concentrated stock solutions should be made up (known as the mother liquor) according to the formulation requirements, such as major elements, microelements, ferric salt, and organic substances other than sugar. Then, the various quantitative concentrated solutions are measured, diluted, and mixed. After adding the sucrose and agar, heat and dissolve all components, and then, determine the constant volume, adjust the pH value, and add to the culture vessels. Finally, sterilize the medium with an autoclave and store the vessels after cooling.

The amount of plant growth regulators used in culture medium is very little, but with huge effect. Because of the properties of various growth regulators, their dissolution and preparation methods differ. In this chapter, several commonly used plant hormones are briefly introduced.

Auxins. IAA is decomposed by light and should be stored in a brown glass bottle and placed in a refrigerator. It cannot tolerate high temperature and high pressure, and it is usually filtered by a microfiltration membrane and added to culture medium. IBA, 2,4-D, NAA are very stable and can be sterilized at high temperatures. This type of substance can be dissolved with a small amount of NaOH of 1 mol/L, and then diluted to the required concentration with distilled water. It can also be dissolved in a small amount of 95% ethanol and then diluted with water. When diluting, a dropper is used to drop the ethanol solution containing auxin into the distilled water slowly, stirring it while dropping, otherwise auxin is easy to separate out. It can be heated slightly when the dissolution is incomplete.

Cytokinins. ZT and its analogues 2-iP are natural products. BA and KT are synthetic substances, which are stable under high temperature and high pressure at 120 ℃; ZT is generally sterilized through filtering. KT decomposes in light and should be stored in a dark place at 4 to 5 ℃. Such substances should be dissolved with a small amount of 0.5 or 1 mol/L HCl and then diluted with water.

Gibberellin. GA_3 cannot be sterilized at high temperatures because of the instability when heating. It dissolves in alcohols and decomposes rapidly in water. When using GA_3, 95% ethanol needs to be used to prepare the mother liquor and it should then be stored in the refrigerator.

Plant growth regulators are generally expensive, and they easily oxidize and deteriorate, so they should not be prepared in small quantities each time. The mother liquor should be placed in a brown reagent bottle and kept in a low-temperature and dark place to avoid deterioration and decomposition.

(3) pH Value of Medium

Because the pH value of the medium directly affects the ion absorption of the explant, and then affects the differentiation and development of the explant, the adjustment of the pH value is a necessary link in the process of culture medium preparation. Generally, 1 mol/L HCl and NaOH are used to adjust the pH value. Generally, the acidity of the medium will increase by about 0.2 after autoclaved sterilization, which should be considered when adjusting the pH value. The pH value of the medium also affects the solidification degree of agar. Sometimes the medium does not solidify after high-temperature sterilization and cooling, which may be attributable to insufficient stirring when adjusting the pH value and too-low pH value, or because the high temperature treatment time is too long. When the pH value of the medium is required to be faintly acid, some agar powder can be added appropriately.

(4) Medium Sterilization

The medium can besterilized by moist heat sterilization. When the volume of the bottled solution is less than 1000 mL, the sterilization effect can be achieved under the condition of 1.1 kg/cm^2 and 120 ℃ for 15 to 20 min. After more than 20 min, high temperature will increase the decomposition of some organic matters (sugars and vitamins, for example), which is harmful to the culture. IAA, GA_3, etc. should be filtered and sterilized by a microfiltration membrane and added into the medium before the culture medium temperature is reduced but not solidified. Shake well.

(5) Medium Storage

The medium should be stored in the shade. If IAA and KT are included, it needs to be placed in the dark. After the newly prepared medium solidifies, there will be water on the surface, and inoculating should be carried out preferably after 2 to 3 days, when the water on the surface is absorbed. If inoculating is in urgent need, sterile pipettes need to be used to suck up the flowing water on the surface to prevent the pollution source from spreading with the water flow. After sterilization, the medium should be used within 15 days, definitely not more than 30 days.

10.3.6 Preparation of Explants

10.3.6.1 Acquisition of Explants

The commonly used materials for tissue culture include bulbs, corms, stem segments, stem tips, flower stalks, petals, petioles, leaf tips, and leaves, of which the physiological state has a great influence on organ differentiation during the tissue culture process. Generally speaking, young seedlings are easier to differentiate than old trees, terminal buds are easier to differentiate than axillary buds, and sprouting buds are easier to differentiate than dormant buds. In addition, immature seeds, ovary, ovule, and mature seeds can be used as materials. Plant material was washed with tap water several times, and after removing the dirt on the surface, it was transferred to a clean beaker.

10.3.6.2 Disinfection of Explants

Disinfection and sterilization kill the microorganisms on the surface of the explant, but at the same time, try not to harm the living cells in the plant materials. Commonly used disinfectants include ethanol, sodium hypochlorite, and mercury bichloride. Type and treatment time of disinfectants vary with plant species, organ types, and growth conditions. Generally, 70% to 75% ethanol can be poured into the beaker, and the washed explant material can be put into the beaker and immersed for less than 1 min; then, the ethanol can be poured out, and the 0.5% sodium hypochlorite solution can replace it for disinfection for 15 to 20 min (the explant material and the disinfectant solution should be transferred to the aseptic beaker and put into the clean bench for operation). Later, the disinfectant solution is poured out and the material should be washed with aseptic water three times, each time for 2 min. For materials with serious pollution, 0.1% (m/v) mercury bichloride (which is highly toxic should be; use as little as possible) can be used for disinfection. Sodium hypochlorite can also be used for disinfection for two consecutive times, or mercury bichloride disinfection can be used after one time of sodium hypochlorite disinfection. After sterilization with mercury bichloride, wash the explant material with aseptic water many times.

10.3.7 Primary Culture Inoculation

10.3.7.1 Explant Inoculation Process

Inoculation needs to be carried out in aseptic conditions, and the ultraviolet lamp in the inoculation room or on the clean bench should be turned on for 30 min at 1 hours before the inoculation. Strong ultraviolet light will hurt people's eyes and skin, and ozone (O_3) produced by oxygen irradiation will hurt the respiratory system, so people must leave after turning on the light. After the 1 hours wait, turn off the UV lamp, and turn on the pump on the clean bench to blow air for 15 min, removing the ozone generated by the UV lamp irradiation and making the clean bench surrounded by sterile air. Before inoculation, workers should wash hands with soap first, then wipe hands with a cotton ball soaked in 70% to 75% ethanol. Wipe the surface of the clean bench with gauze soaked in 0.1% benzalkonium bromide solution, then wipe with absorbent cotton soaked in 70% to 75% ethanol. Then put all sterilizing apparatus on the side of the air passage that does not affect the aseptic filtration and prepare for the aseptic operation.

The explant material that has been sterilized and washed withsterile water should be put in the sterile vessel with filter paper, and a little sterile water should be added to wet the filter paper. Using sterile tweezers, take out the sterilized and washed (with sterile water) plant materials and put it in the inoculation room or on the clean bench, into the aseptic culture dish padded with filter paper. Burn the scalpel on the alcohol lamp flame and cool it, then cut off the part soaked by disinfectant near the end cut. Cut the plant material into small pieces according to the needs of the project and transfer it to the culture dish or conical flask containing the culture medium. These small pieces used as the culture materials are the explants. Use the sealing film to seal the culture dish or conical flask and place it on the culture rack.

The inoculation should be strictly protected from pollution, and attention should be paid to the blowing direction of the sterile air. Non-sterile vessels should not be stored at the air inlet; the operator should wear a mask to prevent the microorganisms discharged from their respiratory tract from entering the culture dish. The surface of the culture medium should be free of water to prevent the microorganisms from diffusing with the water flow; after cutting or transferring several explant materials several times, the scalpel or tweezers should be burned to again sterilize them.

10.3.7.2 Precautions

With tissue culture procedures, the movement should be accurate and agile, but not too fast, so as to prevent air flow and increase pollution opportunities.

Do not touch the sterilized vessels with hands. If it has been touched, it should be sterilized with flame or replaced with backup sterilized equipment.

For the convenience of workflow, the articles on the workbench should have a reasonable layout. In principle, the right-hand articles should be placed on the right side, the left-hand articles on the left side, and the alcohol lamp in the center.

Maintain a certain sequence throughout the work; don't expose tissues or cells to the air too early before handling. Similarly, don't open the bottle too early before using the culture solution; after use, the bottle should be immediately closed.

When sucking up nutrient solution, cell suspensions, and other solutions, pipettes should be used separately and should not be mixed so as to prevent pollution or cross contamination of cells.

Do not speak or cough facing the operation area during the tissue culture work process so as to avoid the pollution caused by the bacteria or mycoplasma brought onto the workbench surface by saliva.

The hand or relatively dirty items cannot pass over the open bottle, which is the most likely item to be polluted. If the tip of the pipette touches the bottleneck when adding the solution, the pipette should be discarded.

10.3.7.3 Culture Conditions

Temperature, illumination, oxygen, and water are the main factors affecting the development of plants *in vitro*; those elements need to be regulated by close human attention.

(1) Temperature

The most suitable culture temperature for most plant tissues is 25 to 28 ℃. Generally, the temperature in a culture room is 25 ℃ ± 2 ℃, and temperatures higher than 35 ℃, lower than 15 ℃, are harmful to growth.

(2) Illumination

For the growth of *in vitro* culture materials, illumination plays an important role in morphogenesis induction; cell, tissue, and organ growth; and differentiation. Generally, dark conditions are beneficial to callus induction. Callus cultured in light conditions has texture and color that differs from callus cultured in the dark conditions; however, the differentiation organ needs light, and the light needs to be strengthened with the growth of buds. In general, 12 to 16 hours of illumination per day and 1000 to 3000 lx of light intensity are required.

(3) Humidity

Two aspects of humidity cultivation are particularly important in tissue culture: One is the humidity in the culture container, which can generally guarantee 100% humidity; the other is the humidity in the culture room, which changes with seasons and weather. Too high humidity will cause the growth of infectious microbes, resulting in a large amount of pollution; too low humidity will cause the loss of water and drying of the culture medium. Or, too low humidity can increase osmotic pressure, which will affect the growth and differentiation of the culture materials. Generally, the humidity of the culture room should be kept at 70% to 80% relative humidity.

(4) Oxygen

In plant tissue culture, oxygen is needed for explant respiration. In general, the air above the culture dish or conical flask is enough for culture material respiration in solid culture; in deep liquid culture, ventilation can be solved by vibration or rotation. The temperature of the culture room should be uniform, so proper indoor air circulation should be required.

Pollution in tissue culture can be caused by three reasons: It was brought in by the explants with incomplete sterilization; brought in by the surrounding air or inoculation tools during inoculation; and/or brought into the culture bottle by the edge of the sealing film after inoculation. After inoculation, observations should be made frequently, and the key period to detect pollution is within 7 days after inoculation. Pollution can be caused by bacteria or fungi. If pollution is found, promptly remove the culture bottle.

10.3.8 Enrichment Culture

Sterilebuds were induced from explants by primary culture. To meet the needs of large-scale production, a large number of sterile buds have to be cultured by continuous enrichment tissue culture. The type of proliferation medium varies with the plant species, varieties, and types, which can be the same as the primary culture medium but can also gradually have adjustments made in the concentration of cytokinin, the proportion of inorganic nutrients according to the possible situation, or have active carbon added to prevent vitrification or browning.

The optimal size and cutting method of the propagules vary with species. Generally speaking, the explants have to reach a certain size in order to obtain uniform and rapid proliferation in subsequent subculture. The elongated stem segments are cut from the original culture for subculture, and there should be 2 to 4 nodes for the subculture explant length. Remove the leaves and insert them vertically into the medium or horizontally on the surface of the medium. Proliferation can be repeated several times to increase sufficient materials and to lay a foundation for rooting and transplantation. However, if too many subcultures are made, the viability of the propagules will decrease, the necrosis rate will increase, and the rooting ability will reduce.

10.3.9 Rooting Induction and Culture

After the stem segment proliferates to a certain amount, transfer it into rooting induction culture in time to allow it to root and to then obtain complete plants for transplantation. Compared with enrichment culture and primary culture, rooting medium has the following characteristics:

(1) Low Concentration of Inorganic Salt

It is generally believed that a higher concentration of mineral elements is beneficial to the development of stems and leaves, and a lower concentration is beneficial to rooting; therefore, 1/2 MS or 1/4 MS medium is often used.

(2) Few or No Cytokinins

In the rooting medium, cytokinins should be removed or only a small amount of cytokinins should be used, and auxin, such as NAA, IBA, IAA, and so forth. should be added appropriately.

(3) Low Sugar Concentration

In the rooting stage, the sugar concentration in the medium should be reduced to 1.0% or 1.5% to enhance the autotrophic ability of plants and to facilitate the formation and growth of intact plants.

In the aspect of environmental control, the light intensity should be strengthened in the rooting stage. It is generally believed that under strong light, plants can grow well and their resistance to water stress is enhanced. The rooting time varies with plant species. Generally, after 2 to 4 weeks of rooting culture, the stem segment can produce roots and become a transplantable complete plant.

10.3.10 Domestication and Transplanting

After rooting, the tissue culture seedlings have to be domesticated and transplanted to the normal greenhouse environment, and to finally grow in the natural environment. For *in vitro* culture, tissue culture seedlings grow and develop in the environment of sterility, supplied nutrition, suitable temperature and illumination, and higher relative humidity. After they are removed from the culture bottle, the growth environment changes greatly. Therefore, it is necessary to meet the needs of tissue culture seedling growth from the aspects of water, temperature, light, culture medium, and management measures.

To adapt to the lower humidity and higher light intensity after transplanting and to complete the transformation from heterotrophic to autotrophic, the test-tube plantlet should be properly hardened before transplanting to make the plants strong and strengthen the adaptability to the external environment, so as to improve the survival rate of transplants. This process is usually realized through the transition of hardening and domestication of tissue culture seedlings.

In general, the domestication in the bottle gradually changes to that out of the bottle. The culture bottle can be placed under a high light condition, the seal can be gradually opened for ventilation, and the seedlings can be domesticated in the bottle for a few days. After being removed from the culture room and then planted to the seedling container or seedbed, the seedlings still need to undergo a period of domestication in terms of moisture and shading. The domestication process should follow the principle of gradual transition. Generally speaking, in the first few days, the domestication conditions should be similar to the environment conditions of *in vitro* culture, whereas in the later period of domestication, the conditions should be gradually transferred to be the same as the natural cultivation conditions. In the process of transplanting, take note of the following problems.

(1) Maintain Water Balance of Tissue Culture Transplant Seedlings

Water balance is the first problem to be solved in domestication transplantation. The cuticula on the surface of stems and leaves of seedlings growing in high-humidity environment in test tubes or culture bottles may be underdeveloped, and the root system is not very developed, which makes it important to maintain their water balance when they are transplanted to the outside of the bottle. Therefore, through spraying, sprinkling water, and/or a covering plastic film or glass cover, the air humidity in the seedling growth environment can be increased, transpiration can be reduced, and the water balance of seedlings can be maintained.

(2) Reasonable Selection of Cultivation Matrix

The culture medium should have loose ventilation, good water retention, and be easy to sterilize; vermiculite, perlite, river sand, and peat are the most commonly used mediums. In practical use, matrixes are often mixed according to a certain proportion, which can only be used after sterilization and disinfection.

(3) Prevention of Breeding Germs

When transplanting, seedlings should be treated with a certain concentration of fungicides (such as chlorothalonil or carbendazim), and the culture medium should be cleaned when seedlings are out of the bottle. Seedlings should be protected from being damaged as much as possible, so as to reduce and prevent pollution and effectively improve the survival rate.

(4) Suitable Light and Temperature Conditions

The adaptability of seedlings is poor, so the best light is diffuse light, which is regulated as the seedlings grow. The habits of plants toward light loving or shade tolerant generally range between 1500 and 4000 lx, or at times even 10 000 lx. In the same way, the use of greenhouse facilities, combined with the appropriate transplant season and appropriate temperature conditions, is an important factor to promote strong growth of seedlings. For most plants, a temperature of approximately 25 °C is suitable. Temperature control also involves water balance, pathogen breeding, and other issues, which should be in combination with other measures that are considered overall.

Keep in mind that water absorption by the root system and self-regulation of tissue culture seedlings is weak after transplanting. Therefore, it is necessary to strengthen management; maintain appropriate temperature, humidity, and light conditions; and strictly control the spread of pathogens, so that the seedlings can have the highest possible survival rate.

Questions for Review

1. What are the main types, characteristics, and application fields of vegetative propagation? Briefly describe.
2. What are cutting seedlings? Briefly describe the characteristics of cutting propagation seedlings.
3. What is the polarity of cuttings and its importance in cuttingpropagation?
4. What are the types and characteristics of cutting propagation?
5. What are the internal factors affecting the cutting propagation?
6. What are the environmental factors affecting the cutting propagation?
7. What are the key problems to be considered in the selection and making of cuttings?
8. What are the characteristics and application of hardwood cuttings and softwood cuttings?
9. What are the methods and implementation basis for promoting the survival of cuttings?
10. How would you improve the survival rate of cuttings for a tree species difficult to root?
11. What are grafting seedlings? Briefly describe the characteristics of grafting propagation.
12. What are the main methods and operation points of grafting propagation?
13. What is the process of grafting survival?
14. What is the value of grafting propagation in seedling production?
15. What are the factors that affect the survival of grafting?
16. What is the importance of selection and cultivation of stock and scion to grafting propagation? What should be paid attention to in this process?
17. What are the characteristics and key links of tissue culture?
18. What is the theoretical basis of plant tissue culture?
19. What are the conditions for plant cells to show totipotency?
20. What is the effect of explant selection on tissue culture?
21. In the process of plant tissue culture, why should a series of disinfection and sterilization be carried out and sterile operation be required?
22. What are the key technology and physiological bases for the success of tissue culture seedling?
23. How would you make comprehensive use of seed propagation and vegetative propagation to achieve efficient breeding of some precious tree seedlings?

References and Additional Readings

CHEN H Y, LIU L W, 2011. Seed and seedling science[M]. Shanghai: Shanghai Jiao Tong University Press.

CHENG F Y, 2012. Nursery of landscape plants[M]. 2nd ed. Beijing: China Forestry Publishing House.

DUMROESE R K, 2009. Nursery manual for native plants: a guide for tribal nurseries[M]. Washington, DC: US Department of Agriculture, Forest Service.

GONG X, GENG L Y, LIU Z L, 1995. Nursery of landscape plants[M]. Beijing: China Architecture & Building Press.

LIU X D, HAN Y Z, 2011. Nursery of landscape plants[M]. Beijing: China Forestry Publishing House.

LIU Z L, SHI A P, LIU J B, 2001. Nursery of landscape plants[M]. Beijing: China Meteorological Press.

SHEN H L, 2009. Seedling cultivation[M]. Beijing: China Forestry Publishing House.

SUN S X, 1992. Afforestation[M]. 2nd ed. Beijing: China Forestry Publishing House.

SUN S X, 2013. Cultivation techniques of forest seedlings[M]. 2nd ed. Beijing: JINDUN Publishing House., pp 134-135

WANG D P, LI Y P, 2014. Nursery of landscape plants[M]. Shanghai: Shanghai Jiao Tong University Press.

WANG X J, CHEN G, LI M J, et al., 2010. Application of plant growth regulators to plant tissue culture[M].

Beijing: Chemical Industry Press.

WHITE P R, 1934. Multiplication of the viruses of tobacco aucuba mosaic in growing excised tomato roots[J]. Phytopathology, 24: 1003-11.

WU S H, 2004. Nursery of landscape plants[M]. Shanghai: Shanghai Jiao Tong University Press.

WU S H, Zhang G, Lv Y M, 2009. Floriculture and seedling science[M]. Beijing: China Forestry Publishing House.

XI R T, 2011. Pomology pandect[M]. 3rd edition. Beijing: China Agriculture Press.

XIAO Y, 1998. Forest cultivation[M]. Beijing: China Agricultural Science and Technology Press.

YU J, 1987. Nursery of landscape plants[M]. Beijing: China Forestry Publishing House.

ZHAI M P, 2011. Theory and techniques of modern forest cultivation[M]. Beijing: China Environmental Press.

ZHAI M P, SHEN G F, 2016. Silviculture[M]. 3rd edition. Beijing: China Forestry Publishing House.

ZOU X Z, QIAN S T, 2015. Technology of seedlings production[M]. 2nd ed. Beijing: China Forestry Publishing House.

Chapter 11 Cultivation of Transplants and Large-size Seedlings

Zheng Yushan, Lu Xiujun

Chapter Summary: With the increasing demand for urban greening, the market demand for large-size seedlings is increasing. Transplantation is a key technical link in the cultivation of large-size seedlings. This chapter introduces the key technologies in the cultivation of bare-root transplantation, soil ball transplantation, large-size container seedling cultivation, and large-size vegetative seedling cultivation based on grafting. The aim is for readers to master the technical processes of modern transplanted seedlings and cultivation of large-size seedlings.

Transplanting refers to removing small seedlings from the original nursery or container and replanting them according to a certain spacing in another designated place of seedling cultivation technology. Transplantation enlarges the growth space of aboveground and underground parts of seedlings, changes the ventilation and light conditions, and produces large-size seedlings that meet the requirements of landscape greening. In addition, the transplant process cuts off part of the main and lateral roots; promotes the development of fibrous roots; and closely concentrates the root system, which is conducive to the growth of seedlings and is especially beneficial to improving the survival rate of landscape plantings. In the process of transplantation, reasonable pruning of roots and crown is necessary to artificially regulate the growth balance between aboveground and underground. Doing so eliminates inferior seedlings and improves seedling quality. Transplantation is an important measure to improve the survival rate of seedlings and is a necessary step in nursery production.

11.1 Bare-root Transplantation

Transplantation of bare-root seedlings is a commonly used transplant method in China. In the transplant process, the soil is not extracted, that is, most of the "soil protecting the seedling" is taken away with this simple operation that is low cost and saves resources in manpower and materials. Bare-root transplantation is ideal for young propagating seedlings of most of the dormant deciduous trees and shrubs.

11.1.1 Transplantation Site Preparation

First, the transplant site should be selected according to the biological characteristics of seedlings along with the land characteristics. Generally, loamy soil with moderate thickness, fertility, loose texture, good drainage, and low groundwater level should be selected.

(1) Land Leveling

Most of the transplant land is empty after the seedlings are lifted. When the seedlings are moved, some of them take the soil boll (the soil around and among the seedling roots), resulting in uneven land, and the plow layer soil is partially lost. In addition, the growth of the previous seedlings consumed nutrients. Therefore, before transplanting back into that land, it is necessary to fill and level the nursery land with good soil.

(2) Base Fertilizer Application and Soil Disinfection

Based on the nutrient status of the nursery, an appropriate amount of organic fertilizer is applied as the base fertilizer to restore the soil potential and ensure the growth of seedlings. The soil is disinfected by applying decomposed organic fertilizer mixed with insecticides, fungicides, and herbicides. For plots with severe underground pests, insecticides such as toxic zinc at a rate of 20 to 25 g/m^2 can be applied to the deep soil along with the cultivation of nursery land, which can effectively prevent and control underground pests such as grubs.

(3) Meticulous Land Preparation

Land preparation work can be done in autumn and spring. Plowing depth depends on the size of seedlings, but generally 20 to 30 cm is suitable. For the first plowing in autumn or when breaking in fallow land, the plowing depth can be deeper, and the second or spring plowing can be shallower. Two times of land preparation, in autumn and spring, are adopted: The first is to promote the germination of weed seeds in the soil over various times to achieve the purpose of eliminating them or to use pesticides and exposure to bright sunlight to kill insects and sterilize the soil. The second is to increase the soil humus, promote the formation of soil granular structure, improve the soil permeability, improve the survival rate and good growing environment for the transplanted seedlings, and facilitate the cultivation of seedlings.

(4) Seedbed or Ridge Preparation

The method of operation is determined according to the characteristics and scale of the transplanted seedlings. For the seedlings of broad-leaved and coniferous trees with slow growth rates and those sensitive to water, the high seedbed operation mode can be adopted, and the seedbed specification can refer to the type of seedlings. For seedlings with a fast growth rate or larger-size seedlings, ridge culture or conventional field planting can be used. The specifications of ridge culture depend on the characteristics of seedlings, and conventional field planting is based on the planting spacing set to dig planting holes at fixed points. The size and direction of the seedbed (or ridge) should be convenient for transplanting and irrigation and should be connected with the nursery road and irrigation system to create good conditions for future management practices.

11.1.2 Transplanting Density and Times

(1) Transplanting Density

The planting density of seedlings refers to the number of seedlings planted per unit area, which depends on the spacing between plants and rows of transplanted seedlings. The planting spacing of seedlings is closely related to the seedling's growth rate, climatic conditions, soil fertility, age and specification of seedlings, cultivation objectives, and cultivation time. Generally, the planting spacing of broad-leaved tree seedlings is larger than that of coniferous tree seedlings; planting spacing of fast-growing tree seedlings is larger than that of slow-growing tree seedlings; planting spacing of long-term cultivated seedlings is larger than that of short-term ones; and planting spacing of seedlings with an open crown and developed lateral roots and fibrous roots should be larger (Table 11-1).

Table 11-1 Spacing of landscape plant transplants

Seedling type	First transplanting (cm × cm)	Second transplanting (cm × cm)	Notes
Evergreen large-size seedlings	150 × 150	300 × 300	Seedlings for the first transplanting can be cultivated to 3-4 m high. Seedlings for the second transplanting can be cultivated to 5-8 m high.
Evergreen small-size seedlings	30 × 20 (seedbed) or 40 × 30	60 × 40 or 80 × 50	The first transplanting is for the small-size seedlings of *Pinus tabuliformis*, *Pinus bungeana*, and others. The second transplanting is for cultivating hedge seedlings of *Pinus bungeana*.
Deciduous fast-growing trees	110 × 90 or 120 × 80	150 × 150	The first transplanting is to cultivate the trunk diameter to 4-6 cm. The second transplanting is to cultivate the trunk diameter to 7-10 cm.
Deciduous slow-growing trees	80 × 50 or 100 × 80	120 × 120 or 150 × 150	The first transplanting is to closely plant to cultivate the trunk or to cultivate the trunk diameter to 4-6 cm. The second transplanting is to cultivate the trunk diameter to 7-10 cm.
Flowering shrub	80 × 80 or 80 × 50		
Fruit trees	120 × 50	120 × 100	
Climbing plants	80 × 50 or 60 × 40		

Source: From Zhang Donglin et al. (2003).

Depending on the cultivation objectives, requirements for density differ at the various stages of seedling cultivation. For example, the tree seedlings need close attention to promote trunk growth, especially for those with poor apical dominance. They can be closely planted at the early stage of seedling cultivation to encourage the height growth of seedlings through density control and to control the development of lateral branches (buds) through density control, which will assist with cultivating the straight trunk.

For nursery land with mechanical transplanting, the transplant density of seedlings should be subject to the mechanical specifications of the equipment used.

In short, to ensure sufficient vegetative area for seedlings, it is necessary to plant in a reasonable density to make full use of the land and improve the yield of seedlings per unit area.

(2) Transplanting Times

Transplanting times of cultivated seedlings in the nursery are related to the growth rate of the tree species and the requirements of the seedling size. The arbor species used in street trees and shade trees for urban greening often require larger-scale seedlings. Generally, the first transplantation should be carried out when the seedlings have grown for one year. Later, according to the growth rate and planting spacing of tree species, the trees should be transplanted every 2 to 3 years, and the planting spacing should be enlarged accordingly. However, the arbor species with slow growth rate, underdeveloped roots, and survival difficulties after transplanting can be cultivated in the seedbed for 2 years after sowing and transplanted in the third year. Then transplant once every 3 to 4 years, and they can be shifted to every 2 to 3 years transplantation after they reach their desired specifications. According to the characteristics of tree species, transplantations for common shrubs should be done only 1 to 2 times, after which they should be outplanted to a permanent location.

It is a trend to use large-size seedlings for afforestation in mountainous areas. Generally, first transplanting should be carried out when the seedling age of sowing or cutting reaches 1 year, and the broad-leaved tree species should be continued to cultivate for another 2 to 3 years. Coniferous tree species should be lifted after 3 to 4 years for permanent afforestation.

11.1.3 Transplanting Time

The transplanting time of seedlings can be determined according to the local climate conditions and the specific tree species characteristics. In general, the transplanting of seedlings should be carried out during the dormancy period of trees, for example, deciduous trees can be transplanted from the beginning of autumn defoliation to before seedling budding the following spring. Generally speaking, evergreen tree species can also be transplanted in the rainy season, and other tree species are transplanted primarily during the seedling dormancy period. In recent years, with the acceleration of urbanization in China, green construction has developed into a phenomenon of full-year development, and out-of-season transplantation of seedlings has been gradually promoted.

(1) Transplanting in the Dormancy Period

Transplanting in the dormancy period is mainly in spring and autumn. Transplanting occurs occasionally in winter, but the cost of transplanting in winter is usually 2 to 5 times higher or even more than that in spring and autumn in the North (such as in Shenyang).

Spring Transplanting. Spring is the most important transplanting season in North China. Because of the cold winter in the North, spring is the best time for transplantation. It is ideally suited for unfrozen soil transplantation and early spring sprouting. At this time, the temperature gradually rises; the soil temperature can meet the requirements of the root growth of the seedlings; the sap flows, and the transpiration capacity of the branches and leaves is weak, which is conducive to water retention in the seedlings; and the survival rate of seedlings is higher. The specific time of transplanting in spring should be arranged according to the time of germination. Generally speaking, the tree species with early germination should be transplanted first, those with late germination should be transplanted later; those with defoliation should be transplanted first, the green ones should be transplanted later; the woody ones should be transplanted first, the perennial herbs should be transplanted later; and the larger-size seedlings should be transplanted first, the small-size seedlings should be transplanted later.

Autumn Transplanting. Autumn is the second good season for seedling transplantation. The aboveground part of seedlings ceases to grow in autumn, and transplanting can be carried out from the beginning of leaf abscission of deciduous trees until the soil freezes. At this time, another growth peak of the root system of the seedlings occurs, which is conducive to root wound healing and recovery growth and improves the survival rate of the transplant. It should be noted that the transplanting time should not be too early, to reduce the water loss in seedlings and affect the survival rate of the transplant.

(2) Transplanting in the Growing Period

This time frame is also called summer transplanting or rainy season transplanting. It should be used carefully for transplanting bare-root seedlings in their growth period. In seedling land that has convenient irrigation or spray irrigation facilities, after transplanting the seedling transpiration can be controlled by spray irrigation. Also, a sunshade net can be set up to control the light to reduce the transpiration. This method can be applied to the small-size bare-root seedlings, but it will increase the management cost of seedling raising correspondingly. Therefore, it is not recommended to adopt large-scale transplanting during the growth period.

11.1.4 Transplanting Method and Technology

(1) Bare-root Lifting

Seedlings should be lifted during windless, cloudy days as much as possible; avoid the bright sunlight days and windy days. If the soil in the nursery is too dry, it should be properly irrigated approximately 1 week before the lift of seedlings to increase the soil moisture content. It is undesirable to lift seedlings when the soil is frozen. There are two kinds of seedling lifting methods: Artificial seedling lifting and mechanical seedling lifting.

A relatively complete root system is one of the important conditions to ensure a high survival rate of bare-root transplanting. Therefore, the basis of bare-root transplanting is to keep the root system undamaged and to prevent water loss. When transplanting propagating seedlings and small shrubs, the quality of the root system must be protected; when transplanting large-size seedlings, the retention length of the root system should be 8 to 10 times of DBH. The length of lateral roots (root width) and axial roots is determined according to the size of seedlings before artificial seedling lifting (Table 11-2, Table 11-3). Make sure that the main lateral roots do not split, or split during the seedling rise, and pay attention to protecting the branches and terminal buds. Shown in Figure 11-1 is a large-size seedling lifting machine.

Table 11-2 Specification requirements for small-size bare-rooted seedling lifting

Seedling height /cm	Lateral root range /cm	Axial root length /cm
< 30	12	15
31-100	17	20
101-150	20	20

Table 11-3 Specification requirements for large-size bare-rooted seedling lifting

Seedling DBH /cm	Lateral root range /cm	Axial root length /cm
3.1-4.0	35-40	25-30
4.1-5.0	45-50	35-40
5.1-6.0	50-60	40-45
6.1-8.0	70-80	45-55
8.1-10.0	85-100	55-65
10.1-12.0	100-120	65-75

(a) (b)

Figure 11-1 Mechanical lifter for large-size bare-rooted seedlings (Schmidt Wholesale Nursery, Oregon, USA).
(a) seedling lifter suitable for large-size seedlings with a range of specifications; and
(b) a closer view of the seedling lifter.

The seedlings should be graded after they are lifted. Grade seedlings to be planted in different areas to reduce the differentiation of seedlings after transplanting, facilitate the management of seedlings, and promote the growth of seedlings evenly and skillfully. Seedling classification is based primarily on seedling height, seedling ground diameter, and seedling root system.

Grading is often followed by proper pruning before or after seedling lifting. For example, shoots and prolonged fibrous roots of young propagation seedlings can be cut off, and transforming steps can be applied in conjunction with the characteristics and the status of seedling species for arbors. After seedling lifting, the damaged roots and redundant fibrous roots should be pruned to ensure that the length and quantity of fibrous roots meet specifications. Seedlings with too-large crowns should be properly pruned. Pruning should also be carried out for branches with diseases, insect pests, and mechanical damages. If the seedlings are not planted immediately after lifting, they should be heeled-in for temporary planting.

(2) Protecting Seedling Vigor after Seedling Lifting

Water loss is one of the main reasons for a low survival rate of bare-rooted seedling transplants. The main concern for the survival of bare-root transplants is how to avoid root water loss, and special attention should be paid to all steps of transplantation, such as seedling lifting, pruning, grading, transportation, and planting as well as the management steps after cultivation. Because of the tender root system, the smaller-size bare-rooted seedlings have to be placed in a moist environment after seedling lifting to prevent water loss of the root system (Figure 11-2); if the larger-size bare-rooted seedlings cannot be transplanted promptly, they must be heeled-in for temporary planting (Figure 11-3) or wrapped and covered with moisturizing materials after seedling lifting. Mud dipping before transplanting can be mixed with some rooting agents and water-retaining agents (such as urea gels) that can assist rooting and water retention, and water needs to be spread and sunk into the land to shorten the time from seedling lifting to transplanting.

Figure 11-2 Spraying water to the seedling root system after seedling lifting (Photo by Li Yajiang).

(3) Seedling Transplanting

A key point of planting technology is that the correct depth of planting must be perfected. Planting depth should be slightly deeper than the depth in the original planting ground. Generally, it can be slightly deeper than 2 to 5 cm. The root system should stretch downward without root folding and root bending. In the process of planting, the soil should be filled in, compacted, and in close contact with the root system. Especially for the transplanted seedlings with a larger root system, be certain the root system is filled with soil. To a certain extent, timely watering after

Figure 11-3　Provisional planting of seedlings (Photo courtesy of Shenyang Youth Nursery).

planting can make up for the loss that occurs during planting.

Dibble planting and trench planting methods can be used. Dibble planting is often used for transplanting large-size seedlings. Dibble digging tools often have measurement indicators on the shaft of the dibble, which assist with digging holes at a consistent depth. The dibble bars are also ergonomically designed to minimize back pain and to speed up the planting process. The method of manually laying wires on fixed points to determine where to plant seedlings in an orderly arrangement and then mechanically digging a hole can greatly improve the working efficiency and planting quality for the plots with flat and loose soil (Figure 11-4). The trench planting method can be adopted for small-size seedlings, that is, the seedlings can be placed into a planting trench dug in the nursery land according to a certain row spacing, covered with soil, and compacted. The larger-size seedlings should be fixed with brackets to prevent them from being blown down by the wind.

Figure 11-4　Planting holes and digging machines for mechanical excavation (Photo courtesy of Shenyang Shengshi Lvyuan Nursery).
(a) overall effect of mechanically digging holes; (b) single planting hole by mechanical excavation; and (c) rotary cultivator of hole-digging machine.

For the nursery, spring is the peak of labor use as it is the season of agricultural spring planting and forestry afforestation. Given the low efficiency of manual transplantation and high labor intensity, the average productivity of most workers is 4000 plants per day. However, the transplanting season is only about 15 days, so large-scale adoption of mechanized seedling transplantation can save manpower and material resources and improve the quality of transplantation (Figure 11-5).

(a)   (b)

Figure 11-5 Seedling transplanter(Photos courtesy of Rocky Mountain Institute of the U. S. Forest Service).
(a) seedling transplanting; and (b) details of mechanical transplanting.

In recent years, the development of foreign seedling transplanting machines has advanced. Except for the changes in an individual structure, accessory parts, and appearance design, there have been no great changes in the operation mode and working principle.

The flexible disc-type planting apparatus is an earlier and simpler one when compared to the transplanting machine. It uses two flexible discs that can be elastically deformed to clamp onto and send seedlings and to then loose the seedlings at the opening of the disc to complete the planting work. When planting, the seedlings should be placed at the opening of the disc manually (or by a conveyor belt), aiming at the center of the disc, and the pressing wheel drives the disc to rotate through the transmission mechanism. When the disc turns to the gathering point, the seedlings are clamped and brought to the opening of the disc, then the seedlings fall into the ditch. At this time, the pressing wheel is covered with soil and suppressed in time to complete the planting process. This kind of planting apparatus can also be used for planting small-size bare-rooted seedlings.

(4) Management after Planting

After planting, at least two to three saturating drenches should be carried out, and pay attention to the soil cultivation work after watering. After the trees germinate, water can be sprayed once every 1 to 2 days, if conditions permit, to accelerate the growth rate of new buds. The new shoots of trees can be easily damaged by insects, so timely observing and spraying, frequent weeding and loosening of the soil, and regular pruning need to be carried out during dry weather. Seedlings should be shaded, and seedlings that did not survive should be replaced in time to ensure the consistent growth of seedlings.

11.2 Transplantation and Cultivation with Soil Boll

When the seedlings can survive by bare-root transplantation, the transplanting method with soil bolls is generally not adopted. The transplantation and cultivation with soil bolls are mainly applicable to some evergreen trees, a few precious deciduous trees, and larger-size seedlings, as well as seedlings difficult to regenerate by the root system and that do not readily survive by bare-root transplantation. The advantage of this method is that the survival rate of planting is high, but the construction cost is also higher.

11.2.1 Transplanting Time

The transplanting time with soil bolls is not as strict as that of bare-root transplanting. According to the range of transplanting times, it can be divided into normal planting season transplantation and abnormal planting season transplantation. In normal planting season, transplantation can be divided into spring transplantation, rainy season transplantation, and autumn transplantation. In the process of seedling transplanting, the most suitable transplanting period should be selected according to the biological habits of seedlings and the actual demand for production. Generally, the most suitable period for transplanting seedlings is in spring, when the sap begins to flow in early spring, but before bud germination. Evergreen tree species can be transplanted in the rainy season. The arbors (trees) after defoliation can be transplanted in autumn.

11.2.2 Preparation before Transplanting

(1) Seedling Selection

The principles of seedling selection include: There should be no obvious diseases and pests; if it is an exotic seedling, it must go through plant quarantine; there should be no obvious mechanical damage; there should be a full crown and good ornamentals; strong plant and normal growth; and site conditions should be suitable for seedling excavation, hoisting, and transplanting.

(2) Seedling Pruning

Before transplanting the seedlings, proper pruning should be carried out according to their biological habits, growth status, transplanting time, excavation method, and cultivation purpose. In the process of seedling lifting, there will be some degree of damage to the root system of the seedlings. Therefore, to improve the survival rate of transplanting, a key point is to keep the relative balance between the above ground part and the underground part of seedlings. Although, it is also one of the necessary measures to reasonably prune the aboveground part of seedlings to improve the survival rate of transplanting.

Pruning before seedling transplantation includes root cutting, cutting the crown back, branch thinning, leaf picking, and cut crown processing. Seedlings with DBH <15 cm and likely to survive can be transplanted in the normal season without root cutting. For seedlings with DBH ≥15 cm and that may have difficulty surviving, root cutting should be carried out before transplanting. That is to say, in the spring or autumn of 1 to 2 years before transplanting, a circular ditch with a width of 30 to 40 cm and a depth of 50 to 70 cm should be dug outward with the trunk as the center and the 6 to 8 times the DBH as the diameter. For deciduous trees such as maple, Chinese ash (*Fraxinus chinensis*), robur, and white birch (*Betula*), one-third to one-half of the whole crown is cut off generally. When *Ginkgo biloba* is transplanted with soil boll in abnormal transplanting seasons, the leaves can be properly removed; for tree species with fast growth and easy crown restoration, such as poplar and willow, the crown can be removed. The crown of evergreen trees should be kept intact as much as possible, and only some dead branches and/or over-dense branches should be properly pruned. Besides, the tree structure and the distribution of retained branches should be considered when pruning. The thick-cut can be coated with Yushang ointment or wrapped with plastic film, vaseline, or paraffin, for example, to protect the cut from water loss or bacterial infection. Generally speaking, for the seedlings with soil bolls that were damaged in the process of seedling lifting, the old roots and rotten roots of the plants should be cut off and wrapped with wet grass and straw bags, and the withered and yellow branches and leaves should also be cut off before loading. Also, according to the quality of the soil bolls, part of the stems can be cut off properly, or even cutting the trunk, or combining with the methods such as lopping and shaping to ensure their survival.

11.2.3 Plant Pit Processing

The size of the plant pit should be determined according to the size of the soil bolls taken while transplanting the seedlings; the pit is generally 30 to 40 cm larger than the soil bolls. The soil at the base of the plant pit should be loose and show a concave shape. The soil in the planting area should have strong permeability and good drainage. If the soil needs to be changed, the loosened soil must be compacted and irrigated in advance; for transplanting in the saline-alkali land, the transplant pit must be treated with salt control.

11.2.4 Seedling Transplantation

11.2.4.1 Artificial Seedling Lifting

Within 3 to 4 days before transplanting, the transplant seedlings are irrigated to allow the roots to fully absorb water. The size of the soil boll is related to the height and diameter of trees and is determined by the DBH of seedlings. Generally, the diameter of a soil boll is 6 to 8 times the DBH of seedlings, and the height of a soil boll is two-thirds to four-fifths of its diameter.

Specific steps are as follows (Figure 11-6).

i. Draw a circle with the trunk as the center to indicate the diameter of the soil boll, which should be slightly larger than the specified size.

ii. Remove the topsoil. After the circle has been drawn, first dig a layer of topsoil in the circle to a depth that does not damage the seedling root on the surface.

iii. Excavate a trench to a width of 60 to 80 cm running vertically outward with the trunk as the center of the circle and slightly higher than the height of the soil boll. When digging, the surface of the soil boll needs to be trimmed, and when excavating to half the depth, the upper half of the soil boll should be trimmed using the spade or shovel to make it smooth and not angular, but be certain that the root of the tree is not exposed. The lower half of the soil boll needs to gradually shrink inward. For nursery soil with deep roots and sandy loam, the soil bolls appear apple-shaped, and for the nursery soil with shallow roots and clayey soil, the soil bolls present oblate. Do not step on the soil boll during operation.

  iv. Undercutting. After the surface of the soil boll has been trimmed well, slowly digging from the bottom to the inside is called "packing". The soil boll with a diameter less than 50 cm can be directly undercut and held outside the pit for "packing", whereas the soil boll with a diameter greater than 50 cm should retain a part of its bottom center to support the soil boll for "packing" in the pit (Figure 11-7).

Figure 11-6 Steps of the artificial lifting of seedlings with the soil boll(Photos by Liu Yufeng).
(a) determining the diameter of soil boll; (b) actual operation of workers; and (c) the trimmed soil boll.

Figure 11-7 Packing method of seedlings with the soil boll(Photo by Li Yajiang).
(a) soil boll packing in the pit (Photo by Wang Bin); and (b) packing with oilcloth and binding with iron wire.

After the soil boll is excavated, it needs to be tightly packed with materials such as cattail bags and tied firmly with straw ropes around the outside, which is called "packing". Before packing, the cattail bags and straw rope should be soaked in water to increase their strength. There are mainly three types of dressing: Orange-shaped packing, Jingzi packing, and pentagon packing (Figure 11-8).

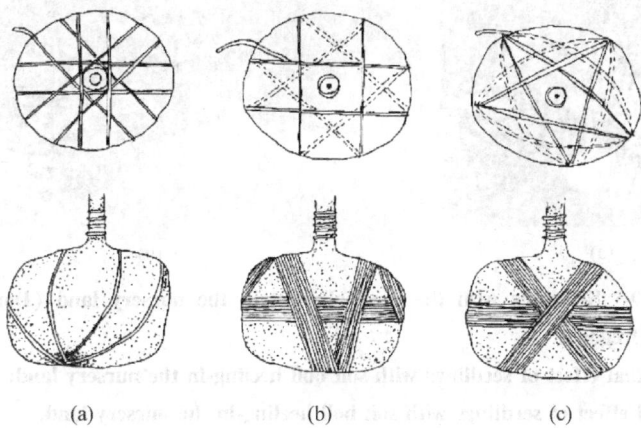

Figure 11-8 Common packing methods of soil bolls.
(a) orange-shaped packing; (b) Jingzi packing (ancient coins-shaped packing); and (c) pentagon packing.

When packing, the soil boll with a diameter less than 50 cm can be packed outside the pit. The loose soil boll should be packed in the pit. The winding strength and density are determined by the diameter of the soil boll. When the diameter of soil boll is less than 40 cm, a straw rope (or iron wire) (see Figure 11-7) is used to wind it once. For the larger soil ball, a straw rope should be used to wind it in the same direction twice, which is called "single-stranded and double shaft". If the soil boll is extremely large, with a diameter of more than 1 m, it needs to be wound with two straw ropes, which is called "double-stranded and double shaft". After winding using the longitudinal straw rope for one circle, it should be tied firmly at the stump base. For the soil boll with a diameter of more than 50 cm, at the end of the longitudinal straw rope tie the straw rope horizontally in the middle of the soil boll for its protection, which is called "tying waist strap". For all the soil bolls packed in the pit, the bottom of the soil bolls should be tightly wrapped with cattail bags and straw ropes after the waist ropes have been tied, which is called "bottom sealing". If the seedlings with the soil boll are not transplanted right away after packing, they should be heeled into the nursery land (Figure 11-9b).

Notes:

When the soil is too loose to ensure the formation of the soil boll, the soil boll can be dug and tied with the straw rope at the same time, which is called "inner waist rope", then packing it outside the inner waist rope.

To ensure that the soil boll is not scattered, no matter the size of the soil boll, no one is allowed to stand on the soil boll during the whole process of digging and packing.

In the rainy season, the soil boll must be lifted out of the pit for transportation to a location

Figure 11-9 Seedlings with the soil boll heel in the nursery land (Don Marjama Nursery, Oregon).
(a) individual effect of seedlings with soil boll heeling-in the nursery land; and
(b) overall effect of seedlings with soil boll heeling-in the nursery land.

where it can avoid being soaked in water.

11.2.4.2 Mechanical Seedling Lifting

At present, the price of the labor force in China is rising, and the aging of the labor force in the greening industry is very serious in that fewer people are available to work. Mechanization development of nurseries is clearly imminent. Especially in recent years, large-scale tree transplantation has generally been carried out in the process of urban greening construction. Given the urban ecological environment, which is more diversified, the survival rate and growth trends after transplantation differ from rural and forested areas. The survival rate of some transplant trees is very low, which not only wastes the social-ecological resources to a certain extent but also affects the goals of urban green space construction. Traditional manual transplantation has a high cost, needs a variety of devices to be used at the same time, and also needs a large number of workers, especially in the peak season of transplantation when the phenomenon of a labor shortage is more apparent and detrimental. In addition, in the process of transplanting, human factors often lead to accidental damage of seedling root systems. The seedling transplanter is a type of mechanical equipment integrating mechanical and hydraulic technology, and it can be used to excavate seedlings, provide transport, and so forth. It can also ensure the integrity of the soil boll at the root of seedlings and minimize the damage to trees.

The research and design of a tree transplanter to improve work efficiency in the short run and the survival rate of seedling transplants in the longer run is of great significance to the industrialization of the seedling base and large-scale excavation and transplantation of seedlings.

With the gradual improvement of greening requirements and the maturity of tree transplanting technology, tree transplanting equipment has gradually developed the advantages of simple mechanical operation, high tree transplanting efficiency, low labor intensity, and high survival rate after tree transplanting. For example, the tree digging group (8 persons) of Rostrup Nursery in

Germany, using a small excavator, in one day can complete the task of digging up to 220 seedlings with a DBH of 8 cm, more than 1000 trees a week. The efficiency of seedling excavation is greatly improved; the specifications of seedling excavation are consistent; the speed of seedling excavation is fast; the operation quality is high; and the survival rate of seedling transplants is effectively improved.

Many kinds of tree transplanters are in use at home and abroad. According to the shape of the shovel, it can be divided into a straight shovel type and a curved shovel type. According to the layout of the shovel, it can be divided into an external shovel type and an internal shovel type. According to the number potential of the shovel, it can be divided into a two-shovel type, three-shovel type, four-shovel type, and six-shovel type. According to the connection mode of the supporting power, it can be divided into a vehicular tree transplanter, a tree transplanter drawn by an automobile trailer, a wheel-type or crawler-type loader, a tree transplanter with tractor rear suspension and side suspension, and a tree transplanter with a multi-bar mechanical-arm-type hydraulic drive.

The study of tree excavation and transplantation began in the 1950s. Existing tools and resources include products from Big-John Company in the United States, Damcon Company in the United Kingdom, and Dutchman Company in Canada. The products of these companies, however, are aimed primarily at transplanting trees with higher DBH, which are bulky, have high requirements for transplanting space, and cannot adapt to the environment very well. Therefore, in the 1970s, Federal Germany, Netherlands, and Japan improved the adaptability of tree transplanters so they can be used in relatively small spaces.

Tree transplantation research in China began in the 1970s. Gu Zhengping and Zhang Yingyan of Beijing Forestry University used a shovel model to study the mathematical model of shovel drag in soil, which provided a valuable theoretical reference for the design of a domestic tree transplanter. In the aspect of research and development of tree transplanting machines, many models are also in China.

The research period on tree transplanters in China is not long. It is mainly based on the introduction of foreign equipment for technical improvement and imitation, which cannot meet the market demand. In addition, in the process of tree excavation, the shovel blade movements of existing types are driven by hydraulic pressure, moving slowly, and it does not cut off the roots in the soil but pulls them off, which causes root injury and loosens soil bolls, resulting in the low survival rate of trees. At present, foreign tree transplanting technology is relatively mature and has corresponding excavating and transplanting machinery and equipment for various tree DBH and planting conditions, but the prices are expensive. At the same time, many problems arise in the use, maintenance, and staffing of imported garden machinery. In addition, the soil properties at home and abroad differ, and the production model of the domestic nursery (many gullies and short planting spacing) will cause the "being unacclimatized" of these machines. According to a research investigation, only some powerful large-scale nurseries use professional imported tree diggers and within a small range, such as the Dutch-Man Tree Spade tree digging shovel made in Canada,

(a)   (b)   (c)   (d)

Figure 11-10 The seedling lifting process using a tree-digging shovel.
(a) tree-digging shovel before operation; (b) working state of a tree-digging shovel during operation; (c) lifting of seedlings; and (d) lifted seedlings being carried away.

introduced by Shenyang Jiayuan Seedling Breeding Company Ltd (Figure 11-10).

11.2.5 Hoisting and Transportation

For the large-size seedlings with the soil boll, the methods of crane loading and unloading and automobile transportation are generally adopted. For only a few seedlings within a short distance, they can be directly lifted and moved by crane (Figure 11-11) or directly transported and transplanted by a mechanical seedling lifter (see Figure 11-10d). When loading seedlings, the crown needs to be toward the rear of the vehicle; the trunk should be wrapped with a straw rope or other materials to protect it, and the crown should be properly wound with soft rope. Loading, transporting, and unloading should ensure that the trunk, crown, and soil bolls at the root are not

(a)   (b)

Figure 11-11 Hoisting and loading of seedlings (Photos by Liu Yufeng).
(a) crane-lifting process; and (b) loading and placing of seedlings with the soil boll.

damaged. For long-distance transportation or transplanting in unsuitable seasons, attention should also be paid to water spraying, shading, windproof and shockproof, and preventing the soil boll from being leached and loosed in heavy-rain days.

11.2.6 Planting of Seedlings with the Soil Boll

Seedlings with the soil boll are planted in holes. The planting spacing is determined according to the size of the trees, anticipated growth speed, cultivation years, and other factors, and holes are dug according to the planting spacing. The tree pit should be standardized and dug in advance, of which the specification needs to be determined according to the size of the root system and the soil boll. The upper and lower sizes of the tree pit should be the same, and the width and depth of the tree pit should be 20 to 50 cm larger than the soil boll. According to the situation of soil diseases and insect pests in the nursery, the pit soil should be sterilized properly before planting (50% carbofuran granular can be mixed with soil to kill insects according to the proportion of 0.1%, and 50% thiophanate methyl or 50% carbendazim powder can be mixed with soil to kill bacteria according to the proportion of 0.1%). The bottom of the tree pit should be backfilled with good soil. If possible, rotting farmyard manure can be applied to the pit to enhance soil fertility and promote the growth of seedlings. When the soil is dry, the pit needs to be soaked in advance so the seedlings can absorb water after planting.

The seedlings must be inspected before planting, and any mechanical damage found should be handled without delay. The planting depth needs to be suitable, and the soil-buried depth is generally 5 to 10 cm higher than the surface of the soil boll. The planting depth of the soil boll is the same, and before planting, the package or rope on the soil boll should be removed to facilitate the combination between the root system and the soil for rooting. More soil should be piled at the base of the tree trunk to make the soil boll stable and unshakable, and the particularly tall trees should be stabilized with brackets. When planting, it is necessary to straighten the trees, and when filling, it is necessary to backfill and compact in layers. When the soil is backfilled to the height of two-thirds of the soil boll, the first watering should be carried out to make the backfilled soil fully absorb water, and the soil should be filled after the water seeps into the soil completely (paying attention not to compact the soil at this time). Finally, a cofferdam is built around the periphery, and the second water is poured over. After watering, attention needs to be paid to observing whether the soil around the trunk sinks or cracks; if any is seen, add soil and fill it without delay.

11.2.7 Management after Transplanting

(1) Setting up Support

For tall seedlings, the support should be set up after planting. Generally, triangle support and fixation method with three-column support (that is, the seedling is supported by three poles and each pole forms a triangle shape with the seedling main stem) is adopted to ensure that trees are stable. The joint between the support and the bark should be padded with straw bags, or something similar, to avoid damaging the bark.

(2) Water Management

The principle of "no watering when the soil is not dry; pouring water over soil when watering" should be grasped in watering. For seedling planting in spring, depending on the soil moisture, watering needs to be carried out once every 5 to 7 days, and repeat 3 to 5 times. For seedling transplanting in the growing season, the interval time of watering should be shortened and the frequency of watering should be increased; in case of particularly dry weather, the frequency of watering needs to be increased. To promote the development of new roots, NAA of 200 mg/L or ABT rooting powder can be added in combination with watering.

(3) Tree Body Moisture

For evergreen trees or broad-leaved, large-size seedlings, high-pressure sprayers or gun-type sprinkler irrigation can be used to spray the trees to achieve their moisture retention after transplanting. Water should be sprayed 2 to 3 times per day, and 1 time per day after 1 week, for 15 days. In addition, a sunshade net can be set up to reduce light and control water consumption to improve water use efficiency and survival rate.

(4) Infusion Promoting Survival

For tree species that have difficulty surviving, the method of an internal water supply after planting can be used to promote their survival. At the base of the plant, three to five infusion holes are drilled with a wood drill at an angle of 40° from top to bottom, deep to the pith. The number and pore size of the infusion holes should be appropriate for the thickness of the trunk and the plug, or the top, of the infusion set. The horizontal distribution of the infusion holes should be even and the vertical distribution should be staggered. Water should be used as the main liquid for infusion; however, micro plant hormones and mineral elements should be added. Also, 0.1 g ABT6 rooting powder and 0.5 g potassium dihydrogen phosphate should be dissolved into a liter of water and applied.

(5) Adjusting Tree Shape

After transplanting and surviving, the tree will sprout a lot of shoots. According to the characteristics of tree species and tree shape requirements, some unnecessary sprouts on the trunk and main branches should be removed in time to form the full crown and achieve the ideal landscape effect.

(6) Loosening Soil and Weeding

Because of the large size and long planting spacing of seedlings with the soil boll, these areas are suitable for mechanical weeding (Figure 11-12). The use of a rotary tiller or special weeder between seedling rows can greatly improve work efficiency.

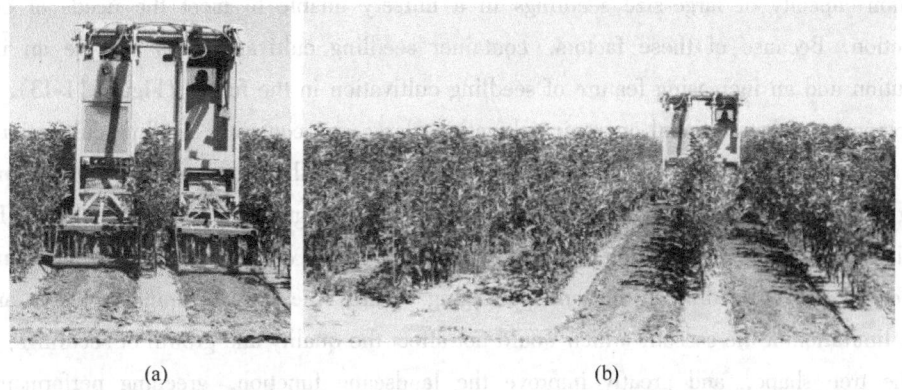

Figure 11-12 Weeding operation of seedling weeder (KCK Farms Nursery, Oregon).
(a) before weeding machine operation; and (b) after weeding machine operation.

11.3 Large-size Container Seedling Cultivation

Container seedling cultivation uses various containers to load nutrient soil or culture medium; adopts the ways of sowing, cutting, or transplanting the seedlings; and adopts water and fertilizer management measures to cultivate seedlings. Container seedling practices were first developed in the forestry nurseries in Nordic countries, such as Sweden, Finland, Denmark, Norway, and others, in the mid-1950s. In the 1960s to 1970s, container seedling production began in the United States, Australia, New Zealand, and many developed European countries. They developed into very mature technologies. Given the differences and diversities in materials, shapes, structures, and specifications of seedling containers, many types of seedlings raising forms have evolved, such as plug seedling, nutrient bowl seedling, net bag seedling, root-control container seedling, and more. Given the differences in propagation methods, cultivation purposes, and cultivation modes, many types of container seedling forms, such as sowing container seedlings, cutting container seedlings, tissue culture container seedlings, and transplanting container seedlings (in which bare-rooted seedlings are transplanted and cultivated by sowing, cutting, grafting, tissue culture, and so on within the container for continuous cultivation). In the forest nursery, for the seedlings cultivated in the above way, after conifer species have been cultivated in the nursery for 3 to 5 years and broad-leaf species have been cultivated for 2 to 3 years, they can be lifted for afforestation. In the urban nursery, however, because of high requirements for seedling specifications, they often need to continue to be cultivated in larger containers, so that they can finally become seedling products that meet the necessary requirements. Therefore, large-size container seedling cultivation is the process of cultivating large-size seedlings in larger containers.

In recent years, with the needs of ecological environment construction and the rise of urban forestry, new requirements have been put forward for greening and beautification, and the market demand for large-size seedlings is growing. It takes a long time to cultivate large-size seedlings, they cover a large area, and the necessary efforts have no obvious short-term benefits, which makes the production capacity of large-size seedlings in a nursery unable to meet the needs of ecological construction. Because of these factors, container seedling cultivation has become an important contribution and an increasing feature of seedling cultivation in the future (Figure 11-13).

Compared with traditional open-air cultivation, large-size container seedling cultivation has the following advantages. First, it is convenient for container seedling management. For example, the spacing between seedlings can be promptly adjusted with the growth status of seedlings. It is also convenient for shaping and pruning. Second, containers can be convenient for transportation, saving the time and cost of seedling lifting and packaging. Third, they can be transplanted at any time, without limitation of the season, which would not affect the quality and growth of seedlings; it would keep the tree shape, and greatly improve the landscape function, greening performance, and survival rate of transplants. Fourth, it is convenient for intensive management, mechanization, and marketing, which can promote commercialization. Therefore, container seedling and container cultivation technology is a change of production and cultivation mode and a product of the modernization of the seedling industry. It has caused great changes in related nursery construction, seedling production, management technology, application, management concepts, and more, and it will become the main trend of nursery production development in the future.

Figure 11-13 Container seedling cultivation in Oregon, USA.
(a) container cultivation of small-size evergreen seedlings; (b) container cultivation of small-size evergreen colored-leaf seedlings; (c) container seedlings of deciduous trees placed according to planting spacing; (d) large-size modeling container seedling cultivation; (e) container cultivation of large-size evergreen seedlings; and (f) container seedlings of deciduous trees placed closely.

Container cultivation technology is widely used in developed countries. In China, the application and research of large-size container seedlings started relatively late, and it took time to absorb, customize, and improve the advanced technology from developed countries in forestry. The large-size container seedling is generally larger, and its cultivation always focuses on the individual cultivation. Large-size seedling cultivation takes a long time, and strong stress effects of the environment affect seedlings. Many technical problems still need to be solved. Based on Chinese national conditions, the cost of seedling management will increase, and the protection or buffering effect of the substrate on the root system of seedlings will be reduced if problems such as suitable substrate materials, formulation, and water and fertilizer management technology for large-size container seedlings are not optimized. Ultimately it will affect application performance and seedling markets for large-size container seedlings. However, as the largest construction market in the world, China has broad prospects for the application of large-scale seedlings, and with the improvement of the cultivation technology of large-scale seedlings, the production mode of container cultivation will become an important factor in the national greening efforts.

11.3.1 Type and Specification of Containers

Nowadays, several container types are used for large-size container seedling cultivation.

(1) Rigid Plastic Container

The rigid plastic container is made of plastic materials by a blow-molding or injection-molding process, which is a landmark event of container seedlings that greatly promote the development of container seedlings. The rigid plastic containers are a commonly used seedling container for large-size container seedlings, and they are easy to manufacture and scale to production needs with relatively low cost (Figure 11-14). Rigid plastic containers are mostly cylindrical or cone-shaped, which is convenient for filling the substrate. Because the materials of containers are impervious to water and air permeability, some treatments would be carried out on the inner wall and bottom of the containers for better use and to expand the functions of the containers, such as opening holes, setting vertical ridges, convex and concave, and so forth.

(a)　　　　　　　　　　　　　(b)

Figure 11-14　Rigid plastic seedling container.
(a) the container buried in the nursery land; and (b) the container placed on the ground.

(2) Rocket-basin, Root-controlled Container

The rocket-basin, root-controlled container was invented by Australian experts in the early 1980s, and then it was introduced and used in Australia, New Zealand, the United States, Japan, the United Kingdom, and other countries. In the 1990s, rocket-basin root-controlled containers were introduced in China and were used to carry out experiments with Chinese arbors and shrubs on the Loess Plateau by the cooperation of the Institute of Soil and Water Conservation of the Ministry of Water Resources of the Chinese Academy of Sciences and the Indak Group of Australia, then the containers were widely promoted throughout China. The rocket-basin root-controlled container was made of polyethylene and composed of three parts: Chassis, sidewall, and inserting rod or rivet to hold the sidewalls together (Figure 11-15). Each part is made independently, and the relevant parts can be assembled when using. A series of containers with different specifications can be assembled by adjusting the size or shape of the components. Generally, the size of containers is within 20 to 60 cm × 21 to 62 cm, which is used to cultivate seedlings with a DBH of 1.5 to 7 cm, and bottomless containers are suitable for seedlings with a DBH greater than 7 cm.

Figure 11-15 Root-control container.
(a) sidewall; (b) chassis; (c) bottom structure; (d) assembled root-control containers; and
(e) container planting of large-size arbor seedlings.

(3) Root-control Bag Container

The root-control bag container is known as a planting bag or physical bag root-controlled container, made of non-woven polypropylene material by special processing. The containers are pervious to water and air permeability, so bags do not have water accumulation that could cause root rot, and the bags allow fine roots to pass through (Figure 11-16). The thickness of the material can be expressed by the weight per unit area, generally ranging from 200 to 400 g/m^2. The container can be thickened or equipped with a handle if needed. Its design and production are an achievement

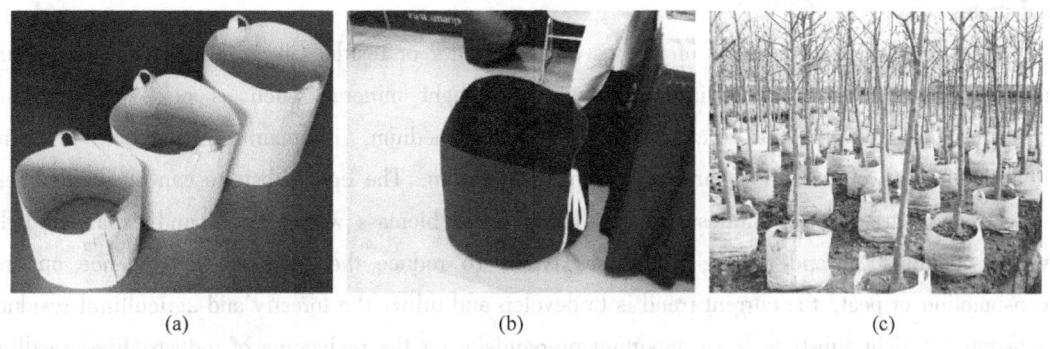

Figure 11-16 Root-control bag container.
(a) white root-control bag; (b) black root-control bag; and (c) root-control bag planting.

from Oklahoma State University that was introduced in the early 2000s.

The above three types of containers can be used alone, or two types of containers can be combined, which can achieve good performance, such as root-control bag + root-control bag, root-control bag + plastic container, plastic container + plastic container, and so forth. Among them, a successful combination used for large-size container seedling cultivation has been plastic container + plastic container mode and related technologies, such as drip irrigation, forming the double containers, or pot-in-pot seedlings. For example, the pot-in-pot seedling technology (Figure 11-17) is to place the container where the seedlings are cultivated into the holder pot in the field for planting, combining the advantages of two cultivation modes of open-air cultivation and container cultivation, with the ability for wind resistance, heat resistance, and freeze resistance. This technology is better conducted in cold areas and dry heat areas with good performance.

Figure 11-17 Pot-in-pot cultivation of container seedlings in Oregon, USA.
(a) holder pot; (b) pot-in-pot of evergreen seedlings; and (c) pot-in-pot of deciduous seedlings.

11.3.2 Seedling Substrates

(1) Substrate Types

The selection of substrates is an important factor for the success of container seedling cultivation. Selection should be based primarily on the biological habits of seedlings, material source, and container type. Generally speaking, according to the composition, texture, and weight per unit area of substrates, it can be divided into three categories: Light, semi-light, and heavy substrates.

The light substrate is made of various organic matter or residues from forestry and agriculture through fermentation or carbonization, mixed with light minerals such as peat, perlite, and vermiculite. Compared with the traditional seedling medium, its main characteristic is being lightweight, as well as other characteristics of the medium. The light substrate can be divided into forestry and agricultural residues type, industrial solid biomass wastes type, and expanded light wastes of industrial and mining enterprises type. To reduce the excessive dependence on and consumption of peat, the current trend is to develop and utilize the forestry and agricultural residues substrate. A light substrate is an important prerequisite for the realization of industrialized seedling production, and the basis of container seedling cultivation using a light substrate net bag. It has been widely used in agricultural seedling production, forest tree container seedling, and other industries.

The semi-light medium has nutrient soil and various organic matter in a certain midrange proportion. Its texture weight is between heavy medium and light medium, with the bulk weight within 0.25 g/cm^3 to 0.75 g/cm^3.

Heavy medium is made up of various nutrient soils as the main components. Its texture is compact, with a higher weight per unit area and the bulk weight greater than 0.75 g/cm^3. This type of medium can be obtained from local areas, so it can reduce the cost of raising seedlings.

At present, light medium and the semi-light medium have been more widely used. Generally speaking, the lighter medium is favorable for the increase of fine root proportions and better clustering of roots; the heavier medium is relatively close to the root system and provides a better buffer to external changes. The medium-type should be determined according to the local situation. In places with good seedling cultivation conditions and good management, the lighter medium can be selected; otherwise, the heavier substrate should be selected.

(2) Medium Selection and Physicochemical Properties

Good medium can provide the rhizosphere environmental conditions with stable and coordinated water, fertilizer, gas, and heat for the growth of seedlings; it can support and fix plants; and maintain water and air permeability. Therefore, the physicochemical properties of the medium are the main reference indicators for the selection and preparation of the cultivation medium. The selection of the medium should be adapted to local conditions and tree conditions, with both applicability and economy. The following characteristics should be considered in the selection of medium materials. First, the raw materials should have wide sources, not be limited by regional resources, and be cheap and easy to obtain. Second, the physicochemical properties should be good, with a high water-absorption rate, strong water-holding capacity, and certain water permeability—total porosity of 35% is best—along with good fertilizer retention and weak acidity. Third, it should have low fertility to allow for adjustments to be made to the growth state of seedlings through the external nutrition supply. Fourth, the medium material should be light in weight, so it is easy to carry and transport. Fifth, the medium should be clean and sanitary, without soilborne diseases, pests, and weeds. The physicochemical properties of the medium can have different effects on the seedlings. Some experiments of the seedling cultivating in the medium have further proved that the physical properties of the medium are more important than the chemical properties for the growth of container seedlings. The commonly used medium materials for container seedlings include peat, vermiculite, perlite, forest humus (wasteland topsoil), yellow soil (yellow-brown soil removing topsoil), untilled mountain soil, river sand, slag, and agricultural and forestry production wastes (straw, chaff, seed coat, fruit shell, forest litter, sawdust). But from the scientific, economic, and practical point of view, peat, rotten bark, and rotten wood scale are the best container culture mediums.

(3) Medium Ratio

Soil structure of the medium is an important factor to determine the cultivation effect of container seedlings. Once it is determined, it cannot be changed in a short time. The components of the medium and their proportion are closely related to the soil structure, and the selection and proportion of the medium components is the basic and key work in container seedling cultivation. The good seedling raising medium is often not only made up of one component but made up of two or more components in a certain proportion.

The medium composition and proportion will change greatly based on tree species, seedling raising conditions, source of raw materials, and region. Peat and vermiculite account for 50%, respectively, and together they make up a medium formula widely used at home and abroad. For example, in Poland, peat and litter or bark powder are used as the substrate; in Finland, peat moss or a mixture of mineral and compost is used as the culture substrate; and in the United States, mixed soil of fine mud and vermiculite in a ratio of 1:2 or 2:1 is used as the substrate. In addition, the mixed media with peat and sawdust or carbide and vermiculite accounting for 50%, respectively, and the mixed media with peat, vermiculite, sawdust or soil, carbide, vermiculite or soil, sawdust, and vermiculite accounting for 33%, respectively, also had good seedling raising effects.

In production, two methods can be used to prepare seedling raising medium: Mechanical mixing and manual mixing. For large-scale or industrialized production of container seedlings, a special substrate processor can be used to control the crushing, screening, and mixing in a quantitative proportion of two or three kinds of substrates; in addition, chemical fertilizer or organic fertilizer can be added to the mix and prepared with the substrate at one time. For small-scale container seedling, the medium components can be broken and sifted by a mobile small crusher, or manually, and fertilizer can also be added to the mix.

The pH value of the prepared substrate should be measured and adjusted according to the requirements of the specific tree species cultivation. Generally, the pH value of conifer species is in the range of 5.5 to 7.0, and that of broad-leaved species is in the range of 6.0 to 8.0. If the pH value is too low, alkaline fertilizers such as calcium nitrate [$Ca(NO_3)_2$] or sodium nitrate ($NaNO_3$) can be added into the base fertilizer; if the pH value is too high, acid fertilizers such as ammonium sulfate [$(NH_4)_2SO_4$] or muriate of ammonia (NH_4Cl) can be added.

In China, many large-size container seedlings are cultivated on the ground, and the container can easily be affected by environmental stress. The medium components with higher bulk weight and stronger stability should be selected to facilitate the long-term growth of seedlings. Therefore, compared with the small-size container seedlings, the content of local soil components, such as yellow soil, forest topsoil, or garden soil, should be increased properly when selecting the seedling raising substrate for the large-size container seedlings. Doing so can reduce the seedling raising cost. Given the development trends of culture media at home and abroad that increasingly consider the economy and environmental protection, the utilization of low-cost, recyclable, and environmentally friendly forestry and agricultural residues will become the main direction of culture media application.

11.3.3 Selection of Nursery Land

The best place for container seedling cultivation is on gently sloping land with sufficient water sources, high terrain, and good drainage. The container seedling is produced by using the cultivation medium, therefore, the selection of the nursery land is not dependent on the soil conditions. These nurseries can make full use of abandoned land or plots that are not suitable for planting, because the nursery owners can cover the ground with covering materials that effectively prevent the growth of weeds and reduce the management cost. For example, in European and American nurseries, the ground management methods such as covering the ground with gravel, sawdust, and ground cloth are often used, and the thickness of the gravel is generally 5 to 10 cm. At present, some domestic greenhouses and advanced nurseries have also adopted ground covering, but there are fewer waste wood chips in China. Instead, some nurseries use gravel, cinder, and other materials for covering, while others use sunshade net instead of ground cloth. However, the sunshade net will age and break after long-time sunlight, resulting in a new round of ground pollution, so it is not recommended to use. Compared with long-term artificial weeding or other ground treatment materials, the horticultural ground cloth is the first choice of ground covering materials, with low price and good quality. At the same time, nursery owners must consider the convenience of transportation, power conditions, and communication to meet the requirements of nursery construction.

11.3.4 Main Cultivation Techniques of Large-size Container Seedlings

(1) Potting (or Changing the Pot) and Placing

In container cultivation, the technique of transplanting the seedlings raised in a container, as well as the small-size seedlings or bare-rooted seedlings cultivated on the seedbed in the nursery, into a container (or into the large-size containers) for further cultivation is called potting (or changing the pot), which is one of the key technical links of large-size container seedling cultivation. In China, the work of potting is a mainly manual operation, which requires a lot of labor and is time-consuming. In foreign countries, it is primarily a mechanized operation or an artificial auxiliary operation under mechanized conditions. In the United States, many types of machines are used for potting or changing the pot (Figure 11-18), but the operation process is basically the same. Generally, a tractor loads the substrate into the feed box of the pot-loading equipment with a loading shovel, and the stirrer device in the assembly machine agitates continuously to discharge the substrate from the discharge port. The workers need only to prepare the seedlings or containers, put them under the discharge port to load the pot, and then transfer the container seedlings after potting or changing the pot to a certain area through a conveyor belt. Then the pots are loaded and transported by special personnel to the nursery for placement. The mechanical filling equipment greatly improves the efficiency of potting and transportation, and the operation quality is high. With the increase of labor costs in China, it is an inevitable development direction to implement mechanized operations in container seedling production.

(a)   (b)

Figure 11-18 Examples of a container medium-filling machine and a manual auxiliary container medium-filling machine.
(a) one container medium-filling machine; and (b) another style of container medium-filling machine.

Container seedling placement is the main technical content of container cultivation. Before placing, the nursery area should be divided according to the type of container seedlings, such as arbor area, shrub area, weeping seedling area, specimen area, and so forth. In each region, consider the characteristics of the seedlings, such as the different requirements of seedlings in terms of water, light, acid, and alkali. Based on that information, distribute seedlings in different plots. The seedlings that require a particular environmental condition should be placed in an area that has adopted the same management measures, which not only facilitates the management but also benefits the growth and development of the seedlings. The row spacing of the placed large-size container seedlings should be set according to the specifications, cultivation years, and rate of growth of the seedlings. For large-size container seedlings, because of the difficulty of moving the containers, enough space should be reserved for the growth of seedlings. Generally, the seedling cultivation period is 2 to 3 years, after which it is best to lift the seedlings. For some slow-growing tree species or relatively small-size container seedlings, the container can easily be moved and replaced, so the reserved space may not be necessary.

According to the containers used and the locale-specific situation, three methods can be used to place the large-size container seedlings: On the ground, half-buried, and fully buried. A given placement form will have a certain impact on the management intensity and root control effect of seedlings. The container seedlings with half-buried and fully-buried placement can reduce the management intensity of seedlings, enhance the buffering effect of seedlings against the external environment, but weaken the root control effect of seedlings. To achieve a better root control effect, the production area is generally paved with gravel or ground cloth and then filled with containers.

After the potted container seedlings grow to a certain size, and the container can no longer meet the requirements of root growth, it is necessary to replace the container with a larger-size one.

(2) Fertilizer and Water Management

Container seedlings differ from bare-rooted seedlings in that they cannot obtain water and fertilizer supplements through the infinite extension of the root system, which has significant limitations. Container seedlings are more dependent on external water and fertilizer supply, so the management of water and fertilizer is an important cultivation technology measure in container seedling cultivation.

The water quality, irrigation method, amount, and times of large-size container seedling irrigation will have an impact on container cultivation. Generally speaking, neutral or slightly acidic water with a low content of soluble salts (a salt content of less than 1.5%) is conducive to the growth of seedlings. It is best to use pipeline irrigation, such as fixed sprinkler irrigation or mobile sprinkler irrigation, for example, although a nursery with good conditions can also adopt drip irrigation. For shrubs and seedlings with a height of less than 1 m, spray irrigation is often used, and drip irrigation is the main method for large-size seedlings with a sparse arrangement. In the United States, computer-controlled sprinkler irrigation is adopted for the large-size container seedlings in many nurseries; in addition, the water recycling system is designed to save water and labor, with even irrigation and good irrigation effects. If automatic drip irrigation is adopted, the water-saving effect is more obvious, and the automatic drip irrigation technology can also combine watering with fertilization. Therefore, considering long-term development and increasing labor costs, the adoption of advanced irrigation technology is the future development trend.

Water consumption of container seedlings is generally higher than that of field-cultivated seedlings. The times of irrigation are adjusted with the change of seasons. Generally, watering should be carried out once every 1 to 2 weeks for large-size containers and once every 3 to 6 days for small-size containers.

Because the root system of container seedlings is concentrated in a limited-space container, it is inconvenient to apply fertilizer, so a more suitable option is to select a slow-release fertilizer. Slow-release fertilizer can control the release of urea, ammonium hydrogen phosphate, potassium chloride, and other rapidly soluble fertilizers; slow down the release rate of nutrients; improve the utilization rate of fertilizer; reduce the intensity of fertilization; and change multiple fertilizations into one time of fertilization, or even fewer depending on how long the seedlings need to be in the containers. Slow-release fertilizer needs to be applied only 1 to 2 times per year, which greatly saves labor. Fertilizer can be directly mixed into the substrate during potting or pot changing, with good effect and even root growth and distribution as a result. In the growth period of seedlings, fertilizer can be directly applied on the surface of the substrate in the container, similar to topdressing, but the utilization rate of fertilizer will be reduced. In the production of large-size green seedlings, the use of slow-release fertilizer is more and more common.

(3) Cold Proof in Winter

In the cold area of North China, container seedlings are more vulnerable to low temperatures in winter compared with the field-cultivated seedlings. The root system of container seedlings will be affected by frostbite in the following year and even die. Therefore, it is necessary to implement cold prevention measures for container seedlings before winter. First, the seedlings can receive an application of antifreeze water; the stem can be brushed with a white or antifreeze solution and wrapped with straw rope; and the roots can be wrapped with film or covered with sawdust, straw, wheat straw, or rice husk. Second, the seedlings can be moved into a greenhouse or plastic shed. To save space, the container seedlings moved into the greenhouse should be placed close together, even in layers. In addition, the container seedlings can be placed with the container half-buried or fully buried underground.

(4) Transpiration Inhibition in Summer

High temperatures in summer can cause stronger transpiration, leading to water loss and wilting of plants, even drying and death. Especially, the root system of container seedlings has limited contact with the substrate, which can only absorb the water provided by the substrate in the container, so the phenomenon of water loss of plants will be more obvious. Therefore, taking some measures to restrain summer transpiration can not only reduce water consumption but also be beneficial for plant growth. These measures include setting up a sunshade net, spraying water to cool down, and using some plant transpiration inhibitors such as mercury benzene acetic acid, long-chain alcohols, silicone, butadiene acrylic acid, and so forth. In addition, some thinning and pruning should be carried out to reduce the transpiration area when summer comes. Choosing a reasonable way of container seedling placement can also play a role in moisturizing and inhibiting the intensity of transpiration.

(5) Fixation and Binding of Container Seedlings

For the fixation of container seedlings, different measures can be taken, but they are basically carried out from two aspects of seedling fixation and container fixation. Container seedlings placed on the ground easily blow down or are moved by the wind, so the first thing to do is to fix the seedlings. Generally, it is necessary to set the column and bind the seedling trunk to the column with plastic tape or rope to ensure that the seedling is upright (Figure 11-19a、b). In the North American nurseries, seedlings are fixed with a small tool similar to a stapler, which is convenient and quick to operate. The columns used for fixing seedlings include bamboo poles, steel bars, and so on, and the short ones are approximately 1 m long with the long ones measuring 2 to 3 m long. For the large-size container seedlings, it is generally necessary to build fixed columns at both ends of the seedling row and erect iron wire or steel wire to fix the seedlings (Figure 11-19c), or fix the containers with specific supports (Figure 11-19d).

Figure 11-19 Binding of container seedlings and container fixation.
(a) seedlings binding on the stand columns; (b) seedlings bound on the stand columns; (c) erecting iron wire to fix large-size seedlings; and (d) fixing the containers with iron wire supports.

Questions for Review

1. What is transplantation? Why is transplanting the key technology for large-scale seedling cultivation?
2. What are the key technologies of bare-root transplanting?
3. Why is the dormancy period the best time for seedling transplantation? How could you improve the survival rate of out-of-season transplanted seedlings?
4. What should be done before transplanting the seedlings with the soil boll? How to improve the seedling lifting quality of seedlings with the soil boll?
5. How to improve the transplanting survival rate of seedlings with the soil boll?
6. What are the advantages of large-size container seedling cultivation compared with open field cultivation?
7. How to choose the medium of large-size container seedling cultivation? What should be paid attention to in medium proportioning?
8. What are the main techniques and key points of large-size container seedling cultivation?
9. What is the purpose of cultivating transplanted seedlings?
10. What is the key technology of transplantation?
11. What is the key technology to ensure the survival in the process of transplanting when cultivating transplanted seedlings?

References and Additional Readings

CHENG F Y, 2012. Nursery of landscape plants[M]. Beijing: China Forestry Publishing House.

DENG H P, YANG G J, WANG Z P, 2011. Research status on cultivating techniques of big container seedlings[J]. World Forestry Research, 24(2): 36-41.

JIANG D S, BAO Z Y, 2004. Production of nursery plants for landscape gardening[M]. Beijing: China Forestry Publishing House.

WANG J F, CHEN Z M, LIU J X, et al., 2016. Effect of different substrates and slow-release fertilizer loading on growth of *Taxus chinensis* var. *mairei* container saplings[J]. Journal of Zhejiang Forest Science & Technology, 36 (2): 74-78.

ZHAI M P, SHEN G F, 2016. Silviculture[M]. 3rd edition. Beijing: China Forestry Publishing House.

ZHANG D L, SHU Y Z, CHEN W, 2003. Garden nursery seedling handbook[M]. Beijing: China Agriculture Press.

ZOU X Z, QIAN S T, 2014. Technology of seedlings production[M]. Beijing: China Forestry Publishing House.

Chapter 12 Seedling Pruning

Zhao Hewen

Chapter Summary: This chapter discusses the biological basis, the principles, and the methods of seedling pruning.

Pruning refers to the adjustment of tree growth in order to make the tree form a certain ideal tree shape. It is the implementation of certain technical measures on the tree plant so that it forms the structure and shape required by the cultivator.

Seedling pruning is important to remove excess growth of the branches or poor aesthetic branches, cultivate strong tree potential, regulate the growth and development of trees, and ensure the health of trees. In addition, seedling pruning can prevent overspread branches and leaves and tree disorder, control tree structure, and cultivate good tree shape. More broadly, pruning can adjust the relationship between trees and the environment, improve the ventilation and light transmittance inside the canopy, reduce the infection of diseases and insect pests, and help rejuvenate old trees.

12.1 Biological Basis of Pruning Seedlings

Understanding the biological characteristics of seedlings helps make clear the effect of pruning on seedling health. This section will cover the interrelationship between the parts of the seedlings.

12.1.1 Root System and Pruning

The root system has 5 functions: Support the aboveground part of the seedlings and maintain the stable function of the seedlings; absorb the minerals and water in the soil; generate some root microorganism organisms, such as rhizobia, benefiting the growth of seedlings; synthesize cytokinin; and store part of the energy generated by photosynthesis in the leaves.

The root system is the most easily forgotten part of seedling pruning. Nowadays, many seedlings are raised in large containers. Sometimes, because of the limitation of space, container seedling raising can cause damage to the root system of the tree, such as the phenomenon of one-sided rooting and loop rooting, resulting in poor stability of the tree body, the possibility of tree collapse or falling (i. e., lodging), and even death of the seedlings (Figure 12-1).

The distribution of roots in the horizontal direction is often 3 times that of the crown (Figure 12-2). More than half of the roots extend beyond the crown projection.

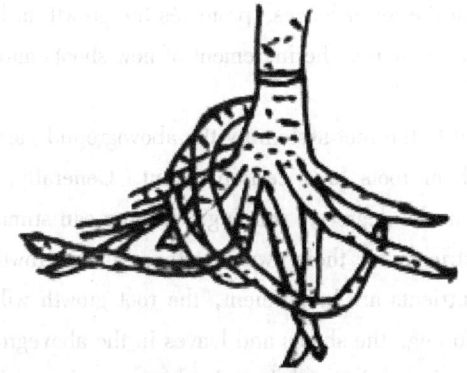

Figure 12-1 Root ring phenomenon caused by a too-small container.

Figure 12-2 Horizontal distribution of root systems of seedlings is 3 times larger than the crown diameter.

Seedlings with defective roots need pruning. For example, for the seedlings raised in a container, the root bending phenomenon will occur when the tree grows to a certain size, and the lateral root is seldom produced outside the bent root. When the seedlings in the container are outplanted into the garden, the root system will tend to grow on one side, and these kinds of seedlings will easily lodge. At that point, appropriate thinning and cutting of the crown is in order to remove the coiling, so as to promote the outward extension of the root system and enhance its wind resistance.

12.1.2 Relative Balance between the Aboveground Part and the Underground Part

Seedlings are a unity of opposites composed of crown and root system. Once pruned, the aboveground part and the root system of seedlings change, breaking the physiological balance between crown and root system. The seedlings have the characteristics of strong regeneration function, and after pruning, the new branches will grow rapidly to obtain a new balance. The vigorous growth of branches and leaves promotes new growth of the root system. The interaction of the two makes the seedlings keep vigorous growth potential.

Pruning cuts off a small part or a large part of the aboveground organs of the seedlings and reduces thephotosynthetic assimilation area, so that the carbohydrate content in the plants changes significantly. By measuring the starch content in the roots of the pruned plants compared with the control (no pruning), the pruned plants were experiencing slow growth for a long period of time. This outcome is because at the early stage of pruning, the photosynthetic assimilation area is reduced, so the materials needed for regeneration and growth after pruning need to use the nutrients stored in the roots. Therefore, the promotion of pruning on seedlings is to reduce the number of branches and leaves, so as to change the distribution relationship between the original nutrition and water in the roots, so that the nutrients are supplied to the remaining axillary buds and adventitious buds. At the same time, pruning improves the ventilation and light transmission conditions of the seedlings, increases the photosynthetic performance of the lower leaves, promotes the growth of buds of this part, prolongs the growth period, and therefore increases the increment of new shoots and the area of single leaves.

After pruning, the energy and nutrients supplied to the root system by the aboveground part are reduced. Therefore, pruning can inhibit the growth of roots to a certain extent. Generally, the increment of root system decreases with the increase in the degree of pruning. Pruning can stimulate the growth of roots when the roots store enough nutrients for their own growth and the growth of aboveground shoots and leaves. When the storage nutrients are insufficient, the root growth will be restrained at the initial stage of pruning, but after pruning, the shoots and leaves in the aboveground part will not only provide nutrients for their own sprouting and growth but also have certain nutrients available for the root growth. At this time, pruning can make the aboveground part rejuvenate while also promoting root growth.

12.2 Principles of Pruning

Pruning modeling is an important part of seedling conservation. In urban greening, any type of trees should be pruned and shaped according to their functional requirements so that they can coordinate with the surrounding environment and display their ornamental traits most effectively. Therefore, pruning is one of the regular work events in seedling planting and conservation, and it is an important measure to adjust the tree structure, restore the vitality of seedlings, and promote the growth balance.

12.2.1 General Operation Procedures

The overall operating procedures of seedling pruning are summarized as "first knowing, second observing, third pruning, fourth taking, fifth processing, and sixth protecting". *Knowing* means that all technical personnel who participate in pruning must master the operative procedures, technical specifications, safety regulations, and special requirements. *Observing* means looking closely at trees before pruning to determine how best to prune specific trees. *Pruning* trees reasonably means according to the principles of adjusting measures to local conditions and the trees themselves. *Taking* means the pruned branches should be removed promptly and transported away collectively to ensure a clean and tidy environment. *Processing* the cut branches should be tended to promptly to prevent the spread of diseases and insect pests. *Protecting* involves thinning and removing big branches and thick branches.

According to the pruning scheme, the branches to be pruned need to be marked. Pruning should be carried out according to the order of the lower part first and the upper part later; the inner shoots first and the peripheral shoots later; and moving from bulk pruning to fine pruning. Generally, pruning starts from thinning, pruning, and removing dead branches, dense branches, and overlapping branches. After that, according to the order of large, then medium, and then small branches, the perennial branches should be cut back. Finally, according to the needs of modeling, short pruning needs to be carried out for the 1-year-old branches. After pruning, check the rationality and results of pruning, for example, if some portion of pruning has been skipped or is wrong, correct it promptly.

12.2.2 Principles of Pruning Modeling

When pruning, we should follow the basic operation principle of "from the whole to the part, from the bottom to the top, from the inside to the outside, removing the weak and retaining the strong, removing the old and retaining the new". Cuts should be smooth and neat, without ponding (surface of the cuts should not collect water) or stumps (do not leave long branches on stem after making cuts). When pruning the bough, avoid the bough falling and tearing the bark. Diseased and insect-ridden branches, dry and withered branches, water sprouts, inverted branches, and shady branches need to be cut off as soon as they appear, and the lopsided crown or overly dense branches should be pruned in time to maintain a balanced and transparent crown.

When pruning, we must observe and then prune seedlings according to the rules of the tree structure. The observation and analysis should be made according to the crown structure, tree vigor, growth of main and lateral branches of the pruned seedlings, and the specific pruning plan should be made according to the pruning purpose and requirements.

Those engaged in pruning should know the biological characteristics of seedlings, technical specifications, and safe operation. When pruning trees, observation and analysis should be done first. If the tree vigor is unbalanced, analyze whether it is strong or weak in the upper part, in the lower part, or unbalanced between main branches. Consider the causes so as to adopt corresponding pruning technical measures. If it is because branches are too abundant, especially too many boughs (meaning the main branches) with strong growth potential, thinning should be carried out. Before thinning, first determine the number of boughs and their positions on the skeleton branch; cut off the useless boughs first. Determine the strongest, primary boughs that should be kept and how they will help to determine the shape of the tree. If small branches and middle branches are cut first, that work may have been useless if you then find main boughs that are weak and should be pruned out. The smaller branchlets should be pruned after the main bough(s) are adjusted well, and pruning should be done from the top of each main branch or each lateral branch in turn. At this time, pay special attention to the cutting-back height of elongated shoots of each main branch or each lateral branch, and balance the branch potential among them. Finally, balance the overall tree potential through the corresponding length of the same type of elongated shoots at all levels.

In the past, the cut off plant debris was generally removed and disposed of elsewhere. Now, a mobile wood chipper is often used to cut and crush the branches into small wood chips on the spot, which can save the transportation cost and work and allow for wood debris to be reused. For example, some nurseries use the wood chips in the walkways between planted rows.

12.2.2.1 According to the Purpose of Urban Greening

Various pruning modeling measures result in outcomes that differ from one another, and different greening purposes have their own special pruning requirements, so the cultivation purpose requirements of the tree have to be clear for pruning modeling. For example, the pagoda tree

(*Sophora japonica*) should be pruned into a cup shape when it is planted as a street tree, but it should be left in its natural tree form if it is planted as a court shade tree. For the Chinese juniper (*Sabina chinensis*), pruning modeling requirements when it is planted on the lawn for viewing differ from what is needed for making hedges, so the specific pruning modeling methods also differ.

12.2.2.2 According to the Environmental Conditions of the Growing Land

Pruning methods differ depending on the ecological environment. For flowers and trees growing in thin soil and higher groundwater level, the trunk should be kept low, and the crown should be small; in the *tuyere*, or windy area, a low trunk and low crown should be adopted and the branches need to be relatively sparse. In addition, saline-alkali land has a limited amount of soil replacement, and the soil layer is correspondingly thin, so when planting trees, the low trunk and low crown should be adopted, otherwise it will not have a good ornamental effect.

Pruning methods also differ dependent on the configuration of the environment. If the flowers and trees grow in a wide area within a large space, the branches can be opened as much as possible without affecting the surrounding configuration, which maximizes the crown of the trees; if the space is smaller, the volume of individual plants should be controlled by pruning to prevent crowding and reduce the viewing effect.

If a particular species of tree is planted in a different configuration environment, the pruning methods will differ. For example, three pruning methods are used for flowering almond (also known as flowering plum) (*Amygdalus triloba*) in Beijing: Plum-pile shape, suitable for collocating beside buildings and rocks; round-head shape with the trunk, collocating in front of evergreen trees and on both sides of garden roads; the clump-oblong shape, best planted on hillsides and lawns.

12.2.2.3 According to the Biological Characteristics of Trees

(1) Growth and Development Period

Appropriate pruning methods and intensities should be adopted for trees of different ages based on their growth vigor and development stage.

Early Stage. This stage should be based primarily on modeling, laying a solid foundation for the growth of the whole life cycle and giving full play to its garden function benefits. The main task of modeling is to match the main and side branches, expand the crown, and form a good tree body structure. The precocity of the flower and fruit trees should also be promoted by proper pruning.

Middle-aged Stage. A complete and beautiful crown should be formed in the middle-aged stage. The purpose of pruning modeling is to maintain the perfect and robust state of the plant; delay the arrival of the aging stage; adjust the competing priorities among growth, flowering, and fruiting; and stabilize the flowering and fruit-bearing time.

Tree Senescence. Because of weakening vigor, the growth increment is decreasing year by year, and the crown is in the stage of centripetal regeneration (meaning the original apex dominance may be changing from one stem to another). Intense pruning is typically adopted, so as to stimulate the germination of dormant buds, to renew and rejuvenate the inner part, to restore the growth vigor, and to achieve the purpose of renewal and rejuvenation by using the elongated shoots.

(2) Position of Bud, Nature of Flower Bud, and Flowering Period

For the flowers and trees blooming in spring, the buds usually differentiate in the summer and autumn of the previous year and grow on the 1-year-old branch. Therefore, when pruning in the dormancy season, pay attention to the location of the buds. The flower bud that grows on the top of the branch is called the top flower bud. The flowers and trees with the top flower bud, for example, Yulan magnolia (*Magnolia denudata*) and Manchu rose (*Rosa xanthina*) cannot be cut back in the dormancy season or before the flowering (except for the purpose of renewing the branch potential). If the flower bud grows in the axil of the leaf, it is called an axil flower bud, and according to the needs, the branch can be cut back before the flowering [e. g., *Amygdalus triloba* and peach blossoms(*Prunus* spp.)].

The trees with axillary pure flower buds, such as weeping forsythia (*Forsythia suspensa*) and peach blossom, should not have flower buds at the cut point when cutting branches back. Because pure flower buds bloom but cannot grow branches and leaves, after flowering there will be a short dry branch left here. If there are too many dry branches, the ornamental effect will be affected. For fruit viewing trees, if there are no branches and leaves above flowers as the source of organic nutrition, they cannot bear fruit after flowering, resulting in the reduction of fruit yield.

For tree species that flower in summer and autumn, flower buds form in the new shoot in the same year, for example, crape myrtle (*Lagerstroemia indica*) and Rose of Sharon (*Hibiscus syriacus*). Therefore, these types of trees should be pruned after leaves fall in autumn and before sprouting in early spring. Since it is cold in winter and dry in spring in Beijing, pruning should be postponed until the temperature rises in early spring, when buds are about to sprout. During pruning, leave 3 to 4 pairs of full buds at the base of 1-year-old branches, which can sprout strong branches after pruning. Although the number of flower branches may be less, because of the concentration of nutrients, larger flowers will be produced. For some flowers and trees that could bloom twice in the same year, the residual flowers can be cut off after flowering, strengthening the management of fertilizer and water. Crape myrtle is also known as "one-hundred-day red" because it can blossom up to 100 days after removing the remnant flowers.

(3) Branch Properties

For the tree species with main axis branches, when pruning, attention should be paid to control the lateral branches, cut off the competitive branches, and promote the development of main branches. For example, trees with a spire-shaped or cone-shaped crown, such as the Lombardy poplar (*Populus nigra* var. *italica*), Chinese white poplar (*Populus tomentosa*), and ginkgo (*Ginkgo biloba*), have strong apical growth potential and an obvious trunk, so it is suitable to adopt the modeling method of retaining the central leading trunk. The tree species with sympodial branches readily form several branches with equal force, showing a multi-forked trunk. To cultivate the trunk, the terminal bud of other lateral branches can be removed to weaken its apical dominance, or the telome can be cut back, with strong buds left at the cut, and at the same time, 3 to 4 lateral branches under the cut can be thinned to accelerate its growth.

For the tree species with pseudo-dichotomous branching, apical buds cannot be formed from the top tip of tree trunks at the later stage of growth, so the opposite lateral buds below have a balanced advantage, which affects the formation of the trunk. For these trees, peeling off one of the buds can be used to cultivate the trunk. For the tree species with polytomous branching, the central main branch can be re-cultured by the method of bud picking or cutting the main branch back. In pruning, we should fully understand the characteristics of all branches, pay attention to the balance between all branches, and master the principle of "intensely pruning for strong main branches, less pruning for weak main branches". This approach is important because the main branch is generally strong, with more new shoots, larger leaf area, and more organic nutrients produced, which makes it grow stronger. By comparison, the weak main branch has fewer new shoots and poor nutrition conditions, which makes it grow weakly. Therefore, pruning should balance the growth vigor of various branches. Lateral branches are the basis of flowering and fruiting, and it is not easy to form flower buds if they grow too strongly or too weakly. Hence, the purpose of less pruning for strong lateral branches is to promote the germination of lateral buds, increase branching, moderate growth vigor, and promote the formation of flower buds. At the same time, the growth and development of flower and fruit inhibit the growth potential of strong lateral branches. The weak branch should be pruned intensely to make it germinate a stronger branch, because the flower bud formed by this branch is weaker with less nutrition consumed.

12.3 Method of Pruning Seedlings

12.3.1 Pruning Season

There are many types of seedlings, with a range of habits and functions. According to the purposes and properties of pruning, although each has its own suitable pruning season, on the whole, trees can be pruned at any time of the year, and the selection of a specific time should be based on the purposes and properties of pruning.

The pruning season of seedlings is generally divided into winter (dormancy period) pruning and summer (growth season) pruning. Dormancy period refers to the time frame from falling leaves to the beginning of sap flow in the early spring of the following year (generally from December to February of the following year); growth period refers to the time from germination to the end of growth of new shoots or auxiliary shoots (generally from April to October).

12.3.1.1 Pruning in Winter (Dormancy Period)

In winter (the dormancy period), the nutrients stored in the tree body are sufficient, and after the aboveground part is pruned, the branches and buds are reduced, which means the stored nutrients can be used intensively. Therefore, the growth of new shoots is strengthened, and the buds near the cut have a long-term advantage. For deciduous fruit trees with normal growth, they should be pruned approximately 1 month after defoliation. When pruning after sprouting in spring, the sprouted shoots have consumed part of the stored nutrients. Once the sprouted shoots are cut off and the lower buds sprout again, growth is delayed and the growth vigor has obviously weakened. When pruning during the entire winter, the young trees, trees with good benefits, trees with poor wintering ability, and trees on dry land should be pruned first. In terms of time management, it is also necessary to first ensure the pruning of trees that present more technical difficulties. For example, bleeding phenomenon, which can occur in grapes, means that the nutrients and water in the tree body flow out at the tree wound, and if the bleeding is too much, the tree vigor will weaken, even the branches will die. Therefore, for these species it is better to carry out pruning during the dormancy season when the bleeding is less and is easier to stop.

12.3.1.2 Pruning in Summer (Growth Period)

Pruning in summer (the growth period) can be carried out during the entire growth season from germination in spring to defoliation in autumn. The main purpose of pruning in this period is to improve the ventilation and light transmittance of the tree crown. Generally, less pruning is used to avoid adverse effects on seedlings, which can be caused by cutting off a large number of branches and leaves. Pruning in summer is an easier time to adjust the light and shoot density and to judge diseases and insect pests. Dead and weak branches can be removed to trim the crown into an ideal shape. Pay more attention to pruning in summer for young tree modeling and controlling vigorous growth.

12.3.2 Tree Types That May Need to Be Pruned

12.3.2.1 Deciduous Trees

For tree species, especially those with strong apical dominance, such as poplar, metasequoia, and larch seedlings, pay attention to the timely thinning of the root sucker and the lateral branches below 1.8 m of the trunk. Then, with the continuous increase of the trunk, the branches at the lower part of the middle trunk will be thinning year by year, and the overly dense branches in the crown and the branches disturbing the tree shape will be thinning at the same time. For tree species with weaker apical dominance and stronger germination, such as *Sophora japonica*, corkscrew willow (*Salix matsudana*), and plane trees (*Platanus*), the trunk is often cultivated by cutting trunk (i.e., cutting the main stem). When the plant height reaches 2.5 to 3.0 m, the required height, it is cut at 1.8 to 2.0 m to culture the main trunk and branches. In later stages, pay attention to the protection of the elongated shoot of the trunk, and pinch the lateral branch to promote the growth of the elongated shoot of the trunk. (See more on pinching later in this chapter under 12.3.5. Pruning Methods, section (6) Other Methods.) Select 3 to 5 branches evenly distributed around the tree as main branches, which should be cut back at 30 to 40 cm in the next year to promote the growth of lateral branches and form the basic tree shape. Branches in the lower part of the trunk need to be thinned year by year; at the same time, the overly dense branches in the crown and the branches disturbing the shape of the tree should be thinned.

12.3.2.2 Evergreen Trees

For the evergreen tree species such as cedar and spruce, trees have a strong trunk and obvious central leading branch, and the trunk is upright and does not bifurcate, so it is necessary to maintain the dominance of the central leading trunk growing upward. If competing branches are on the trunk, one strong branch should be chosen as the central leading trunk, and the other branch (es) should be shortened and retracted for cutting back in the second year. The main stem of some seedlings is bent or weak, which is bound to affect the normal growth of plants. In the following years, pruning will mainly affect dead branches, diseased and insect-ridden branches, overly dense branches, or a small number of branches with unsuitable azimuth (meaning they are growing in an undesired direction or angle). Pay close attention to protect the top tip of the trunk.

For the evergreen tree species with weaker apical dominance and stronger germination, such as southern magnolia (*Magnolia grandiflora*), Chinese banyan (*Ficus microcarpa*), and rubber fig (*Ficus elastica*), the vigorous and upright branches need to be selected as the trunk, and the overly dense lateral branches can be removed. The rest of the branches should be removed according to specific conditions, such as removing the thick competition branches, water sprouts, upright branches, drooping branches, and dead branches at the top of the crown. In summer, the vertical and strong branches shall be removed from the pruning, and according to the principle of "pruning the strong and keeping the weak, removing the straight and keeping the flat, pruning more for the upper part of the crown and pruning less for the lower part", pruning should be carried out by batch. In the following years, the branches at the lower part of the trunk, the overly dense branches in the crown, and the branches disturbing the shape of the tree should be thinned.

12.3.2.3 Shrubs

For bushes with strong sprouting and pruning resistance, such as white dogwood (*Swida alba*), bridewort (*Spiraea salicifolia*), and Japanese kerria (*Kerria japonica*), it is advisable to prune them with garden shears or a hedge trimmer; that is to say, prune them more when they are young to promote the development of new shoots, and then prune them for daily maintenance. From the beginning of winter to early spring every year, necessary thinning and cutting back should be carried out for dense plants and excessively dense branches, disease and pest branches, water sprouts, and overly weak branches. In addition, the branches need to be evenly distributed and formed. For low shrubs, such as the China rose (*Rosa chinensis*), rhododendron, and boxwood (*Buxus sinica* var. *parvifolia*), pruning can be divided into shrubby pruning and tree pruning according to different applications. In terms of shrubby pruning, the flower residue is cut in winter, with more axillary buds left in order to facilitate more new shoots sprouting in the early spring. Branches in the upper part of the trunk grow stronger and more buds should be left; branches in the lower part of the trunk are weaker and fewer buds should be left in place. After flowering in summer, the inner buds should be left in the low shrubs with expansive growth and the outer buds should be left in the low shrubs of upright growth. With regard to tree pruning, when the new trunk reaches the appropriate height, pinching should be carried out, and 3 to 4 axillary buds below the trunk cut are selected as the main branch to be cultured. Remove other axillary buds from the trunk. During the growth period, the main branch is pinched to promote the secondary branch germination. In the growing period, the flower residue and overly dense branches should be cut off early in the season. For shrubs with large tree shapes, such as crape myrtle, flowering peach, and sweet osmanthus (*Osmanthus fragrans*), most of them are ornamental shrubs, and many types of reshaping, such as cup-shaped, natural open heart-shaped, and multi-trunk cluster shaped, can be applied to the seedlings according to their specific use.

12.3.2.4 Liana

Lianas such as wisteria, Chinese trumpet vine (*Campsis grandiflora*), and rambling rose are pruned twice a year, generally in summer and winter. The thick branches need to be selected as the main vine to be cultured, and the unnecessary base tillering should be cut off at any time, so that the nutrients are concentrated in the main vine to promote its continuous thickening. Then, among the lateral vines sprouted from the main vine, 2 to 3 strong branches should be selected as the main lateral vine to pull to attach to the pillars, and a part of a small lateral vine is left as the auxiliary vine. In summer, the auxiliary vine is pinched to promote growth of the main lateral vine, and in winter, the main lateral vine is cut back to the strong bud, which promotes sprouting of the strong vine in the next year and blooming on the top. For some small lateral vines on main branches, attention should be paid to the proper pruning and keeping a certain distance to form the main lateral vine structure with clear primary and secondary layers.

12.3.3 Basic Techniques of Pruning Modeling

12.3.3.1 Tree Structure

The main parts of a tree (Figure 12-3) include crown, trunk, middle trunk, main branch, lateral branch, blossom bud branch, and extension branch.

Crown: A general term for the branches and leaves above the trunk.

Trunk: The portion of the tree from the first branch point to the ground.

Middle stem: An extension of the trunk in the crown.

Main branch: The main branch growing on the middle stem.

Lateral branch: A main branch growing from the main branch.

Flowering branch group: A group of branches composed of flowering branches and growing branches.

Branch group: A group of branches formed by branching from lateral branches.

Elongated branch: The elongated part of the apex of the backbone branch at all levels.

Backbone branches: The permanent branches that make up the crown skeleton, such as trunk, middle trunk, main branch, and lateral branch.

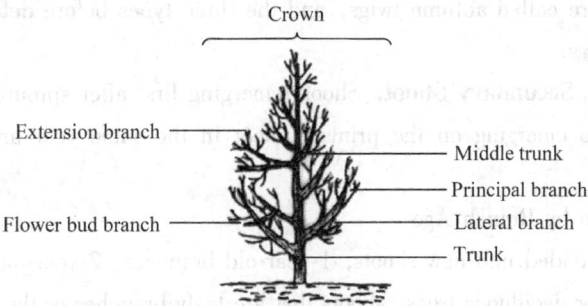

Figure 12-3 Main parts of trees.

12.3.3.2 Type of Branch

In the aspect of pruning modeling, the types of branches are studied and analyzed from the following aspects.

(1) Classification According to the Position of Branches on the Tree

Classifications can be divided into trunk, middle trunk, main branch, lateral branch, and elongated branch.

(2) Classification According to the Posture of Branches and Their Interrelation

These classifications can be divided into vertical branch, oblique branch, horizontal branch, droopy branch, inward branch, overlapping branch, parallel branch, whorled branch, cross branch, adnate branch, and so forth.

Vertical Branch, Oblique Branch, Horizontal Branch, Droopy Branch, and Retrograde Branch. All branches that grow upright and are perpendicular to the ground are called vertical branches; the branch with a certain angle to the horizontal line is called an oblique branch; the branch parallel to the horizontal line is called a horizontal branch; the branch with a downward growth tip is called a droopy branch; and the branch with an inverted position is called a retrograde branch.

Inward Branch. A branch growing to the crown.

Overlapping Branch. Two branches overlapping each other in the same vertical plane.

Parallel Branch. Two branches growing parallel to each other on the same horizontal plane.

Whorled Branch. If the growing points of multiple branches are close to each other, they can emerge from one point and radiate outward to the surrounding area.

Cross Branch. Two branches crossing each other.

Adnate Branch. Two or more branches grow from a point or a bud of the node order. (That is, when a branch grows from a usually unlike part.)

(3) Classification According to the Period and Order of Emergence in Growing Season

Growth can be divided into spring twig, summer twig, and autumn twig; primary shoot, secondary shoot, and so forth.

Spring Twig, Summer Twig and Autumn Twig. Twigs emerging from dormant buds in early spring are called spring twigs; twigs emerging from July to August are called summer twigs; twigs emerging in autumn are called autumn twigs, and the three types before defoliation are collectively referred to as new twigs.

Primary Shoot, Secondary Shoot. Shoots emerging first after sprouting in spring are called primary shoots; shoots emerging on the primary shoot in the same year are called the secondary shoots.

(4) Classification by Branch Age

Growth can be divided into new shoots, 1-year-old branches, 2-year-old branches, and so on.

New Shoots. For deciduous trees, shoots that are leafy branches or the current year's branches before defoliation are called new shoots; for evergreen trees, shoots that emerge from spring to autumn in the same year are called new shoots.

1-year-old Branch and 2-year-old Branch. Branches emerging in the same year from the defoliation to the germination in the next spring are called 1-year-old branches; the 1-year-old branch since sprouting till spring in the next year is called the 2-year-old branch.

(5) Classification by Nature and Use

Nature and use can include categories such as vegetative shoots, elongated shoots, leafage branches, flowering branches (fruiting branch), regeneration branches, and auxiliary branches.

Vegetative Shoot. This term refers to all growth branches, and includes long, medium, and short growth branches, leafage branch, and elongated shoot.

Elongated Shoot. These branches have particularly vigorous growth, thick shoots, and large leaves, long internodes, small buds, high water content, and non-full tissues, which often grow uprightly and are called the elongated shoot.

Leafage Branch. These short branches with short internodes and dense leaves, often show a rosette shape and are called the leafage branch.

Flowering Branch (Fruiting Branch). Branches with the blossom buds are called flowering branches for ornamental flowers and trees and are called fruiting branches for fruit trees. According to the length of the branch, it can be referred to as long flowering branch (long fruiting branch), middle flowering branch (middle fruiting branch), short flowering branch (short fruiting branch), and bouquet-shaped branch. Tree species differ in their dividing criteria.

Regeneration Branch. This new branch that replaces the old branch is called a regeneration branch.

Auxiliary branch. This branch helps the tree with nutrition, such as when the weaker branch left on a young tree trunk contributes to providing nutrition and makes the trunk full. When reserved temporarily, it is called an auxiliary branch.

12.3.4 Basic Modeling Mode

The modeling methods of seedlings vary greatly depending on the planting purpose, configuration mode, and environmental conditions. In practical application, these modeling shapes are the primary modes.

12.3.4.1 Natural Modeling

This modeling refers to a natural tree shape formed on the basis of the unique shape of the tree species, with a little manual adjustment and intervention according to the growth and development habits of the tree itself. The tree shape not only reflects the natural beauty of seedlings but also conforms to the growth and development habits of trees, which is beneficial to the long-term maintenance and management of trees. Street trees, court shade trees, and general landscape trees basically adopt natural modeling, for example, an oblong shape, such as the Yulan magnolia (*Magnolia denudata*), Chinese crab tree (*Malus spectabilis*); a circular spherical shape, such as the Manchu rose (*Rosa xanthina*), flowering plum (*Prunus triloba*); oblateness, such as a locust tree (*Robinia* spp.), peach (*Prunus* spp.); an umbrella shape, such as Persian silk tree (*Albizia julibrissin*), *Amygdalus persica* cv. Pendula; an oval shape, such as apple, purple-leaf plum; and an arch shape, such as weeping forsythia (*Forsythia suspensa*), and winter jasmine (*Jasminum nudiflorum*).

12.3.4.2 Artificial Modeling

According to the special requirements of urban greening, trees are sometimes pruned into regular geometric shapes, such as squares, circles, and polygons, or irregular shapes, such as birds and animals. This type of modeling is against the natural law of the growth and development of trees, with strong inhibition intensity. The plant materials adopted to this modeling require strong germination and branching abilities, for example, Oriental arborvitae (*Platycladus orientalis*), boxwood, elm, genista, plum pine (*Podocarpus* spp.), June flower, privet, yew, coral, Japanese photinia (*Photinia glabra*), and an ash tree (*Fraxinus hupehensis*). Dead branches should be cut off immediately. Dead plants need to be replaced immediately to keep the planted trees in order, so this kind of modeling is often used to meet special ornamental requirements.

(1) Modeling Method of Geometry

Pruning modeling is carried out according to the composition standard of geometric form, such as spherical, hemispherical, mushroom shape, cone, cylindrical, cube, cuboid, gourd shape, castle type, and so forth.

(2) Modeling Method of Non-Geometry

Wall Type. The purpose of this shape is to create a vertical greening wall near the garden and buildings, which is often seen in classical gardens in Europe. Common forms include U-shaped, cross-shaped, and rib-shaped.

In this way, the main trunk is kept low as it is meant to be short, and the main branches are arranged symmetrically or radially on the left and right sides of the trunk and are kept in the same vertical plane.

Sculpture Type. According to the intention of the gardener and/or artist, all types of forms can be created. Note that ideally the form of the trees should be coordinated with the surrounding landscape, and the lines should not be too complicated. It is preferable to have a clear and concise outline.

12.3.4.3 Natural and Artificial Mixed Modeling

Central Leading Trunk Shape. In this model, there should be a strong, central leading trunk, on which scattered main branches are retained, primarily with a semi-circular crown. If the main branch is layered, it is called a scattered stratified form. The first layer comprises 3 to 4 main branches relatively adjacent to each other. The second layer comprises 2 to 3 main branches. The third layer also has 2 to 3 main branches. After that, 1 to 2 main branches are left in each layer until reaching a total of 6 to 10 main branches. The distance between the main branches of each layer is successively reduced moving upward. In this tree form, the growth advantage of the central leading branch is strong, and the crown can be expanded outward and upward. The main and lateral branches are evenly distributed, with good ventilation and light transmission, which allows them to enter the flowering and fruiting period earlier than those trees without pruning and they will have a high yield (Figure 12-4).

Cup Shape. This shape is often called "three-ply, trigeminal, and twelve branches". The tree has no central trunk, but 3 main branches are left at a certain height of the trunk to stretch in 3 directions. The angle between each main branch and trunk is approximately 45°, and the angle between the 3 main branches is about 120°. Two primary lateral branches are left on each main branch, and 2 secondary lateral branches are left on each primary lateral branch, and so on, forming a cup-shaped crown similar to a false dichotomous branching (Figure 12-5). This type of modeling method is used primarily in tree species with a weaker trunk.

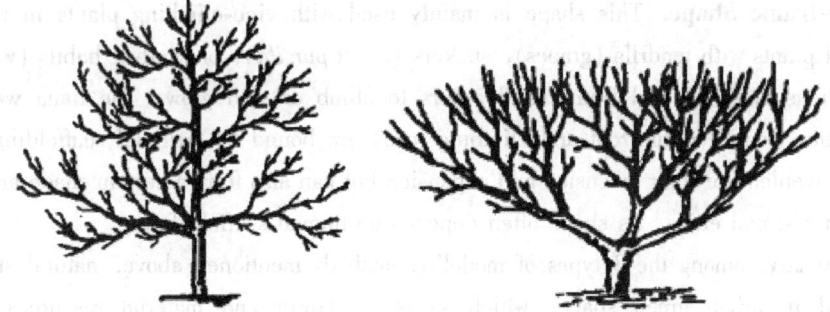

Figure 12-4 Central leading trunk shape.   Figure 12-5 Cup shape.

Natural Open Heart Shape. This shape is improved from the cup shape. It has no central trunk, and it is not hollow in the center compared with cup shape, but the branching is relatively low. The 3 main branches are scattered and distributed in a certain interval, with the trunk radiating out to the 4 sides, extending in a straight line, and opening from the center. However, the lateral branches divided by the main branches are not false dichotomous branching, but instead are scattered on the left and right, so the crown is not arranged completely in a 2-dimensional structure (Figure 12-6). This type of tree has a large area of flowering and fruiting, a strong structure of growth branches, and can make good use of space. The crown is full of sunlight, which is conducive to flowering and fruiting. Therefore, this shape is often used for pruning and modeling of peach, plum, pomegranate, and other ornamental trees in the garden.

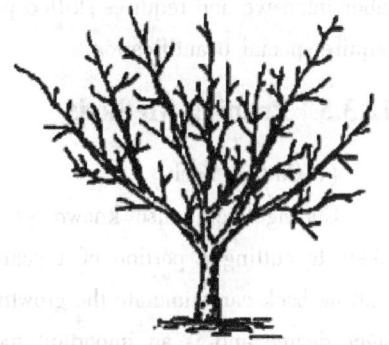

Figure 12-6 Natural open heart shape.

Multi-leading Trunk Shape. There are 2 to 4 leading trunks left on the tree, on which the lateral main branches are arranged by layers, forming a uniform crown. This tree shape is suitable for vigorous tree species, most suitable for flower-viewing trees and shade tree modeling. The crown is beautiful, and it can blossom early and prolong the life of small branches.

Thicket-spherical Shape. The main trunk of this type of modeling is relatively short, and the main and lateral branches at different levels are arranged in clumps, with a thick leaf layer and good afforestation and beautification effects. For the most part, this shape is used for the modeling of small trees and shrubs, such as boxwood, waxberry, and pittosporum.

Umbrella Shape. Often used at both sides of a building exit and entrance, or at the exit and entrance of regular green space, two pairs of plants are used as guide tips. The pool side, the corner of the road, and other places can be embellished adopting this kind of modeling, with good effect. It is characterized by an obvious trunk, and all the lateral branches are bent down and drooped while the upper buds continue to extend outward and expand the crown year by year, forming an umbrella shape, such as what is found in a Chinese pagoda tree, weeping large-leaved early cherry (*Cerasus subhirtella* var. *pendula*), weeping trident maple, weeping elm (*Ulmus pumila* L. cv. Tenue), weeping Japanese apricot (*Prunus mume* var. *pendula*), and *Amygdalus persica* cv. Pendula.

Fence-frame Shape. This shape is mainly used with vine-climbing plants in urban green spaces. All plants with tendrils (grapes), suckers (*Ficus pumila*), or winding habits (wisteria) can rely on various forms of grid frames and arbors to climb on and grow. The liana without these characteristics (e.g., costus root and climbing rose) are bound by artificial scaffolding, which is not only convenient for their extension and expansion but can also form a certain shade area in which tourists can rest and enjoy. Its shape often depends on human scaffold forms.

In summary, among the 3 types of modeling methods mentioned above, natural style is most widely used in urban green space, which saves manpower and material resources and grows successfully. The second most-often used growth habit is a combination of natural and artificial modeling, which is a modeling method for the purpose of making flowers large and dense or fruit rich and plump. The method is more labor intensive and needs to be properly coordinated with other cultivation technical measures. As for the artificial modeling, generally speaking, because it is very labor intensive and requires skilled personnel, it is used only in part of the garden or in places that require special beautification.

12.3.5 Pruning Methods

(1) Cutting Back

Cutting back, also known as short pruning, refers to cutting a portion of 1-year-old branches. Cutting back can stimulate the growth of branches to some degree and is an important method to adjust the growth potential of branches. Within a certain range, the more the short pruning, the more vigorous the germination. According to the degree of short pruning, it can be referred to as light short pruning, medium short pruning, heavy short pruning, and extremely heavy short pruning (Figure 12-7).

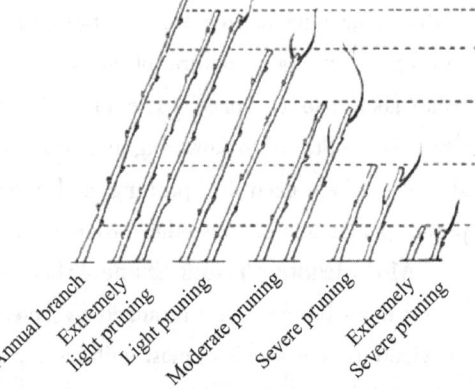

Figure 12-7 Effect of the degree of short pruning on flower bud differentiation.

Light Short Pruning. One-quarter to one-third of the branch tip is cut off, that is, only a light pinching. Because the pruning is light and many buds have been left, the reaction after pruning is to produce a few medium-long branches below the cut, and to then develop many short branches downward. Generally, the growth potential is moderate, which is conducive to the formation of fruiting branches and the promotion of flower bud differentiation.

Medium Short Pruning. In general, approximately one-half of the total length of the branch is cut off when the branch is full of buds. The response after pruning is that more vigorous branches are produced below the cut, a few medium short branches sprout downward, and the short branch quantity is less than that produced by light short pruning. Therefore, the pruning can promote branching to strengthen the branch potential, and the continuous medium short pruning can delay the formation of flower buds.

Heavy Short Pruning. Pruning is carried out below the full buds on the branch, and more than two-thirds of the branch is cut off. After pruning, because fewer buds are left, it grows strongly because of its lower branching ability. It can alleviate the growth potential.

Extremely Heavy Short Pruning. Prune back to the ring mark, or prune so that only 2 to 3 shriveled buds are left at the base of the branch. Only 1 to 3 weak branches can emerge after pruning, which can reduce the position of branches and weaken the growth of vigorous branches, elongated shoots, and upright branches, so as to diminish the branch potential and promote the formation of flower buds.

(2) Retraction

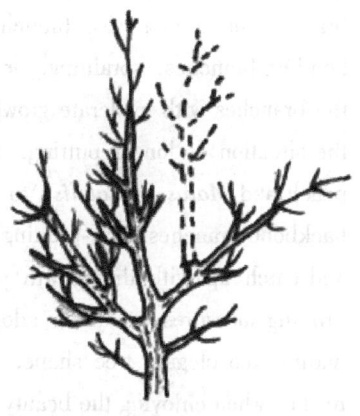

Figure 12-8 Back spune.

Retraction, also known as back spune (Figure 12-8), refers to the pruning of branches for 2 years or more. Generally, large amounts of pruning can stimulate regeneration and rejuvenation. Back spune is mostly used for regeneration of a branch group or backbone branch and control of crown auxiliary branches. The reaction is related to the degree of back spune, the strength of the remaining branches, and the size of the cut. If the strong branches and upright branches remain during pruning and the cut is small, the proper back spune can promote growth; otherwise, growth can be inhibited. The former is used primarily for regeneration and rejuvenation, and the latter is mainly used for controlling the crown or auxiliary branches.

(3) Thinning Pruning

Thinning pruning refers to pruning branches at the mitogenetic point, that is, at the base of the branch. It is generally used to prune dead branches, diseased and insect-ridden branches, overly dense branches, elongated branches, competitive branches, weak branches, droopy branches, cross branches, overlapping branches and adnate branches, and so forth and is the pruning method to reduce the number of branches in the crown. For example, 1-year-old branches pruned at the base, and more than 2-year-old branches pruned at their meristems, are all called thinning pruning.

When thinning pruning, although the non-objective branches that have hindering or shadowing effects on the future growth will be removed eventually, they should be kept temporarily during the young tree period so as to nourish the tree. But to prevent these types of branches from growing too vigorously, they may be left unpruned. In particular, a lower branch of the same tree stops growing earlier and consumes fewer nutrients than an upper branch, which supplies more nutrients for the root and other necessary growth areas. Therefore, it is better to keep those branches rather than remove them too early.

The application of thinning pruning should be used appropriately, especially for young trees, otherwise the tree shape will be disturbed, which will cause trouble for later pruning. For the plants with overly dense branches, thinning pruning should be carried out year by year.

(4) Non-pruning

Vegetative branches are not pruned, which is called non-pruning. Non-pruning uses a natural law that the growth vigor of a single branch decreases year by year. If more buds are left on non-pruning branches, more shoots emerge, causing the nutrients to disperse in the early growth period and more medium and short branches to form; but, accumulating more nutrients in a later growth period can promote flower bud differentiation and fruiting. However, when the vegetative branches have been experiencing non-pruning, the branches will thicken faster, especially the upright branches. Therefore, if non-pruning is used improperly, there will be the phenomenon that trees grow on trees, which should be prevented. In general, non-pruning is not adopted for the upright branches on a crosswise branch, and if non-pruning is used, other pruning measures such as bending branches, spraining, or girdling should also be used. Non-pruning is generally applied to the branches with moderate growth to promote the formation of flower buds, which will not lead to the situation of longer putting, more vigorously growing. Generally, for flowering wood, such as peach and *Malus spectabilis*, to balance the tree vigor and increase the growth potential of weak backbone branches, non-pruning measures are often taken to make the branches rapidly grow coarser and catch up with the growth potential of other backbone branches; for clustered shrubs, non-pruning measures are often adopted. For example, when pruning *Forsythia suspensa* to form a natural and elegant tree shape, 3 to 4 long branches are often put aside above the crown. As a result, when enjoying the beauty of the tree, the long branches swing in the wind with an impressive effect.

(5) Injury

Introducing an injury can be an intentional technique rather than a mistake. The method of damaging the phloem and xylem of a branch by various methods to weaken the growth potential of the branch and to alleviate the tree vigor is called injury. It is usually carried out in the growing period, which has a greater impact on the local part and a lesser impact on the growth of the whole tree. It is one of the auxiliary measures for pruning modeling. The main methods are as follows.

Girdling. A knife is used for girdling the bark of a certain width at the appropriate part of the stem or the branch base to prevent the downward transportation of carbohydrates in the branch tip for a period of time, which is conducive to the accumulation of nutrients and flower bud differentiation of the branches above the girdling. It is applicable to the branches with a small amount of blossoming and fruiting at the peak of development.

Note that the girdling width should be determined according to the thickness of the branch and the callus ability of the tree species, which is approximately one-tenth (2 to 10 mm) of the diameter of the branch. Wounds that are too wide do not heal well, and wounds that are too narrow heal too early to achieve the purpose of girdling. The depth of girdling is best if it reaches the xylem, but if the wounds are too deep it will injure the xylem and cause the girdling branches to break or die. Wounds that are too shallow will make the phloem residual with a non-obvious girdling effect. Enough leaves and branches should be reserved above the girdling branches for normal photosynthesis.

Girdling is a temporary pruning measure applied in the growing season (Figure 12-9) and is generally carried out after flowering or fruiting ends. The part above the girdling area should be gradually cut off during winter pruning, so girdling is not used on the trunk, middle trunk, and main branch. It is generally not adopted for the trees with strong bleeding and a tendency for gummosis (when sap oozes from wounds or cankers), such as fruit trees.

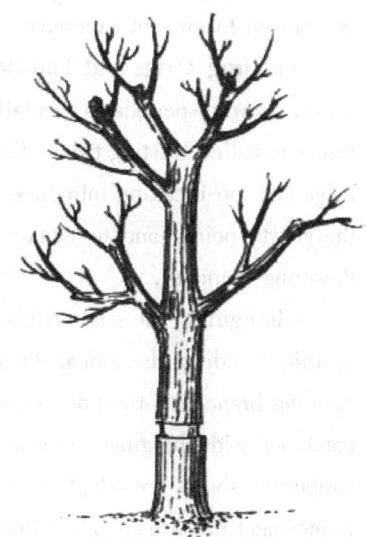

Figure 12-9 Girdling.

Cutting Injury. This is a method of breadthwise cutting (or lengthwise cutting) the upper (or lower) side of the bud (or branch) with a knife deep to the xylem. It is often used in the dormancy period in combination with other pruning methods. The main methods are as follows.

Eye injury refers to the injury on the top of a bud or branch, the shape of which is similar to the eye, and the xylem is injured to prevent the water and mineral nutrients from continuing to transport upward, so as to sprout shoots in the ideal position. On the contrary, when the injury is below the bud or branch, the growth vigor of that bud or branch will be weakened, but the accumulation of organic nutrients is beneficial to the formation of flower buds.

Lengthwise injury refers to the method of cutting the branches lengthwise deep to the xylem with a knife. The purpose is to reduce the mechanical binding force of the bark and promote the thickening growth of the branches. Cutting injury should be carried out before the trees start to grow in spring. The hardened part of the bark should be selected during the implementation, and one lengthwise injury can be applied to the twigs and several lengthwise injuries can be applied to the thick branches.

Breadthwise injury refers to the method of breadthwise cutting the trunk or thick main branch for several times. Its function is to block the downward transport of organic nutrients and to promote the branches to be full and facilitate the flower bud differentiation so as to promote the flowering and fruiting. The mechanism of this action is the same as that of girdling, but the intensity is lower.

Jackknifing. To bend branches to make them form various artistic forms, jackknifing is often carried out at the beginning of budding in early spring. First, a knife is used to cut obliquely to a depth that is one-third to two-thirds of the diameter of the branch, then to carefully bend the branch, using the inclined plane at the xylem fracture for support and positioning. Often the wound is wrapped to prevent excessive water loss.

Twisting Twigs and Top-breaking. These techniques are usually used for overgrown branches during growing periods, especially elongated branches on the back of the twigs. Twisting without injury is called twisting twigs. Bending twigs without breaking them is called top-breaking. Twisting twigs and top-breaking introduce conduction tissue damage to block water, to transport nutrients to the growth point, and to weaken the growth vigor of branches to facilitate the formation of short flowering branches.

Changing. The goal of this modeling measure is to change the direction and angle of branch growth, to adjust the apical dominance, and to change the crown structure with methods such as bending branches, twisting branches, pulling branches, and lifting branches. They are usually combined with the growth season pruning, and the technical measures such as twisting, binding, or supporting shoots are adopted. Vertical inducement can enhance growth potential, while horizontal inducement has a moderate inhibition effect. It is easy to form flower buds by enriching the tissue, but it has strong inhibition effect by inducing downward flexion. However, strong new shoots readily sprout on the upper part of the back of branches, and they should be removed in time to avoid counterproductive effects.

(6) Other Methods

Pinching. Pinching is a measure to remove the growing part of the top of new shoots, which weakens the apical dominance of branches, changes the transporting direction of nutrients, and facilitates the flower bud differentiation and fruiting. Removal of the terminal bud can promote the germination of the lateral buds, thus increasing the branching and promoting the early formation of the crown. But timely pinching can allow branches and buds to still receive enough nutrition, making them full and plump, and can improve cold resistance (Figure 12-10).

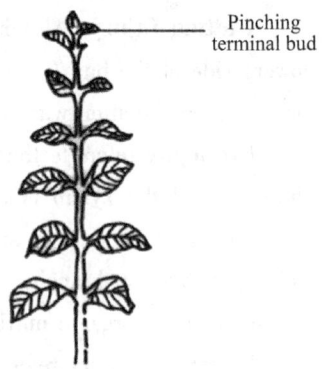

Figure 12-10 Pinching.

Bud Picking. The method of picking the excrescent buds at the base is called bud picking or bud removing. This measure can improve the nutrient supply of the retained buds and enhance their growth vigor. For example, on one hand, bud picking can make the main trunk of the street trees straight without branching to avoid affecting the traffic; on the other hand, to reduce unnecessary nutrition consumption and ensure the healthy growth of the street trees, the cryptoblasts germinated on the main trunk should be removed every summer. For example, for tree peony (*Paeonia suffruticosa*), the lateral buds are usually picked before flowering to concentrate nutrients on the terminal buds, so that the flowers at the top will bloom abundantly and colorfully. To inhibit the over-strong growth vigor of the top or to delay the germination period, the main buds are picked for some species, and the germination of the auxiliary buds or the cryptoblast is promoted.

Leaf Picking. Cutting off leaves with petioles is called leaf picking, and it can improve the ventilation and light transmission conditions in the crown. For trees with ornamental fruits, it can make the fruit fully light visible and well colored, increase the aesthetic degree of the fruit, and thus improve the ornamental effect. For a crown with overly dense branches and leaves, leaf picking can prevent the occurrence of diseases and insect pests.

Removal of Tiller (or Removal of Unwanted Sprouts). To avoid disturbing the tree shape and to reduce the ineffective consumption of the tree body nutrients, sprout tillers should be removed at any time during the growing season for the seedlings that take root easily, such as *Amygdalus triloba* and *Rosa chinensis*. To avoid interfering with the tree shape and affecting the normal growth of scion crown, it is necessary to remove the sprout tillers before grafting.

Disbudding. In essence, disbudding is an early measure of flower and fruit thinning, which can effectively adjust the quantity of flowers and fruits and improve the quality of the remaining flowers and fruits. For example, for hybrid tea rose, the lateral buds are usually removed before flowering, so that the main buds receive enough nutrients to blossom beautifully; for floribunda roses, the lateral buds or overly dense small buds are usually removed to concentrate the florescence and make the flower large and regular and the ornamental effect enhancing.

Root Excision. Root excision is the measure of cutting off all or part of the root system of the plant in a certain range, which can stimulate the root to produce new fibrous roots. Therefore, when transplanting precious trees or mountain trees, root excision is usually carried out 1 to 2 years before transplanting to promote the development of new fibrous roots in a certain range, which is conducive to the survival of the transplant.

12.3.6 Pruning Considerations

(1) Cut State

The cut is slightly inclined to the opposite side of the lateral bud, so that the upper end of the slope is basically flush with the bud end or slightly higher than the bud tip by approximately 0.6 cm, and the lower end is basically flat with the base of the bud. In this way, the cut area is small and the injury surface is not too large, so it will easily heal; the growth of the bud is also better. If the slope of the cut is too high, the scar area is large and the water evaporation is excessive, and the supply of nutrients and water to the cut bud is affected, thus inhibiting the growth of the cut bud. The growth potential of the subjacent bud will be strengthened, and this type of cut is generally used only when weakening the growth potential of the tree. However, when a small section of pile (part of a cut branch) is left above the cut bud, because the nutrients are not suitable to flow into the small pile, this type of cut is difficult to heal, causing withering and affecting the viewing effect of the growing tree, so it is generally not suitable for use.

(2) Selection of the Cut Bud

If the strength and position of the cut bud differ, the strength and position of sprouted branches differ and the posture of the sprouted branches also differ. If a strong bud is left at the cut, a strong branch is produced; if a weak bud is left at the cut, a weak branch is produced.

The buds on the upward part on a horizontal branch are easy to develop into strong branches, and the buds under the back will develop moderate branches. When the cut bud is left on the outside of the branch, it can expand the crown outward, but when the cut bud is facing inward, it can fill the cavity. To restrain overgrown branches, leave weak buds as cut buds, and when turning weak branches into strong ones, be certain to leave full strong buds on the back at the cut.

(3) Removal of Large Branches

When pruning dead branches or unnecessary old branches, disease-and insect-ridden branches, and so forth, cut from the upside of the branch point downward obliquely to shrink the wound as much as possible. The wound of the raised part underneath the residual branch point is not large and will heal. If a part of the residual branch is left, a section of residual pile will become withered and rotted in the future, gradually getting into its tissue with its mother branch growing up. This process can result in a non-healing wound, likely becoming a nest of disease and insects.

When retracting perennial large branches, elongated shoots usually sprout. To prevent a large number of elongated shoots from sprouting, thinning and heavy, short pruning can be done first to weaken the growth vigor and then back spune is carried out. At the same time, the weak branch should be left under the cut, which will help to alleviate the growth potential and reduce the occurrence of elongated shoots. If the perennial branch is thick and must be sawed off, it should be shallowly sawed from the bottom first, and then sawed from the top, so as to avoid jackknifing in half-way sawing due to the down weight of the branch itself, causing the wound to become too large and not easy to heal. As the sawed-off branches may have large wounds and rough surfaces, they need to be smoothed with a knife to facilitate healing. To prevent water evaporation from the wound or the decay of the wound caused by the invasion of diseases and insect pests, a protective agent or plastic cloth should be applied.

12.3.7 Pruning Wound Protection

Tree pruning is essential and important work that should be based on the principle of no or minimal damage to healthy tissues, to meet the requirements of a smooth damaged surface, symmetrical contour, and protection of the natural defense system of trees.

12.3.7.1 Wound Treatment and Dressing

The purpose of wound treatment and dressing of trees is to promote the formation of callus, speed up wound closure, and prevent infection by pathogenic microorganisms.

(1) Dressing of Damaged Bark

If only the bark is damaged, and the cambium is not damaged and still has the ability of meristem activity, the bark should be pasted back on to the exposed cambium and fastened down with flat nails or tied tightly with rubber plastic tape. Generally, wound coating is not used, but the bark should be covered with a 5 cm layer of wet and clean water mosses and then covered with white plastic film. The upper and lower ends should be sealed with asphalt paint to prevent the entrance of water and other moisture. The covered plastic film and water moss should be removed within 3 weeks.

For wounds in thick bark, but with only surface damage and no obstruction to the movement of cambium, immediately cover the area with a clean rag or polyethylene film. Such wounds can generally heal quickly. If the cambium or even the xylem is damaged, it should be trimmed as much as possible according to the natural shape of the wound, and be trimmed into a circle, ellipse, or shuttle (fusiform) shape so as to avoid damaging the healthy cambium as much as possible. When the cambium of the wound surface is old, the dead or loose bark should be removed from the edge of the wound, and damage to the healthy tissue should be avoided. When the trunk or a big branch is damaged by freezing, burning, or lightning, it is not easy to determine the scope of the wound. It is better to trim it at the end of the growing season when it is easier to judge the damage.

(2) Dressing of Thinning-out Cut

Thinning-out is a method of cutting off non-target branches from the trunk or mother branch. On the premise of protecting the branch collar, the final cut of removing the big branch should be properly close to the trunk or the mother branch, never leaving long piles or projections, and the cut should be flat and not torn, otherwise, it will accumulate water and rot, making it difficult to heal. The upper and lower ends of the wound should not be cut horizontally, but instead should be cut as an ellipse or circle with the long diameter parallel to the long axis of the branch (trunk), otherwise the wound will have difficulty healing.

In addition, to prevent the wound from forming a high-surrounding and low-central water basin because of the development of callus, the xylem in the center of the wound should be trimmed into a convex, spherical surface when dressing the larger wound, so as to prevent the decay of the xylem.

12.3.7.2 Function and Type of Wound Dressing

There is no complete agreement on the role of dressings. Some people think that although the coating has played a certain role in promoting the formation of callus and wound closure, it has no great value in reducing the infection and spread of pathogenic microorganism. Some research results show that although some coatings, such as lanolin, can promote the formation of callus, they have little effect on the prevention of wood-parasitic microorganisms infecting deeply into the tree. Many coatings can stimulate the formation of callus, but the formation of callus has nothing to do with the decay process. Big wounds in trees are rarely completely closed. Other wounds seem to be closed on the outside, but there may still be very thin cracks. Research supports that the wound coating can rarely be maintained for more than 1 year and will eventually crack and weather given the action of wind, sun, and rain.

In addition, the performance and painting quality of the coating become the key as to whether to use a wound coating and how to manage the role of the coating. The ideal wound coating should be able to disinfect the treated wound surface, prevent the invasion of wood-rotting fungus and wood cracking, and promote the formation of callus. The coating should also be easy to use allow for the excessive water in the wound to permeate through the coating and evaporate, so as to keep the wound relatively dry. And the paint film should be resistant to weathering and cracking after drying. The coating quality for the wound needs to be improved, with thin, dense, and uniform paint film that never misses coating or blistering because the paint film is too thick. Coatings and bitumen that damage living cells should not be used directly in the cambium area. After application, regular inspection should be carried out, and remedial measures should be taken immediately in case of missing areas of coating, blistering, or cracking, so as to achieve better effects.

Commonly used wound coatings are as follows.

(1) Wound Disinfectant and Hormone

After dressing, the wound should be disinfected with 2% to 5% copper sulfate solution or 5% lime sulphur solution. If 0.01% to 0.1% α-NAA (naphthaleneacetic acid) was applied to the cambium, the formation of wound callus could be promoted.

(2) Lac Varnish

Lac varnish does not harm living cells and has good waterproof performance. It is often used in the cambium area adjacent to the bark and sapwood around the wound, and it is safe to use. The alcohol solution of lac is a good disinfectant. Lac paint alone, however, is not durable, and it should be covered with other tree coating agents.

(3) Creosote Coating

Creosote is the best coating to deal with large, internal wounds of the tree holes that have been attacked by fungi, but it is harmful to living cells, so special care should be taken when using it on any new wounds on the surface. Common commercial creosote is the best material to eliminate and prevent wood-rotting fungus, but most coatings do not easily adhere to it except for coal tar or hot-melting asphalt. Like the creosote coating, creosote oil is harmful to living tissue and is used primarily for the treatment of heartwood.

In addition, the balanced mix of creosote oil and asphalt is also a kind of coating, of which toxicity to living tissue is not as harmful as creosote oil alone.

(4) Grafting Wax

Grafting wax is very effective to treat a small, damaged surface. The solid grafting wax is prepared by boiling 1 part animal oil (or vegetable oil), adding 4 parts of rosin and 2 parts of beeswax, fully melting, and then pouring it into cold water. This type of grafting wax needs to be heated when using, so it can be inconvenient to use. Liquid grafting wax is prepared by heating and melting 8 parts of rosin and 1 part of Vaseline (or lard) at the same time, and then cooling slightly. Next add alcohol till foaming, but not too many bubbles, to make the sound of "zi-zi", and then add 1 part of turpentine. Finally add 2 to 3 parts of alcohol with stirring. This type of grafting wax can be directly applied with a brush, will dry when there is wind, and is easy to use.

(5) Bituminous Coating

This coating is more toxic to tree tissues than water emulsion coating, but it is slow to dry and weather resistant. The composition and preparation method follows: Melt each kilogram of solid asphalt on a slight fire, add about 2500 mL of turpentine or oil, stir it fully, and then cool it.

(6) Lanolin Paint

The application of lanolin as the main ingredient of tree coatings has been widely developed throughout the world. It can protect cambium and cortex, and it makes callus form and expand smoothly.

(7) House Coating

Exterior wall house paint is a mixture of lead and zinc oxides and linseed oil, with good brushing effect. It is not as durable as asphalt coating, however, and it is harmful to tender tissue. Therefore, lac paint should be applied in advance of using this type of paint.

Note that the large wound of wood sun cracks (and frost cracks) is a detrimental entrance for the invasion of wood-rotting fungus. For this kind of wound, especially intact xylem wounds, in addition to the normal dressings, oilcloth should be cut into small pieces larger than the wound, which should be firmly nailed to the surrounding healthy bark, so as to further prevent the wood from cracking and leading to decay.

After the application, regardless of the quality of the paint, to ensure the better effect of the tree paint, the wounds should be regularly inspected. Generally, check and recoat once or twice a year. Measures should be taken to avoid blistering, cracking, or peeling of paint before these events take place. When repainting an old wound, it is better to remove all paint bubbles and loosen paint skin with a brush. Except for the callus, other exposed surfaces should be recoated once.

12.3.7.3 Cut Protection Agent

Wounds caused by pruning on the trunk, especially the precious tree species, should be protected by a protective agent in the main part of the tree body. At present, two kinds of protective agents are widely used.

(1) Solid Protectant

Use 4 parts rosin, 2 parts beeswax, and 1 part (by weight) animal oil. First put the animal oil in a pot and melt it with fire or another high-temperature heat source. Then remove from the heat, immediately add the rosin and beeswax, and re-heat it slowly, stirring it fully, and taking it away from the heat after condensation is gone. After cooling, put the protectant in a plastic bag for sealing. When using, just slightly heat it so as to soften it, and then smear it on the wound with a putty knife. It is generally used to seal and smear large wounds.

(2) Liquid Protectant

Raw materials include 10 parts rosin, 2 parts animal oil, 6 parts alcohol, and 1 part turpentine (by weight). First put rosin and animal oil into the pot to heat. Stop the fire or other heat source immediately after melting (or remove it from the fire immediately). After cooling slightly, pour alcohol and turpentine in at the same time. Stir evenly and pour mixture into a storage bottle with good sealing qualities to prevent alcohol and turpentine from volatilizing. It can be applied with a brush. This liquid protectant is suitable for small wounds.

Questions for Review

1. What is the purpose of seedling pruning modeling?
2. What are the methods of seedling pruning modeling?
3. What are the specific methods of natural and artificial mixed modeling?
4. What is the biological basis of seedling pruning modeling?
5. What are the main types of seedling wound protectants? How are they prepared?

References and Additional Readings

CHEN Q B, 2007. Landscape plant cultivation[M]. Beijing: China Agriculture Press.

CHENG F Y, 2012. Nursery of landscape plants[M]. Beijing: China Forestry Publishing House.

DUMROESE R K, 2009. Nursery manual for native plants: A guide for tribal nurseries[G]. Washington, DC: US Department of Agriculture, Forest Service.

GUO X W, BAO M Z, 2004. Seedling planting and maintenance[M]. 2nd ed. Beijing: China Forestry Publishing House.

HU C L, 1996. Illustrated handbook of ornamental flowers and trees shaping and pruning[M]. Shanghai: Shanghai Scientific & Technical Publishers.

LI Q W, 2011. Seedling shaping and pruning[M]. Beijing: China Forestry Publishing House.

LIU Z L, 2005. Nursery of landscape plants[M]. Beijing: China Meteorological Press.

SHEN H L, 2009. Seedling cultivation[M]. Beijing: China Forestry Publishing House.

SU J L, 2003. Nursery of landscape plants[M]. Beijing: China Agriculture Press.

SUN S X, 1992. Silviculture[M]. 2nd ed. Beijing: China Forestry Publishing House.

SUN S X, 2002. Techniques of forest seedling raising[M]. Beijing: JINDUN Publishing House.

ZHAI M P, SHEN G F, 2016.Silviculture[M]. 3rd edition. Beijing: China Forestry Publishing House.

ZHANG X Y, 1999. Shaping and pruning of ornamental flowers and trees[M]. Beijing: China Agriculture Press.

Chapter 13 Seedling Quality Evaluation and Seedling Lifting

Zhu Yan, Wei Xiaoli

Chapter Summary: This chapter introduces the seedling quality evaluation index, evaluation method, and seedling quality control system; seedling quality sampling method; and the procedures of seedling lifting, grading, packaging, transportation, and storage in the processing of nursery stock. Given the diversity of seedling type demands of afforestation, urban forest construction, and environmental greening, seedling types and sizes vary from bare-root to container and from small to large, respectively, which lead to a complexity of seedling quality evaluation and seedling raising techniques. In the nursery production practice, specific analysis should be made according to the specific situation. Nursery practitioners need a flexible use of knowledge to solve practical problems and a timely grasp of new trends at home and abroad in seedling quality evaluation and seedling quality control.

13.1 Seedling Quality Assessment

Seedlings are the material basis of afforestation. The task of a nursery is to cultivate a large number of seedlings with high quality and of appropriate varieties for forest construction and landscaping. Seedlings with high quality are usually called strong seedlings. Specifically, strong seedlings refer to those with good genetic quality, robust growth and development, strong resistance to adversity, and high survival rate of transplanting or afforestation. Compared with the seedlings for common forest construction, the species and types of seedlings for urban forest cultivation and landscaping have greater variety, and the requirements for the quality of seedlings differ. Therefore, specific quality indicators and evaluation methods need to be put forward according to species' needs.

13.1.1 Purpose and Significance of Seedling Quality Assessment

The purpose of seedling quality assessment is to ensure the production and application of high-quality seedlings. Included in that goal are tasks such as understanding and mastering the quality status of seedlings, so as to explain the seedling status to the users; determining the measures for seedling lifting and storage; evaluating whether the nursery cultivation measures are reasonable; determining suitable site conditions for the current batch of seedlings; and formulating appropriate seedling processing and planting measures. In addition, nursery managers must avoid losses caused by improper use of seedlings, determine the sequence of seedling planting, and analyze the reasons for unsuccessful planting and the most important quality factors affecting the success of planting. Through seedling quality assessment, we can judge whether the genetic and sowing quality of propagation materials are good in seedling cultivation; whether the various techniques and management measures of seedling cultivation are appropriate, which need to be abandoned, which need to be maintained, and which need to be improved; whether the evaluated seedlings can be used for afforestation for various purposes and under what conditions, as well as what effect will be produced after application; and what control measures should be taken to ensure the quality of seedlings in all aspects of seedling breeding. All of these situations and problems can be determined by seedling quality assessment.

In a long-term afforestation practice, people gradually realize that the quality of seedlings determines the success of afforestation and the effect and function of afforestation to a large extent. As a result of intensive management, the cost of seedling purchase, land preparation, and early tending increased dramatically. To reduce the cost and improve the quality of forests, researchers and nursery managers have realized the importance of scientific management of seedling quality. With the development of forest genetic improvement, the quality of seedlings has also become a concern in the forestry field. The study of seedling quality assessment and control technology has become one of the research hotspots, and impressive progress has been made. To unify the understanding and scientific assessment of seedling quality, IUFRO held the first Seedling Quality Assessment Technology Symposium in New Zealand in 1979, at which the role of seedling quality in afforestation, seedling quality assessment technology, and factors affecting seedling quality were discussed. After the meeting, the forestry research departments of the United States, Canada, Germany, Japan, and other countries paid more attention to the research of seedling quality assessment technology, and some countries carried out the research of improving seedling technology and seedling quality. In 1994, three working groups of the International Union of Forest Research Organizations (IUFRO), including seedling production, plant material characteristics, and tree physiology, jointly held an academic conference in Ontario, Canada, with the theme of seedling quality assessment. Results from the conference summarized various methods of seedling quality assessment proposed by predecessors, analyzed the complexity of seedling quality assessment more comprehensively and objectively; and put forward the measurement standard and application range of various methods, which provided further basis for the standardization and scientific seedling quality assessment.

Research on the relationship between seedling quality and afforestation in China began in the 1950s. Since the early 1980s, under the leadership of the Ministry of Forestry and through fixed-point research, the first seedling quality standard for main afforestation species in China was published: *Tree Seedlings of Major Species for Afforestation* (GB 6000—1985). Standards were formulated by using morphological quality indicators; on this basis, according to the actual environments, provinces have formulated local standards to guide seedling work and to test the quality of seedlings, which have played a certain role in the improvement of seedling technology and the promotion of seedling quality assessment research. In 1999, based on a great deal of research and practice, a new national standard, *Tree Seeding Quality Grading of Major Species for Afforestation* (GB 6000—1999), was revised and formed. Compared with the standard formulated in 1985, the new standard added the quality index of roots, and seedling quality assessment paid attention to the importance of roots in the survival of seedlings, which indicates that the seedling quality assessment in China had entered a new stage. A primary concern of seedling quality assessment in China is the seedlings for afforestation, most of which are 1-year-old seedlings. With the rise of urban forestry, requirements on the varieties, types, and specifications of seedlings were put forward in urban forest cultivation and landscaping. The evaluation standard GB 6000—1999 and the local seedling quality standard formulated by each region are suitable only for large-area suitable land afforestation, but they are not suitable for seedling quality assessment in urban forest cultivation and landscaping in terms of the index system, methods, and standards. Therefore, it is urgent to establish the seedling quality standards for urban landscaping. At present, a set of unified seedling quality standards for landscaping are not available in our country, but some industry standards for main greening species and landscaping seedling standards applicable for local uses have been issued successively. For example, Beijing has formulated *Woody Seedlings of Plant Materials for Landscaping* (DB11/T 211—2017); Shenzhen has formulated *Tree Seedling Grading for Urban Greening* (DB440300/T 28—2006); and many for the tree species, such as *Seedling Quality Grading of Major Bamboo Species for Urban Greening* (LY/T 2345—2014), *Virescence Seedling Cultivation Regulation and Quality Grading of Camphor tree* (LY/T 1729—2008), *Planting Technical Specification and Produce Quality Grade for Ornamental Palmae—Part 1: Field Production* (LY/T 1734.1—2008), *Planting Specification and Quality Grade of Ornamental Palmae—Part 2: Containers Cultivation* (LY/T 1734.2—2008), and others. These standards enrich and improve the content of seedling quality assessment in China.

13.1.2 Index System of Seedling Quality Assessment

Seedling quality refers to the degree to which the seedlings meet the afforestation target under the specific site conditions in terms of type, age, morphology, physiology, and vitality. Quality refers to the specific site conditions and operation purposes. Generally speaking, two types of indicators are used to describe the quality of seedlings: One is the morphological or physical measurement of seedlings; the other is the measurement of physiological or internal quality of seedlings. Since the 1980s, the study of seedling quality has gradually changed from a single morphological quality index to the combination of morphological and physiological indexes, extended to the molecular level. Seedling quality assessment has also extended from seedling raising process to the entire process including seedling lifting, storage, planting, and early growth after outplanting. The measurement results of physiological quality of seedlings are not visible or stable, and may even be destructive to the seedings, especially if they need special instruments or laboratory procedures. Therefore, those indexes are only suitable for research, not for production and application. To find the correlation between various physiological and morphological indexes, determine the primary and auxiliary morphological indexes that can best represent the seedling quality of each species and seedling age type in each region, as well as the physiological indexes that are convenient for measurement and application, and then apply them to the production practice. These tasks are important steps in seedling quality assessment research.

13.1.2.1 Morphological Index

(1) Morphological Index of Small-size Seedlings

Small-size seedlings refer to the seedlings, including container seedlings and asexual propagation seedlings, that have been cultivated or transplanted in the original seedbed for no more than 3 years and are used primarily for afforestation in patches. For most of the tree species, seedlings are lifted at 1-year-old, but 2- or 3-year-old seedlings are lifted for a few tree species. The main indexes used for small-size seedling quality assessment include seedling height, root collar diameter, height-diameter ratio, root system index, weight index, stem-root ratio, terminal bud status, and comprehensive quality index. The morphological indexes are easy to operate in production, can be measured with simple instruments, and easy to control directly. Moreover, the morphological indexes are related to the physiological and biochemical conditions, biophysical conditions, vigor conditions, and other conditions of seedlings. Therefore, morphological index is always the seedling quality index especially concerned in research and production.

Seeding Height, Root Collar Diameter, and Height-diameter Ratio.

i. Seedling Height. Which refers to the stem length from the root collar (soil mark) to the base of the terminal bud, which is the most intuitive morphological index that is easy to measure. Seedling height can reflect its genetic advantage to a certain extent and also reflect the seed sowing quality, the superiority of living microenvironment, and the appropriateness of cultivation measures. Generally speaking, for a single seedling, seedling height reflects the foliage volume, photosynthetic capacity, and transpiration area. Therefore, seedling height can well reflect the seedling increment. After the survival of seedlings, the seedlings with high initial height grow faster.

However, seedling height is not closely related to the survival rate of afforestation. In some adverse conditions, the taller the seedlings, the lower the survival rate. The taller the better is not the case for seedlings; rather, they need a height commensurate with the diameter. Each type of tree species has their own suitable height, and within the range of suitable height, the survival rate and growth can be considered. In production, for example, Scots pine (*Pinus massoniana*) afforestation shows that the survival rate of afforestation is high when the seedling height is more than 15 cm and the root collar diameter is more than Grade II level seedlings. Proper seedling height saves labor and time. But the survival rate is low for super tall seedlings in afforestation. The reason is that given the same site conditions, afforestation specifications and planting methods are more productive with the survival of general qualified seedlings rather than super seedlings. For super seedlings, nursery workers must increase the land preparation specifications and fine-tune a more precise planting. , If the planting hole is small, the planting depth will not be deep enough, and the soil cover thickness will be insufficient, affecting the root system rooting and water balance of seedlings. Therefore, afforestation technology should be robust.

The suitable seedling height cannot be generalized. It should be determined according to tree species, afforestation site conditions, afforestation time, and other factors. The general principle is that on the premise of ensuring the survival of afforestation seedlings, the taller the seedling height, the better. But for the whole seedlings used in the same forestation land, in order to prevent the strong differentiation of stands in the future, the seedling height should be in the given range and as neat as possible.

ii. Root Collar Diameter (RCD). Which refers to the diameter of the root stem junction (soil mark) of seedlings, which is the abbreviation of root collar diameter. Among all morphological indexes, RCD is one of the primary indexes to evaluate the quality of seedlings, not only because it is simple and easy to measure but also because of the large amount of information that RCD contains in various morphological indicators. The RCD is closely related to seedling height, root status, weight, mineral nutrition and carbohydrate content, and stress resistance of seedlings. In addition, it is highly related to the survival rate of afforestation and early growth after afforestation (Figure 13-1). Therefore, the reliability of RCD to reflect the quality of seedlings is very high. Most studies have shown that the RCD of seedlings was positively correlated with the survival rate of afforestation. In the practice of seedling production and afforestation, researchers have found that "it is better to be short and fat than tall and thin".

However, RCD is limited in improving the survival rate of afforestation. When the RCD increases to a certain extent, the survival rate of afforestation increases slowly. It can be sure that with the further increase of the RCD, the survival rate will increase only a little, or change based on a horizontal line, or even decrease. The study of Mulin and Svaton (1972) also confirmed that the effect of RCD on the survival rate of afforestation and tree increment was similar to that of seedling height; specifically, the relationship curve between RCD and survival rate of afforestation was much smoother than that between seedling height and survival rate.

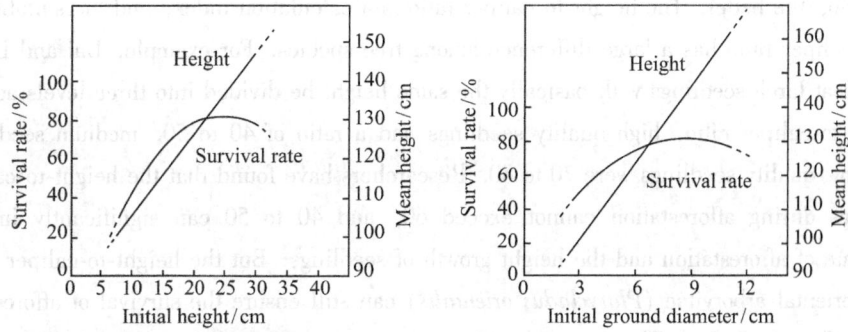

Figure 13-1 Relationship between initial seedling height, root collar diameter and tree height, survival rate of white spruce after 10 years of afforestation (Modified from Mulin and Svaton, 1972).

As we all know, the survival rate of afforestation depends not only on the quality of seedlings but also on the planting technology and quality, as well as the environmental conditions. When the RCD increases to a certain extent, it is no longer the limiting factor of survival. Therefore, the relation curve between the root collar diameter and the survival rate tends to be gentle. At this time, the survival rate is mainly subject to the planting conditions and environmental conditions of seedlings. In addition, the various parts of the seedlings are interrelated and in a certain proportion, and there are few "short and fat seedlings" absolutely. The increase of the RCD will inevitably lead to the increase of other parts, so seedling height and the parts above the ground will also expand correspondingly, and the negative effects such as the increase of transpiration and the rapid loss of water will become more and more prominent. If the sufficient root biomass and the timely recovery of the root system after afforestation cannot be guaranteed, low survival rate of afforestation is inevitable. Seedlings that are too thick are adverse to lifting, packing, storage and transportation, and planting. Therefore, on the premise of ensuring the survival rate of afforestation, the larger RCD of seedlings, the better.

iii. Height-to-caliper Ratio. Which measures the ratio of seedling height to root collar diameter. Because height-to-caliper ratio is only a ratio, there is no unit. The unit of seedling height and root collar diameter differs across countries, so the ratio also differs. For example, in the United States, centimeter is used for seedling height and millimeter is used for root collar diameter, so the ratio is mostly a 1-digit value; in China, centimeter is used for seedling height and root collar diameter, so the ratio is mostly a 2-digit value.

The height-to-caliper ratio reflects the balance between seedling height and thickness of seedlings. The combination of seedling height and root collar diameter is a useful index to reflect the stress resistance of seedlings and the survival rate of afforestation. Generally, the higher the height-to-caliper ratio is, the thinner and higher the seedlings are, with weak stress resistance and lower survival rate of afforestation. By comparison, the smaller the height-to-caliper ratio, the coarser and shorter the seedlings, with stronger stress resistance and high survival rate of afforestation. Generally speaking, as long as the seedling height meets the requirements expected, the lower the height-to-caliper ratio, the better. The height-to-caliper ratio is a calculation index, and the suitable range of height-to-caliper ratio has a large difference among tree species. For example, Dai and Liu (1992) proposed that larch seedlings with basically the same height be divided into three levels according to the height-to-caliper ratio, high-quality seedlings had a ratio of 40 to 50, medium seedlings were 60, and low-quality seedlings were 70 to 80. Researchers have found that the height-to-caliper ratio of seedlings during afforestation cannot exceed 60, and 40 to 50 can significantly improve the survival rate of afforestation and the height growth of seedlings. But the height-to-caliper ratio of 70 to 80 for oriental arborvitae (*Platycladus orientalis*) can still ensure the survival of afforestation.

Root System Index. The root system is an important organ of trees, and it plays a decisive role in the survival and early growth of afforestation. At present, the root system indexes used in production primarily include root length, root width, and number of lateral roots, while in scientific research, root system indexes also include root weight, root volume, total root length, and root surface area index.

Root length refers to the natural length of the main root measured from the root base near the surface to the root tip. It is the length of a root system that should be preserved when the seedling is lifted, and it is the basis of controlling the lifting depth of a seedling; root width refers to the length measured from the base of the main root near the surface to the lateral root, and it is also the amplitude of retaining the lateral root when the seedling is lifted. The number of lateral roots generally refers to the number of all lateral roots that meet the requirements of a certain length (such as 1 cm, 5 cm, 10 cm, and so on). The total length of roots refers to the total length of all lateral roots that meet the requirements of a certain length (such as 1 cm, 5 cm, 10 cm, and so forth); the root surface area index is the product of the number of lateral roots that meet the specifications of a certain length (such as 1 cm, 5 cm, 10 cm, and so on) and the total length of lateral roots. Root volume refers to the volume of all roots. Among the above root system indexes, the number of lateral roots is most closely related to the survival rate and early growth status of afforestation. The number of lateral roots specified in the current national standards in China refers to the number of primary lateral roots>5 cm.

Seedling Weight and Shoot-root Ratio.

i. Seedling Weight. Also known as biomass, it refers to the dry weight or fresh weight of seedlings, and it is more accurate when using dry weight. The dry weight reflects the dry matter accumulation of seedlings and is a good index to measure the quality of seedlings. Seedling weight can be the total weight of seedlings, or the weight of all parts, such as root weight, stem weight, and leaf weight. Generally speaking, the seedlings with high biomass have luxuriant branches and leaves and good seedling quality, otherwise they are not good. The total biomass or total dry weight of seedlings is the best indicator to reflect the competitiveness of seedlings, because it can reflect the photosynthetic area (dry leaf weight), root size (dry root weight), stem and branch size (dry stem and branch weight), and comprehensively reflect the competitiveness of seedlings. In addition to the total biomass of seedlings, the weight of each part of the seedlings is also very important. It reflects the distribution of biomass in each part and plays an important role in the evaluation of seedling quality. However, in the actual production, the dry weight of seedlings can be used only for sample investigation to estimate the quality of the whole batch.

ii. Shoot-root Ratio. This ratio uses the weight or volume of the aboveground part to the underground part of the seedlings. Some researchers also use the root-shoot ratio, which is the reciprocal of the shoot-root ratio. The shoot-root ratio reflects the balance of stem and root, which is actually the balance of water and nutrients. It is usually the dry weight ratio. Most studies have shown that the lower the shoot-root ratio is, the more favorable the survival of afforestation seedlings. But for the shoot-root ratio, the lower the better does not hold true. Each species of seedlings has its own suitable shoot-root ratio, and if the ratio is over the suitable range, it will have a negative impact on the growth and survival of seedlings after afforestation. The suitable shoot-root ratio of different species of seedlings needs further study.

Terminal Buds. The size and presence of terminal buds are very important for some conifer trees with weak germination ability. The normal and full development of terminal buds is an important condition for qualified seedlings, such as *Pinus massoniana*, Japanese cedar (*Cryptomeria fortunei*), and other seedlings. Because the larger the terminal bud, the higher the number of protophylls in the bud; the higher the vigor of the seedlings, the greater the increment of the post-afforestation. But for most broad-leaved trees and some fast-growing conifer trees, the terminal bud has little to do with the quality of seedlings, such as Chinese arborvitae, slash pine (*Pinus armandii*), and loblolly pine (*Pinus taeda*).

Quality Index. A single morphological index of seedlings often reflects only a certain side of the seedling, and the influence on the survival and initial growth of afforestation is often the comprehensive function of each index, which needs the coordination and balance of each part of the seedling. Therefore, Dickson et al. (1960) proposed the seedling quality index (QI), which is calculated as follows:

$$QI = \frac{\text{dry weight of seedlings, g}}{(\text{seedling height, cm/root collar diameter, mm}) + (\text{dry stem weight, g/dry root weight, g})}$$

The formula shows that the lower the height to caliper ratio and the shoot-root ratio, the higher the total dry weight and *QI*, the higher the quality of seedlings. Foreign research showed that the quality index can better reflect the quality of seedlings. Research results of Chinese scholars, however, indicate that this index has some limitations (Liu, 1999).

(2) Morphological Indexes of Medium- and Large-size Seedlings

Medium- and large-size seedlings refer to seedlings cultivated or transplanted in the original seedbed for more than 3 years, and which have a specific crown, trunk, and compact root lump after some shaping, pruning, root cutting, and other measures. The seedlings are used primarily for urban landscaping, landscape forest construction, and so forth, including sowing seedlings, transplanted seedlings, and transplanted container seedlings. According to the shape, the seedlings used for urban landscaping can be divided into cluster seedlings, single stem seedlings, multi-stem seedlings, and creeping seedlings; according to the height of seedlings, the seedlings used for urban landscaping can be divided into small trees, medium trees, and large trees.

Landscaping requires high-quality seedlings that have the following conditions: Developed and complete root system, short and straight main root, more lateral roots and fibrous roots within a certain range close to the root neck, no split of large root system after seedling lifting; thick and straight main trunk of seedlings (except liana), with a certain suitable height, and no excessive growth; and uniform distribution of main and lateral branches, which can form a perfect crown. In production, classification of the quality indexes of medium- and large-size seedlings usually includes the diameter of the trunk, the diameter of the soil boll, clear bole height (from ground to the first branch), seedling height, the diameter of the crown, the number of branches (or the number of layers of lateral branches).

Seedling Height. The commonly used indexes of medium- and large-size seedlings include plant height, clear bole height, clear stem height, and shrub height. Plant height refers to the vertical height of the plant from the ground surface to the highest point of the plant under natural conditions; clear bole height refers to the vertical height of the tree from the ground surface to the lowest branch point (live branch); clear stem height refers to the height of the palm trees from the ground surface to the lowest leaf sheath; shrub height refers to the vertical height from the ground surface to the top of the normal growth of the shrub. Different types of greening land have different height requirements for trees. For example, the main quality indexes of arbor seedlings for street trees are as follows: The stem diameter of deciduous arbors is not less than 7.0 cm; the height of evergreen arbors is more than 4.0 m; there are 3 to 5 main branches, and the clear bole height is not less than 2.5 m (under special circumstances, it can be stipulated separately).

Diameter at Breast Height (DBH), Meter Diameter, Base Diameter, and Root Collar Diameter.

i. DBH refers to the diameter of the arbor trunk at 1.3 m from the ground surface, also known as the stem diameter;

ii. Meter diameter refers to the diameter of the trunk at 1.0 m above the ground;

iii. Base diameter refers to the diameter of the trunk of seedlings at 0.2 to 0.3 m from the ground surface; and

iv. Root collar diameter refers to the diameter of the root neck (soil mark) of the tree.

According to the specific situations of tree species and uses, the datum points of tree diameter measurement differ. The required branching points of street trees are higher than other tree types, and the single trunk is straight, therefore, the DBH is often used as the datum point of measurement for street trees. The meter diameter is often used as the datum point for isolated trees and ornamental trees. The root collar diameter is often used as the datum point of measurement for trees with lower natural branching points (lower natural growth branching). Some plants, however, such as the pineapple palm (*Phoenix canariensis*), whose leaves are whorled and clustered, and without the obvious main stem, can often be measured by the base diameter.

Crown Diameter and Canopy Diameter.

i. Crown diameter refers to the diameter of the vertical projection plane of the crown of the arbor; and

ii. Canopy diameter refers to the diameter of the vertical projection surface of shrubs.

Soil Boll Diameter and Soil Boll Height.

i. Soil boll diameter, also known as the diameter of the mud ball, refers to the diameter of mud ball at the root when transplanting seedlings, usually expressed as "D".

ii. Soil boll height refers to the vertical height from the bottom of the soil boll to the base of trunk. Generally, the diameter of arbor soil boll is 7 to 10 times the DBH of the tree, and its height is about 2/3 the diameter of the soil boll; the size of a shrub soil boll is based on 1/4 to 1/2 of its crown diameter.

Main Vine Length Refers to the Length of the Main Stem of Liana.

Comprehensive Indexes Include Terminal Bud, Leaf Color, and Branching.

Types of seedlings have different quality assessment standards. Trees are classified according to DBH, clear bole height, plant height, crown diameter, number of branches, and diameter of soil boll. Palm and cycad are classified according to the base diameter, plant height, clear stem height, crown diameter, number of branches, number of leaves, and diameter of soil boll; terminal buds must be well preserved. Bamboo plants are classified according to the base diameter, number of branches per cluster, trunk (or bole) cutting height, and diameter of soil boll. Shrub height, canopy diameter, and diameter of the soil boll are as the classification indexes for shrubs. Root collar diameter, length of main stem, number of branches, and diameter of soil boll are the classification indexes of woody lianas.

13.1.2.2 Physiological Attributes

In the seedling quality assessment, people used to pay attention to morphological attributes and paid less attention to physiological attributes. In fact, the function of a physiological index in seedling quality assessment is far greater than the morphological index. More and more, research shows some limitations in seedling quality assessment by morphological index classification.

The relationship between morphological index and physiological index of seedlings is that the morphological index is the external manifestation of various physiological responses of seedlings to the nursery conditions. When the growth of seedlings is normal, generally morphological indexes can best reflect the quality of seedlings, but when the seedlings are affected by some external factors, the physiological status of seedlings has changed, but it cannot be promptly expressed in terms of morphology. During this time frame, limitations arise in using morphological indexes to evaluate the quality of seedlings. Therefore, morphological classification of seedlings is only the perceptual stage of seedling quality assessment, and what really determines the survival rate and future growth potential of seedling afforestation is the inherent physiological characteristics of seedlings. Using physiological indicators to evaluate seedling quality is the developmental direction of seedling quality assessment in the future.

(1) Water Status of Seedlings

The water status of seedlings is closely related to the quality of seedlings. A large number of studies and nursery practices have also proved that one of the important causes of seedling death after afforestation is the imbalance of seedling water. Many indicators reflect the water status of seedlings, such as water content (roots and leaves), water potential, pressure and volume (P-V) curve, and its parameters of water dynamic change.

Water Content. Water content of seedlings refers to the percentage of water content in the dry weight of seedlings. Researchers have found that within a certain range of water content, a linear correlation occurs between the water status of seedlings and the survival rate of afforestation. With the gradual loss of water in seedlings, the survival rate of afforestation decreased. For example, for every 1% decrease in root water content of Chinese white poplar (*Populus tomentosa*) and black locust (*Robinia pseudoacacia*), the survival rate of afforestation decreased by 1.5% and 8.67%, respectively. However, using water content to measure the quality of seedlings has some limitation. Before the complete loss of water in the seedlings, their physiological activities have been greatly affected, sometimes the seedlings have died, even though water is still in the body, so it is not accurate to use water content to measure physiological activities. In addition, this index cannot distinguish the normal seedlings from the dead ones that still contain enough water.

Water Potential. Water potential is one of the most important indicators to reflect the water status of plants. It can sensitively reflect the changes of water content of seedlings under drought stress; it can also help to explain the law of water movement in the soil-plant-atmosphere continuum. According to Kramer and Kozlowski (1979), the water potential in a system is the chemical potential difference per mole volume between water and pure water at the same temperature, in Pa (the abbreviation for Pascal, the basic unit of pressure in the International System of units). The water potential (φ_w) of seedlings is composed of osmotic potential (φ_s) and pressure potential (φ_p), that is,

$$\varphi_w = \varphi_s + \varphi_p$$

The pressure potential is a positive pressure on the expanded cell wall, just as the pressure produced by the surface of a balloon on the air inside the balloon, and as the cell loses water, the pressure will weaken. Pressure potential is an important indicator to measure the water status of seedlings. It is very sensitive to water stress. If the pressure potential drops to a certain level and stays at that low level for a long time, it may cause permanent damage to seedlings. The osmotic potential is a negative pressure that is caused by adding solute (such as sugar or salt) and other substances into pure water with zero potential energy. With the increase of solute concentration, the osmotic potential decreases. The osmotic potential of pure water is zero.

The relationship between the above three potentials changes with the water absorption and loss of seedlings. When the seedlings fully absorb water (water content is 100%), their water potential is zero. In this case, the values of pressure potential and osmotic potential are equal but their signs are opposite. With the loss of water, the cell membrane allows only water to pass through, while the solute is left in the cell, so the solute concentration in the cell increases, and the osmotic potential decreases. At the same time, because the cells lost their original volume, the pressure potential decreases, and finally water potential decreases so the water stress of seedlings increased.

The biggest problem in practical application is the measurement of water potential. Common methods include small liquid flow method, conductivity method, ice point depressurization method, and pressure chamber method (Figure 13-2). But these methods are difficult to adapt to production processes and are used only in scientific research at present. The common method is to dry the seedlings for a period of time, measure the water potential of seedlings in the process of water loss and the afforestation survival rate of seedlings with different degrees of water loss, and measure the critical water potential value that determines the survival rate of afforestation. Song et al. (1993) put forward a simple method (Table 13-1) to measure the physiological grade of seedlings according to π_0 (osmotic potential during incipient plasmolysis) and π_{100} (osmotic pressure when the seedlings are saturated with water). The dividing standards are as follows:

$\varphi_w \geqslant \pi_{100}$ I-level seedlings (with the survival rate>80%)

$\dfrac{\pi_{100}+\pi_0}{2} \leqslant \varphi_w < \pi_{100}$ II-level seedlings (with the survival rate of 40%-80%)

$\pi_0 < \varphi_w < \dfrac{\pi_{100}+\pi_0}{2}$ III-level seedlings (with the survival rate < 40%)

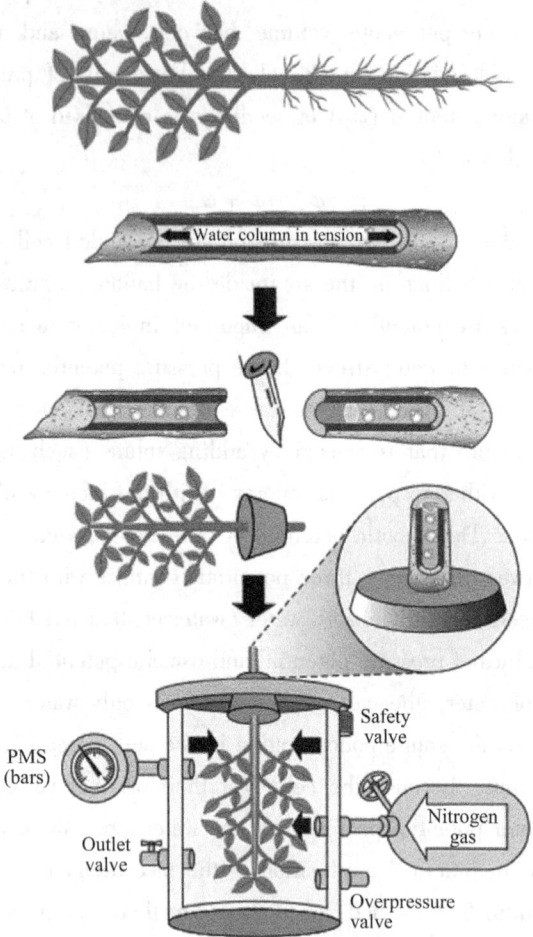

Figure 13-2　Water potential measurement with pressure chamber (From Landis et al., 2008).

Table 13-1　Relationship between water potential and quality level of types of seedlings

Seedling types		Water potential/MPa		
		I-level seedlings	II-level seedlings	III-level seedlings
		Survival rate>80%	Survival rate between 40%-80%	Survival rate < 40%
Pinus sylvestris var. mongolica	1-0.5 bare-root seedling	≥ -1.45	[-1.83, -1.45]	[-2.20, -1.84]
	2-0.5 bare-root seedling	≥ -1.30	[-1.65, -1.30]	[-2.00, -1.65]
	1-1 bare-root seedling	≥ -1.65	[-2.08, -1.65]	[-2.50, -2.08]
	2-1 bare-root seedling	≥ -1.70	[-2.15, -1.70]	[-2.60, -2.15]
Larix gmelinii	1-0.5 bare-root seedling	≥ -1.55	[-1.78, -1.55]	[-2.00, -1.78]
	1-1 bare-root seedling	≥ -2.45	[-2.80, -2.45]	[-3.10, -2.80]
	2-1 bare-root seedling	≥ -2.35	[-2.60, -2.35]	[-2.85, -2.60]

Source: From Song et al. (1993).

***P-V* Curve and Its Water Parameters.** *P-V* curve established by pressure chamber method in the process of gradual water loss of seedlings is very beneficial to the study of the dynamic change law of water content in seedlings. *P-V* technique can also be used to test dead and rotten seedlings that fully absorbed water again, which is a problem that water content method and single water potential index cannot solve. According to the research of Yin (1992), the cell structure of the root system of dead seedlings has been destroyed, and the water control ability of the semipermeable membrane has been lost. With a little pressure in the pressure chamber (2-5 kg/cm^2), the water is almost discharged at one time (accounting for 76% of the total water displacement), which shows that the water potential of the dead seedlings after absorbing enough water is more than −5 to −2 kg/cm^2, and if the pressure continues to increase, almost no more water can be discharged; however, the cell membrane of the roots of the high-quality seedlings with sufficient water absorption is complete and has strong water control ability, with the water potential of −10 to −5 kg/cm^2, which is significantly lower than that of the dead seedlings. In addition, the initial discharge of water in the pressure chamber is very little (only 7.7% of the total water volume). With the increase of pressure, the water can be discharged more evenly for several times. When the test results are reflected in the *P-V* curve, the curve of dead seedlings is almost perpendicular to the horizontal axis, indicating that the inside and outside of cells can almost freely pass water through without hindrance, and almost all water is discharged at one time with a little pressure, and almost no water is discharged by re-pressurization (Figure 13-3), which can be judged as dead seedlings.

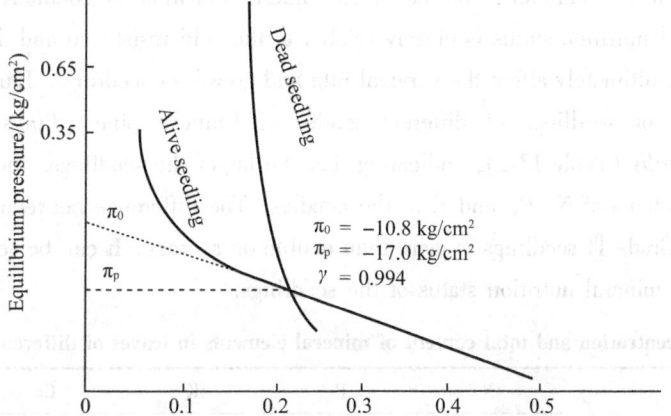

Figure 13-3 *P-V* curve of root system of normal and dead seedlings absorbing enough water of Korean pine (From Yin et al., 1992).

(2) Electro-conductibility

The theoretical basis of measuring seedling quality with electro-conductibility is that the water condition of plant tissue and the damage of membranes are closely related to the electro-conductibility of tissue. The higher the water content, the stronger the tissue electro-conductibility. Any adversity will cause damage to plant cell membranes, increase membrane permeability, and lead to electrolyte leakage. The resistance to stress and the degree of injury can be determined by measuring the electro-conductibility of exudate.

Two methods can be used to measure the electro-conductibility of seedlings: One is to measure the electrical conductivity of exudate of plant tissue; the other is to measure the electrical resistivity of plant tissue. The conductivity of exosmotic solutions of plant tissue can be measured by a special Conductivity Meter, and the conductivity of root leaching solutions can best reflect the vitality of seedlings; the resistivity of plant tissue can be measured by a Resistance Meter, from which an electrode is inserted into plant tissue, and the resistivity quickly displays.

Measuring seedling quality with electro-conductibility is simple, fast, does not damage the seedlings, and the result is reliable. However, the electrical conductivity of seedlings is affected by tree species, season, temperature, water, and measuring position. When evaluating the quality of seedlings, the relationship between the electrical conductivity of seedlings and the survival rate of afforestation should be measured carefully according to the specific tree species, seasons, and environmental conditions, and the relevant curve between the electrical conductivity of seedlings and the survival rate of afforestation should be drawn, correcting as necessary according to the temperature when taking the reading. Only in this way can the seedling quality be evaluated correctly.

(3) Other Indicators

Mineral Nutrition. Mineral nutrition status in seedlings is closely related to the quality of seedlings. At present, research has shown that at least 17 kinds of nutrient elements are involved in the growth and development of seedlings, and the lack of any one element will cause adverse reactions of seedlings. Sufficient and balanced mineral nutrition is conducive to the growth of seedlings. Mineral nutrition status is closely related to the cold resistance and drought resistance of seedlings and will ultimately affect the survival rate and growth of seedlings. Liu (1999) studied the mineral elements of seedlings of different grades of Chinese pine (*Pinus tabuliformis*) and *Platycladus orientalis* (Table 13-2), indicating that the larger the seedlings, the higher the grades, the higher the contents of N, P, and K in the needles. The difference between Grade I and Grade II, Grade II and Grade III seedlings is more than double on average. It can be seen that the seedling grade reflects the mineral nutrition status of the seedlings.

Table 13-2 Concentration and total content of mineral elements in leaves of different grades of seedlings

Seedling type		Grade	N		P		K		Ca		Mg	
			Concentration /%	Total content /mg per seedling	Concentration /%	Total content /mg per seedling	Concentration /%	Total content /mg per seedling	Concentration /%	Total content /mg per seedling	Concentration /%	Total content /mg per seedling
Pinus tabuliformis	1.5-0	I	1.99	37.6	0.28	5.3	0.67	12.7	0.12	2.3	0.12	2.3
		II	1.72	17.0	0.26	1.6	0.71	7.0	0.11	1.1	0.11	1.1
		III	1.42	6.4	0.18	0.8	0.65	2.9	0.15	0.7	0.13	0.6
Platycladus orientalis	1.5-0	I	2.46	46.5	0.29	5.5	1.38	26.1	0.49	9.3	0.39	7.4
		II	2.30	15.6	0.29	2.0	1.37	9.3	0.47	3.2	0.34	2.3
		III	1.72	5.7	0.18	0.6	1.12	3.7	0.58	1.9	0.33	1.1

Source: From Liu (1999).

For a long time, researchers have paid much attention to the increase of mineral nutrition storage in seedlings, especially the accumulation of nitrogen. They have explored the effects of conventional fertilization, steady-state nutrition loading, and controlled release fertilizer on mineral nutrition accumulation of seedlings (Quoreshi and Timmer, 2000; Li et al., 2011). Steady-state nutrition loading is considered the best measure to meet the demands for the growth and development of seedlings for nutrient laws, and to fix nutrients in the seedlings as much as possible to form a nutrient pool, promote the root growth of afforestation seedlings, and promote survival (Grossnickle, 2012; Pokharel et al., 2017). Some studies have shown that the diameter (Jackson et al., 2007) and height growth (Jackson et al., 2012) of longleaf pine (*Pinus palustris*) seedlings with additional nitrogen reserves increase in the field. However, there are also contrary cases, for example, the growth of loblolly pine seedlings with fall fertilization in the nursery has not increased on sandy land (South and Donald, 2002), that is to say, the seedlings with additional nutrient reserves do not always show a positive response after afforestation, because the inherent nutritional status of the seedlings is better than the additional loading of nutrition (Hawkins, 2011), nutrient availability of forested land (Andivia et al., 2011), or other site factors that limit growth (such as water stress) (Wang et al., 2015). Because of the uncertainty of the mineral nutrition and the performance after afforestation of the seedlings, the influence of the mineral nutrition elements on the physiology of the seedlings is more complex. In addition, the determination of the mineral nutrition content is complex and time-consuming, and needs some instruments and equipment, so many obstacles arise when evaluating the quality of the seedlings with the content of the leaf nutrients.

Carbohydrate Content. Carbohydrate is an important nutrient in seedlings, which provides energy and raw materials for the growth of seedlings. From seedling lifting and planting to the beginning of photosynthesis, the seedlings rely on the carbohydrate stored in their bodies to maintain growth and respiration. If the carbohydrate stored in their bodies cannot meet their needs, they will die. Therefore, the relative content of carbohydrate in seedlings can be used as part of a physiological quality index. To determine carbohydrate, refer to related books on the process.

Chlorophyll Content and Chlorophyll Fluorescence. Chlorophyll is the important pigment for photosynthesis of plants. The amount of chlorophyll can reflect the strength of photosynthetic capacity, so it can quantitatively reflect the health status of seedlings, such as changes in seedling morphology and nitrogen content. The relationship between them should be determined through research.

The red light reflected from the chloroplast membrane is related to the main process of photosynthesis, including light absorption, activation of energy conversion, and photochemical reaction of photosystem II. Chlorophyll fluorescence reaction is an indicator of plant photochemical reaction, which is related to species, season, environment, sample conditions, and other factors affecting physiological effect of plants. The changes of chlorophyll fluorescence can be measured to reflect the quality of seedlings.

Chlorophyll fluorescence is a direct measurement of the physiological status of the chloroplast membrane, which can be combined with the measurement of conductance, root growth potential, and stress-induced volatile matter. It takes a short time to determine the physiological status of seedlings in the growth stage, with the characteristics of reliability, instant results, and no damage to the seedling. Chlorophyll fluorescence measurement has great potential in the following aspects: Determining the time of seedling lifting; determining the vigor of seedlings after storage; monitoring the effect of environmental conditions on photosynthesis; and measuring the difference of photochemical effects of conifer species.

Root Activity. Root activity refers to the comprehensive performance of root system absorption, synthesis, and growth. Root activity was measured by the 2, 3, 5-triphenyltetrazolium chloride (TTC) method. The aqueous solution of tetrazole is colorless, and when the roots of seedlings are immersed in the solution, the hydrogen produced by the dehydrogenase in the living cells of fibrous roots make the tetrazole reduce to a stable, insoluble, non-transferring, and non-diffusing red substance—TTC. The reduction amount of tetrazole is directly proportional to the root activity of seedlings; that is, the deeper the solution is dyed, the higher the seedling activity. In addition, root activity can also be measured by α-naphthylamine method.

Bud Dormancy. Seedling dormancy is a type of self-protection for seedlings to adapt to the external environment conditions (such as temperature, water, light), and seedlings with timely dormancy have strong stress resistance. However, seedlings during non-dormant periods are vulnerable. According to this natural law, there is a theoretical basis to use the dormancy state of buds as an index to evaluate the quality of seedlings. The methods of bud dormancy determination include bud opening speed, dormancy release index (BRI), total hours of low temperature, oscilloscope technology, mitosis index, dry weight ratio, plant hormone analysis technology, and resistivity technology. Among them, the bud opening rate method is most accurate and reliable.

In addition, plant growth and regulatory substances, enzymes and proteins, photosynthesis and respiration, and plant temperature have a certain role in reflecting the physiological status of seedlings.

(4) Functional Indicators

The functional indexes of seedlings refer to the ability of seedlings to survive and grow under specific environmental conditions. All of the abovementioned morphological and physiological indexes are various manifestations of seedling vigor, but any single morphological or physiological index cannot fully reflect the vitality of seedlings. Consider the functional indexes of seedlings to be the most representative indexes of seedling vigor, because the functional indexes are the comprehensive performance of morphology and physiology of seedlings under certain environmental conditions. In addition, the resistance of seedling planting under adverse conditions is a good indicator of the vitality of seedlings.

Root Growth Potential. Also known as RGP, root growth potential refers to the rooting ability in the most suitable growth environment for seedlings. It not only depends on the physiological status of seedlings but is also closely related to the morphological characteristics of seedlings, biological characteristics of tree species, and growth season. It can predict the vitality of seedlings and survival rate of afforestation better than the morphological indexes. Therefore, since Stone (1955) put forward this concept, RGP has been widely used in seedling quality assessment, and it is one of the most reliable methods to evaluate seedling quality at present. The disadvantage is that its measurement time is long, the method is complicated, and it is inconvenient to promote in production.

RGP determination method: Remove all white root tips from the seedlings. Plant them in a container with mixed matrix (peat and vermiculite), sandy soil or river sand. Culture them in the most suitable environment for root growth, such as 25 ℃ ± 3 ℃ in the daytime, 12 to 15 hours in the light, 16 ℃ ±3 ℃ in the night, 9 to 12 hours in the dark, and at 60% to 80% relative humidity of the air, to maintain the required moisture (water once every 2 to 4 days). After 28 days, carefully take out the seedlings, wash off the sand, and count the number of new root growth points (white color).

RGP can be expressed in many ways, such as the commonly used number of new root growth points (TNR), new root number>1 cm ($TNR>1$), total new root length>1 cm ($TLR>1$), new root surface area index ($SAI = TNR>1 \times TLR>1$), and the fresh weight and dry weight of new roots. Different indexes reflect various physiological processes in the process of rooting. TNR reflects the situation of rooting, and $TNR>1$, $TLR>1$, SAI, fresh weight of new roots, and dry weight of new roots reflect the situation of root elongation. In the past decade, the appearance of different types of root analyzers and root analysis systems and other advanced instruments has provided convenience for RGP measurement. These instruments can automatically analyze the cleaned root image with multiple parameters and batches.

In principle, the determination time of RGP is 28 days (4 weeks), but tree species differ in their rooting time frame. For example, using TNR representing RGP, *Platycladus orientalis* roots relatively quickly, so the results can be seen in 1 to 2 weeks, while *Pinus tabuliformis*, Scotch pine (*Pinus sylvestris*), and Gmelin larch (*Larix gmelinii*) need at least 2 to 3 weeks.

Resistance of Seedlings. Seedlings may encounter various types and levels of adversity after planting. Therefore, the resistance of seedlings can be determined by measuring their tolerance to adversity. The determination of stress resistance of seedlings is a method proposed by Oregon State University (1984), and it is also called the OSU activity test. The theoretical basis is that seedlings after planting experience some degree of adverse conditions, resulting in decreased vitality, weakened vital forces, and may be damaged until death. In the OSU method, the seedlings are exposed to artificial adversity first and are then monitored in a human-controlled environment. If the growth and survival of the seedlings progress well, it shows that these seedlings are robust, vigorous, with high afforestation survival rate and growth potential. On the contrary, if the seedlings die, it means that the resistance was poor and the quality was poor. The results of the stress resistance test were consistent with the survival rate of afforestation.

The most commonly encountered adversity after seedling planting is cold and drought. Therefore, seedlings can be planted under specific cold and drought adverse conditions to determine the frost resistance and drought resistance of seedlings. However, to do so takes too long to obtain results of the OSU activity test—usually 2 months. Therefore, its popularization and application are limited, but it is widely used in scientific research.

13.1.3 Comprehensive Evaluation and Control of Seedling Quality

In the evaluation of seedling quality, it is very difficult to fully reflect the quality of seedlings by only one or two morphological, physiological, or functional indicators. In addition, seedlings are greatly affected by the environmental factors of forest land after planting, which increases the difficulty of evaluation. From the definition of seedling quality, it can be seen that seedling quality is aimed at site conditions and specific cultivation purposes. Therefore, the evaluation of seedling quality should consider the site conditions of afforestation land and the dynamics of seedling quality, adopt multiple indicators, and establish a complete, comprehensive evaluation and assurance system of seedling quality, which is the problem to be solved in current and future seedling quality research. To do well in the comprehensive evaluation and control of seedling quality, we can refer to the following aspects.

(1) Suitable for Land and Seedlings

On the basis of site investigation and site classification, when suitable tree species, geographical provenance, and ecological type have been determined, planting suitable seedlings in suitable land refers to selecting the most suitable seedling type, age, size, and physiological condition for afforestation according to site conditions.

Planting suitable seedlings into the appropriately suitable land is to give full play to the advantages of different seedlings, so that they can better adapt to the site conditions of the afforestation site. At the same time, the afforestation work should be combined with cultivating seedlings. In the afforestation design, the stock type, age, size, and physiological status of the seedlings are clearly defined. When nursery seedlings are raised, the seedling cultivation is oriented according to the requirements, so as to provide qualified seedlings for afforestation.

Afforestation effect is the starting point and basis of seedling quality assessment; that is to say, the quality of seedlings depends on their adaptability to afforestation land. When evaluating the seedling quality according to the site conditions of the afforestation site, stock type is an important aspect of seedling quality. Different types of seedlings are produced by propagation materials and methods that may differ from one another, and there will be great differences in morphology, physiology, and adaptability. Therefore, the selection of stock type is the first step to seedlings successfully adapting to the suitable land. Second, it is necessary to choose the right size of seedlings. There are various morphological indexes of seedlings, and different indexes reflect different aspects of growth and development of seedlings. Select seedlings of appropriate sizes according to the site conditions of the afforestation site. For example, seedlings with developed root systems, low height to caliper ratio, low shoot-root ratio, and strong resistance should be selected for afforestation in arid areas; seedlings with a high degree of lignification and full terminal buds should be selected for afforestation in areas with frost damage; for landscaping, attention should be paid to the trunk shape, crown shape, and root lump of seedlings. In addition, seedlings with the same morphological index grade should be selected with a good physiological index to ensure the vitality of seedlings during afforestation.

(2) Multi-index Comprehensive Evaluation of Seedling Quality

The indexes of normal growing seedlings are related to each other to some extent, but when seedlings are in abnormal conditions such as drought, cold, lack of nutrients, and so forth, the results of each test method are the reflection of the seedlings in only a certain one-sided way, not the comprehensive reflection of the quality of the seedlings. Therefore, use multi-index and multi-aspect approaches to evaluate the quality of seedlings. The following should be taken into account: Provenance and seed quality, stock type and age, morphological index, physiological status, and functional performance of seedlings.

(3) Dynamics of Seedling Quality and Stages of Quality Evaluation

A seedling is a living organism whose morphology, physiology, and vitality are constantly changing. Therefore, the quality of a seedling is dynamic. The evaluation of seedling quality cannot be done by static methods or by a one-time test. Instead, the quality of a seedling should be evaluated and controlled by stages according to the changing characteristics of seedling quality and the characteristics of each index. Evaluations should be carried out in every stage of seedling cultivation, before seedling lifting, before outplanting, and before afforestation, but the emphasis of controlling and evaluating seedling quality varies in each stage.

The key point before seedling lifting is to promote seedling growth and support it so that it reaches the required standard. Therefore, the establishment of the standard curves of height, root collar diameter, root system, and mineral element content of the seedlings of the main afforestation species is the basis of scientifically raising seedlings and performing quality control. When the morphology, hardening stage, and mineral element content of the seedlings meet the requirements, the seedlings can be lifted. The change rule of RGP should be mastered in the nursery, and if seedling lifting is carried out when RGP reaches the highest quality, the seedlings can be guaranteed to have high vitality and resistance.

The classification of seedlings after lifting is determined by seedling height, root collar diameter, and root system. The protection of seedling vigor should be considered when seedlings leave the nursery. In this stage, the determination of physiological indexes should be emphasized to provide important basic data and information for afforestation and storage.

(4) Establishment of Seedling Quality Control System

At present, the seedling quality assessment is only to evaluate the quality of the seedlings that have been bred and decide to eliminate them individually or in batches, which is a negative countermeasure in quality management. In fact, the quality of seedlings is a concentrated reflection of the cultivation measures and conditions applied in the process of seedling cultivation, and causal relationships occur in all stages of development. Therefore, in each stage of seedling cultivation, we can consciously determine the most suitable seedlings according to the site conditions of the afforestation site, and through environment regulation and the adoption of appropriate cultivation technical measures, we can regulate the physiology and morphology of seedlings, achieve high-quality seedlings with consistent morphology and physiology, and make the production process of seedlings a high-quality assembly line of seedlings. This multi-step process is similar to the total quality management in industrial production, and it is the management target for seedling quality in every seedling stage which shifts the control of seedling quality from passive to active. We should start from the following aspects.

Seed Quality Control. Seed quality control is the first step toward seedling quality control. To ensure the success of seedling cultivation, the first thing is to ensure the genetic quality of seedling raising materials, and the second thing is to guarantee the quality of the varieties, starting from seed collection, processing, storage, transportation, germination (or pre-treatment), and other aspects, so as to make the improved varieties in the optimal state.

Seedling Raising Environment and Cultivation Technology Control. A suitable seedling cultivation environment and advanced and practical seedling cultivation technology control are the guarantees of successful seedling cultivation. Intensive and dynamic regulation should be carried out on the abiotic and biological environment of seedling cultivation. Abiotic environment control includes suitable nursery land selection, soil cultivation, soil nutrient, water control, and so forth, while biological environment control includes pest control, reasonable density, mycorrhizal inoculation, and rhizobia. These measures are needed throughout the whole process of seedling cultivation to meet the needs of every type of seedlings in each stage of development.

Outplanting Control. It is very important to control the quality of seedlings before and after they leave the nursery. The quality control of outplanting includes seedling lifting, grading, packaging, transportation, heeling-in, and young forest tending. The quality of each step is directly related to the survival and growth performance of seedlings after afforestation.

13.2 Seedling Investigation Method

13.2.1 Investigation Purpose, Time, and Requirements

To make outplanting and production plans for the next period and to accurately determine the quantity and quality of seedlings, a seedling investigation must be carried out before outplanting. The investigation is usually conducted from the time when the seedlings stop growing to the time when the seedlings leave the nursery, or during the seedling growth period. To explore the growth and development of seedlings and to understand the impact of environmental conditions and seedling cultivating technology on the growth and development of seedlings, regular surveys should be conducted during the seedling growth period. On the one hand, developing a seedling investigation can help to comprehensively understand the seedling yield and quality and to satisfactorily complete all work before outplanting to supply required seedlings for afforestation. On the other hand, a seedling investigation can also judge the effects of various seedling cultivation measures, so as to summarize the experience and lay a foundation for further improvement of seedling cultivation technology.

The seedling investigation requires scientific sampling methods to carefully investigate the seedling yield and determine the quality indicators. The investigation requires 90% reliability, with the yield accuracy reaching 90% and the quality (root collar diameter and seedling height) accuracy reaching 95%. The percentage of Grade I and II and unqualified seedlings, as well as the total yield of qualified seedlings, should be calculated at the same time.

13.2.2 Sampling Method for Seedling Investigation

The reliability of the yield and quality data obtained from the seedling investigation depends primarily on the representativeness of the sampling and the measurement accuracy of the seedlings. The principle of mathematical statistics, using sample data to estimate the overall situation, has to be applied in sampling. Common sampling methods are as follows.

(1) Simple Sampling

Simple sampling includes mechanical sampling and random sampling. Mechanical sampling refers to randomly determining a starting point in the nursery and investigating rows or seedling beds according to a predetermined interval. There should be equal distance between sample plots or points and even distribution. Random sampling refers to using a random number table to determine the location of the sample plot, with an equal sampled probability of all seedlings.

(2) Stratified Sampling

When large differences in the density and growth of seedlings occur in the nursery, it is difficult to obtain the true results of seedling yield and quality by simple sampling. Therefore, stratified sampling has to be adopted. Stratified sampling refers to dividing the production area of the surveyed seedlings into good, medium, and poor levels according to the density and quality of the seedlings, and sampling respectively. The average value, standard error, and standard deviation of each level are calculated, and then the total average value, standard deviation, and standard error are calculated according to the weighted method. Finally, the yield and quality accuracy of the investigated seedlings are calculated by the total standard deviation.

13.2.3 Investigation Steps

13.2.3.1 Division of Investigation Area

Any tree species, seedling cultivation method (or seedling type), seedling age, operation method and main technical measures for seedling cultivation (sowing method, fertilization time, fertilization amount, irrigation frequency and irrigation amount) that are the same can be divided into an investigation area. Seedbeds in the same investigation area should be uniformly numbered.

13.2.3.2 Determination of the Type and Specification of the Sample Plot

The commonly used sample plot types include quadrat and sample section, and sometimes the sample circle is also used. The quadrat is a rectangular or square section, with the survey unit of 1 m × 1 m. The sample section uses a linear section as the survey unit (sample line), which is suitable for drilling, cutting, transplanting, and spot sowing seedlings. To improve the accuracy and reduce the workload of investigation and calculation, the sample group can be a set number and configuration. A sample plot group is composed of one main quadrat plus two or more quadrats, and the sample group is taken as a statistical unit. This method is suitable for the nursery with seedling density that changes greatly and/or has irregular growth.

The size of the sample plot depends on the seedling density, seedling cultivation method, and the number of seedlings that must be measured. For example, a small sample plot is suitable for seedlings planted with a higher density and vice versa; a small sample plot for sowing seedlings, and a large sample plot for cutting and transplanting seedlings; a small sample plot for fewer seedlings and a large sample plot for more seedlings. Generally, 20 to 50 seedlings are required for one quadrat, with coniferous tree seedlings needing 30 to 50; rooted cutting and transplants need 15 to 30. The principle for determining the sample section length is the same as determining the sample plot size.

13.2.3.3 Determination of the Sample Plot Number

The number of sample plots directly affects the accuracy and workload of the survey. The survey results have higher precision when more sample plots are used, but with more workload; fewer sample plots may obtain a low accuracy that cannot meet the requirements, and it is often necessary then to supplement the sample plots, which increases the workload. Therefore, it is very important to determine the number of sample plots. This determination is affected by the evenness of seedling density and quality. If the seedlings are even and grow neatly, the number of sample plots can be small.

Two methods determine the number of sample plots.

(1) Empirical Data Method

Long-term seedling investigation shows that, in general, 20 to 50 sample plots designed initially can meet the requirement of 90% yield accuracy and 95% quality accuracy. If the variation coefficient is less than 25%, 20 sample plots can reach the requirements; if the variation coefficient is 25% to 40%, more than 40 sample plots can meet the requirements.

Actually, if little difference is seen for the density and growth of seedlings, the initial sample plots of no less than 20 can generally meet the accuracy requirements. If the requirements cannot be met, additional sample plots should be set up, and the number of additional sample plots should be calculated according to the following formula:

Actual number of sample plots required: $n = \left(\dfrac{t \times c}{E}\right)^2$

Where t is the reliability index, $t = 1.7$ for reliability of 90%, $t = 1.96$ for reliability of 95%; c is the variation coefficient, calculated by the survey results; E refers to the allowable error percentage ($E = 5\%$ for precision of 95%).

(2) Range Estimation Method

After determining the size of the quadrat, the sections with dense seedlings (N_{dense}) and sparse seedlings (N_{sparse}) in the production area need to be investigated according to the area of the quadrat (except for the densest and thinnest sections), and the number of seedlings and the range in the sections with the same area as the quadrat should be calculated.

Estimated range: $R = N_{dense} - N_{sparse}$

Estimated standard deviation: $S = \dfrac{R}{5}$

Estimated average: $\bar{N} = \dfrac{N_{\text{dense}} + N_{\text{sparse}}}{2}$

Estimated coefficient of variation: $C\% = \dfrac{S}{\bar{N}} \times 100\%$

Estimated number of sample plots: $n = \left(\dfrac{t \times c}{E}\right)^2$

According to the normal distribution characteristics, the range can be approximately considered as 5 times the standard deviation. Since the quality accuracy of seedlings is required to be 95%, $E = 5\%$.

13.2.3.4 Setting of the Sample Plot

The number of sample plots with rough estimation should be determined in the survey area. The key to setting sample plots is to distribute them evenly and representatively. Mechanical sampling refers to the mechanical distribution of points, with one sample point set at a predetermined number of certain beds or rows. The steps are as follows.

(1) Calculate the Sample Plot Spacing (d)

Total seedling row length in the survey area = average seedling row length per bed × average seedling rows per bed × seedling bed number

If the total length of the seedbed in the survey area is 200 m, the number of sample rows is 15 with 30 seedlings per row, and the sample row length is 1 m, spacing d = total length of seedling rows/number of sample rows = 200/15 ≈ 13.3 m, that is, one row is measured every 13 m.

(2) Calculate the Number of Spacing Seedlings (n)

n = average seedling number of sample rows × number of sample rows ÷ planned measured seedlings − 1 = 30×15÷200−1 = 1.25 (planned measured seedlings = 200)

That is, in the sample row, one seedling is measured at a spacing of every other one. If the previous row is insufficient, it can be combined with the next row.

Two methods can set the quadrat: One is to first calculate the number of spacing seedbeds, N = the total number of seedbeds/number of sample plots, and then calculate the number of spacing seedlings according to the above method. The other method is to set the diagonal, calculate the total diagonal length, and calculate the spacing.

For Example, When the total diagonal length is 180 m, the number of quadrats is 20, the average seedling number of quadrats is 60, and the predicted number of seedlings is 300, the calculation is as follows:

Spacing d = 180/20 = 9 m, that is, one quadrat is set every 9 m, and the quadrat area is 1 m^2.

Number of interval seedlings n = 60×20÷300−1 = 3, that is, 1 seedling is measured for every 3 in the quadrat.

(3) Determination of Starting Point of Quadrat

To avoid subjective fixed-point, a random fixed-point method is generally adopted. For example, a small stone can be thrown randomly to the center of the survey area, and where it falls is the starting point. If the stone falls at 100 m of the sample row, a sample row is set at 100 m, and then a sample row is set at a specific certain length (such as 13 m), and so on. A random number table can also be used to determine the starting point.

13.2.3.5 Investigation on Yield and Quality of Seedlings

The number of all seedlings in each quadrat or samplerow needs to be counted, and the seedlings with diseases and pests, mechanical damage, malformations, and double terminal buds should be recorded in the notes of the seedling questionnaire, so as to count the percentage of various seedlings. Then the quality indexes of seedlings are measured according to the number of interval seedlings. The height of seedlings below 2 m should be measured with the steel tape, and the height above 2 m with the ruler. The shoot height measurement should be from the root collar to the tip of the terminal bud within an accuracy of 0.1 cm. The root collar diameter is measured by the vernier caliper, with a measurement accuracy of 0.05 cm (sometimes with 0.02 cm, and an electronic vernier caliper with 0.01 cm). For sowing seedlings and transplanted seedlings, root collar diameter should be measured at the soil mark, and the upper part needs to be measured if there is a expanded soil mark. For vegetative propagation seedlings of deciduous tree species, root collar diameter should be measured at the base of the newly germinated trunk above the cuttings, or at the normal starting position of the expanded trunk at the base. The diameter at the soil mark needs to be measured for evergreen tree cuttings. At the same time, the length and width of main roots and the number of primary lateral roots ≥5 cm in length are measured by completely digging the sample plants. The survey data are filled in on a form such as Table 13-3.

Table 13-3 Record form of seedling investigation

Tree species: _____ Seedling type: _____ Number: _____
Sample plot ID: _____ Seedling number: _____ Seedling age: _____
Seedling production unit: _____

Number	Root collar diameter /cm	Seedling height /cm	Root length /cm	Number of primary lateral roots ≥ 5 cm	Root width /cm	Overall quality status	Quality level		
							I	II	Unqualified

13.2.3.6 Calculation of Seedling Yield and Quality Precision and Total Seedling Yield

(1) Calculation of Seedling Yield and Quality Precision

Step 1: Calculate the average $\bar{X}$ and standard deviation S of each index according to the specific measured value (X_i) and sample number (n):

$$\bar{X} = \frac{\sum_{i=1}^{n} X_i}{n}$$

$$S = \sqrt{\frac{\sum_{i=1}^{n} X_i^2 - n\bar{X}^2}{n-1}}$$

Step 2: Calculate the standard deviation ($S_{\bar{X}}$) and error percentage ($E\%$) of each index:

$$S_{\bar{X}} = \frac{S}{\sqrt{n}}$$

$$E\% = \frac{t \cdot S_{\bar{X}}}{\bar{X}}$$

Step 3: Calculate the survey precision of each index (P):

$$P = 1 - E\%$$

If the precision does not meet the specified requirements, additional sample plots are required. The number of additional sample plots should be calculated according to the above method, and the precision needs to be recalculated to meet the precision requirements.

(2) Calculation of Seedling Yield

If the precision meets the requirements, the average value of each quality index is the seedling quality index. According to the quality index data of the investigated seedlings, the percentage of Level I and II seedlings should be calculated according to the national standards or local standards. The average density of sample plots can be used as the seedling density of the survey area, and the seedling yield in the survey area can be calculated accordingly. The seedling yield is calculated as follows:

Gross area = survey area = length × width × 1/667 (mu)

Net area = average bed length × bed width × number of beds × 1/667 (mu)

Quadrat area = quadrat length × width (m²)

Total seedling yield = $\frac{\text{net area}}{\text{quadrat area}}$ × average seedlings of quadrat

Seedling yield per mu = total seedling yield ÷ gross area (seedling per mu)

Yield of qualified seedlings = level I seedling percentage × total seedling yield + level II seedling percentage × total seedling yield

The above yield and quality investigation methods are mainly used for the investigation of small-sized afforestation seedlings. The investigation of small-sized vine seedlings and shrubs for landscaping can be conducted in the same way, but the quality indicators need to be determined according to the morphological indicators of vines and shrubs. For large-size seedlings of garden greening plants and precious tree species, the standard section or standard plot method can be adopted for investigation. The standard section is 50 to 100 m, and the standard plot is more than 100 m^2. The corresponding morphological indicators, such as DBH (or base diameter, meter diameter), plant height (bare stem height or height with crown), height under branches, crown width (crown diameter or canopy diameter), can be measured according to the seedling characteristics, and then the yield and quality of seedlings in the whole production area can be calculated.

For the investigation of seedlings with less quantity or that are more precious, the number of seedlings is often counted according to the planting line, and the quality indicators of seedlings are sampled and measured to master the quantity and quality of seedlings.

13.3 Seedling Lifting, Packing, and Transporting

Outplanting is the final harvest stage and the end stage of seedling raising. If this work is done well, it can ensure the quality of the seedlings and the yield of qualified seedlings. Otherwise, it will seriously reduce the quality of seedlings and the yield of qualified seedlings, and may even lead to a large number of abandoned seedlings, resulting in high yield but bad harvest. Therefore, the outplanting work is also a key step of seedling raising. We should take care of all links related to outplanting work (seedling lifting, seedling classification and statistics, seedling packaging, transporting, and so forth), with the focus on the protection of seedling vitality.

13.3.1 Water Physiology of Seedlings after Seedling Lifting

The activity of seedlings depends on the water condition of seedlings to a great extent. The growing seedlings normally absorb water from the soil, carry out photosynthesis, dissipate water into the atmosphere through transpiration and evaporation, and form the soil-plants-atmosphere continuum (SPAC system). To maintain the normal life activities of plants, it is necessary to maintain the internal water balance and the normal water movement of the SPAC system. However, after the seedling is lifted, the system is damaged, and the water in the seedling can only go out, and the speed and amount of water loss determine the time that the seedling can maintain its vitality.

The water-losing parts of seedlings include the aboveground part and the underground part. The aboveground part includes the stem and leaf of the seedlings. To protect the seedlings from the impact of the drastic changes of the external environment, the stem and leaf have formed certain adaptive characteristics in the morphological anatomy during the long-term adaptation process, such as regulating the tissue through transpiration of cork layer, cuticle, and stomata, and inhibiting the water loss in the seedlings. The underground part, that is, the root system has been living in the soil with relatively small changes in environmental conditions, sufficient moisture, little temperature change, and no wind, so the water transpiration regulation tissue is not as developed as the stem and leaf. For example, to make the soil water enter into the root tissue, the epidermal cells of the root are very thin; the cell arrangement is loose and its membrane is relatively thin, thus the water moves easily by way of transpiration. When the seedlings are in the open air after lifting, the root hairs composed of parenchyma cells and the non-lignified fibrous roots are likely to lose water, wither, and fall off. The water loss of roots leads to the decrease of water potential, and because of the water transport from high water potential to low water potential, water in the stems and leaves with high water potential flows back into the roots, causing the seedlings to lose water and die. The root system is the main path of water loss after seedling lifting.

The essence of root water loss is the damage of cell membrane structure, an increase in membrane permeability, and the weakened ability of controlling water and ions. From the perspective of water loss potential, more fibrous roots indicate a larger root system, which has a higher water loss rate, while the water loss potential of coarse roots is less than that of fine roots. Therefore, the key to the protection of seedling vitality is to protect the root system and reduce the water loss of the root system, and to properly reduce the number of fibrous roots and fine roots by pruning, so as to maintain the water content of the seedlings. According to the research, if the root water content is reduced to 50% of the original root water content, the vitality of seedlings will drop sharply. If it is reduced to 30%, insufficient water content will cause the seedlings to lose their regeneration ability.

13.3.2 Seedling Lifting

Attention should be paid to protection of the root activity of seedlings, keeping the root integrity as much as possible, and reducing damage to the root system. Generally speaking, afforestation with seedling lifting can ensure the root activity of seedlings and improve the survival rate of afforestation. However, the time of seedling lifting and afforestation often cannot coincide with each other, so it is necessary to protect the root system of seedlings in the process of seedling lifting in advance of afforestation.

13.3.2.1 Seedling Lifting Season and Time

The time of seedling lifting is generally from dormancy in autumn to germination in spring. To determine the suitable seedling lifting season, pay attention to the connection with afforestation time, production arrangement of nursery land, local climate conditions, storage tolerance, storage effects on seedlings, and seedling survival after outplanting.

(1) Spring Seedling Lifting

Most of the seedlings are suitable to be lifted in early spring and either transplanted or outplanted immediately after lifting. The seedlings do not need to be stored, so they can keep their vitality and have a high survival rate. Some tree species that are not suitable for lifting and transplanting in autumn are more suitable for lifting in spring, such as evergreen trees and larger-size seedlings that are not easy to be heeled in and stored. The disadvantage of spring seedling lifting is that if the arrangement of nursery fallow is not reasonable (for example, the land may be needed for another species at the same time), passive seedling production may occur in spring. Lift seedlings in spring before germination, otherwise, the survival rate of afforestation will be affected.

(2) Autumn Seedling Lifting

In autumn, the seedling lifting should be carried out after the aboveground part of the seedlings is dormant and leaves of deciduous trees fall off. Two kinds of conditions support seedling lifting in autumn: First, planting with seedling lifting, which is suitable for the warm area in the South. The aboveground part of seedlings stops growing in autumn, but the soil temperature is still high after seedling lifting and planting, which is beneficial for root system recovery and growth of seedlings, and it creates favorable conditions for rapid growth during the second year. Second, storing after seedling lifting, which is suitable for the North area. The small-size seedlings can be cellared after they are lifted, while the large-size seedlings can be heeled in for overwintering. If the storage conditions are good and the storage work is done well, generally speaking, there is no significant difference in the effect of the deciduous tree species storage after autumn seedling lifting and spring seedling lifting on seedling vigor, but the evergreen tree species are not easy to be stored. Some of them will experience a decrease in the seedling vigor and are not suitable for overwintering storage. Some seedlings with high root water content, such as Chinese wingnut (*Pterocarya stenoptera*) and Chinese paulownia (*Paulownia chinense*), are not suitable for overwintering storage. The advantage of seedling lifting in autumn is that it is beneficial to the reuse of nursery land, so seedling production can be arranged.

(3) Seedling Lifting in Rainy Season

In areas with severe drought in spring, it is not easy to survive for afforestation. Some seedlings can be lifted for afforestation in rainy seasons, for example, *Platycladus orientalis*, *Pinus tabuliformis*, *Pinus massoniana*, Yunnan pine (*Pinus yunnanensis*), Manchurian walnut (*Juglans mandshurica*), and camphor tree (*Cinnamomum camphora*).

Specific seedling lifting time: Whether for deciduous tree species or for evergreen tree species, it is better to lift seedlings in the dormancy period of seedlings, which can be determined according to the bud sprouting. It is better to lift seedlings on a windless, overcast day, when the water potential of seedlings is high and the water loss rate is slow. In addition, the soil moisture content should be considered, because too high soil moisture content is inconvenient to work with and will cause damage to the soil structure. When the soil moisture content is 60% of the saturated water content, the seedling lifting effect is better. If the soil is too dry, it should be watered properly one week before seedling lifting to moisten the soil.

13.3.2.2 Seedling Lifting Specification

Seedling lifting specification refers primarily to the seedling lifting depth and width determined according to the root quality requirements or DBH size of the seedlings. The root system of seedlings is an important organ of seedlings. An injured and incomplete root system will affect the growth and survival of seedlings, and the seedling quality will be reduced if the seedling lifting specification fails to meet the requirements. The various types and specifications of seedlings will differ in their specifications for seedling lifting.

(1) Bare-root Seedling Lifting

Bare-root seedling lifting is most suitable for small-size seedlings. For 1- to 3-year-old, small-size seedlings, the depth of seedling lifting should be favorable to ensure the quality of the root systems of seedlings is up to the standard. Generally, the seedling lifting depth is 2 to 5 cm longer than the root system of qualified seedlings. Seedling lifting depth of coniferous seedlings is 18 to 28 cm, and that of broadleaf seedlings, cutting seedlings, and transplanting seedlings is 25 to 40 cm. The seedling lifting root range affects whether the number of lateral roots and fibrous roots of seedlings is up to standard. Generally, the seedling lifting width of coniferous seedlings is 20 to 30 cm; that of broadleaf seedlings is 25 to 35 cm; and that of cutting seedlings and transplanting seedlings is 40 to 60 cm (Table 13-4). If the medium- and large-size seedlings are lifted with bare roots, the depth and width of the seedling lifting should conform to the Technical Specification for Seedling Raising in Urban Garden (Table 13-4).

(2) Seedling Lifting with Soil Boll

Seedling lifting with soil boll is suitable for medium- and large-size seedlings for landscaping. This kind of seedling lifting specification can be carried out in accordance with the seedling lifting specification in the Ministry of Urban and Rural Construction and Environmental Protection StandardTechnical Specification for Seedling Raising in Urban Garden (Table 13-4).

Table 13-4 Specifications for digging depth based on root size of landscape seedlings

Small-size seedlings		
Seedling height/cm	Root length needed to be reserved/cm	
	Lateral root (width)	Taproot
< 30	12	15
30-100	17	20
101-105	20	20
Large- and medium-size seedlings		
Seedling DBH/cm	Root length needed to be reserved/cm	
	Lateral root (width)	Taproot
3.1-4.0	35-40	25-30
4.1-5.0	45-50	35-40
5.1-6.0	50-60	40-45
6.1-8.0	70-80	45-55
8.1-10.0	85-100	55-65
10.1-12.0	100-120	65-75
Seedlings with soil boll		
Seedling height/cm	Soil boll specification/cm	
	Transverse diameter	Vertical diameter
< 100	30	20
101-200	40-50	30-40
201-300	50-70	40-60
301-400	70-90	60-80
401-500	90-110	80-90

For the large-size seedlings with longer root extension or the large-size seedlings without transplanting, the absorbing root group extends beyond the projection range of the crown, so the soil boll often does not bring a large number of fibrous roots. It is necessary to cut the roots and shrink the soil boll. The size of the soil boll can also be determined according to the DBH of the trees, in addition to the regulations of the Technical Specification for Seedling Raising in Urban Garden. Generally, the diameter of the soil boll is 5 to 10 times the DBH of the tree, and its height is approximately 2/3 of the diameter of the soil boll; the size of the soil boll of the shrub is based on 1/4 to 1/2 of its crown size. How to shrink the soil boll: 1 to 2 years before seedling lifting, dig a ditch around the trunk, with a distance between the ditch and the trunk that is approximately 5 to 10 times the DBH of the tree for deciduous tree species. Because the fibrous roots of the evergreen tree species are more concentrated than the deciduous tree species, the radius of the distance can be slightly shorter. The shape of the ditch can be square or round, but the perimeter should be divided into 4 to 6 equal parts. The width of the ditch should be wide enough to make it easy to dig and work in, usually 30 to 40 cm. The depth of the ditch should be determined according to the depth of the root, usually 50 to 70 cm. The root system exposed in the ditch should be cut with a sharp knife, so that the roots are level with the inner wall of the ditch. The wound should be smooth to avoid unevenness. Coat the large wound with preservative. Then fill the ditch with fertile soil (preferably sandy loam or loam), compacted layer by layer, and water it regularly, so that many fibrous roots will grow in the ditch. In the spring or autumn of the second year, the same method should be used to excavate the other two opposite sections. In the third year, when the ditches all the way around the tree are full of fibrous roots, the seedlings can be removed (Figure 13-4). The excavation should be carried out from the outer edge of the ditch, and the time of root cutting is different according to the climatic conditions of the tree's location.

Figure 13-4 **Schematic diagram of tree root cutting and shrinking.**

13.3.2.3 Seedling Lifting Method

(1) Artificial Seedling Lifting

Attention should be paid to reducing root damage when using artificial seedling lifting. If the soil is too dry before seedling lifting, apply water for 2 to 3 days to first moisten the soil of the root distribution layer, which will benefit the seedling lifting process.

Bare-root Seedling Lifting. Bare-root seedling lifting can be used for seedlings of most deciduous and evergreen trees. Before the seedlings are lifted, dig a ditch along the direction of the seedling line to the specific distance of the seedlings; the distance depends on the size of the seedlings and the root system, including outside the distribution area of the main root system. The ditch depth should be the same as that of the main root system. This method can be used for a trial excavation before formal seedling lifting, and the distance to the seedling and the ditch depth can be observed and determined. A chute is dug on one side of the seedling side of the ditch wall, and the root system is cut off according to the required length of the root system, and then a shovel is used vertically from the other side of the seedling to cut off the too-long root system, pushing the seedlings into the ditch. When the root system has been completely cut off, the seedlings can be lifted out of the ditch, but the seedlings cannot be pulled too hard, because the root system could be damaged (Figure 13-5).

The method of seedling lifting for large-scale bare-root seedlings is basically the same as the above method. However, with a large root system, the distance and depth of trenching should be increased. The ditch should surround the seedling stem, cutting off the redundant roots, and then using the shovel from one side obliquely to cut off the main root as required. While lifting out the seedling, take care to prevent the root from splitting. The root system of the seedlings is large, with much soil that can be knocked down gently, but avoid thumping it, so as to avoid damaging the root system. Part of the original soil should also be reserved.

Seedling Lifting with Soil Boll. For the larger-size seedlings of evergreen trees, precious tree species, and large shrubs, in order to improve the survival rate of planting it is necessary to lift seedlings with the soil boll. For the specific methods, refer to 11.2.4.1 in Chapter 11.

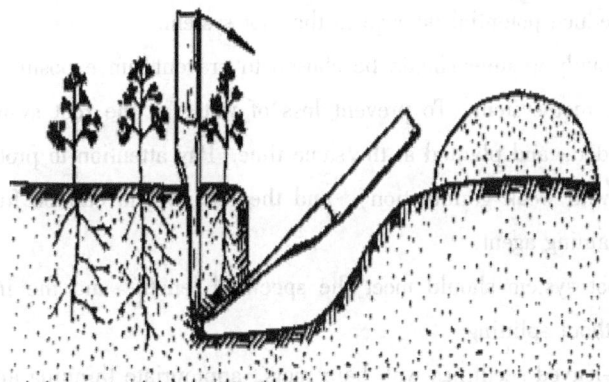

Figure 13-5 Schematic diagram of small-size bare-root seedling lifting.

(2) Mechanical Seedling Lifting

Mechanical seedling lifting has high efficiency and provides high-quality results. At present, the commonly used U-shape plow or specially designed seedling lifter (Figure 13-6) can reduce labor intensity, improve work efficiency by ten to dozens of times, the seedling lifting quality is good, and the root length is consistent with the root width.

See 11.2.4.2 of Chapter 11 for the mechanical lifting of large-size seedlings.

Figure 13-6 Two types of seedling lifters.

(3) Precautions for Seedling Lifting

In case of drought and dry soil, water should be applied 2 to 3 days before seedling lifting to loosen the soil and reduce potential damage to the root system.

Windless and cloudy weather should be chosen to prevent sun exposure that will cause drying and shriveling of the root system. To prevent loss of water in the root system of seedlings, they should be lifted, graded, and planted at the same time. Pay attention to protecting the apical buds (especially conifers with weak germination), and the root system can be further protected with a mud dip or water retaining agent.

The seedling root system should meet the specified length, and the interface of thick roots should be smooth without splitting.

For large broad-leaved seedlings with bare roots, appropriate thinning and short cutting should be carried out first to reduce water loss.

13.3.3 Seedling Grading, Packaging, Transporting, and Storage

13.3.3.1 Seedling Grading

(1) Small-size Seedlings

The purpose of seedling classification is to make the outplanting reach the national standard, to ensure the use of strong seedlings for afforestation, to reduce the differentiation of seedlings after afforestation, and to improve the survival rate and growth of trees. *Tree Seedling Quality Grading of Major Afforestation* (GB 6000—1999) has been formulated in China, in which according to the seedling height, root collar diameter, root system (root length, >5 cm lateral root number and root width), and comprehensive control index, seedlings are divided into I and II grades. The specific grading regulations are as follows.

When grading, first look at the root system index, the seedling level is determined based on the level achieved by the root system. If the root system meets the requirements of level I seedling, the seedling can be level I or level II; if the root system reaches only level II, the highest level of the seedling is only level II. If the root system fails to meet either of these requirements, it is an unqualified seedling. After determining the root system level, grading looks at the root collar diameter and seedling height index.

Qualified seedlings are divided into I and II grades, which are determined by the two indexes of root collar diameter and seedling height. When the seedling height and root collar diameter do not belong to the same grade, the grade of the root collar diameter shall prevail. The quality grading of container seedlings should be carried out in accordance with the provisions of *Technical Regulations of Containerized Seedlings* (LY/T 1000—2013).

During the process of classification, take protective measures for seedling vitality. Specifically, work in the shade, leeward of the wind, but indoor conditions that can keep low temperature and humidity are the best; the classification speed should be fast, and the time of seedling root exposure should be minimized to prevent water loss. After classification, seedlings should be packaged or stored immediately.

(2) Classification of Medium- and Large-size Seedlings

In production, the quality indexes of medium- and large-size seedlings usually include the diameter of the trunk, diameter of the soil boll, clear bole height, height of the seedling, diameter of the crown, and number of branches (or the number of layers of lateral branches). However, the various types of seedlings differ in their grading indexes and standards. At present, no uniform grading standard is available for medium- and large-size seedlings, although some standards are formulated on the local level, and the quality of seedlings specifies only the basic requirements, without formal grading.

Quality Requirements for Arbor Seedlings. The trunk diameter, tree height, crown diameter, main branch length, branch point height, and transplanting times are taken per the specified indexes. Included should be a trunk with a main shaft and three to five main branches that are evenly distributed. The slow-growing trunk diameter of large deciduous trees should be more than 5.0 cm, and the fast-growing trunk diameter more than 7.0 cm; the trunk diameter of small deciduous trees should be more than 3.0 cm; the height of evergreen trees more than 2.5 m; the grafting time of top-grafting trees should be more than 3 years, and the junction should be flat and firm (Table 13-5, Table 13-6).

Table 13-5 Main specifications and quality standards of deciduous trees commonly used in urban landscaping in beijing (excerpts)

Tree species (varieties)	Scientific name	Trunk diameter (≥cm)	Main branch length after pruning (≥m)	Crown diameter (≥m)	Branching point height (≥m)	Transplanting times (≥times)
Dawn redwood	Metasequoia glyptostroboides	—	—	1.2	—	3
Chinese ash	Pterocarya stenoptera	7	0.4	—	—	2
Cork oak	Quercus variabilis	5	—	1.2	—	3
White elm	Ulmus pumila	7	0.5	—	—	2
Zelkova	Zelkova schneideriana	5	0.4	—	—	3
Echium	Pteroceltis tatarinowii	5	0.4	—	—	2
Yulan	Magnolia denudata	4	—	1	—	3
Magnolia biondii	Magnolia biondii	5	—	1	—	3
Magnolia soulangeana	Magnolia × soulangeana	4	—	1	—	3
Cortex eucommiae	Eucommia ulmoides	7	0.4	—	—	2

Table 13-6 Main specifications and quality standards of evergreen trees commonly used in urban landscaping in Beijing (Excerpts)

Tree species (varieties)	Scientific name	Tree height (≥m)	Trunk diameter (≥m)	Crown diameter (≥m)	Branching point height (≥m)	Transplanting times (≥times)
Cedar	Cedrus deodara	4	—	2	—	3
Chinese pine	Pinus tabuliformis	4	—	1.5	—	3
White bark pine	Pinus bungeana	3	—	1.5	—	3
China Armand pine	Pinus armandii	3	—	1.5	—	3
Oriental arborvitae	Platycladus orientalis	3	—	1.2	—	2
Chinese juniper	Sabina chinensis	4	—	1	—	3
Dragon juniper	Sabina chinensis ' Kaizuca'	2.5	—	1	—	2
Privet	Ligustrum lucidum	—	4	1.2	—	2

Quality Requirements for Shrub Seedlings. Bushy shrubs are required to be plump, with main and lateral branches evenly distributed. The number of main branches is not less than five, and the average height of main branches is more than 1.0 m; creeping shrubs are required to have more than three main branches of more than 0.5 m; single-trunk shrubs are required to have main branches with uniform branches, and the base diameter is more than 2.0 cm, and the height of trees is more than 1.2 m; hedgerow shrubs should have a full crown, even branches, non-bare lower branches and leaves, and a seedling age of more than 3 years (Table 13-7).

Table 13-7 Main specifications and quality standards of shrubs commonly used in urban landscaping in Beijing (Excerpts)

Tree species (varieties)	Scientific name	Main branch number (≥branch)	Canopy diameter (≥m)	Seedling age (≥a)	Shrub height (≥m)	Main twig length (≥m)	Base diameter (≥cm)	Transplanting times (≥times)
Peony	Paeonia suffruticosa	5	0.5	6	0.8	—	—	2
Berberis thunbergii	Berberis thunbergii	6	1.5	3	0.8	0.8	—	1
Wintersweet	Chimonanthus praecox	—	—	—	1.5	—	—	1
Spiraea	Spiraea	5	0.8	4	1	—	—	1
False spiraea	Sorbaria kirilowii	6	0.8	4	1.2	1	—	1
Chaenomeles speciosa	Chaenomeles speciosa	5	0.8	5	1	—	—	1
Rosa xanthina	Rosa xanthina	6	0.8	4	1.2	1	—	1
Kerria japonica var. pleniflora	Kerria japonica var. pleniflora	6	0.8	6	1	0.8	—	1

Quality Standard of Liana Seedlings. Which requires that the number of branches shall not be less than three, the diameter of the main vine shall be more than 0.3 cm, and the length of the main vine shall be more than 1.0 m (Table 13-8).

Table 13-8 Main specifications and quality standards of liana commonly used in urban landscaping in Beijing (Excerpts)

Types	Tree species	Scientific name	Seedling age (≥a)	Branching number (≥branch)	Main vine diameter (≥cm)	Main vine length (≥m)	Transplanting times (≥times)
Evergreen liana	Euonymus fortunei	Euonymus fortunei	4	3	1	1	1
	Euonymus fortunei var. radicans	Euonymus fortunei var. radicans	3	3	1	1	1
	Hedera	Hedera	3	3	0.3	1	1
	Fagopyrum esculentum	Fagopyrum esculentum	2	4	0.3	1	1
	Rose	Rosa multiflora	3	3	1	1.5	1
	Rosa multiflora var. albo-plena	Rosa multiflora var. albo-plena	3	3	1	1.5	1
Deciduous liana	Costustoot	Rosa banksiae	3	3	1	1.2	1
	ClimbingRose	ClimbingRose	3	3	1	1	1
	Chinese wistaria	Wisteria sinensis	5	4	2	1.5	2
	Gelastrus orbiculatus	Gelastrus orbiculatus	3	4	0.5	1	1
	Amur grape	Vitis amurensis	3	3	1	1.5	1

Quality Standard of Bamboo Seedlings. The seedling age, the number of bamboo leaves, the size of the soil lump, and the number of bamboo stalks are taken as the specified indexes; the parent bamboo is 2 to 5 years old; the scattered bamboo seedlings require that the large- and medium-sized bamboo seedlings have one to two bamboo stalks, small bamboo seedlings have more than five bamboo stalks, and the clumped bamboo seedlings have more than five bamboo stalks per clump (Table 13-9).

Table 13-9 Main specifications and quality standards of bamboo commonly used in urban landscaping in Beijing

Tree species	Scientific name	Seedling age ($\geq$a)	Mother bamboo branching number ($\geq$branch)	Bamboo rhizome length ($\geq$m)	Bamboo rhizome number ($\geq$rhizome)	Bamboo rhizome bud eye number ($\geq$bud eye)
Phyllostachys propinqua	*Phyllostachys propinqua*	3	2	0.3	2	2
Black bamboo	*Phyllostachys nigra*	3	2	0.3	2	2
Bambosa vulgaris var. *striata*	*Bambosa vulgaris* var. *striata*	3	2	0.3	2	2
Phyllostachys aureosulcata	*Phyllostachys aureosulcata*	3	2	0.3	2	2
Indocalamus	*Indocalamus tessllatus*	3	2	0.3	2	—

13.3.3.2 Seedling Test Method and Label

(1) Sampling

The quality test of seedlings should be carried out in a batch of seedlings, and the random sampling method according to the rules in Table 13-10 should be adopted. For the seedlings in bundles, sample bundles shall be sampled first, and then 10 plants shall be sampled in each sample bundle, and for the seedlings without bundles, sample seedlings shall be sampled directly.

Table 13-10 Sampling quantity of seedling test

Seedling number	Testing number
500-1000	50
1001-10 000	100
10 001-50 000	250
50 001-100 000	350
100 001-500 000	500
$\geq$ 500 001	750

(2) Test

Generally, the afforestation seedlings should be tested according to the method specified in *Tree Seeding Quality Grading of Major Species for Afforestation* (GB 6000—1999), and the large-size seedlings of landscape afforestation can be tested according to the local standards. Within the allowable range of a seedling test, among the same batch of seedlings, the number of seedlings with a grade lower than the specified grade shall not exceed 5%. If the test result does not meet the above requirements, it should be rechecked. The seedling test certificate should be filled in at the end of test. All the seedlings out of the nursery should have the seedling test certificate attached, and the seedlings transported out of the county are subject to quarantine and have the quarantine certificate attached.

(3) Seedling Label

Seedling sales must have "two certificates and one label", which refers to the "seedling test certificate", "seedling quarantine certificate", and "seedling origin label". These must be held in the process of outplanting, transportation, and use of forest seedlings. The "seedling test certificate" is issued by the quality inspector of the forest seedling management institution at or above the county level to provide seedling users with accurate information about the seedling quality. This test proves that the quality of the seedlings is guaranteed. The "seedling quarantine certificate" is issued by the local forest pest control and quarantine department for the seedlings that need to be transported to other counties (cities and districts), which can effectively prevent the spread of dangerous diseases, insects, and weeds. The "seedling origin label" includes seedling type, tree species or variety name, origin, quality index, seedling age, plant quarantine certificate number, quantity, seedling production license or business certificate number, production date, producer or operator name, and address.

13.3.3.3 Seedling Packaging

After grading, the seedlings shall be packed promptly. Generally, the seedlings shall be packed according to a certain quantity according to their respective grades.

(1) Root System Treatment of Seedlings before Packaging

The purpose of root system treatment before packaging is to maintain the water balance of seedlings for a long time, to create a better water conservation environment for seedlings before storage or transportation to planting, and to prolong the vitality of seedlings as much as possible. The root treatment method is as follows.

Dipping in Mud. Dip the root system into the mud to form a moist protective layer for the root system, which will protect the vitality of the seedlings.

Water-retaining Agent Treatment. Water-retaining agent is a type of functional polymer material with strong water absorption capacity and is non-toxic and harmless. It can release and absorb water repeatedly, with strong water-retaining capacity. At present, 2 water retention agents are used at home and abroad: One is acrylamide-acrylate copolymer (polyacrylamide, sodium polyacrylate, potassium polyacrylate, ammonium polyacrylate), and the other is starch graft acrylate copolymer (starch graft acrylate). Usually, one part of a water-retaining agent and 400 to 600 times of water are stirred into gel, and then the roots are immersed in the gel so the gel adheres to the surface of the root, forming a protective layer that prevents evaporation.

(2) Packaging Materials and Methods

At present, the commonly used packaging materials include straw bags, gunny bags, nylon bags, plastic bags, and cartons. The packaging process is generally to i. lay the packaging materials on the ground; ii. put wet straw, such as wheat straw, on it; iii. place 2 groups of seedlings so that one group heads outward with roots inward, and the other group heads in the opposite direction, so that the roots are on the opposite end to the roots in the other group of seedlings; iv. add moisture materials between the roots; v. roll the seedlings into bundles; and vi. bind them with ropes, although not too tight when binding so as to facilitate air permeability.

Container seedlings are usually packed in cartons, plastic baskets, or plastic bags. The baskets (cartons or bags) shall be placed in layers, and the extrusion between seedlings shall be avoided to avoid damaging the terminal bud and branches and leaves. If the transportation time is long, water should be properly applied to avoid losing water and scattering lump.

Large-size seedlings should be well packed when they are lifted. The key point is to pack the roots with straw ropes, straw bags, or gunny bags. The packing methods include orange-shape bags, pound-sign-shape bags, and pentagon bags.

13.3.3.4 Seedling Transportation and Storage

(1) Seedling Loading and Transportation

When loading bare-root seedlings, use straw bags and cattail bags on the bottom plate of the vehicle to avoid damage to the seedlings and maintain water content. Small-size seedlings can be tied directly to the packaging car.

As the container seedlings must be transported with containers, the compartment capacity space is relatively tight. To ensure that the container seedlings are not damaged during transportation (especially for long-distance transportation), the containers should be stacked carefully. Pay attention to not damaging seedlings and try to maximize the use of the space to reduce transportation costs. The specially designed container seedling transport vehicle can be equipped usefully, and the vehicle is designed with a layered frame structure, which can effectively prevent the damage of seedlings in the transportation process and can facilitate loading and unloading. Attention should be paid to the orderly loading of seedlings to prevent them from being overly squeezed together. During transportation, slow down and avoid weather with strong wind and rain.

When loading the seedlings with soil bolls, the seedlings with heights less than 2.0 m can be loaded vertically, and the high seedlings must be laid down, which can place them horizontally or obliquely. Generally, the soil boll is forward and the treetop is backward. If the shoot is too long, it should be surrounded and hoisted with rope instead of being dragged on the ground; and the tree crown should be fixed with brackets to avoid damage caused by friction between the tree crown and the vehicle. The number of stacking layers is determined according to the specifications of soil bolls. Generally, seedlings with the soil boll of a diameter of more than 50 cm are placed in only one layer, and those with smaller soil bolls can be stacked in 2 to 3 layers. Soil bolls must be closely stacked to prevent them from being scattered by the swing of the vehicle. Do not stand or place heavy objects around the soil boll during transportation. The transport vehicles should be covered with tarpaulins to block out sunlight and rain, as well as cold and freezing weather events.

No matter what type of seedlings, there must be a designated person to escort the seedlings during transportation with the quarantine certificate from the local quarantine department. Attention should be paid to inspection on the way, especially for long-distance transportation, for example, check whether the cover is blown away by the wind and whether the root system is losing water. If necessary, water should be carried along to maintain moist root systems.

(2) Seedling Storage

The commonly used ways to store seedlings include heeling-in, cellaring, pit storage, stack storage, and low-temperature storage. This section introduces heeling-in and low-temperature storage.

Heeling-in. With this method, the roots of seedlings are temporarily heeled-in with moist soil to prevent the roots from drying and to protect the vitality of seedlings. Heeling-in is divided into temporary and overwintering heeling-in. The short-term heeling-in after seedling lifting but before afforestation is called temporary heeling-in. All seedlings that are lifted in autumn and cannot be planted for afforestation in the same year, need to be stored for overwintering, which is called overwintering heeling-in.

Bare-root seedling heeling-in: A site with high terrain, good drainage, and leeward from the wind should be selected for heeling-in. The heeling-in ditch should be dug on flat ground, with a depth of 20 to 100 cm (depending on the size of seedlings), a width of 100 to 200 cm, and wet soil. Seedlings of broad-leaved trees are arranged in the ditch with the same number in each row, so as to make them easier to count, and the shoots should incline in the downwind direction. The lower part of the seedling trunk and the root system should be well buried with moist soil and compacted. Master the requirements of "thin arrangement, deep burying, and compacting" to prevent dry wind from invading and negatively affecting the heeling-in area. Fifty or one hundred coniferous seedlings are in a bundle, placed in the heeling-in ditch in order, and with the roots separated by sand; for overwintering heeling-in, the seedlings shall also be covered with 10 to 30 cm of soil to prevent wind drying and mildew (Figure 13-7).

(a)　　　　　　　　　　(b)　　　　　　　　　　(c)

Figure 13-7　Three types of heeling-in.

Heeling-in of seedlings with soil boll: The trenching method of seedlings with soil boll is the same as that of bare-root seedlings, but the depth and width of the trench should be appropriately increased according to the size of the soil boll. Attention should be paid to the plant spacing when heeling-in. The plant spacing should not interfere with the lateral branches of the trunk, nor with the maintenance management, access, and loading during the period of heeling-in. During temporary heeling-in, the soil shall be earthed up to the height of 1/3 of the soil boll, and the soil boll shall be firmly patted with the shovel. Remember not to bury the whole soil bolls to prevent the packaging materials from rotting. Pillars should also be set up to prevent trees from skewing and to keep them upright. During the overwintering heeling-in, the seedlings should be arranged in a regular way, with the crowns close to each other, vertically placed in ditches, and the thickness of the covering soil should just cover the soil boll. Watering should be applied after soil covering. If strong winds occur in winter, use straw bags and loose straw to cover the aboveground part of the heeled-in seedlings. The stem should also be protected to prevent it from freezing.

After heeling-in, insert a sign to indicate the tree species, seedling age, and quantity within that area. During the period of heeling-in, check the seedlings frequently, especially when they are unable to leave the nursery in early spring. It may be necessary to take cooling measures to inhibit germination.

Low-temperature Storage. Low temperatures can keep the seedlings dormant, reduce the intensity of physiological activity, and reduce the consumption and loss of water. It not only keeps the vigor of seedlings but also delays the germination of seedlings and prolongs the afforestation time. Low-temperature storage should be controlled at 0 to 3 ℃, and the relative humidity of the air kept at 85% to 90%, with ventilation facilities provided. The best way to store seedlings at low temperatures is in cellars and low-temperature storehouses.

Questions for Review

1. What is the relationship between morphological index and physiological index of seedlings?
2. How do you comprehensively evaluate the quality of seedlings?
3. How do you grade the quality of small-size seedlings?
4. What are the morphological indexes for the medium- and large-size seedling quality assessment?
5. How do you ensure that the vitality of seedlings will not be affected in the process of outplanting?
6. What measures should be taken to protect the vitality of seedlings during the process from seedling lifting to afforestation?

References and Additional Readings

ANDIVIA E, FERNANDEZ M, VAZQUEZ-PIQUE J, 2011. Autumn fertilization of *Quercus ilex* ssp. *ballota* (Desf.) Samp. nursery seedlings: Effects on morpho-physiology and field performance[J]. Annals of Forest Science, 68(3): 543-553.

PRC FORESTRY INDUSTRY STANDARD, 2013. Technical regulations of containerized seedlings: LY/T 1000—2013[S]. Beijing: Standards Press of China.

DAI J X, LIU J L, 1992. Relationship between seedling density and seedling quality of larch[J]. Forestry Technology Newsletter, 7: 12-13.

DICKSON A, LEAF A L, HOSNER J F, 1960. Quality appraisal of white spruce and white pine seedling stock in nurseries[J]. Forestry Chronicle, 36(1): 10-13.

GROSSNICKLE S C, 2012. Why seedlings survive: Influence of plant attributes[J]. New Forests, 43(5-6): 711-738.

HAWKINS B J, 2011. Seedling mineral nutrition, the root of the matter[A]. In: Riley L E, Haase D L, Pinto J R, technical coordinators. National Proceedings, Forest and Conservation Nursery Associations, 2010. General Technical Report RMRS-P-65[C]. Fort Collins, CO: U.S. Department of Agriculture, Forest Service. p 87-97.

JACKSON D P, BARNETT J P, DUMROESE R K, et al., 2007. Container longleaf pine seedling morphology in response to varying rates of nitrogen fertilization in the nursery and subsequent growth after outplanting[A]. In: RILEY L E, HAASE D L, PINTO J R, technical coordinators. National Proceedings, Forest and Conservation Nursery Associations, 2010. General Technical Report RMRS-P-50[C]. Fort Collins, CO: U.S. Department of

Agriculture, Forest Service. p 114-119.
JACKSON D P, DUMROESE R K, BARNETT J P, 2012. Nursery response of container *Pinus palustris* seedlings to nitrogen supply and subsequent effects on outplanting performance[J]. Forest Ecology and Management, 265(1): 1-12.
KRAMER P J, KOZLOWSKI T T, 1979. Physiology of woody plants[M]. New York: Academic Press.
LANDIS T D, TINUS R W, MCDONALD S E, et al., 2008. The container tree nursery manual. Volume 7: Seedling processing, storage, and outplanting. Agriculture Handbook 674[C]. Washington DC: US Department of Agriculture, Forest service. p31.
LI G L, LIU Y, ZHU Y, et al., 2011. Study of techniques for steady-state nutrition supply of seedlings[J]. Journal of Nanjing Forestry University (Natural Science Edition), 35(2): 117-123.
LIU Y, 1999. Seedling quality regulating theory and techniques[M]. Beijing: China Forestry Publishing House.
MULLIN R E, SVATON J, 1972. A grading study with white spruce nursery stock[J]. Commonwealth Forestry Review, 51(1): 62-69.
NATIONAL FLOWER STANDARDIZATION COMMITTEE, 2008. Planting technical specification and produce quality grade for ornamental Palmae—Part 1: Field production: LY/T 1734.1—2008[S]. Beijing: Standards Press of China.
NATIONAL FLOWER STANDARDIZATION COMMITTEE, 2008. Planting Specification and Quality Grade of Ornamental Palmae—Part 2: Containers Cultivation: LY/T 1734.2—2008[S]. Beijing: Standards Press of China.
POKHAREL PREM K, JIN-HYEOB C, SCOTT X, 2017. Growth and nitrogen uptake of jack pine seedlings in response to exponential fertilization and weed control in reclaimed soil[J]. Biology and Fertility of Soils, 53(6): 701-713.
QUORESHI A M, TIMMER V R, 2000. Early outplanting performance of nutrient-loaded containerized black spruce seedlings inoculated with *Laccaria bicolor*: A bioassay study[J]. Canadian Journal of Forest Research, 30: 744-752.
PRC FORESTRY INDUSTRY STANDARD, 2014. Seedling quality grading of major bamboo species for urban greening: LY/T 2345—2014[S]. Beijing: Standards Press of China.
PRC NATIONAL STANDARD, 2000. Tree seeding quality grading of major species for afforestation: GB 6000—1999[S]. Beijing: Standards Press of China.
SHEN G F, ZHAI M P, 2011. Silviculture[M]. 2nd edition. Beijing: China Forestry Publishing House.
SHEN H L, DING G J, 2009. Seedling cultivation[M]. Beijing: China Forestry Publishing House.
SONG T M, ZHANG J G, LIU Y, et al., 1993. Study on seedling vigor of main conifer species in Great Khingan Range[J]. Journal of Beijing Forestry University, 15(Supplement 1): 1-17.
SOUTH D B, DONALD D G M, 2002. Effect of nursery conditioning treatments and fall fertilization on survival and early growth of *Pinus taeda* seedlings in Alabama U.S.A.[J]. Canadian Journal of Forest Research, 32(7): 1171-1179.
STONE E C, 1955. Poor survival and the physiological condition of planting stock[J]. Forest Science, 1: 90-94.
SUN S X, 1992. Silviculture[M]. 2nd edition. Beijing: China Forestry Publishing House.
PRC National Standard, 2001. Technical specification for seedling raising in urban garden: CJ/T 23—1999[S]. Beijing: Standards Press of China.
SHENZHEN QUALITY AND TECHNICAL SUPERVISION, 2006. Tree seedling grading for urban greening: DB440300/T 28—2006[S]. Shenzhen: Shenzhen Quality and Technical Supervision.
PRC FORESTRY INDUSTRY STANDARD, 2008. Virescence Seedling cultivation regulation and quality grading of camphortree: LY/T 1729—2008[S]. Beijing: PRC Forestry Industry Standard.
WANG J, LI G, PINTO J R, et al., 2015. Both nursery and field performance determine suitable nitrogen supply of nursery-grown, exponentially fertilized Chinese pine[J]. Silva Fennica, 49(3): article 1295. 13 p.

BEIJING QUALITY AND TECHNICAL SUPERVISION, 2017. Woody seedlings of plant materials for landscaping: DB11/T 211—2017[S]. Beijing: Beijing Quality and Technical Supervision.

YIN W L, WANG S S, et al., 1992. Study on the physiological index evaluating seedling quality and the development of the plant vigor tester[R]. Beijing: Beijing Forestry University.

Chapter 14 Nursery Management

Lin Na, Zhang Gang

Chapter Summary: Nursery management involves nursery science, including nursery goal management, human resource management, production management, sales management, and archive management. Whether a nursery is large, medium, or small in size, its success depends on careful planning and good management, the capacity to produce high-quality seedlings, and the ability to meet the needs of forest construction and landscaping. Managing a nursery is a science as well as an art, and the art side means learning from experience. Observational skills, a flexible management style, and a willingness to take responsibility for seedlings are important characteristics of successful managers. An important aspect of good management is having clear responsibilities in a well-structured organization with capable managers. Customer and employee feedback, information gained from trials and daily records, and ongoing research and education will always be the basis for understanding and improving nursery production.

Management plays an important role in the organizational activities that enable an enterprise to achieve its objectives and develop its functions. The main task of scientific management is to study and implement the phenomenon and laws of organizational activities, to guide the development of an organization.

The majority of nurseries are businesses engaged in managing seedling production. Most of the national nurseries are government-affiliated institutions, with a major goal of active production of seedlings. With government-affiliated institutions, economic accounting should be carried out in accordance with enterprise management. The tasks of a nursery include reproduction, excellent tree species popularization, demonstration and guidance of seedling production, and the launch of many types of operations. Nursery management includes organizational leadership and management of planning, finance, labor, technical practices, maintenance, and so on. In addition to government-affiliated nurseries, there are joint stock nurseries and private nurseries.

Nursery management is a creative activity, geared toward specific targets and with a wide range of responsibilities to fulfill through effective integration of owned resources.

14.1 Nursery Positioning and Target Management

14.1.1 Nursery Positioning

Enterprise positioning refers to enterprise strategic positioning, and for nursery enterprises, it encompasses the overall goals of the nursery. Three questions should be addressed in enterprise positioning: Who I am, where I am now, and where I am going. The characteristics of a nursery include nursery scale, tree species and varieties, and the combination of long and short terms of nursery production. The distinguishing features of a nursery are determined on the basis of market research, combined with its own actual situation. The production capacity of seedlings is limited by the nursery's economic situation. The most attention should be paid to tree species, the selection of varieties, and the determination of a long-term and short-term combination strategy for seedling production.

The manager must determine and fix the position of the nursery by its functions. Only in this way can a manager and the business reach their targets. The most suitable site for a nursery is near the main nursery planting areas. Generally, the central provinces are concentrated in the provincial capital cities, while the coastal provinces are not suitable given their remote counties or locations far away from the seedling planting areas.

14.1.2 Target Management of the Nursery

Target management is the top leadership of the organization. They formulate the general objectives according to the specific situations and the social needs. The target system consists of a range of departmental objectives, which are taken as the basis for assessment for the staff. The main advantage is to improve the management of the nursery at all levels, which means the managers organize and handle the departments depending on their scheduled goals. The staff of the departments in a nursery make their plans and make every effort to achieve all their goals.

14.1.2.1 Factors to Consider in Setting Targets

Is the target specific? For example, how many types of seedlings are produced by the production department? How many specifications are there? How many seedlings of various specifications? Can these targets be achieved? How to achieve these targets? When will these targets be achieved? Can these targets be verified?

Setting up the overall objectives is the origin of the target management system. The sub-objectives are generated from the overall objective, forming an objective net.

14.1.2.2 Basic Works in Target Management

(1) Improve Employees' Understanding of Target Management

Make sure the employees make accurate and meaningful development targets. Make sure the employees understand that setting a reasonable target system is the basis for ensuring the target management goes well. Make sure the employees realize that target achievements are for their own use and success. Make sure the employees understand the basic theory and methods of target management.

(2) Strengthen Standardized Work and Lay a Qualified Foundation

Criteria are divided into technical and managerial according to the contents. Technical criteria refer to the quality of seedling production and the standards formulated for the maintenance and use of various mechanical equipment in the nursery; managerial criteria refer to the standards formulated for maintaining the normal operation of the nursery, which are divided into general and departmental principles.

(3) Strengthen Statistical Work and Lay a Qualified Informative Foundation

Establish the necessary statistical system for collecting and analyzing data. Specify unified statistical statements across the nursery departments. Improve the statistical quality of original records.

14.1.2.3 Nursery Target Setting

(1) Contents of Target Setting

Specify the Policies. The policies are directional, overall, incentive, and other characteristics.

Develop Targets. According to the targets' nature of the nursery plan, they can be divided into strategic and tactical. According to the time needed to achieve the goal, it can be divided into long-, medium-, and short-term targets.

Setting Target Values. Target values indicate the standard each target should reach, which is further concretization and quantification of each target.

(2) Principles of Target Setting

Targets include the all-round principle, the emphasis principle, the advanced principle, the feasibility principle, and the flexibility principle.

(3) Basis for Target Setting

Target setting should consider the national policies, economic development, the specific situation of the nursery, and it should not go against environmental conditions.

14.2 Human Resource Management of Nursery

14.2.1 Organization and Design

The principle and structure of the organization include target; administrative levels and structure; an artificial system; and the operation process.

The organizational design is the process of activities and structure. Its purpose is to coordinate the relationship between people and situations; among people, encouraging their full initiative and enthusiasm; improving efficiency; and achieving the goals.

The principles of organizational design are integration; unity of command; responsibility of right correspondence; and effective management range.

The basic models of organizational structure include a linear system, functional system, division system, simulation decentralization system, matrix system, super division system, new matrix system, multidimensional structure system, and more.

The organization of a nursery can be relatively simple, for example, a larger nursery could adopt a linear system (Figure 14-1), a shareholding nursery often adopts a division system (Figure 14-2), and the private nurseries do not necessarily have a fixed organizational system.

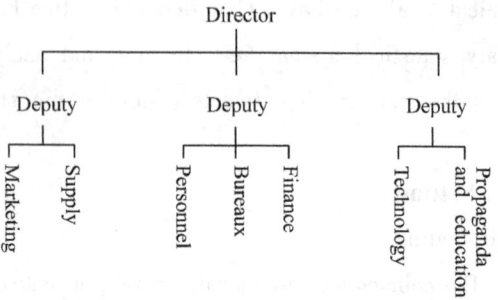

Figure 14-1   Organization structure sketch of a linear system (Bureaux = administrative office).

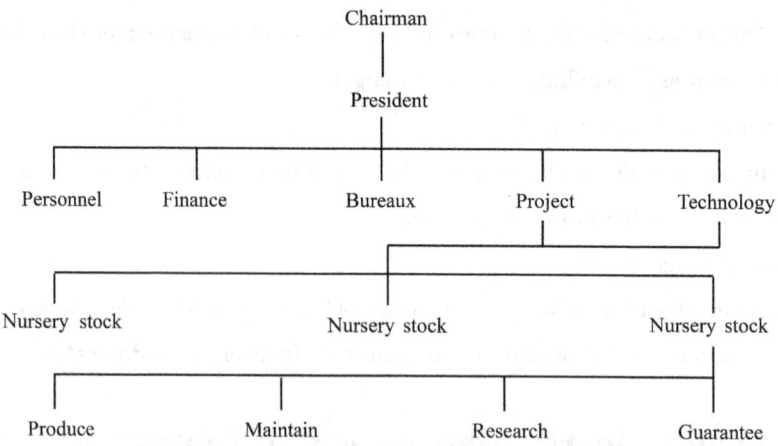

Figure 14-2   Organization structure sketch of a division system (Bureaux = administrative office).

14.2.2   Human Resource Management

Human resource management includes skill management and consciousness management. Skill management is for operators, and it is closely related to several humanistic variables, such as physique, specialty, experiences, and personal coverage. Consciousness management is to ensure and promote the management efficiency, the workmanship, the quality of technological processes, the dispatching, and progress level. It is closely related to several humanistic concerns, such as education, qualifications, actual achievement, adaptability, and so on.

14.2.3 Talent Management

Talent management is the procedure formulation, implementation, and adjustment of talents to substantially promote the growth of the effective production, significantly reduce the invalid consumption (meaning costs not related to production), and enhance the amount of effectiveness. Talent management includes discovery, job placement for each worker, and control of the talents. Talent discovery is a dynamic talent selection mechanism, which creates a competition mechanism within the company and ensures the talent could be selected in a fair circumstance.

Compared to talent discovery, talent usage is even more important. Effective matching of talent with responsibilities is also an art and a science, and retaining good talent is critical to the stability of the workplace. Talent control includes institutional constraints and encouragement of competition. Better control of talent requires the manager to be the top of his or her profession. Make perfect, scientific, and reasonable rules and regulations, or sign relevant contracts with new talent to make clear their responsibilities, and managers should do all they can to prevent the outflow of talents. Regarding talent, the manager should encourage competition to build the talent echelon.

14.3 Production Management of Nursery

14.3.1 Tasks and Content of Nursery Production Management

14.3.1.1 Tasks of Nursery Production Management

(1) Production Management of Nursery

The target of the nursery is to produce types, specifications, and quantities of seedlings that meet the needs of the society.

(2) Essential Productive Factors Management

Essential productive factors include people, property, objects, and information. The essential productive factors management refers to include reasonable allocation of those essential productive factors to maximize the nursery benefits.

(3) Production Process Management (PPM)

PPM includes the formulation of a production plan, the purchase of production goods, the propagation and cultivation of seedlings, and the expansion of reproduction.

(4) Feedback Management

In the process of implementing the production plan, any problems that arise should be solved, the production plan should be adjusted and implemented, and the production goals should be smoothly realized.

14.3.1.2 Contents of Nursery Production Management

Production management includes the preparation of nursery production goods, formulation of the production plan, and management of the production process (Figure 14-3).

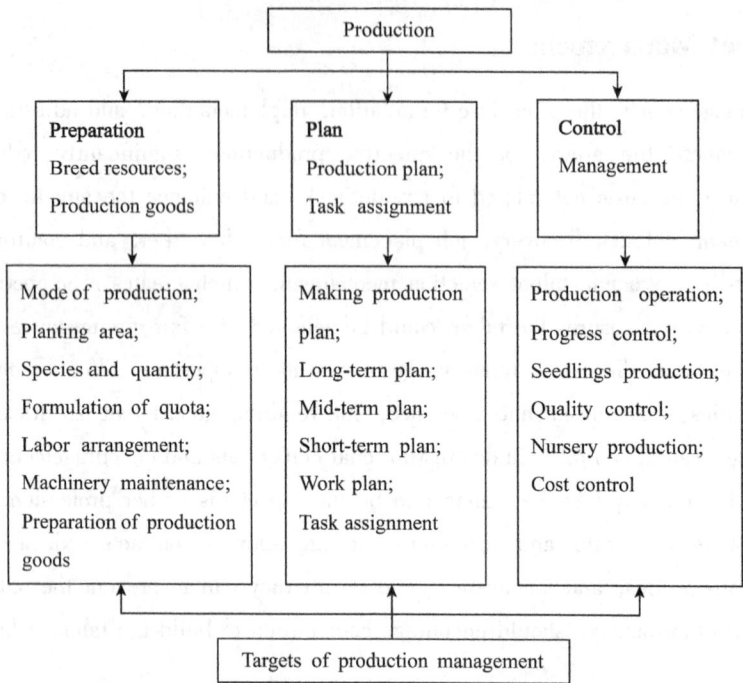

Figure 14-3 Content diagram of modern nursery production management (From Han, 2008).

14.3.2 Quality Management of Nursery Seedling Production
14.3.2.1 Overall Quality Management

(1) Quality Management of All Staff

The enthusiasm and creativity of all staff should be inspired, and qualities of the staff should be promoted. All the staff need to have concern for the quality of the seedlings and be involved with management.

Train all the staff to enhance their quality awareness and encourage participation in management goodwill. Formulate the quality liability mechanism, clarify the responsibility and authority of each staff in the quality responsibility system, ensure every staff performs their own duties, and every staff should closely cooperate. Carry out mass quality management activities in various forms, to inspire all the staff to develop their own talents and interests.

(2) Whole Process Quality Management

This term includes the whole process of quality management from market investigation, tree species (varieties) design, development, production, marketing, and service. Control all links and relevant factors across the entire process of seedling quality formation, focus on prevention, combine prevention and inspection, eliminate unqualified seedlings in the formation process, and take preventive measures.

(3) Whole Nursery Quality Management

All managers at all levels of the nursery must have a clear understanding of quality management and what that entails. Management emphasizes quality decision-making, quality policymaking, quality targets, quality plans, along with strong supervision and inspection. Nursery staff shall carry out production in strict accordance with the standards and regulations, cooperating with one another in the division of labor, making suggestions to improve the work, and continuously improving their work.

(4) Multi-method Quality Management

Multi-method refers not only to material factors but also people, technology, management, and internal or external factors that may affect the quality of seedlings. Put all the affected factors together and manage them well in a comprehensive way. We must widely and flexibly use a variety of modern management methods to solve the quality problems according to the specific situations and influencing factors. No matter which method is adopted, the working procedure of planning, execution, inspection, and summary should be followed.

14.3.2.2 Problems to Notice in Total Quality Management

(1) Strictly Control the Quality of Seeds

First of all, improve seed viability and seedling vigor. Choose the best breeding material to ensure seedling quality. Pay attention to the temporality and regionality of the improved seed viability. Plant the latest cultivated, improved varieties in a suitable place to ensure the quality of seedlings. Second, test the sowing quality of the seeds, and do a good job in seed selection and grading before sowing. Seeds of different grades should be seeded and raised separately to ensure high emergence rate, neat emergence, and robust growth. Third, test the vitality of the cuttings, that is, whether they lose water, whether their clarity is basically the same, and improve the uniformity of seedling emergence.

(2) Carry Out Quality Management in the Whole Process of Seedling Raising

The whole production process, which includes soil preparation, sowing (cutting or grafting), watering, fertilizing, weeding, thinning, pest and insect control, seedling investigation, and outplanting, must be strictly in accordance with the production technical regulations for intensive management. Strictly control the density of seedlings in each period, strengthen the measures for thinning seedlings, ensure a reasonable production number of seedlings, and improve the emergence rate of level-I seedlings. According to the soil conditions and the characteristics of the seedlings, a reasonable irrigation system should be established.

(3) Pay Attention to Standardization

Standards include seedling standards, inspection standards, operation standards, and other technical standards, as well as management methods, procedures, rights and responsibilities, and other management standards.

14.3.3 Internal Control of Seedling Production

Control refers to the process of monitoring all activities to ensure that all important details are carried out as planned and/or corrected as necessary. The key points of nursery control include plan control, quality control, and cost control. To achieve effective control, we must do the following.

(1) Create and Follow a Scientific Plan

The plan limits the activities of the nursery to a certain extent, which is a kind of control in itself. But the plan is not equal to control.

A nursery plan is the overall target of control. The ultimate goal of control is to complete the plan and achieve the target. The more detailed, clear, and feasible the plan, the easier and more effective the control. Management control itself should also be part of the plan. To implement effective control, first establish control standards and procedures; and second, define the key points, methods, and objectives of control work.

(2) Collect Accurate Information on Time

Information is the basis and premise of control. Only through the timely transmission and feedback of information, can control be carried out. And only accurate, reliable, and timely information can achieve the purpose of control.

(3) Establish a Clear Responsibility System

A clear responsibility system should be established for effective control. Everyone should know their responsibilities and the standards to be achieved, so they can conscientiously perform their duties in the work and complete the task according to the standards.

(4) Build a Strict Organization

Make clear who is in charge; that is, implement who controls and solves the problem of vacancy (meaning no one is in charge of the work). According to the whole operation and management process of a nursery, the control work can be divided into the following three stages.

Pre-control, Also Called Prior Control. To avoid losses afterward, the potential problems and deviations that may occur in the implementation can, for the most part, be predicted and estimated before the implementation of the plan. Preventive measures should be taken to eliminate problems before they occur.

Field Control, Also Called Real-time Control or Process Control. This concept refers to control in the process of production and operation. When a problem arises, it should be remedied in time, by supervising effectively and implementing a prompt correction.

Post-control, Also Called Achievement Control. This step compares the implementation results with the original expected plans and standards. Then, by analyzing, evaluating, and taking measures, improvements can be made to the plan. Post-control cannot redeem losses that may have occurred in the earlier stages, but it can help prevent future irreparable losses.

14.3.4 PDCA Working Method

PDCA is an abbreviation made from the first letter of "Plan", "Do", "Check", and "Action", which is a common management method. "Plan" means why and what you are planning to do, where to do it, when to do it, who will do it, how to do it, and the aim is to achieve a goal. "Do" refers to the specific organization, implementation, and execution of the plans and measures that have been formulated. "Check" refers to the comparison of the implementation results with the plan objectives and the inspection of the implementation of the plan. "Action" is to sum up experiences and lessons, consolidate the successful cases and form standards, formulate preventive measures for the failed cases, and turn any remaining problems into the next cycle. PDCA working method is a work process with four stages of repeated circulation. To meet prescribed goals, the four stages should not be lost, the sequence should not be changed, and the cycle should be continued.

To operate PDCA scientifically, effective measures should be made and a "5W1H" working method should be adopted for seedling production management. The abbreviation "5W1H" stands for 5 "W" and 1 "H" considerations, that is, Why (why to develop these measures or means); What (what purpose should the implementation of these measures achieve); Where (which process and department should these measures be implemented for); When (when must each measure be completed); Who (by whom will the measures be implemented); and How (how to implement these measures). The 5W1H working method ensures the realization of PDCA, improving the production efficiency of a nursery, guaranteeing the quality of seedlings, and achieving the goals of nursery production management.

14.3.5 Nursery Management Cases

14.3.5.1 Who Is Responsible

A nursery manager should have the following characteristics: Keen observation; flexible management style (dispatch is not necessarily rigid but adapts to the changing needs of plants growing day by day); the ability to "think like plants" (people who have a "sense" of plants, may probably do a better job than a manager who treats plants more strictly from an engineering point of view); and willing to be responsible for the plants in the nursery.

14.3.5.2 What Needs to Be Done

(1) Weekly Plan

The strategic plan for weeks or months include the following tasks: Summarize what needs to be done; check the seedling growth schedule, and facilities arrangement schedule and its deadline; evaluate seedling development and required maintenance; evaluate potential problems; establish the schedule for one week and one month; reasonably arrange the tasks; assign the tasks; follow up to ensure the completion of the tasks; summarize and plan the next task; carry out long-term planning (future vision, objectives, and steps).

Looking up weekly logs, plant growth and development records, and other observations will help to prioritize the work. What's going on with the seedlings? Which growth stage are they in: Survival stage, fast-growing stage, lignification stage? Are all stages on schedule? What needs to be done next: Transplanting, changing the amount of fertilizer applied? Any potential problems that may arise should be identified, such as the emergence of potential pests. Do customers need to update their seedlings and schedule progress? Once the list is completed, the tasks of the list are arranged from the most important to the least important, and appropriate roles and tasks of employees are assigned.

(2) Daily Tasks

Watering; maintenance (such as weeding, pest control, fertilize, and so on); seedling inspection and monitoring; paying attention to the progress of seedlings daily and weekly and making notations in the plant development record; and recording the regular observations and activities in the log.

Keeping a daily log is an essential nursery work. The manager should be quite familiar with the daily log and should know exactly: What was done today? Have all the supplies been purchased? What is the time spent on a certain seedling? How much time does it take to manage work, such as irrigation? What work needs to be arranged for the next seedling? What observations are required for seedlings or nursery management? What kind of seedlings can be sold and what is the price? As the nursery increases in size and complexity, it will be easier to trace this information by having the daily log in computers (even simple spreadsheets).

The following information is essential for making decisions: Budget; production schedule; determining the maximum benefits from labor-saving equipment; analyzing nursery expenses; improving profits or production; and continue to produce seedlings that have been successfully cultivated.

The record of plant growth and development is another key tool that can improve conscientiousness and develop good observation skills. Daily observation of seedlings ensures that potential problems can be averted before they arise.

Observation includes: Plant physiology: How do these seedlings look? Does the shoot-root ratio meet the standard at this stage? Are root or leaf nutrients lacking or disease symptoms visible? Can the presence and benefits of symbiotic microorganisms be seen in the root system? Are seedlings growing as expected and entering the transplanting time? Smell: Experienced planters can detect a disease, such as the fungi *Botrytis cinerea*, and can prevent widespread outbreaks. Overheated electrical machinery, broken fans, and other factors can also be detected by smell. Noise: Does the engine need to be refueled? Does the water runoff? When or where should runoff not occur? Touch: Is the temperature and humidity in the normal range? Is the root water suitable or excessively wet?

(3) Weekly or Quarterly Seedling Production Tasks

Check the schedule of sowing, transplantation, fertilization, shipments, and facility arrangements; Assign missions (such as preparing growth substrate, sowing, inoculate mycorrhizae, and so forth); Tasks at the early stage of a fast-growing period (such as fertilizing or measuring); Tasks during the lignification period (such as changing fertilization and lighting); Updating customers on plant development; Packaging and transportation; and Classification and cleaning.

The seedling production process includes: Understanding the three growth stages (survival stage, fast-growing stage, and lignification stage) and the clear requirements of each stage; Formulating the seedling growth plan for the detailed change of growth cycle from seed procurement to afforestation; Listing the space, labor, equipment, and supply materials to support seedlings in the three growth stages; Keeping logs and plant development records; Developing accurate breeding programs so that a successful experience of seedlings can be repeated next time.

(4) Record Keeping

Two main records should be kept for a nursery: A production record and a financial record.

(See more on creating and completing nursery files later in this chapter starting at 14.5.)

Production Records. Keep daily records (such as environmental conditions, labor, and daily activities), which are recording the development of each kind of seedlings, establishing and updating the seedling growth schedule and facility arrangement schedule, updating and modifying the seedling plan, evaluating and updating the stock of seedlings (including all seedlings in the nursery numbered according to the seedbed and the current development stage of seedlings), and delivery details (site, customer name, provenance, and expected delivery date).

Financial Records. Required labor and time, material costs, indirect costs (e.g., utilities costs), cost estimation, income monitoring, detailed list of production and maintenance materials (e.g., growth substrate, fertilizer, containers and trays, irrigation parts, and more), and seasonal cleaning (removal of remaining seedlings, cleaning of seedbeds and seedling platforms, cleaning and disinfecting containers, checking and repairing equipment and infrastructure).

The key to maintaining the long-term prosperity of a nursery is to keep detailed financial records. For example, be sure to record: Seedling specification; Growing time; Required labor at each stage (in person/hour); Required materials and their expenses (such as seeds and growth substrate); Needs of custom culture (such as special containers); Indirect operating costs (such as equipment); Cost inflation over time; and Typical losses (proportion of unqualified seedlings).

(5) Seasonal Cleaning

Every 2 to 6 months, after shipping a large number of orders, or at the end of a season, there is the opportunity to complete the following tasks: Clean up the unwanted seedlings; clean and flush the floor and table (if there are no plants, dilute bleach or other cleansers can be used); clean and disinfect containers; flush irrigation system; check and repair other equipment; and if necessary, replace the plastic roof of the greenhouse.

(6) Training, Experimenting, and Problem Solving

Training. The staff participates in trainings and meetings, learns from other nurseries, is assigned and participates in field work, and reads published literature (such as local plant journals).

Experimenting. Identify the most important issues in order to determine priority research, design, and process; evaluate experimental data and results; troubleshoot and solve problems. To systematically solve problems: Identify problems, analyze problems; know whom to ask for help, such as other nurseries, soil experts, pest control experts, and irrigation experts; stimulate creative approaches; develop and test assumptions; and implement solutions.

Problem Solving. A five-step system approach can be adopted for problem solving (Figure 14-4).

Step 1: Confirm the problem. Is it a real problem? Is there anything wrong?

Step 2: Analyze the problem. What happened? When did it start?

Step 3: Stimulate creative thought processes. Identify potential sources of problems, find the solutions or seek help from literature, other nurseries, staff, or external help such as promotional agents or experts who can gather the information.

Step 4: Develop and test assumptions. Find the base of the problem and take action.

Step 5: Implement the solutions. Determine the solution to this problem, observe the results carefully; if the problem is not solved, start over from Step 2.

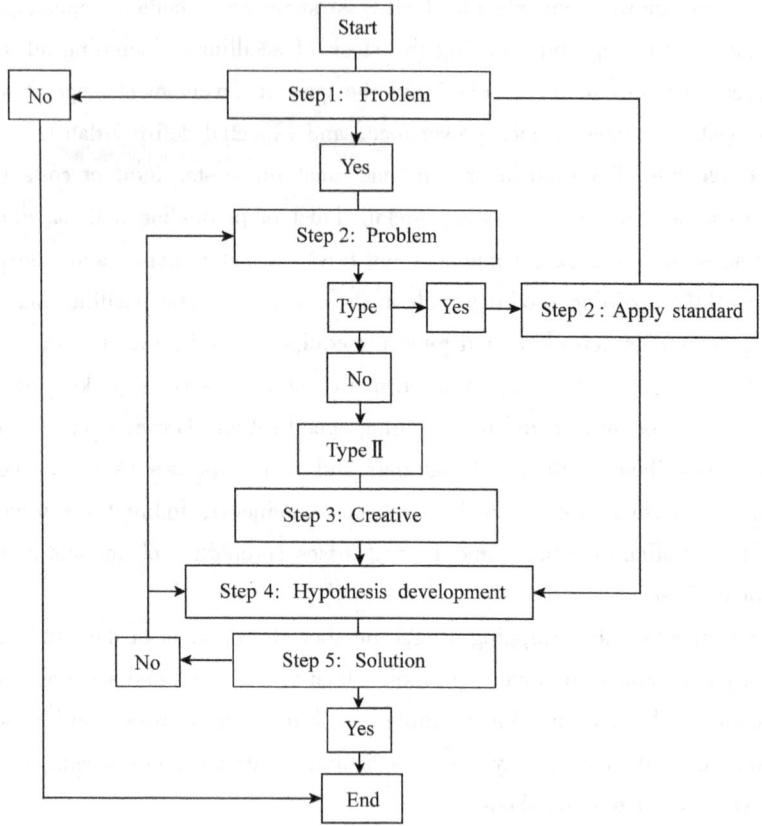

Figure 14-4 Problem solving overview (From Dumroese, 2009).

14.4 Marketing Management

14.4.1 Factors Affecting Consumers' Purchase

14.4.1.1 Internal Factors Affecting Consumers' Purchase

(1) Consumption Demand

Consumption demand is a type of psychological tension state from the sense of absence and expectation. Consumption demands influence consumers to generate purchase actions to solve or buffer the sense of absence. According to the demand level, the influence can be divided into 5 categories.

Physiological Demands. Physiological demands refer to the demands for survival, thus purchasing could solve and relieve the sense of absence. For this group, the edible flower and fruit trees are recommended for purchase.

Security Demands. Security demands refer to the demands from people who want a better life, thus purchasing could meet their need. Planting some tree fence, thorn fence, or flower fence in a company or private garden can meet the security demands.

Social Demands. When the physiological demands and security demands are met, people start to search for the meaning of life, thus social demands are proposed, mainly in terms of seeking beauty. Flowers and bonsais play an important role in people's social lives.

Esteem Demands. Esteem demands refer to a kind of seeking that can attract attention, respect, and admiration from others. The finest flowers and bonsais are the purchase objects of these high-income, esteem-seeking people.

Self-fulfillment Demands. Self-fulfillment demands a are also a form of seeking, in this case to fulfill their abilities and realize their ideals. Many bonsai enthusiasts obtain their own satisfied "works" through their own creative modeling, which is the self-fulfillment demand.

(2) Consumers' Personalities

Human personalities include talent, temperament, and character, which can be divided into the following categories according to the pattern of manifestation.

Reputation Type. This category includes famous brand reputation type, enterprise reputation type, and business reputation type. Relying on science and technology, a nursery can create size effect, establish reputation, and create brand awareness.

Habit Type. In the marketing process for seedlings and flowers, we should provide good quality, appropriate price, and high-quality service to the market and customers. Let new customers be satisfied with your products and form consumption habits to become "return customers".

Emotional Type. In the marketing of seedlings, we impress decision-makers by the meticulous work, wonderful design, and high-quality service, which removes the obstacles that arise during project implementation and marketing processes.

Purchase Type. It is rational shopping to impress users with excellent product quality, preferential price, and thoughtful service.

Random Type. This category is easily affected by the sales environment. The success of product marketing lies in creating a suitable shopping environment so that random customers who weren't necessarily looking to make a purchase become enthused by the environment and become interested in a purchase.

Impulsive Type. For such customers, the characteristics and functions of the products, especially the application scope and disadvantages of the products, should be explained clearly in advance. Try to sign a contract for the larger "sale" first, so as to make it "unrepentant" (i. e., non-refundable).

Execution Type. The "executor" cannot be ignored. Although they do not have the right to decide whether to buy or not, they can transmit information and influence decision-makers.

14.4.1.2 External Factors Affecting Consumers' Purchase

(1) Family

Wealthy families can decorate their houses with a large number of bonsai and even build a landscape-style garden by using a large amount of flowers and trees.

(2) Reference Group

Merchants, for example, can provide consumption modes, provide evaluation information, strengthen the purchase confidence of consumers, and generate the pressure of "consistency", so that people can pursue the trend and promote consumption. From Brazilian wood (*Dracaena fragrans*), Malaba chestnut (*Pachira glabra*) to golden barrel cactus (*Echinocactus grusonii*), and lucky bamboo (*Dracaena sanderiana*), it is the reference group effect.

(3) Social Class

Different classes of people differ in their preferences and pursuits. Businessmen are happy to buy *Pachira glabra*. Literati like asparagus fern and lotus flower, and politicians prefer flowers such as peony and *Dracaena sanderiana*.

(4) Culture

Culture is the synthesis of national and social customs, habits, art, morality, religion, belief, law, and other ideologies. The wide range of nationalities and religions have a wide range of preferences for the color and variety of seedling products.

(5) Sales Promotions

Through sales promotions, consumers can receive information frequently, which has a strong stimulating effect on consumers' purchase motivation and shows the potential demand for products. For example, all types of seedlings, flower exhibitions, and trade fairs have made a great impact on society and played a good role in promoting the products.

14.4.2 Marketing Strategy of Seedlings

14.4.2.1 Marketing Research and Collection of Market Information

(1) Marketing Research

Marketing research methods include, for example, the following methods.

Observation Method. Direct observation and measurement.

In-depth Interview. Invite several people selectively to discuss an enterprise, product, service, marketing, and other topics.

Investigation Method. Includes case-investigation method, key investigation method, sampling investigation method, expert investigation method, comprehensive investigation method, typical investigation method, and school investigation method.

Experimental Method. Collect and analyze experimental data to understand the market through small-scale market experiments, such as packaging experiments, new product experiments, price experiments, and more.

(2) Collection of Market Information

Market information is the true reflection of the development and change and characteristics of the factors that make up the macro- and micro-environment in which the enterprise is located. The term is a general designation of all kinds of information, materials, data, and intelligence, reflecting the nursery's actual situation, characteristics, and correlativity.

14.4.2.2 Market Forecast Method

(1) Qualitative Prediction Method

This method allows a business to make a comprehensive analysis, judgment, and prediction through a social survey, which is made up of a small amount of data and intuitive materials, combined with experience. The advantages are simple and easy to popularize. Methods employed include a buyer's intention survey, salesmen's opinion synthesis, expert's opinion, test marketing method, and more.

(2) Quantitative Prediction Methods

Time Series Forecasting Method. This method is a series made by statistical values in a chronological order. The development trend combines the relevant curve analysis, such as a horizontal development trend, linear change trend, quadratic curve trend, logarithmic linear trend, correction index curve trend, and so on.

Causal Forecasting Method. This method is based on the interrelation and interdependence among things. On the basis of qualitative research, the main factors (independent variables) that affect the prediction objects (dependent variables) are determined, the regression equation is established, and the change of dependent variables is inferred from the change of independent variables.

14.4.2.3 Market Competition Strategy

The successful forms of market competition strategies are as follows.

(1) Differentiated Market Strategy

In the sales activities of similar products, the distinctive features of products or sales are fully displayed, which has a strong attraction to consumers.

(2) Market Strategy of Seeking Difference

This strategy adopts completely different strategies from other enterprises, highlights distinctive features, and establishes competitive strength in market competition.

(3) Pioneering Market Strategy

Pioneering adopts the same sales strategy, strives to be the first, and improves the popularity and competitiveness of enterprises. For example, it could highlight the new, novel, and special ornamental plants.

(4) Induced Market Strategy

Induced adopts a certain sales model to achieve the effect of guiding consumption, stimulating demand, and creating a market. For example, when selling seedlings, a small number of products will be given away for free, which will induce a certain amount of consumers to buy them.

(5) Timeliness Market Strategy

Timeliness grasps some market demand, organizes production in time, sends the products into market before the peak demand, and uses the time difference to gain the profits.

(6) Offensive Market Strategy

Go on the offensive: Break some objective market restrictions and release some imprisoned market demand.

(7) Filling Market Strategy

Enterprises use their own products and services to fill in market demand gaps.

(8) Roundabout Market Strategy

When enterprises encounter some insurmountable obstacles in their business activities, they should take some measures to avoid such restrictions, so that the goals of enterprises can be achieved.

(9) Reverse Market Strategy

Under the condition of a certain product backlog, some enterprises change production or file for bankruptcy; enterprises survive by beating the difficulties.

(10) Hungry Market Strategy

To maintain the sales advantage of their products in the traditional market, enterprises take the initiative to reduce the sales in the traditional market appropriately, so that they are at a state of "hunger", while at the same time, they are constantly exploring new markets.

(11) Joint Market Strategy

Some market demands have special requirements for products, or some products are sold in a certain place with adverse factors. This strategy can be considered to remove sales obstacles and open up a new sales market, such as supporting sales or joint sales of related products.

14.4.3 Network Information Sales System

A nursery should build a suitable nursery stock and sales information system, putting the production, stock, and sales into a variety of data models and integrating them into an information system, simulating and controlling the movement direction of data flow according to the production process. By doing so, nursery managers can realize the functions of browse, query, statistics, and reports of inventory and distribution conveniently and quickly; improve the efficiency of nursery data processing; meet the needs of the nursery in many ways; and speed up the pace of information construction within the nursery.

Nursery stock and sales make up a software system developed for a single large nursery to achieve the main purpose of inventory and sales information processing, which can effectively improve the efficiency of nursery data processing. Nursery stock is the sum of all the seedlings and flowers in the nursery, which is the sum of the number of commodities that can be sold without involving other stocks, such as raw materials.

The stock sale system is based on the specifications and quantities of the existing seedlings and flowers in the nursery at distinct growth periods. Nursery managers can plan the actual production process scientifically, integrating production, stock, and sales into a system; monitor the dynamics of them; and store and back up the important dynamics, so as to achieve nursery informatization.

The principles of system development include integrity, practicability, advancement, development scalability, security and confidentiality, and economy.

The overall goal of the system is to establish a practical information management system. It provides a tool for nursery production and management, improves the efficiency of the nursery operation and management, and speeds up the process of nursery informatization. According to the characteristics of a management information system and the needs of users, the goal of the system design is reliability, maintainability, user friendliness, and high efficiency.

In short, in the era of an internet economy, online sales are an approach to achieve a certain sales target. E-commerce reduces the intermediate links and is convenient and efficient. The nursery should utilize the internet, which has large amounts of information, coverage, low cost, high efficiency, and sound effects, to achieve the goal of popularization and marketing.

14.4.4 Signing the Contract with Customers

The client and the representative of a nursery need to sign a contract, and each party should keep one copy. The terms of the contract shall include the following contents: Description of the seedlings to be provided (e. g., type, container type, seedling specifications); expected schedule; quantity of seedlings that can be provided; unit price and total price of the order; when and how to pay; and what will happen in some cases, for example, if the customer defaults on payment, if the customer receives seedlings late or does not receive the seedlings at all, if the nursery cannot provide the described seedlings, and so forth.

The contract should meet the needs of a specific nursery. A good protocol is to consult lawyers to ensure that the contract protects the rights and interests of nurseries and complies with local laws and regulations.

14.5 Nursery Files Management

Seedlings are the most basic means of production for urban forests, landscaping, and ecological environment construction. To improve the management level of nurseries, we should build and perfect the management files of nurseries and carry out standardized and information-based management.

14.5.1 Definition of a Nursery Technology File

A nursery technology file is a continuous record, analysis, and summary of the utilization of nursery land, labor, machines and tools, materials, medicine, fertilizer, seeds, etc.; the application of various seedling technical measures; the growth and development of various seedlings; and other business activities of the nursery. Such files are the basis of production and marketing activities' records, observations, and research. They are also the central element of the nursery that will support and ensure the progress and innovation of nursery technology.

14.5.2 Significance of Establishing Nursery Technology Files

The technology files record the production, experiments, and management of a nursery. From the beginning of the construction of a nursery, the technology files should be established. From these files, we could know the seedling species, quantity, and quality; the growth and development of seedlings; and analyze and summarize the seedling raising technology experience. We could also explore the use of land, labor, machinery and materials, medicine, fertilizer, seeds, and so on; we could carry out labor organization and management, establish the production quotas, and implement scientific management effectively.

14.5.3 Establishment of Nursery Technology Files

14.5.3.1 Basic Requirements for Establishing Nursery Technology Files

The records of nursery production, experiment, and management must be adhered to for at least a decade, maybe even a century; be practical and realistic; and ensure systematic processes, integrality, and accuracy.

At the end of each year, all kinds of recorded data shall be collected for sorting and statistical analysis so as to provide accurate data and reports for production and operation for the next year.

Full-time and part-time clerks should be hired to be specifically responsible for the nursery technology files. The clerks shall be stable, ideally long-term employees; if any worker changes occur, try to arrange for the transfer of workers at slower times in the annual workload, that is, not at the end of a year.

14.5.3.2 Contents of Nursery Technology Files

(1) Basic Information File

This file should include the nursery location, area, operation conditions, natural conditions, topographic map, soil distribution map, nursery compartment map, fixed assets, instruments and equipment, machines and tools, vehicles, production tools, personnel, organizations, and any other details that may be pertinent to a given nursery.

(2) Nursery Land Use File

This file looks at the operation area as a unit, and it records the area of each operation, species of seedlings, cultivation methods, soil preparation, soil improvement, irrigation, fertilization, weeding, pest control, seedling growth quality, and other basic information so as to analyze the relationship among the changes of soil fertility and cultivation, fertilization history, and to provide a basis for the implementation of reasonable rotation, scientific fertilization, and soil improvement (Table 14-1).

Table 14-1 Nursery land use

Operation area number: Operation area: Soil quality:

Year	Tree species	Seedling raising method	Operation mode	Land preparation	Fertilization	Weeding	Irrigation	Pest	Seedling quality	Notes

Completed by:

(3) Nursery Work File

This work file is updated daily, to record the daily production activities for convenient consulting and summarizing. According to the work file, the labor volume of tree species, the utilization of machines and tools, and the use of materials, medicine, and fertilizer can be counted for calculating the cost, formulating the reasonable quota, strengthening the plan management, and better organizing the production (Table 14-2).

Table 14-2 Nursery work diary

Date:

Species	Operation area number	Seedling raising method	Operation mode	Operation projects	Labor	Machines and tools		Operation quantities		Materials usage amount			Work quality	Notes
						Name	Quantity	Unit	Quantity	Name	Unit	Quantity		
Sum														
Record														

Completed by:

(4) Files of Seedling Technical Measures

This file records the cultivation process of all seedlings, that is, from dealing with the seeds, cuttings, scions, and other propagation materials; then field cultivation to seedling lifting, heeling-in, packaging, and outplanting; from container seedling production to pot replacement and maintenance or sales (Table 14-3).

(5) Investigation Files of Seedling Growth and Development

These files record the regular observation of growth and development of all types of seedlings and their growing process to establish the development cycle, the impact of natural conditions, and the cultivation management on the growth and development of seedlings. Review of these details can assist in determining timely and effective cultivation technical measures (Table 14-4, Table 14-5).

(6) Meteorological Observation Files

These files record the meteorological conditions and their duration each day at the nursery site, for example, temperature, humidity, wind direction, and wind force (Table 14-6). (The observation data of a local meteorological station can be a good source of this information.)

(7) Files of Scientific Experiments

These files record the experiment purpose, experiment design, experiment method, experiment results, analysis, and annual summary report.

(8) Seedling Marketing Files

These files record the type, specification, quantity, price, date, purchasing institution, and use of seedlings sold in each year.

Table 14-3 Seedling raising technology measures

Tree species:				Seedling raising year:		
Seedling raising area:			Seedling age:		Fore-rotating seedling:	
Propagation method	Seedlings	Provenance	Storage method	Storage time	Pregermination method
		Sowing method	Sowing quantity	Soil covering thickness	Mulch
		Covering beginning and ending dates	Emergence rate	Thinning time	Reserved seedling density
	Cutting seedlings	Cutting source	Storage method	Cutting method	Cutting density
		Survival rate			
	Grafting seedlings	Stock name	Source	Scion name	Source
		Grafting date	Grafting method	Binding materials	Uncording date
		Survival rate			
	Transplanting seedlings	Transplanting date	Transplanting seedling age	Transplanting times	Transplanting planting spacing
		Transplanting seedling source	Transplanting survival rate		
Land preparation		Tillage date		Tillage depth	Bedding date

Tree species: _____　　　　Seedling raising year: _____　　　　(Continues)

		Fertilization date	Fertilization types	Fertilizing amount	Fertilization method
Fertilization	Base fertilizer				
	Top-dressing				
Irrigation		Times		Date	
Intertillage		Times		Date	Depth
Diseases and insect pests		Name	Date of occurrence	Prevention date	Agentia name	Concentration	Method	Effect
	Diseases pests							
	Insect pests							
Outplanting		Date	Area	Unit area yield	Qualified seedling rate	Seedling lifting method	Packaging
	Seedlings						
	Cutting seedlings						
	Grafting seedlings						
New technique application condition								
Existing problems and suggestions for improvement								

Completed by:

Table 14-4 Seedling growth summary (____ year)

Tree species   Sowing (cutting, grafting, transplanting) period   Sowing quantity (kg/hm², grain/m²)

Seed germination method   Germination date: From ____ to ____   Maximum germination period: From ____ to ____   Tillage method

Soil   pH   Thickness   Slope aspect   Slope gradient   Fertilization type

Fertilization amount (kg/hm²)   Fertilization time

Survey order	Survey date	Sample plot			Total plants in each point in previous survey	Number of damaged plants				Loss of work	Existing plant number	Growing status									Disaster development notes
		Number of rows	Sample plot	Total area		Diseases	Insect pests	Thin-ning				Seedling height			Seedling diameter			Seedling root		Crown breadth		
												Hig-her	Gen-eral	Lo-wer	Thic-ker	Gen-eral	Thin-ner	Root length	Root width	Wider	Gen-eral	Narr-ower	

Completed by:

Table 14-5 Seedling growth and development questionnaire

Seedling raising year:

Tree species:		Seedling age:		Propagation method:					Transplanting times:	
Starting emergence				Emergence in a large amount						
Bud expanding				Bud unfolding						
Terminal bud forming				Leaf color changing						
Starting defoliation				Complete defoliation						
Increment										
	Day/month	Day/month	Day/month	Day/month	Day/month	Day/month	Day/month	Day/month	Day/month	Day/month
Seedling height										
Root collar diameter										
Root system										

Level		Grading standard		Per unit yield	Total yield
Outplanting	The first level	Height			
		Root collar diameter			
		Root system			
		Crown breadth			
	The second level	Height			
		Root collar diameter			
		Root system			
		Crown breadth			
	The third level	Height			
		Root collar diameter			
		Root system			
		Crown breadth			
	Substandard level				
	Other				
			Sum		

Notes

Completed by:

Table 14-6 Meteorological records

Date:

Month	Mean air temperature (℃)				Mean surface temperature (℃)				Evaporation (mm)				Precipitation (mm)				Relative humidity(%)				Sunlight			
	Mean	First 10 days	Middle 10 days	Last 10 days	Mean	First 10 days	Middle 10 days	Last 10 days	Mean	First 10 days	Middle 10 days	Last 10 days	Mean	First 10 days	Middle 10 days	Last 10 days	Mean	First 10 days	Middle 10 days	Last 10 days	Mean	First 10 days	Middle 10 days	Last 10 days
Full year																								
January																								
...																								
December																								

The annual frost ___ days, the initial frost in ___ day ___ month, the late frost in ___ day ___ month, the final ice in ___ day ___ month; the annual extreme high temperature ___ ℃ in ___ day ___ month; the surface temperature ___ ℃ in ___ day ___ month, the extreme low temperature ___ ℃ in ___ day ___ month, the annual accumulated surface temperature ___ ℃ in ___ day ___ month; the annual temperature stable through 10 ℃, the initial date ___ day ___ month, the final date ___ day ___ month, final date ___ day ___ month. temperature greater than 10 ℃ is ___ ℃; passing through 15 ℃, initial date ___ day ___ month, final date ___ day ___ month; passing through 20 ℃, initial date ___ day ___ month, final date ___ day ___ month.

Completed by:

14.5.4 Requirements for Establishing Seedling Technology Files

All nursery files must be compiled with continuity and integrity by a dedicated manager; files should be practical and realistic, timely and accurate. They should include summary, arrangement, statistics, and analysis, so as to reveal the trend of production and sale of seedlings. Files should be sorted, bound, and catalogued. All the file keeping should conform to standardization and informatization; management personnel should remain stable.

14.5.5 Informatization of Nursery Management

The management information system requires people and computers to collect, transmit, store, maintain, and use the information to know the operation situations, to predict the future by using historical data from the past, to assist the enterprise in decision-making from the perspective of the overall situation of the enterprise, to control the behavior of the enterprise by using information, and to help enterprises achieve their planning goals.

Developed countries introduced nursery management informatization systems into practical application in 1980s. The introduction of these informatization systems can strengthen the management of a nursery and increase the seedling yield of the nursery. This nursery management system contributes significantly across four aspects: Purchase, daily inside work of nursery, outside work, and inventory, which can provide faithful data displays and analyses for managers, with the end goal of improving work efficiency.

The informatization of a nursery is a dynamic process that transforms a traditional single productive nursery into an information-rich nursery, which is embodied in digital management, electronic mapping, network previews, and so much more. It covers all personnel, processes, businesses, and departments in the nursery, as well as stakeholders in the normal operation of the nursery.

Questions for Review

1. What is the target management of a nursery? What factors should be considered in target management?
2. What are the tasks and contents of nursery production management?
3. What are the methods and problems of total quality management of a nursery?
4. What is the five-step system method to solve nursery problems?
5. What are the internal and external factors influencing consumers' purchases?
6. What are the market forecasting methods? Be able to apply these methods to market forecast.
7. What are the contents of nursery technical archives?
8. Briefly, what are the problems existing in the management of a nursery in China and what are the methods to solve them?

References and Additional Readings

CHENG F Y, 2012. Nursery of landscape plants[M]. Beijing: China Forestry Publishing House.
DUMROESE R K, 2009. Nursery manual for native plants: A guide for tribal nurseries[C]. Washington, DC: US

Department of Agriculture, Forest Service.

HAN Y L, 2008. Study on the production and management of modern garden nursery [M]. Beijing: China Agriculture Press.

QI T, 2005. Establishment of the system of stock and selling for the North Stated base of forest seedling demonstration[D]. Beijing: Beijing Forestry University.

SHEN H L, 2009. Seedling cultivation[M]. Beijing: China Forestry Publishing House.

SU J L, 2013. Nursery of landscape plants[M]. 2nd ed. Beijing: China Agriculture Press.

TLIEWUHAN K, 2016. Information management system of three-dimensional nursery based on GIS[D]. Beijing: Beijing Forestry University.

WANG M, 2015. Nursery management[M]. Beijing: China Forestry Publishing House.

WANG X L, 2013. Design and implementation of nursery garden management information system[D]. Dalian: Dalian University of Technology.

YU B C, 2002. Design and implementation of the Beijing seedling management information system[D]. Beijing: Beijing Forestry University.